T. G. [illegible]

Bradford 1984.

Ultrastructure of reproduction

ELECTRON MICROSCOPY IN BIOLOGY AND MEDICINE

Current Topics in Ultrastructural Research

SERIES EDITOR: P.M. MOTTA

Already published in this series:

Motta, P.M. (ed.): Ultrastructure of Endocrine Cells and Tissues.
ISBN: 0-89838-568-7

Ultrastructure of Reproduction

Gametogenesis, Fertilization, and Embryogenesis

Edited by

Jonathan Van Blerkom, Ph.D.
Department of Molecular, Cellular and Developmental Biology,
University of Colorado, Boulder, Colorado, USA

and

Pietro M. Motta, M.D., Ph. D.
Department of Anatomy, Faculty of Medicine,
University of Rome, Rome, Italy

1984 **MARTINUS NIJHOFF PUBLISHERS**
a member of the KLUWER ACADEMIC PUBLISHERS GROUP
BOSTON / THE HAGUE / DORDRECHT / LANCASTER

Distributors

for the United States and Canada: Kluwer Boston, Inc., 190 Old Derby Street, Hingham, MA 02042, USA
for all other countries: Kluwer Academic Publishers Group, Distribution Center, P.O. Box 322, 3300 AH Dordrecht, The Netherlands

Library of Congress Cataloging in Publication Data

Main entry under title:

Ultrastructure of reproduction.

(Electron mioroscopy in biology and medicine ; v. 2)
Includes index.
1. Gametogenesis. 2. Embryology--Vertebrates.
3. Fertilization (Biology) 4. Ultrastructure (Biology)
I. Van Blerkom, Jonathan, 1947- . II. Motta, Pietro. III. Series. [DNLM: 1. Germ cells--Ultrastructure. 2. Embryo--Ultrastructure. 3. Fertilization. 4. Microscopy, Electron--Methods. WQ 205 U47]
QL964.U38 1983 599'.32 83-2437
ISBN 0-89838-572-5

Copyright

PRINTED IN THE NETHERLANDS

Preface

Advances in the development and application of electron microscopic techniques have occurred recently such that the electron microscope has evolved to become an essential tool in both basic and clinical research. Use of this instrument has contributed significantly to the formation of new perspectives and concepts concerning cell fine structure. These structural perspectives are now being integrated with specific functional, biochemical and pathophysiological events and processes of cells and tissues.

Most recently, utilization of innovative electron microscopic techniques such as freeze-fracture, freeze-etching, and scanning and high-voltage electron microscopy offers both the basic and clinical scientist potentially fundamental insights into many morphodynamic processes related to the activities of cells and tissues. Such an approach has been especially rewarding when applied to the dynamic events of gametogenesis and early embryonic development.

The chapters comprising this book have been selected and edited with the aim of providing an up-to-date and comprehensive account of the most important aspects of vertebrate gamets and embryos as revealed by the integration of several different submicroscopic methods. The organization of the chapters is designed to indicate present gaps in our knowledge of the developmental and reproductive biology of gametes and the developing embryo and possible lines of research which may lead to a lessening of these gaps.

The contributions to this volume have been prepared by experts in the field. Each chapter has been composed with the intent of offering the reader not only a concise history of the topic but also emphasizing research problems that must be solved in the future. Much of the information has been presented in an illustrative format, because only in this way will it be apparent how the integration of ultrastructural and biochemical techniques contribute to a more comprehensive understanding of gametogenesis and embryonic development.

This volume should prove useful to advanced students of reproductive and developmental biology, researchers and clinicians. It will be of specific interest to developmental, cell and reproductive biologists involved in the research and teaching of embryology, histology and physiology in both veterinary and medical schools.

We express our thanks to the many authors, not only for the quality of their contributions but also for their patience in responding to exacting and often tedious editorial demands. Our grateful appreciation also is extended to Mr. Hobart Bell for his insightful and very helpful suggestions in the preparation of this book.

October, 1983

Jonathan Van Blerkom, Boulder, Colorado
Pietro M. Motta, Rome, Italy

Contents

Preface V

List of contributors IX

PART ONE: GAMETOGENESIS

1. Origin of the germ cell line, *by* E.M. Eddy 1
2. Ultrastructure of primordial germ cells in the early chick embryo, *by* A. Ukeshima and T. Fujimoto 12
3. Germ and somatic cell interactions during gonadal morphogenesis, *by* H. Larios-Merchant . . 19
4. Germ cell differentiation and intercellular bridges, *by* B. Gondos 31
5. Spermiation – the sperm release process: Ultrastructural observations and unresolved problems, *by* L.D. Russel 46
6. Fine structural features of Sertoli cells, *by* L. Plöen and M. Ritzén 67
7. Membrane organization and differentiation in the Guinea pig spermatozoon, *by* D.S. Friend . 75
8. Comparative aspects of acrosomal formation, *by* J.S. Kaye 86
9. Changes in the sperm surface during maturation in the epididymis, *by* G. Olson. 97
10. The human spermatozoon, *by* B. Baccetti 110
11. Structure and evolution of the nucleolus during oogenesis, *by* A. Coimbra and C. Azevedo . . 127
12. Ultrastructural changes in ovarian oocytes induced by ionizing radiations, *by* T.G. Baker and L.L. Franchi 140
13. Association of oocytes with follicular cells, and oocyte chromatin during maturation, *by* R.W. McGaughey 156

PART TWO: FERTILIZATION AND EMBRYOGENESIS

14. Problems in the analysis of mammalian fertilization, *by* D.M. Phillips 166
15. Ultrastructure of human fertilization, *by* M. Dvořák, J. Tesařík and V. Kopečný 176
16. The cell surface of the mammalian embryo during early development, *by* L.M. Wiley 190
17. The mosaic organization of the preimplantation embryo, *by* M.H. Johnson, C.A. Ziomek, W.J.D. Reeve, H.P.M. Pratt, H. Goodall and A.H. Handyside 205
18. Virus-like particles and related expressions in mammalian oocytes and preimplantation stage embryos, *by* Y. Yotsuyanagi and D. Szöllösi 218

19. Crystalline inclusions in embryonic and maternal cells, *by* L.H. Hoffman and G. Olson . . . 235

20. Cellular aspects of implantation, *by* D.J. Chávez 247

21. The fine structure and function of the trophoblast during normal and delayed implantation, *by* O. Nilsson 260

Index 269

List of contributors

Azevedo, Carlos. Department of Cell Biology, Institute of Biomedical Sciences, University of Oporto, 4000, Oporto, Portugal

Baccetti, Baccio. Institute of Zoology, University of Siena, Via Mattioli 4, 53100 Siena, Italy

Baker, Terry G. School of Medical Sciences, University of Bradford, Bradford, BD7 1DP, United Kingdom

Chávez, Daniel J. Department of Anatomy, School of Medicine, Southern Illinois University, Carbondale, IL 62901, USA

Coimbra, Antonio. Institute of Histology and Embryology, Faculty of Medicine, University of Oporto, 4200, Oporto, Portugal

Dvořák, Milan. Department of Histology and Embryology, J.E. Purkyně University, tř. Obráncu miru 10, 662 43 Brno, Czechoslovakia

Eddy, Edward M. Department of Biological Structure, University of Washington School of Medicine, Seattle, WA 98195, USA

Franchi, Leslie L. Department of Anatomy, University of Birmingham, Birmingham, B15 2TJ, United Kingdom

Friend, Daniel S. Department of Pathology, University of California School of Medicine, San Francisco, CA 94143, USA

Fujimoto, Toyoaki. Department of Anatomy, Kumamoto University Medical School, Kumamoto, 860, Japan

Gondos, Bernard. Department of Pathology, University of Connecticut Health Center, Farmington, CT 06032, USA

Goodall, H. Department of Anatomy, Cambridge University, Cambridge, CB2 3DY, United Kingdom

Handyside, A. Department of Anatomy, Cambridge University, Cambridge, CB2 3DY, United Kingdom

Hoffman, Loren H. Department of Anatomy, Vanderbilt University, Nashville, TN 37232, USA

Johnson, M.H. Department of Anatomy, Cambridge University, Cambridge, CB2 3DY, United Kingdom

Kaye, Jerome S. Department of Biology, University of Rochester, Rochester, NY 14627, USA

Kopečný, Václav. Department of Histology and Embryology, J.E. Purkyně University, tř Obráncu miru 10, 662 43 Brno, Czechoslovakia

Mc Gaughey, Robert W. Department of Zoology, Arizona State University, Tempe, AZ 85287, USA

Merchant-Larios, Horacio. Instituto de Investigaciones Biomédicas, Universitad Nacional Autónoma de

México, Apartado Postal 70228, Ciudad Universitaria 04510 Mexico, D.F., Mexico

Nilsson, Ove. Department of Human Anatomy, University of Uppsala, Biomedical Center, Uppsala, Sweden

Olson, Gary E. Department of Anatomy, Vanderbilt University, Nashville, TN 37232, USA

Phillips, David M. The Population Council, Center for Biomedical Research, 1230 York Avenue, New York, NY 10021, USA

Plöen, Leif. Department of Anatomy and Histology, Faculty of Veterinary Medicine, Swedish University of Agricultural Sciences, S-75007, Uppsala, Sweden

Pratt, H.P.M. Department of Anatomy, Cambridge University, Cambridge, CB2 3DY, United Kingdom

Reeve, W.J.D. Department of Anatomy, Cambridge University, Cambridge, CB2 3DY, United Kingdom

Ritzén, Martin. Pediatric Endocrinology Unit, Karolinska Sjukhuset, S-10401, Stockholm, Sweden

Russell, Lonnie D. Department of Physiology, School of Medicine, Southern Illinois University, Carbondale, IL 62901, USA

Szöllösi, Daniel. Station centrale de Physiologie Animale, I.N.R.A. 78350 Jouy-en-Josas, France

Tesařík, J. Department of Histology and Embryology, J.E. Purkyně University, tř. Obráncu miru 10, 662 43 Brno, Czechoslovakia

Ukeshima, Atsumi. The College of Medical Science, Kumamoto University, Kumamoto 862, Japan

Wiley, Lynn M. Department of Human Anatomy, University of California, Davis, CA 95616, USA

Yotsuyanagi, Yoshio. Centre de Génétique Moléculaire, C.N.R.S. 91190 Gif-sur-Yvette, France

Ziomek, C.A. The Worcester Foundation for Experimental Biology, Shrewsbury, MA 01545, USA

CHAPTER 1

Origin of the germ cell line

EDWARD. M. EDDY

1. Introduction

The definitive germ cell line usually is considered to begin with the appearance of primordial germ cells (PGCs). As we shall see, for a few selected examples, PGCs are identified first in the posterior blastoderm in certain insect embryos, in the floor of the blastocoel in anuran frogs, in the anterior extraembryonic endoderm in chicks and in the endoderm of the yolk sac in mammals. It is tempting to make the generalization from these findings that the germ cell line of diverse animals arises within the endoderm or its equivalent prior to organogenesis. However, PGCs usually are identified on the basis of morphological features. This criterion alone may be insufficient for defining the beginning of the germ cell line. Cells which do not particularly look like PGCs but which have the developmental potential of PGCs, or express gene products specific to PGCs, might be present at earlier times and at other sites during embryogenesis. If this were shown to be the case, one might argue that by these criteria such cells should be called PGCs. One of the main problems, then, in discussing the origin of the germ cell line is in defining a germ cell, particularly one at an early point in the lineage. Is there a marker or a set of markers that can be used to reliably identify precursors of PGCs before they acquire the overt morphological characteristics of germ cells? As will be seen, the answer is yes for some species, but no for others. For the latter species, the question then becomes: Are there markers which have not yet been identified or are there no markers because there is no germ cell line before PGCs can be recognized? There have been several reviews on various aspects of the origin of the germ cell line in different organisms in the last few years (1–5). This Chapter takes a comparative approach but with a focus on mammals, considering in particular the site of origin of the germ cell line and the possible role of germ plasm in germ line determination.

2. Animals with a continuous germ line

Certain nematodes have been found to have determinative cleavage with the egg dividing asymmetrically four times, each time giving rise to a larger somatic tissue precursor cell and to a smaller P cell. At this point, the P_4 cell divides, giving rise to two germ cell line precursor cells (6, 7). It has been found recently that some lots of rabbit antisera identify cytoplasmic granules within cells of the germ cell lineage of *Caenorhabditis elegans* (8). At the first cleavage, the granules are located at the posterior end of the embryo where the P cells form. During the next three cleavages, the granules lie at the side of the cell where the next P cell will form and through this process the granules become segregated into the single precursor cell for the germ cell line. The granules are similar in distribution and size to specific structures previously observed by electron microscopy in some P cells and in subsequent germ cells (9). In other studies, exudation of substantial amounts of cytoplasm from different regions of the uncleaved egg, after puncturing the egg capsule with a laser beam, seldom resulted in sterile embryos, indicating that the putative determinants for the germ cell line were probably not present in the cytoplasm of uncleaved eggs in an unbound form (10). However, neither the cytoplasm that was lost not that remaining was examined for the presence of the granules characteristic of cells in the germ cell lineage.

It was realized by early investigators that the pole cells which bud off the posterior tip of certain insect embryos prior to formation of the cellular blastoderm give rise to PGCs (e.g., 11). Some of the features of pole cells in *Drosophila* are that they are derived clonally from 8–10 nuclei, they divide 1–3 times in the early embryo but then arrest until after gastrulation, they interdigitate within the blastoderm prior to gastrulation with some reaching the boundary of the yolk, and they are loosely attached to the

Van Blerkom, J. and Motta, P.M. (eds.), Ultrastructure of reproduction. ISBN 0-89838-572-5.

midgut rudiment during gastrulation (12). Shortly after gastrulation, most pole cells migrate to the lateral mesoderm where the gonads form and those pole cells which remain outside the gonads degenerate (13).

A characteristic feature of pole cells in *Drosophila* and some other insects is the presence of 'polar granules' in the cytoplasm. A number of morphological studies, initially at the light microscope level (14), and more recently using electron microscopy (e.g., 15), have traced the polar granules from the posterior cytoplasm of the oocyte or mature egg to the pole cells and subsequently into definitive germ cells. Because of this location, Hegner proposed that the polar granules are either germ cell determinants or the visible sign of the germ cell determinant region (14). This hypothesis has been supported repeatedly by results of experimental studies. Surgical ablation of the posterior cytoplasm (16), ultraviolet irradiation of the posterior tip of the mature egg (17, 18), or displacement of polar granules by centrifugation (19), all result in insects lacking germ cells. More recently, it has been shown that transplantation of posterior cytoplasm from donor eggs to eggs exposed to ultraviolet irradiation restores the ability of the treated eggs to form germ cells (20). In other experiments, posterior egg cytoplasm transplanted to the anterior tip of the embryo resulted in cells there which contained polar granules. When those cells were grafted in turn to the posterior pole of other embryos, some of the progeny of the resulting adults bore genetic markers from the donor embryos, showing that posterior cytoplasm can cause anterior blastoderm cells to become PGCs capable of forming gametes (21). A subcellular fraction enriched for polar granules has been examined by SDS polyacrylamide gel electrophoresis and found to contain a major basic protein with a molecular weight of approximately 95,000 that was not present in other cells (22). Additional studies are needed to determine the role of this protein in polar granule structure and function.

In anuran amphibians, cells having the appearance and behavior of PGCs are present in the endoderm of neurula stage embryos. In larval stage embryos, these cells move to the dorsal endoderm and migrate through the dorsal mesentery to enter the genital ridges (Fig. 1) (23). These cells contain distinctive cytoplasmic granules which have been traced from uncleaved, fertilized eggs of *Rana temporaria* where they are aligned along the plasma membrane at the vegetal pole, into a few of the vegetal blastomeres during cleavage (Fig. 2), subsequently to cells in the presumptive endoderm forming the floor of the blastocoele during gastrulation and finally into the definitive PGCs. Bounoure termed these granules the *cytoplasme germinal* (germ plasm), and suggested that they were responsible for determining the germ cell line (23). Similar observations have been made since in other species of amphibians (reviewed in 2, 24). Furthermore, fine structural studies (Fig. 3) have established that the cytoplasmic granules consist of aggregates of fibrogranular material lacking a surrounding membrane (e.g., 25).

Ablation and restoration experiments have given further evidence that the germ cell line is determined by factors from the region of the amphibian egg and early embryo containing the germ plasm. It has been shown that surgical removal of the vegetal pole cytoplasm from eggs, 2-cell or 4-cell embryos results in animals with few or no germ cells (26, 27), with the amount of germ plasm that remains determining the number of PGCs that form (28). It was also shown that irradiation of the vegetal pole of fertilized eggs with ultraviolet light also results in sterility in frogs which develop from treated eggs (29). This effect was found to be dose-dependent and stage-dependent (30) and to vary between batches of eggs (31). The most convincing evidence came when Smith (30) injected subcortical cytoplasm from the vegetal pole of unirradiated eggs (about 3% of the total egg volume) into the same region of eggs which had been irradiated and found that 47% of the injected eggs developed into larvae containing PGCs. Since larvae which developed from irradiated eggs that either were uninjected or injected with animal pole cytoplasm lacked PGCs, it seems likely that vegetal pole cytoplasm, containing germ plasm is responsible for determining PGC formation in these embryos. Thus, for animals from three different phyla there appears to be a marker for the germ cell line which can be traced from the fertilized egg to definitive germ cells. Furthermore, there is experimental evidence in *Drosophila* and in anuran amphibians that the marker is either the determinant of the germ cell line or is at least spatially associated with the determinant.

3. Animals not shown to have a continuous germ line

In consideration of some animals in which a continuous germ cell lineage has not been demonstrated, there are conflicting reports and opinions on the origin of PGC in urodelean amphibians. They are less thoroughly studied than anurans, and although germ plasm is apparently present in fertilized *Ambystoma* eggs (31) and PGCs (32), the germ plasm has not been traced from eggs through early development

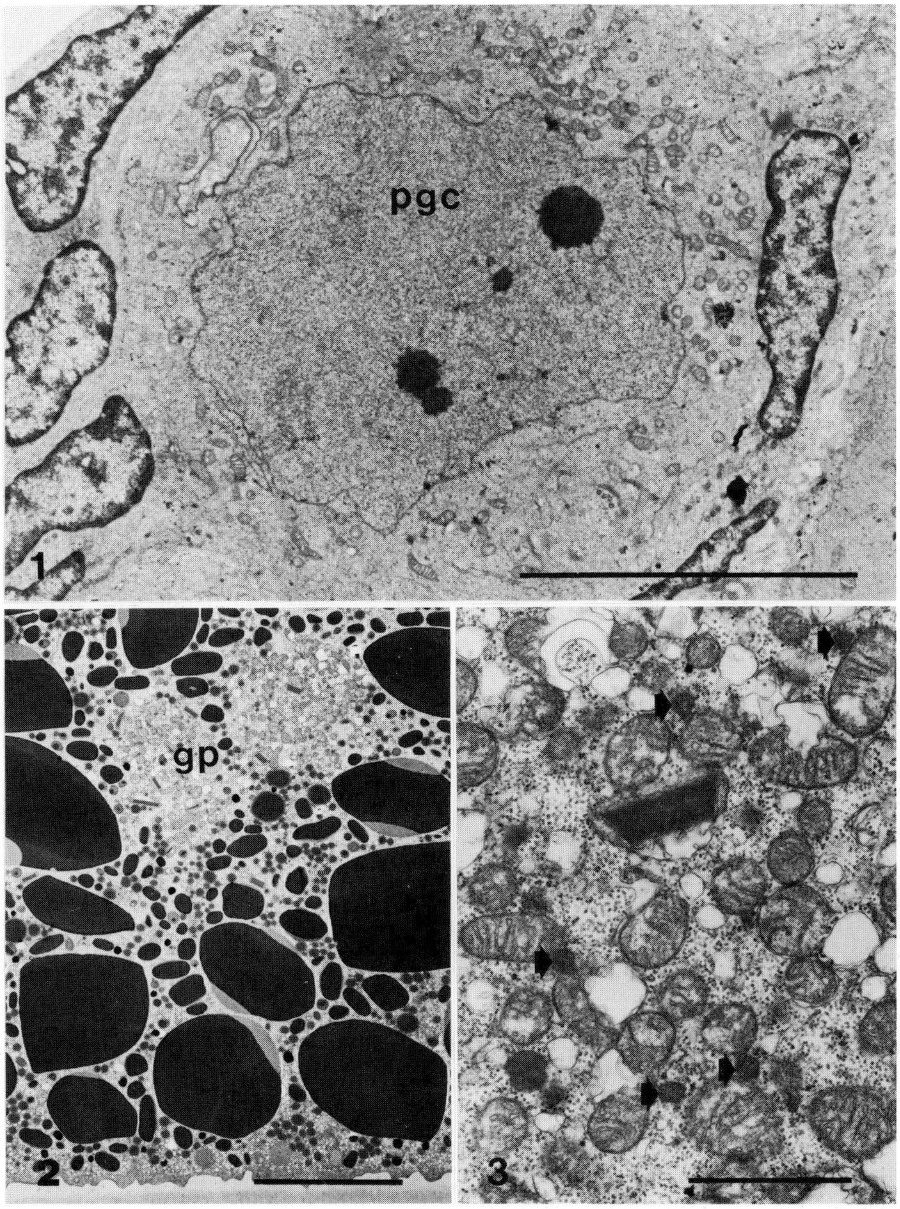

Figs. 1–3. (1) A PGC in the genital ridge of a *Rana pipiens* tadpole is recognizable as a large cell with an irregularly shaped nucleus containing round, dense nucleoli and sparse heterochromatin. Scale line equals 10 μm. (2) A region of the vegetal cytoplasm of a 2-cell stage *Rana pipiens* embryo is shown. Yolk granules of various sizes are present surrounding regions of cytoplasm containing germ plasm (GP). Scale line equals 10 μm. (3) An area from Figure 2 is shown at higher magnification. The fibrogranular germ plasm is indicated by arrows. Scale line equals 1 μm. (Figs. 2 and 3 reprinted by permission (2)).

to PGCs. Some investigators have concluded from morphological or experimental studies that PGCs have an endodermal origin in urodeles similar to that in anurans (33–35). Other investigators have concluded that a germ plasm is not involved in determination of PGCs in urodeles, but rather that they arise epigenetically from totipotent cells in the mesoderm under the inductive influence of the endoderm (3). Although the experimental evidence is consistent with the latter hypothesis, it does not exclude the possibility of an endodermal origin as in anurans. Additional studies clearly will need to be carried out in urodeles to determine if there is a developmental continuity of germ plasm and if it has a role in the determination of the germ cell line.

In the chick, PGCs are first recognized in the hypoblast layer of the extraembryonic endoderm in the region of the germinal crescent (36). They appear there at the head process stage, around 18 hr after fertilization, when the primitive streak has formed but before somites are present (37). The PGCs are rich in PAS-positive material (38), being more abundant in later stages than during earlier times or during PGC migration (39). As development proceeds the PGCs accumulate at the junction of the ectoderm and endoderm and enter the blood islands when the mesoderm invaginates to the germinal crescent area (40). Following this, the PGCs enter the circulation, are transported to the gonadal precursors and then leave the circulation to colonize the gonads (41). There is as yet no morphological or experimental evidence of a germ plasm associated with germ cell formation in the chick.

The time and site of origin of PGC precursors in the chick, as opposed to when PGCs are first recognized, has not yet been well established. Different segments of the blastoderm have been taken from chick embryos and incubated through the time period corresponding to when PGCs are first recognized. One study indicated that PGC precursors originate in the primary hypoblast of the posterior region of the blastoderm and are transported anteriorly to the germinal crescent by morphogenetic movements (42). However, another study found that the usual number of PGCs could arise in anterior halves of the blastoderm as well as in posterior halves (43). Similar results have been obtained from studies using even earlier stages, before blastoderm formation (44), suggesting that PGC formation in the chick may occur in the preblastoderm period throughout the region destined to become extraembryonic endoderm. However, x-irradiation of either the anterior or posterior half of the blastoderm prior to incubation leads to a reduction in the number of PGCs, while x-irradiation of the posterior half after a few hours of incubation causes only a slight reduction in number of PGCs. It has been suggested that presumptive PGCs may be distributed throughout the endoderm during the preblastoderm and early blastoderm stages, but accumulate in the anterior half during the first few hours of incubation upon segregation of the primary hypoblast from the epiblast (44).

Early workers debated whether the germ cell line in mammals arose in the gonads or was of extragonadal origin (45). Morphological evidence for the extragonadal origin was gained with the finding that PGCs could be distinguished from surrounding somatic cells by their high alkaline phosphatase activity (Fig. 4) (46). It was soon demonstrated that PGCs could be found in the vicinity of the yolk sac endoderm and the root of the allantois on the 8th day of gestation in the mouse at about the time somites first appear (47–49). In addition to the presence of alkaline phosphatase on their surface, PGCs in mammals are recognizable as being large, round cells (Fig. 5) with blunt pseudopodia and containing abundant free ribosomes (Fig. 6) but relatively few other organelles (50–52). There is also good experimental evidence that PGCs have an exclusive extragonadal origin in mammals and are not supplemented by cells arising in the gonads. When genital ridges from mouse embryos are transplanted to beneath the kidney capsule in adults before germ cells appeared, they fail to develop germ cells even though other structures develop, including mesonephric tubules, mesonephric ducts and Müllerian ducts (53). Also mutations which result in sterility in homozygotes have been found to result in embryos with normal numbers of extragonadal PGCs in early embryos but with few PGCs in the gonads at later stages (48). These and other morphological and experimental studies clearly indicate that PGCs in mammals have an extragonadal origin prior to the appearance of the gonadal anlagen and that they migrate to the gonads at a later stage in development (see Merchant, Chapter 3).

4. Nuage and germ cells

In addition to those animals in which a visible marker has been traced from the egg cytoplasm through cells of the early embryo to lie in definitive germ cells, a larger number of animals have been reported to contain structures morphologically similar to the germ line markers in their germ cells (2). The term 'nuage' (54) has been applied to these structures. Although some of the reports came from light microscopic studies, most of the observations occurred after the advent of electron microscopy. Nuage usually is found to be a discrete, dense, fibrous cytoplasmic organelle which lacks a surrounding membrane and often lies in association with mitochondria (Fig. 7) or adjacent to nuclear pores. Although nuage is strikingly similar in form and distribution to polar granules of *Drosophila* and germ plasm of amphibians, neither nuage nor the germ line markers have been sufficiently well characterized biochemically or functionally to determine if they are related.

The earliest point in the development of the germ cell line in mammals that nuage has been reported is in PGCs in the hindgut epithelium of 9-day mouse embryos (51) and of 10-day rat embryos (55, 56). Nuage is present in the subsequent stages of germ cell formation in both males and females, including

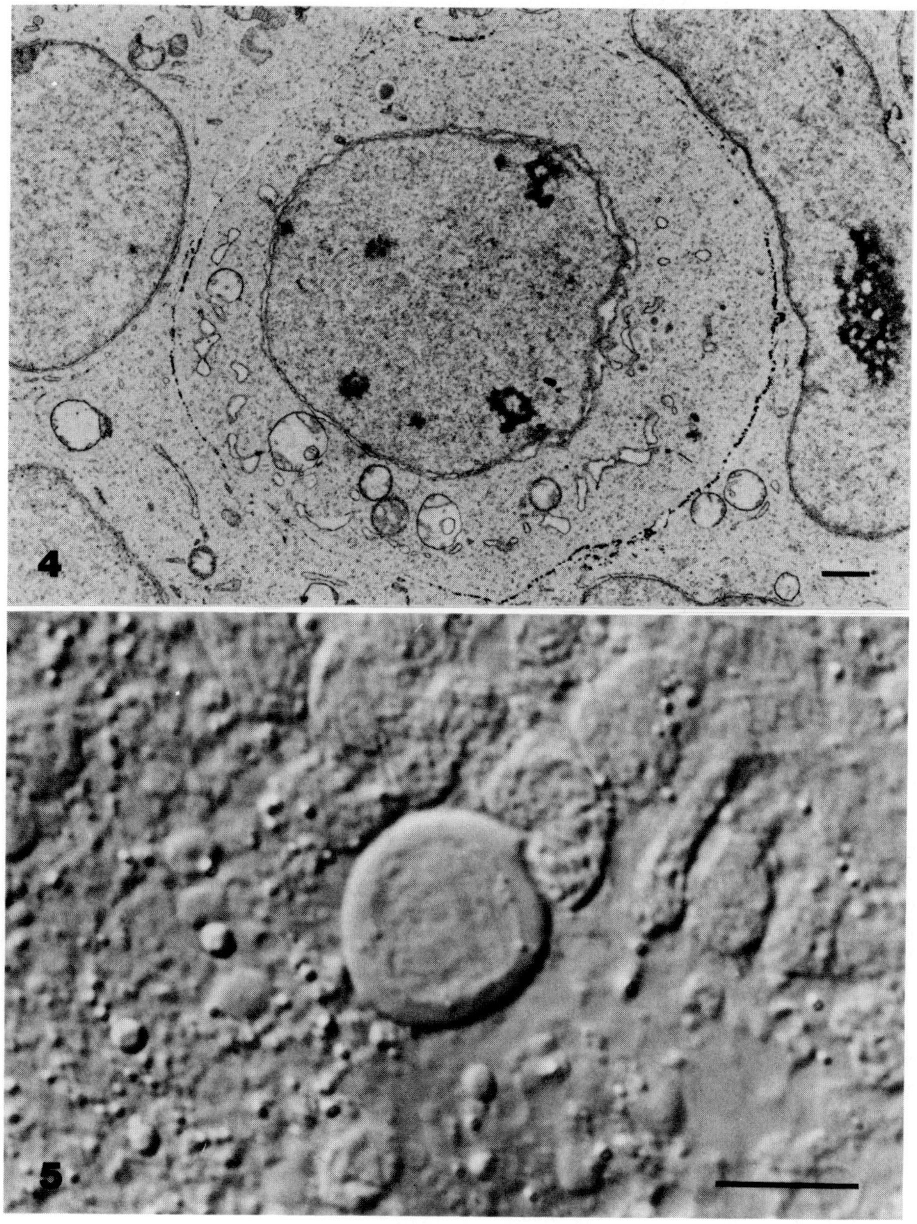

Figs. 4–5. (4) A PGC in the genital ridge of a 14-day rat embryo that has been stained for alkaline phosphatase. The histochemical reaction product is deposited along the plasma membrane of the PGC. Scale line equals 1 μm. (5) A living PGC is seen by Nomarski optics in a squash preparation of a genital ridge from a 12-day rat embryo. PGCs are recognizable at the light microscope level because of their large size, rounded profile and usual solitary distribution. Scale line equals 10 μm.

oocytes, in several mammalian species (reviewed in 2). It is particularly abundant in some species such as the golden hamster (57), but unfortunately is rather sparse in the mouse, the favorite animal for studying early mammalian development. Several ultrastructural studies on cleavage and blastulation in this species did not report nuage (e.g., 58–59). However, these studies were not specifically intended for detecting nuage and it is possible that by using a different species in which nuage is more prominent or by using selective probes that recognize nuage, it might be possible to trace nuage throughout early embryogenesis in mammals. Indeed, there is one report containing convincing electron micrographs of nuage in early stages of embryogenesis of the rabbit (60).

Clearly, it is not possible either to exclude or to implicate nuage in the determination of the germ cell line in mammals based on the limited information presently available. However, it is possible to make some predictions of where nuage should be during

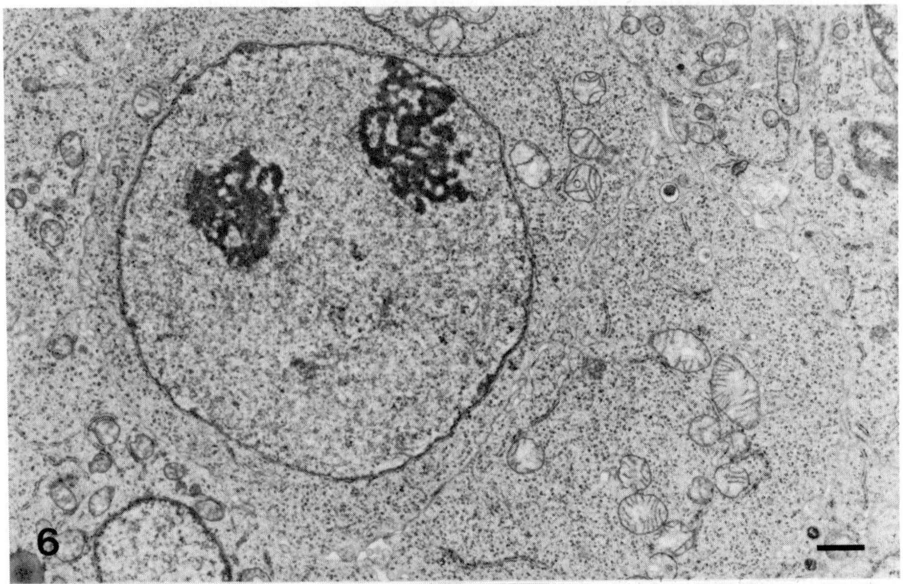

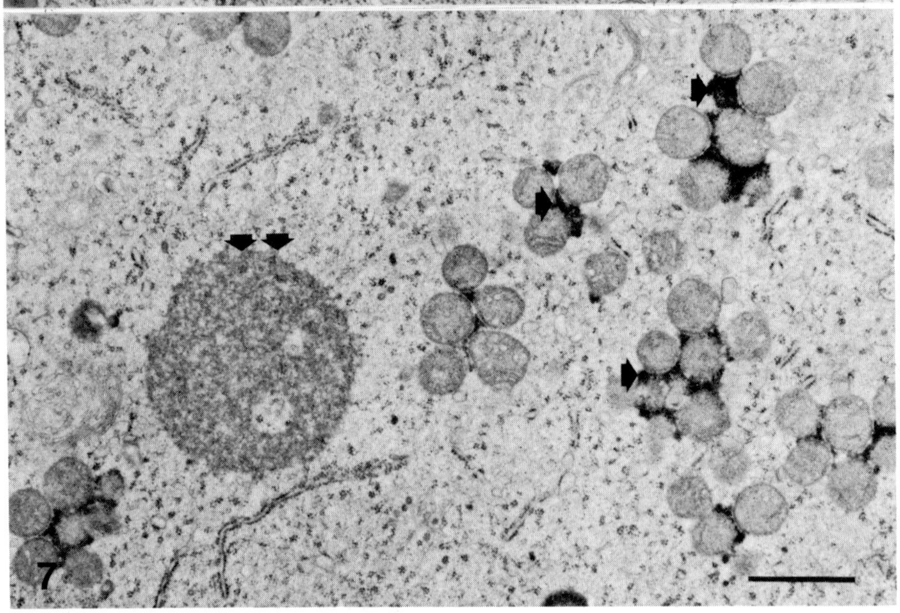

Figs. 6–7. (6) A PGC in the genital ridge of a 13-day rat embryo. The nucleus contains little heterochromatin while the cytoplasm contains abundant ribosomes but relatively few profiles of ER. The scale line equals 1 μm. (7) A region of the cytoplasm of an oocyte from a 13-day postnatal golden hamster. Two forms of nuage are present, the solitary, rounded structure (indicated by the double arrows) and the more electron-dense accumulations lying in the interstices of mitochondrial clusters (single arrows). The scale line equals 1 μm. (Fig. 7 reprinted by permission (5)).

embryogenesis if it is a marker for the germ cell line. For example, nuage would have to be present in all of the first 8 blastomeres, since each can apparently give rise to somatic and germ cells. Likewise, some or all of the cells in the ICM of $3\frac{1}{2}$-day embryos, in the epiblast of $4\frac{1}{2}$-day embryos, and in the posterior end of the primitive streak of 7-day embryos should contain nuage because of their apparent dual somatic cell/germ cell potential (see below). Finally, the nuage should be present in the alkaline phosphatase positive PGCs when they are first detected in 8-day embryos.

5. Cell potential and origin of the germ line

The purpose of the germ cell line is to give rise to cells with the capacity to develop into a new individual by the formation of ectoderm, endoderm and mesoderm, and thence the derivative tissues and organs of these germ layers. At some point during this process, it is necessary to produce the germ cell line in order to assure the next generation. We have seen that in some animals the cells which give rise to the germ cell line are identifiable from the very beginning of development. In these animals, cytoplasmic granules are often present in the cells which are responsible for giving rise to the germ cell line. A variety of studies indicate that these cytoplasmic bodies either contain or serve as markers for the germinal determinant. However, in some other animals such cytoplasmic bodies have not been found consistently throughout the early stages of development. There has not been a demonstration of continuity of transmission of specific components from the egg cytoplasm to precursors of the germ cell line in blastulae and then into PGCs identifiable in gastrula stage embryos. This failure to identify a marker for precursors to the germ cell may be either because the germ cell line is not continuous in these animals, PGCs arising *de novo* by epigenesis (3), or because we have not yet learned how to recognize the germ line markers at all stages. Indeed, it is only quite recently that such germ line markers were found in nematodes (8). Because of this it remains to be seen when the precursors to the germ cell line are formed in such animals as urodeles, chicks and mammals. However, in mammals, recent studies on cell potential in embryos provide information which allow one to predict where such germinal determinants should be located throughout early development, if indeed they are involved in the origin of the germ cell line,

The experimental approach commonly used to determine the developmental potential of specific cells from mouse embryos is to isolate those cells from embryos of a strain carrying particular genetic markers and to test the ability of those cells to form various tissues containing those markers when transplanted into embryos with other genetic markers at those loci. Such cell transplantation studies have indicated which portions of the embryo have the potential to form particular cells and tissues and have suggested where the cells that give rise to the germ cell line are located.

It appears that all of the early blastomeres are totipotent (i.e., can give rise to all components of an embryo). It has been known for some time that each cell from a 2-cell mouse embryo can by itself develop into a mouse (61). More recent studies have shown that the descendants of an individual blastomere isolated from a 4-cell or 8-cell stage mouse embryo and combined with the blastomeres of a host embryo can colonize all tissues in the chimeric mouse that develops (62). Furthermore, these studies have indicated that the germ cell line can arise from any of the blastomeres randomly selected from early cleavage stage mouse embryos, suggesting that the germ cell line precursors are not set aside from somatic cell precursors during this initial period of development. However, it has also been pointed out that these experiments disturb the process of cytoplasmic segregation in the egg that occurs in a regular manner in normal cleavage (63). It remains a formal possibility that experimental manipulation may redirect labile developmental processes that would otherwise result in earlier selection of developmental pathways leading to restriction of cell potential.

The first obvious dichotomy of cell fate in the mouse embryo occurs between the 8-cell and 16-cell stages of development. It has been suggested that microenvironmental differences are responsible for influencing the determination of blastomeres during this period (64) with those on the outside becoming more likely to form trophectoderm cells, which give rise to extraembryonic tissues, and those enclosed by other cells becoming more likely to form the inner cell mass (ICM) which forms the embryo proper. This has been named the 'inside/outside hypothesis' (65). Serial sectioning has demonstrated that inside cells do indeed first appear between the 8-cell and 16-cell stage of development (66), and other studies have shown that the separation of the embryo into trophectoderm and ICM is completed by the 64-cell stage (reviewed in 67). Transplantation experiments using radioactively or genetically labeled cells from cleavage stage embryos have shown that cells placed on the outside tend to become trophectoderm and cells placed on the inside tend to become ICM (66, 68). Although it has not been shown whether determination of trophectoderm and ICM is gradual or abrupt (69), it appears that the trophectoderm cells undergo a specialization or restriction in potential while the ICM cells remain relatively unrestricted. It has been found that single trophectoderm cells from $3\frac{1}{2}$-day embryos (early blastocyst stage) transplanted into host blastocysts have only the potential to form trophectoderm, while ICM cells transplanted under the same conditions were able to produce both embryonic and extraembryonic tissues (70).

The next separation in cell fate during embryogenesis is apparent in the $4\frac{1}{2}$-day mouse embryo (late blastocyst stage), the segregation of the ICM into epiblast and primitive endoderm. Cells transplanted from epiblasts of $4\frac{1}{2}$-day embryos into $3\frac{1}{2}$-day blastocysts have given rise to tissues normally derived from either ICM or trophectoderm. However, primitive endoderm cells transplanted in the same manner give rise only to extraembryonic tissues (71). These results indicate that primitive endoderm cells are of restricted potential, but that epiblast cells retain nearly the same broad potential of earlier stage ICM cells, being able to form all of the tissues of the fetus as well as the extraembryonic mesoderm of the allantois and yolk sac. In other studies it has been found that genetic markers carried by single cells transplanted from the epiblast of $4\frac{1}{2}$-day embryos to other embryos are present both in somatic cells and in germ cells in resultant chimeras (72). This finding suggests that the definitive germ cell line has not yet been determined at this stage and that cells with the potential to give rise to either germ cells or to somatic cells are still present.

In a different approach to the examination of cell fate, fragments of tissue have been removed from embryos and cultured for 24–48 hr *in vitro*. One such fragment was a small piece from the caudal end of the primitive streak of a 7-day embryo. Snow found that after 24 hr the fragment contained large, round, alkaline phosphatase-positive cells (73) like those identified as PGCs lying at the base of the allantois in the intact 8-day embryo (49). It was further found that not only did the fragment form germ cells, but that the number of germ cells in the fragment was comparable to that in the intact embryo, while the remaining portion of the egg cylinder was nearly devoid of germ cells.

Teratocarcinoma cells are the stem cells of teratomas, tumors which usually occur in the gonads and characteristically contain haphazardly organized tissues derived from all three embryonic germ layers. Teratomas arise spontaneously from PGCs in testes of 11- to 12-day fetuses of the 129/Sv mouse strain (74) and can be experimentally induced in several mouse strains by transplantation of either embryos or fetal testes to the testis capsule of adult syngeneic hosts (75). Teratomas also occur spontaneously in the ovaries of mice of the LT/Sv strain (76) and the teratocarcinoma cells from various tumors can be maintained in ascites forms or as *in vitro* cell lines. It has been suggested that teratomas arise by parthenogenetic development, starting with PGCs attempting to form an embryo, but instead producing a disorganized mass of tissues as well as teratocarcinoma cells. Because teratocarcinoma cells are derived from PGCs, or from their precursors (72, 77, 78), information gained from the study of these cells should be helpful in understanding the origin of the germ cell line. Single teratocarcinoma cells transplanted into a subcutaneous site in an adult host can give rise to teratomas containing tissues derived from all three embryonic germ layers (79) indicating that like germ cells, teratocarcinoma cells have broad developmental potential. This potential was shown dramatically when teratocarcinoma cells were injected into blastocysts. The chimeric mice which resulted were found to contain the genetic markers of the teratocarcinoma cells throughout all the tissues of the body (80–83) including germ cells (82, 84). Not only can teratocarcinoma cells be integrated into embryos, but pleuripotent embryonic stem cell lines isolated from preimplantation mouse embryos can give rise to teratomas (85, 86).

If teratocarcinoma cells are derived from PGCs or their precursors, it follows that the portion of the embryo which can give rise to teratomas should contain precursors to PGCs. It had previously been demonstrated that entire embryos from early cleavage to day 7 of development could produce teratomas containing teratocarcinoma cells when transplanted to ectopic sites (76). In a more recent set of experiments, 6-day mouse embryos were separated into fragments consisting of epiblast, extraembryonic ectoderm, or primitive endoderm and the fragments grafted to the testis capsule of histocompatible hosts (87). The epiblast grafts formed teratomas containing tissues characteristic of all three germ layers as well as teratocarcinoma cells. However, the extraembryonic ectoderm gave rise only to invasive trophoblastic giant cells and the primary endoderm cells failed to form other tissues, giving only primitive endoderm and occasional Reichert's membrane. These experiments strongly suggest that at 6 days only cells present in the epiblast have the potential to give rise to PGCs.

Thus, it appears that cells with the potential ultimately to form germ cells in mice probably include all the blastomeres through the 8-cell stage of development, the inside cells of compacted morula embryos, the cells of the ICM of early blastocysts and cells of the epiblast of late blastocysts. Cells with this potential subsequently lie in the epiblast of 6-day embryos and in the caudal portion of the primitive streak of 7-day embryos.

6. Gene products of the germ line

A few studies have provided evidence suggesting that particular gene products might serve as markers for the germ cell line. They began with the finding that antisera against teratocarcinoma cells recognized cell surface antigens of early embryos and of germ cells (88). These initial observations were extended in a series of studies in which antiserum raised in mice against F9 teratocarcinoma cells was shown to recognize a surface antigen present on blastomeres of mouse embryos shortly after fertilization, but restricted to the cells of the ICM and of the epiblast at later stages (89). In 8-day embryos, the surface antigen was present on 80–90% of the embryonic ectoderm cells, but was detected on only 40–45% of the embryonic ectoderm cells from 9-day embryos. Although positive cells were not detected in 10-day embryos, about 10% of the cells from genital ridges of 13- to 14-day embryos carried the F9 antigen on their surface (89). The same antigen was detected on gonocytes of newborn male mice and on cells in the subsequent stages of spermatogenesis (90) and it was suggested that this antigen is a marker for the germ line in the male (89).

A monoclonal antibody has been reported which recognizes a Forssman-like surface antigen on teratocarcinoma, embryonic and germinal cells (91). However, it appears to recognize an antigen different from the one on F9 cells. The antigen first appears on cells of the ICM at $3\frac{1}{2}$ days and is present on cells of the epiblast and of the primitive endoderm of $4\frac{3}{4}$- to $5\frac{1}{2}$-day embryos. In 6- to $7\frac{3}{4}$-day embryos the antigen was detected on parietal and visceral endoderm cells but not on embryonic ectoderm cells. Germinal cells of 11- to 14-day embryos have the antigen but in 16-day embryos only nongerminal cells were positive. However, shortly after birth the antigen appeared on gonocytes of males and was present on their descendants, including sperm.

Several studies have compared the protein synthetic patterns of early embryonic cells and teratocarcinoma cells (92–94). There are protein similarities and differences between specific regions and stages of embryos and between embryo cells and teratocarcinoma cells, but, in general, teratocarcinoma cells appear to be more like embryonic ectoderm cells than other cells. Other studies have focused on biochemically detectable gene products of PGCs and their possible relationship to gene products of embryonic and teratocarcinoma cells (91). PGCs from genital ridges of 12-day mouse embryos were compared to 5-day embryos, to embryonic ectoderm and endoderm cells from 6- and 7-day embryos, and to teratocarcinoma cells. Of the approximately 1000 polypeptides examined, 19 showed variations between the different cell types with PGCs being similar to embryonic ectoderm but lacking some polypeptides present in epiblast or teratocarcinoma cells, possibly concerned with differentiation of these cell types (91).

These immunological and biochemical studies suggest that there are particular gene products associated with the germ cell line that are also associated with embryonic cells which are precursors to the germ cell line. However, the significance of these observations will depend upon whether it will be possible to learn the relationships of such particular gene products to functions and to characterize them sufficiently well to determine if they are specific markers for the germ cell line. Furthermore, the evidence cited is open to criticism. For example, the F9 antiserum is a multicomponent reagent which may be recognizing different antigenic determinants from the various cell types examined. Also, the Forssman-like antigen is present on other cells beside germ cells and those of the embryo (95). Finally, the nature of the biochemically detected polypeptides is unknown and they have not been shown to be restricted to germ cells or to their precursors. Although these various findings are interesting and suggest that particular gene products may be present which are specific to the germ cell line, this possibility remains far from being firmly established at the present time.

7. Concluding remarks

The evidence currently available indicates that pleuripotential cells are present throughout mammalian development, from early cleavage to primitive streak formation. PGCs are recognizable shortly thereafter and appear to be derived from pleuripotential cells sequestered in the embryonic ectoderm at the caudal end of the primitive streak. Since PGCs and their progeny also have broad developmental potential, being able to give rise to a wide array of tissues to form embryos or teratomas, they could be considered to be a stage in a pleuripotential cell lineage. How does this relate to the origin of the germ cell line in other animals? A cytoplasmic marker has not been traced through this lineage in mammals and cannot be used to identify germ cells before more obvious features arise, as in some other animals. Furthermore, cells in this lineage appear to be able to form either germ cells or somatic cells in mammals and are not committed to becoming germ cells, in contrast to the P cells in nematodes or the pole cells in *Drosophila*. These observations seem to be consistent with the hypothesis that PGCs arise from pleuripotent stem cells not yet committed to other developmental fates (72, 77). However, it also seems possible that PGCs themselves are a stage in the pleuripotential cell lineage present throughout development, making the separation between stem cells and germ cells somewhat artificial. Until better markers can be found for the germ cell line in mammals, this point will remain uncertain and the answer to the question of whether the germ cell line in the mammal appears during development or is continuous during development will depend upon how the origin of the germ cell line is defined.

References

1. Beams HW, Kessel RG: The problem of germ cell determinants. In: International review of cytology. Bourne GH, Danielli JF (eds). New York, Academic Press, 1974, Vol. 39, pp 413–479.
2. Eddy EM: Germ plasm and the differentiation of the germ cell line. In: International review of cytology, Bourne GH, Danielli JF (eds). New York, Academic Press, 1975, Vol. 43, pp 229–280.
3. Nieuwkoop PD, Satasurya LA: Primordial germ cells in the chordates. Cambridge, Cambridge University Press, 1979.
4. McLaren A: Germ cells and soma: A new look at an old problem. New Haven, Yale University Press, 1981.
5. Eddy EM, Clark JM, Gong D, Fenderson BA: Origin and migration of primordial germ cells in mammals. Gamete Res 4: 333–362, 1981.
6. Boveri T: Die Blastomerenkune von Ascaris megalocephala und die Theorie der Chromosomen-individualität. Arch Zellforsch 3: 181–286, 1909.
7. Laufer JS, Bazzicalupo P, Wood WB: Segregation of developmental potential in early embryos of Caenorhabditis elegans. Cell 19: 569–577, 1980.
8. Strome S, Wood WB: Cytoplasmic localization of germ cell markers during embryogenesis in C. elegans. J Cell Biol 91: 184a, 1981.
9. Krieg C, Cole T, Deppe U, Schierenberg E, Schmitt D, Yoder B, von Ehrenstein G: The cellular anatomy of embryos of the nematode Caenorhabditis elegans. Analysis and reconstruction of serial section electron micrographs. Develop Biol 65: 193–215, 1978.
10. Laufer JS, von Ehrenstein G: Nematode development after removal of egg cytoplasm: Absence of localized unbound determinants. Science 211: 402–405, 1981.
11. Huettner AF: The origin of germ cells in Drosophila melanogaster. J Morphol 37: 385–424, 1923.
12. Mahowald AP, Allis CD, Karrer KM, Underwood EM, Waring GL: Germ plasm and pole cells in Drosophila. In: Determinants of spatial organization, 37th Symposium of Society for Developmental Biology. Subtelny S, Koningsberg IR (eds), New York, Academic Press, 1979, pp 127–146.
13. Underwood EM, Caulton JH, Allis CD, Mahowald AP: Developmental fate of pole cells in Drosophila melanogaster. Develop Biol 77: 303–314, 1980.
14. Hegner RW: The germ cell cycle in animals, New York, Macmillan Co., 1914.
15. Mahowald A: Polar granules in Drosophila. III. The continuity of polar granules during the life cycle in Drosophila. J Exp Zool 176: 329–344, 1971.
16. Hegner RW: The effects of removing the germ-cell determinants from the eggs of some Chrysomelid beetles. Biol Bull 16: 19–26, 1908.
17. Geigy R: Action de l'ultra-violet sur le pole germinale dans l'oeuf de Drosophila melanogaster. Rev Suisse Zool 38: 187–288, 1931.
18. Hathaway DS, Selman GG: Certain aspects of cell lineage and morphogenesis studied in embryos of Drosophila melanogaster with an ultra-violet microbeam. J Embryol Exp Morphol 9: 310–325, 1961.
19. Geyer-Duszynska I: Experimental research on chromosome elimina-ination in Cecidomyidae (Diptera). J Exp Zool 141: 391–447, 1959.
20. Okada M, Kleinman IA, Schneiderman HA: Restoration of fertility in sterilized Drosophila eggs by transplantation of polar cytoplasm. Develop Biol 37: 43–54, 1974.
21. Illmensee K, Mahowald AP: Transplantation of posterior polar plasma in Drosophila. Induction of germ cells at the anterior pole of the egg. Proc Nat Acad Sci USA 71: 1016–1020, 1974.
22. Waring GM, Allis CD, Mahowald AP: Isolation of polar granules and the identification of polar granule-specific protein. Develop Biol 66: 197–206, 1978.
23. Bounoure L: L'Origin des cellules reproductrices et la problème de la lignée germinale, 1939 Paris, Gauthier-Villars.
24. Smith LD, Williams MA: Germinal plasm and determination of the primordial germ cells. In: The developmental biology of reproduction. 33rd Symposium of the Society for Developmental Biology. Markert CL, Papaconstantinou J (eds), New York, Academic Press, 1975, pp 3–24.
25. Mahowald AP, Hennen S: Ultrastructure of the 'germ plasm' in eggs and embryos of Rana pipiens. Develop Biol 24: 37–53, 1971.
26. Librera E: Effects on gonad differentiation of the removal of vegetal plasm in eggs and embryos of Discoglossus pictus. Acta Embryol Exp Morphol 7: 217–223, 1964.
27. Gipouloux JD: Effets de l'extrusion totale ou partielle du cytoplasme germinal au cours des premiers stades de la segmentatiuon sur la fertilité des larves d'Àmphibiens Anoures. C R Acad Sci, Paris (Serie D) 273: 2627–2629, 1971.
28. Tanabe K, Kotani M: Relationship between the amount of the 'germinal plasm' and the number of primordial germ cells in Xenopus laevis. J Embryol Exp Morphol 31: 89–98, 1974.
29. Bounoure L: Le sort de la lignée germinale chez la Grenoville rousse après l'action des rayons ultra-violet sur le pôb inférior de l'oeuf. C R Acad Sci Paris 204: 1837–1839, 1937.
30. Smith LD: The role of a 'germinal plasm' in the formation of primordial germ cells in Rana pipiens. Develop Biol 14: 330–347, 1966.
31. Blackler AW: The integrity of the reproductive cell line in the amphibia. In: Current topics in developmental biology. Moscona AA, Monroy A (eds), New York, Academic Press, Vol. 5, 1970, pp 71–87.
32. Ikenishi K, Nieuwkoop PD: Location and ultrastructure of primordial germ cells (PGCs) in Ambystoma mexicanum. Develop. Growth Differ 20: 1–9, 1978.
33. Abramowicz H: Die Entwicklung der Gonadenanlage und Entstehung der Gonocyten bei Triton taeniatus (Schneid.). Morphol Jahrb 47: 593–644, 1913.
34. Takamoto K: The development of entoderm free embryo. J Inst Polytech Osaka City Univ Ser D 4: 51–60, 1953.
35. Blackler AW: Embryonic sex cell in amphibia. In: Advances in reproductive physiology. McLaren A (ed), London, Logos Press, pp 9–28, 1966.
36. Swift CH: Origin and early history of the primordial germ-cells in the chick. Am J Anat 15: 483–516, 1914.
37. Reynaud G: Transfert de cellules germinales primordiales de dindon a l'embryon de poulet por injection intravesculaire. J Embryol Exp Morphol 21: 485–507, 1969.
38. Meyer DB: Application of the periodic acid-Schiff technique to whole chick embryos. Stain Technol 35: 83–89, 1960.
39. Fujimoto T, Ukeshima A, Kiyofuji R: The origin, migration and morphology of the primordial germ cells in the chick embryo. Anat Rec 185: 139–154, 1976.
40. Clawson RC, Domm LV: Origin and early migration of primordial germ cells in the chick: A study of the stages definitive primitive streak through 8 somites. Am J Anat 125: 87–112, 1969.
41. Simon D: La lignee germinale des oiseaux et la migration des gonocytes primaires. In: L'Origine de la lignée germinale chez les vertébrés et chez quel ques groupes d'invertébres. Wolff E (ed), Paris, Germann, 1964, pp 237–262.
42. Dubois R: Localisation et migration des cellules germinales du blastoderme non incubé de poulet d'apprès les résultats de culture *in vitro*. Arch Anat Microsc Morphol Exp 56: 245–264, 1967.
43. Fargeix N: Les cellules germinales du Canard chez des embryous normaux et des embryous de régulation. Etude des jeaunes stades au développment. J Embryol Exp Morphol 22: 477–503, 1969.
44. Eyal-Giladi H, Kochav S: From cleavage to primitive streak formation: a complementary normal table and a new look at the first stages of the development of the chick. Develop Biol 49: 321–337, 1976.
45. Everett, NB: The present status of the germ cell problem in vertebrates. Biol Rev 20: 45–55, 1945.
46. McKay DG, Hertig AT, Adams EC, Danziger S: Histochemical observations on the germ cells of human embryos. Anat Rec 117: 201–219, 1953.
47. Chiquoine AD: The identification, origin and migration of the primordial germ cells in the mouse embryo. Anat Rec 118: 1315–146, 1954.
48. Mintz B, Russell ES: Gene induced embryological modifications of primordial germ cells in the mouse. J Exp Zool 134: 207–237, 1957.
49. Ozdzenski W: Observations on the origin of primordial germ cells in the mouse. Zool Pol 17: 367–379, 1967.
50. Zamboni L, Merchant H: The fine morphology of mouse primordial

germ cells in extragonadal locations. Am J Anat 137: 299–336, 1973.
51. Spiegelman M, Bennett D: A light- and electron-microscopic study of primordial germ cells in the early mouse embryo. J Embryol Exp Morphol 30: 97–118, 1973.
52. Clark JM, Eddy EM: Fine structural observations on the origin and association of primordial germ cells of the mouse. Develop Biol 47: 136–155, 1975.
53. Everett NB: Observational and experimental evidences relating to the origin and differentiation of the definitive germ cells in mice. J Exp Zool 92: 49–91, 1943.
54. Andre J, Rouiller C: L'ultrastructure de la membrane nucleaire des ovocytes de l'araignée (Tegenaria domestica Clark). In: Proc. European conf. electron microscopy, Stockholm. Sjostrand FS, Rhodin J (eds), New York, Academic press, 1956, pp 162-164.
55. Eddy EM: Fine structural observations on the form and distribution of nuage in germ cells of the rat. Anat Rec 178: 731–758, 1974.
56. Eddy EM, Clark JM: Electron microscopic study of migrating primordial germ cells in the rat. In: Electron microscopic concepts of secretion. Ultrastructure of endocrine and reproductive organs. Hess M (ed), New York Wiley and Sons, 1975, pp 151–167.
57. Weakley BS: Granular cytoplasmic bodies in oocytes of the golden hamster during the post-natal period. Z Zellforsch 101: 394–400, 1969.
58. Enders AC: The fine structure of the blastocyst. In: Biology of the blastocyst. Blandau RJ (ed), 1971, Chicago, University of Chicago Press, pp 71–94.
59. Nadijcka M, Hillman N: Ultrastructural studies of the mouse blastocyst substages. J Embryol Exp Morphol 32: 675–695, 1974.
60. Motta P, Van Blerkom J: Présence d'un matériel caractéristique granulaire dans le cytoplasma de l'ovocyte et dans les premières stages de la differenciation des cellules embryonnaires. Bull. Assoc Anat 58: 947–953, 1974.
61. Tarkowski AK: Experiments on the development of isolated blastomeres of mouse eggs. Nature (Lond) 184: 1286–1287, 1959.
62. Kelly SJ: Studies of the potency of the early cleavage blastomeres of the mouse. In: The early development of mammals. Balls M, Wild AE (eds), Cambridge, Cambridge University Press, 1975, pp 97–105.
63. Kelly SJ: Investigations into the degree of cell mixing that occurs between the 8-cell stage and the blastocyst stage of mouse development. J Exp Zool 207: 121–130, 1979.
64. Mintz B: Experimental genetic mosaicism in the mouse. In: Preimplantation stages of pregnancy. Wolstenholme GEW, O'Connor M (eds) London, Churchill, 1965, pp 194–207.
65. Tarkowski AK, Wroblewska J: Development of blastomeres of mouse eggs isolated at the 4- and 8-cell stage. J Embryol Exp Morphol 18: 155–180, 1967.
66. Barlow PW, Owen D, Graham CF: DNA synthesis in the preimplantation mouse embryo. J Embryol Exp Morphol 27: 431–445, 1972.
67. Gardner RL: Analysis of determination and differentiation in the mammalian embryo using intra- and inter-species chimeras. In: The developmental biology of reproduction. 33rd Symposium of the Society for Developmental Biology, Markert CL (ed), New York, Academic Press, 1975, pp 207–236.
68. Hillman N, Sherman MI, Graham CF: The effect of spatial arrangement on cell determination during mouse development. J Embryol Exp Morphol 28: 263–278, 1972.
69. Gardner RL, Rossant J: Determination during embryogenesis. In: Embryogenesis in mammals. Ciba Foundation Symposium 40 (New Series). Amsterdam, Elsevier, 1976, pp. 5–18.
70. Gardner RL: Microsurgical approaches to the study of early mammalian development. In: Birth defects and fetal development: Endocrine and metabolic factors. Moghissi KS (ed), Springfield, Ill. Thomas, 1974, pp 212–233.
71. Garner RL, Papaionannou VE: Differentiation in the trophectoderm and inner cell mass. In: The early development of mammals. Balls M, Wild AE (eds), Cambridge, Cambridge University Press, 1975, pp 107–132.
72. Papaioannou VE, Rossant J, Gardner RL: Stem cell in early mammalian development. In: Stem cells and tissue homeostasis. 2nd Symposium of British Soc. for Cell Biology. Lord BI, Patten CS, Cole RJ (eds), London, Cambridge University Press, 1978, pp 49–69.
73. Snow MHL: Autonomous development of parts isolated from primitive-streak stage mouse embryos. Is development clonal? J Embryol Exp Morphol 65(Suppl): 269–287, 1981.
74. Stevens LC: Origin of testicular teratomas from primordial germ cells in mice. J Nat Cancer Inst 37: 859–861, 1967.
75. Stevens LC: The development of transplantable teratocarcinomas from intratesticular grafts of pre- and post-implantation embryos. Develop Biol 21: 364–382, 1970.
76. Stevens LC, Varnum DS: The development of teratomas from parthenogenetically activated ovarian mouse eggs. Develop Biol 37: 369–380, 1974.
77. Martin GR: Advantages and limitations of teratocarcinoma stem cells as models of development. In: Development in mammals. Johnson MH (ed), New York, North Holland Publishing Co, 1978, Vol III, pp 225–265.
78. Mintz B, Cronmiller C, Custer RP: Somatic cell origin of teratocarcinomas. Proc Nat Acad Sci USA 75: 2834–2838, 1978.
79. Kleinsmith LJ, Pierce CB: Multipotentiality of single embryonal carcinoma cells. Cancer Res 24: 1544–1552, 1964.
80. Brinster RL: The effect of cells transferred into the mouse blastocyst on subsequent development. J Exp Med 140: 1049–1056, 1974.
81. Papaioannou VE, McBurney MW, Gardner RL, Evans MJ: Fate of teratocarcinoma cells injected into early mouse embryos. Nature (Lond) 258: 70–73, 1975.
82. Mintz B, Illmensee K: Normal genetically mosaic mice produced from malignant teratocarcinoma cells. Proc Nat Acad Sci USA 72: 3585–3589, 1975.
83. Illmensee K, Mintz B: Totipotency and normal differentiation of single teratocarcinoma cells cloned by injection into blastocysts. Proc Nat Acad Sci USA 73: 549–553, 1976.
84. Stewart TA, Mintz B: Successive generations of mice produced from an established culture line of euploid teratocarcinoma cells. Proc Nat Acad Sci USA 78: 6314–6318, 1981.
85. Martin GR: Isolation of a pluripotent cell line from early mouse embryos cultured in medium conditioned by teratocarcinoma stem cells. Proc Nat Acad Sci USA 78: 7634–7638, 1981.
86. Evans MJ, Kaufman MH: Establishment in culture of pluripotent cells from mouse embryos. Nature 292: 154–156, 1981.
87. Diwan SB, Stevens LC: Development of teratomas from the ectoderm of mouse egg cylinders. J Nat Cancer Inst 57: 937–939, 1976.
88. Edidin M, Patthey HL, McGuire EJ, Sheffield WD: An antiserum to 'embryoid body' tumor cells that reacts with normal mouse embryos. In: Conference and workshop on embryonic and fetal antigens in cancer. Anderson NG, Coggin JH, Jr. (eds), Oak Ridge, Oak Ridge National Laboratory, 1071, pp 239–248.
89. Jacob F: Mouse teratocarcinoma and embryonic antigens. Immunological Rev 33: 3–32, 1977.
90. Gachelin G, Fellous M, Guenet J-L, Jacob F: Developmental expression of an early embryonic antigen common to mouse spermatozoa and cleavage embryos, and to human spermatozoa: Its expression during spermatogenesis. Develop Biol 50: 310–320, 1976.
91. Evans MJ, Lovell-Badge RH, Stern PL, Stinnakre MG: Cell lineages of the mouse embryo and embryonal carcinoma cells: Forssman antigen distribution and patterns of protein synthesis. In: Cell lineage, stem cells and differentiation, INSERM Symposium No. 10. Le Douarin N (ed), Amsterdam, Elsevier/North Holland Biomedical Press, 1979, pp 115–129.
92. Martin GR, Smith S, Epstein CJ: Protein synthetic patterns in teratocarcinoma stem cells and mouse embryos at early stages of development. Develop Biol 66: 8–16, 1978.
93. Dewey MJ, Filler R, Mintz B: Protein patterns of developmentally totipotent mouse teratocarcinoma cells and normal early embryo cells. Develop Biol 65: 171–182, 1978.
94. Failly-Crépin C, Martin GR: Protein synthesis and differentiation in a clonal line of teratocarcinoma cells and in preimplantation mouse embryo. Cell Diff 8: 61–73, 1979.
95. Fox N, Damjanov I, Martinex-Hernandez A, Knowles BB, Solter D; Immunohistochemical localization of the early embryonic antigen (SSEA-1) in postimplantation mouse embryos and fetal and adult tissue. Develop Biol 83: 391–398, 1981.

Author's address:
Department of Biological Structure
University of Washington
School of Medicine
Seattle, WA 98195, USA

CHAPTER 2

Ultrastructure of primordial germ cells in the early chick embryo

ATSUMI UKESHIMA and TOYOAKI FUJIMOTO

1. Introduction

Primordial germ cells (PGCs) of vertebrates appear in the extragonadal (in amniotes, extraembryonic) site at very early stages of development and migrate to the gonadal anlagen with the advance of embryonal development (e.g., 1–4). PGCs are identifiable by morphological or histochemical properties. These properties provide effective means by which their migratory pathway to the gonads may be traced. As to the morphological properties, PGCs are large in size and have a prominent nucleolus, and large amount of glycogen and lipid. Of the histochemical properties, periodic acid-Schiff (PAS) and alkaline phosphatase reactions have been utilized. For example, the PAS staining for glycogen has been effective for the identification of PGCs in chicks (5–11), turtles (4), lizards (12) and humans (13). Alkaline phosphatase activity has been used for identification of these cells in mice (14–18) and humans (19, 20).

The migratory modes of PGCs from the site of their origin to the forming gonads vary according to the species. In vertebrates, these modes are classified into three types; interstitial, vascular, or a combination of interstitial and vascular (21–23). In chicks, PGCs migrate to the gonads mainly via the vascular network.

In this chapter, the migration process of chick PGCs is discussed in three categories: The first is the *endodermal phase*, in which PGCs originate in and separate from the extraembryonic endoderm, and are contained in the space between endoderm and ectoderm; in the second phase, designated as the *migratory phase*, PGCs accumulating between endoderm and ectoderm enter the forming blood vessels, and circulate through the vascular network and ultimately escape from vessels in the vicinity of the gonadal anlagen; Lastly, the *gonadal phase* involves PGCs which arrive and settle down in the gonad.

Recently, transmission electron microscopic studies (TEM) on PGCs have been reported for various species, such as amphibians (24–26), reptiles (4, 27), birds (9, 10), mice (17, 18, 28, 29), rats (30) and humans (20, 31, 32). In these species, fine structure of PGCs is basically equivalent, although differences appear in some respects. Moreover, scanning electron microscopic studies (SEM) on PGCs in some species have been initiated (33–35). However, these SEM studies only show the surface morphology of PGCs, and further investigation should be expected.

In this chapter, ultrastructure and morphological changes of PGCs following migration are mentioned. In addition, some comparisons of chick PGCs with those in other vertebrates are discussed.

2. General morphology of chick PGCs

Because PGCs are very large compared with the surrounding somatic cells in the chick, as well as in other vertebrates, they are easily distinguished from somatic cells. An exception occurs in the endoderm where PGCs and endodermal cells are equally large. The size of chick PGCs measures 14 μm in mean diameter, but frequently reaches 20 μm just after separation from the endoderm.

The surface morphology of PGCs changes according to their migrating stages. When PGCs are located between the endoderm and ectoderm, or in blood vessels, they exhibit a relatively smooth surface presenting variable numbers of microvilli. In the mesenchyme they exhibit amoeboid features and have many cytoplasmic processes and pseudopodia as well as microvilli (33).

The nucleus usually is vesicular and measures 8 μm in mean diameter. Their nucleocytoplasmic ratio is large. The nucleus frequently is situated eccentrically and has a distinct nuclear membrane. The nucleoplasm is homogenous and has low electron-density. This means that PGCs are cells which have not yet differentiated, because in general a nucleus of an

Van Blerkom, J. and Motta, P.M. (eds.), Ultrastructure of reproduction. ISBN 0-89838-572-5.

immature cell does not show the condensation of chromatin. For example, this applies to endodermal cells prior to stage 7 when PGCs escape from the endoderm, but not to mesenchymal cells from stage 7 onward. Consequently, PGCs display striking profiles among the mesodermal cells after segregation from the endoderm.

The nucleolus of chick PGCs consists of multiple electron-dense bodies which range from 150 to 300 nm in diameter (Figs. 2, 4, 7), and is termed a 'fragmented nucleolus' (9, 10). This is clearly different from that of other amniotes, such as humans (20, 32), mice (17, 28) and reptiles (4, 27). Because this fragmented nucleolus in the chick is found even in PGCs of more developed (12 days) embryos (36), it seems to be a specific feature of chick PGCs.

The cytoplasm of chick PGCs contains abundant glycogen particles, lipid droplets and varying amounts of yolk. Glycogen particles are scattered evenly, or often eccentrically in the cytoplasm of PGCs, and their amount varies depending on developmental stage. Glycogen in PGCs tends to be sequestered in some degree before segregation of PGCs from the endoderm. It increases during the migration period and decreases after settlement in the gonads. By light microscopy, PGCs are easily identified by PAS staining owing to their glycogen content. In reptiles, PGCs also contain abundant glycogen (4, 12), though PGCs of rats (30) or mice (28) rarely have glycogen in their cytoplasm. In human PGCs, glycogen is accumulated in large amounts (20, 32).

Lipids are somewhat localized to one side of PGCs and do not show quantitative changes before or after migration. Like glycogen, lipid inclusions are useful for the identification of PGCs during the migratory period. These inclusions are also found in endodermal cells prior to segregation of PGCs. A high lipid content is also one of characteristics of PGCs in reptiles (4, 12) and humans (20). The abundance of lipids, however, has not been demonstrated in mice (28) or rats (30).

Yolk platelets are present in the cytoplasm of PGCs in the earlier stages of development, especially during the period of PGCs association with endoderm. Viewed with the light microscope, they are roundish structures staining with PAS, which remain undigested by amylase. By TEM, they appear as highly electron-dense structures (Figs. 1, 2). As in chicks, the presence of yolk material is specific to PGCs of oviparous vertebrates, such as amphibians (25, 26), or reptiles (4, 12).

The cytoplasm of PGCs contains numerous free ribosomes, which are often present as polysomes. The endoplasmic reticulum and Golgi apparatus in PGCs are rudimentary until migration period, but progressively they develop after settlement of PGCs in the gonads. The endoplasmic reticulum changes from a vesicular form to a lamellar form as development progresses. Mitochondria are abundant and are either roundish or rod-like. Mitotic figures in PGCs are seen throughout PGCs migration, but are especially evident when PGCs are located between endoderm and ectoderm. Microtubules and microfilaments are sparse and randomly arranged.

3. Migration of PGCs in the chick

3.1. Endodermal phase (stage 4–12)

In the chick, PGCs are first recognized in the endoderm located at the so-called 'germinal crescent' which corresponds to the border of area pellucida and area opaca at stage 4, i.e. in the primitive streak stage (staged according to Hamburger and Hamilton, 37). In this stage endodermal cells exhibit similar feature to PGCs. For example, they are large and contain large numbers of yolk platelets and lipid droplets, as well as glycogen granules. This appearance is conspicuous in the cells of the area opaca.

PGCs are derived from the germinal crescent endoderm as development progresses, and locate in the space between endoderm and ectoderm, and later between endoderm and developing mesoderm. The process of segregation of PGCs from the endoderm has been observed in detail by Clawson and Domm (8). According to these investigators, segregation is initiated by the separation of the junction between the endodermal cells and the apical portions of the PGCs. As a result, intercellular space develops and gradually surrounds the PGCs. Therefore, PGCs are delineated from endodermal cells. TEM reveals that cytoplasmic processes are present in the intercellular space between PGC and neighbouring endodermal cells (Fig. 1). Thus with the separation of these portions the segregation of the PGCs is virtually completed.

The segregation of PGCs from endoderm begins at stage 4 and continues to stage 8. PGCs released from the endoderm are located between endoderm and ectoderm until blood vessels are formed. The distribution of detached PGCs is not limited to the germinal crescent region but widely expands laterally to the embryo proper, although these PGCs are only a fraction of the total. It is possible that the wide dispersion area of PGCs is partly the result of amoeboid movements by PGCs themselves (34).

The number of PGCs ranges from 100 to 200 at the

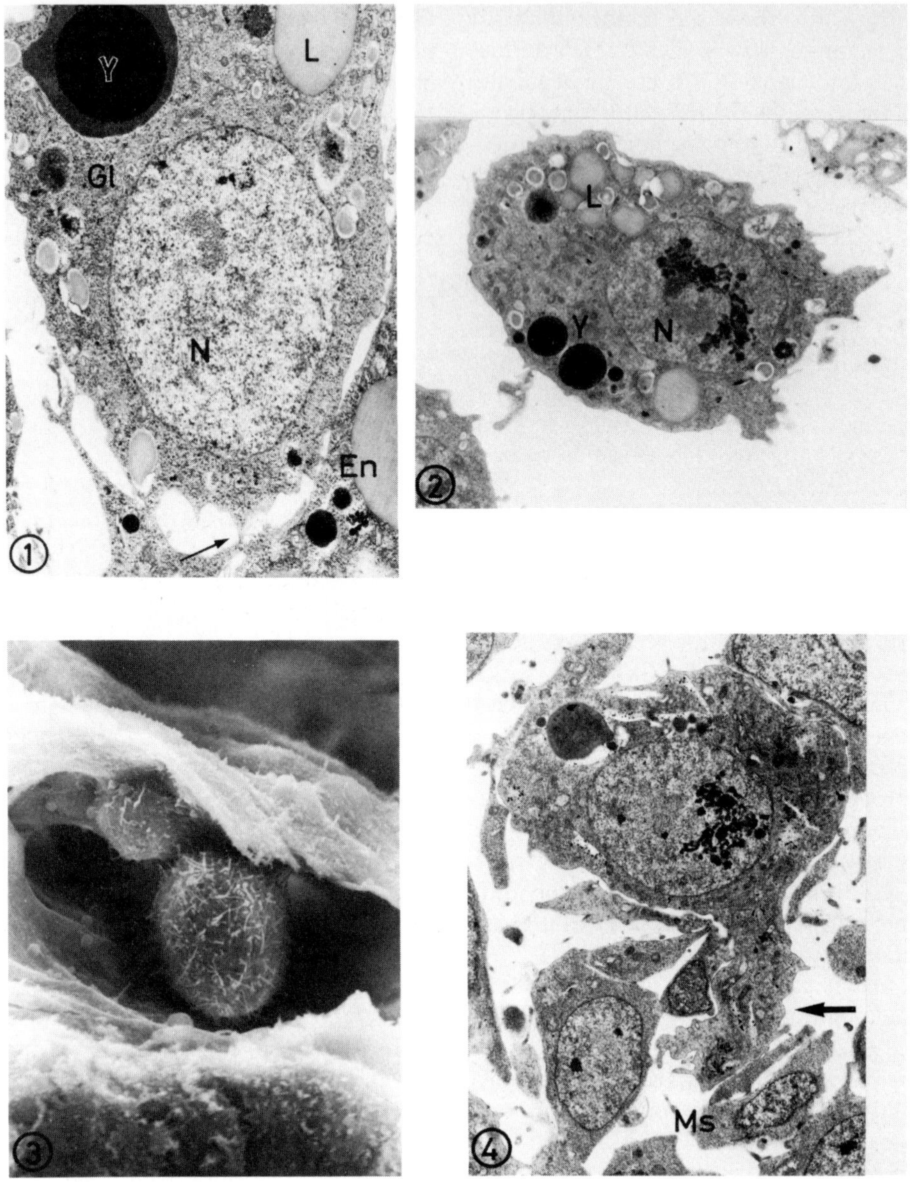

Figs. 1–4. (1) A PGC just separating from the endoderm (En) at stage 7. The cell is almost detached from the endoderm, but still displays thread-like processes (arrow) at the lower end of the cell. The cytoplasm contains yolk platelets (Y), lipid droplets (L) and glycogen particles (Gl). N: Nucleus (× 4,500) From Fujimoto et al. (9). (2) A PGC floating in a space between the endoderm and ectoderm at stage 7. This PGC shows a somewhat irregular shape and cytoplasmic projections. In the cytoplasm, yolk platelets (Y) and lipid droplets (L) are present. Note a peculiar feature of the nucleolus composed of small fragments. N: nucleus (× 3,000) From Fujimoto et al. (10). (3) SEM of a PGC in the primitive blood vessel at stage 10. The cell shows a spherical profile and some microvilli. (× 1,980) From Ukeshima & Fujimoto (33). (4) Amoeboid feature of a PGC in the mesenchyme of a 3-day embryo. An elongated pseudopodium (arrow) is 'creeping' among the mesenchymal cells (Ms). (× 3,000) From Fujimoto et al. (9).

period of PGCs segregation, varying depending on the embryos (8, 10). PGCs located between endoderm and ectoderm are generally round or oval and sometimes possess irregular projections (Fig. 2). The usual sizes of the PGCs range from 14 μm to 16 μm in diameter. However, larger ones measure up to 20 μm and smaller ones are only about 10 μm.

By SEM, the surface of PGCs freed from the endoderm is relatively flattened, although it possesses a small number of blebs and microvilli. The former coincides with the existence of yolk.

PGCs detached from the endoderm are clearly distinguished from the mesodermal elements because at this stage they contain a large amount of yolk and lipid, and many glycogen particles (Fig. 2). Yolk platelets are of variable size and are extremely electron-dense structures. However, some yolk is of low electron-density and others have fibrillar lamellae at their periphery. It is up to this period that PGCs are observed to contain yolk platelets. No PGCs in the migratory phase are observed to have yolk.

Throughout the endodermal phase, mitochondria are abundant and somewhat aggregated within the cytoplasm. The Golgi apparatus and endoplasmic reticulum are poorly developed, the latter often appearing to have vesicular cisternae. Numerous free ribosomes and polysomes are present in the cytoplasm.

3.2. Migratory phase (2–3 days of incubation)

PGCs located in the space between endoderm and ectoderm after segregation soon become surrounded by mesodermal elements expanding from the postero-lateral regions of the embryo. Subsequently, the mesodermal cells form the primitive blood vessels. At this stage, they are only cavities with no blood corpuscles, though PGCs are present (Fig. 3). Dubois (38) examined isolated hypoblast cells in culture, and considered the entrance of PGCs into blood vessels to be the result of active movements of the PGCs themselves. However, because the appearance of the PGCs is followed by formation of the blood vessels, and because most PGCs do not show amoeboid features during this period, the entrance of PGCs into vessels is considered to proceed rather passively, and to be partly complemented by the movements of PGCs themselves.

Almost all PGCs are contained in the vascular network by stage 10. The majority of PGCs are still localized in the area anterior and lateral to the embryo proper, but a small number of them are present within the embryo proper. Once the flow of blood occurs, PGCs circulate with the blood corpuscles throughout the entire embryo via the vascular network. Therefore, PGCs are found in the head, heart, and other tissues once this distribution is expanded into the entire area vasculosa of the blastodisc, including the embryo proper. Circulating through the vascular system, PGCs progressively arrive in the gonads by 3 days of incubation.

Singh and Meyer (39), who applied the PAS technique to circulating blood taken from the chick embryo to investigate the change in the number of the PGCs in the blood at various stages, indicated that the PGCs included in the blood stream attained a maximum number at stage 15 (55 h of incubation). Evidence has appeared recently (40) which indicates that PGCs may escape from the capillaries or small vessels of the splanchnopleure that composes the future dorsal mesentery, posterior to the omphalomesenteric artery. These cells may migrate from the vessels in the manner of leucocytes emerging through the capillary wall.

PGCs in the vascular system are spherical or oval and their surfaces appear to be relatively simple, presenting few microvilli (Fig. 3). The size of PGCs in this period becomes somewhat smaller than those in endodermal phase, measuring 12 μm in mean diameter. A large and spherical nucleus of PGCs and its unique nucleolus are still conspicuous in the blood vessels. PGCs in this period contains many lipids and glycogen particles, but yolk material is rarely observed. Fujimoto et al. (41) examined the circulating blood collected from the 2-day chick embryo (stage 14) by a staining technique specific for lipids, and claimed that the presence of a large amount of lipid in PGCs in this stage may be used as a criterion for distinguishing PGCs from blood corpuscles. After their migration through the wall of the vessels, the PGCs migrate to the gonadal anlagen aided by active amoeboid movement (Fig. 4).

Coelomic epithelium of the presumptive gonad begins to thicken during the migrating period, changing from squamous to columnar epithelium. PGCs penetrate the epithelium by active locomotion through the mesenchyme in the gonadal area. According to observations in humans (20, 32), PGCs often show close associations such as desmosome with the mesenchymal cells during the migration period. Although similar observations are lacking in the chick, PGCs are in contact with mesenchymal cells by means of cytoplasmic projections. Possibly, the mesenchyme may function as a substratum for the amoeboid movement of PGCs. Compared to the other migratory phase, PGCs display their most irregular forms at this stage, extruding pseudopodia or cytoplasmic processes. The nucleus is usually round irrespective of the shape of PGCs. It is considered that the direction of active movement of PGCs is controlled by a chemotactic substance, which is released from the epithelia of the germinal ridges (42, 43).

3.3. Gonadal phase

After their arrival in the gonads, PGCs invade the developing coelomic epithelium. Even in the gonadal area, including the mesenchymal medullary tissues, the migrating PGCs show amoeboid features (Fig. 4). The surface of the PGCs is complex, presenting microvilli, many irregular cytoplasmic protrusions, and occasional pseudopodia (Fig. 5).

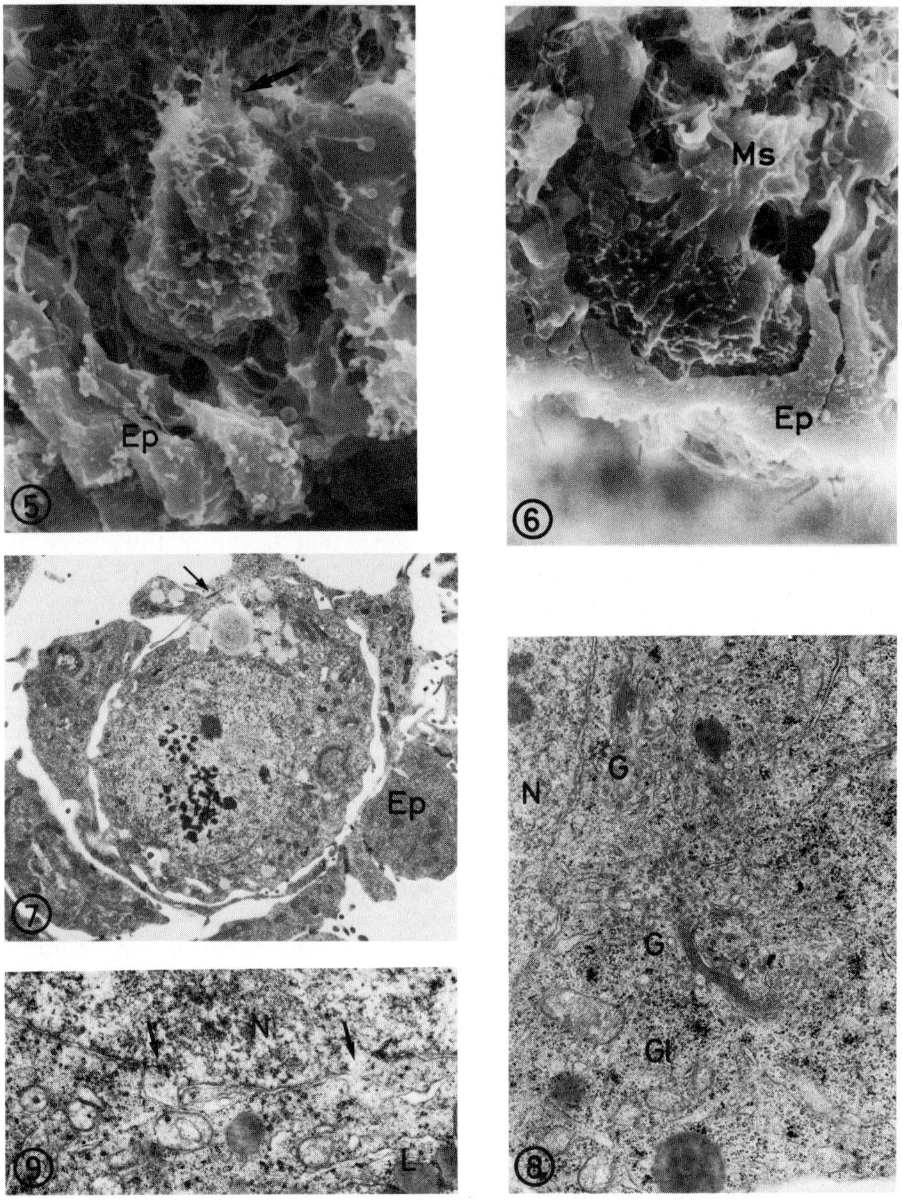

Figs. 5–9. (5) A PGC invading the epithelium (Ep) of the germinal ridge of a 3-day embryo. The cell exhibits a rugged surface with short microvilli. Note an amoeboid feature of the cell with the tail (arrow). (× 3,300) From Ukeshima & Fujimoto (33). (6) SEM of a PGC shortly after settlement in the epithelium (Ep) of the germinal ridge of a 2.5-day embryo. The cell surface still displays an undulated profile and has a small number of microvilli. A lamellipodia of a mesenchymal cells (Ms) adheres to the PGC (× 3,900). (7) TEM of a PGC soon after its arrival at the germinal ridge of a 3-day embryo. The cell is ellipsoidal and surrounded by somatic cells. Note a close association of the PGC with a mesenchymal cell (arrow). Ep: epithelium of the germinal ridge (× 3,300) From Fujimoto et al. (9). (8) Enlargement of a PGC in a 5-day embryo. In this figure, well developed Golgi complexes (G) are found. Some glycogen particles (Gl) are still observed in this PGC. N: nucleus (× 12,000). (9) Enlargement of a part of the PGC in a 5-day embryo. The nucleoplasm protrudes into the cytoplasm (arrows) and the nuclear membrane is continuous with the endoplasmic reticulum. L: lipid droplets N: nucleus (× 12,000).

After invading the epithelia of the germinal ridges, the PGCs return to a round configuration, and are surrounded by the epithelial cells. Therefore, they never face the coelomic cavity directly. The cell surface undergoes radical change. The surface presents a few microvilli, and the long cytoplasmic processes disappear (Fig. 6). The amount of lipids remains unchanged in contrast with the pre-migratory phase, but glycogen is diminished compared to the migratory phase. This supports the concept that glycogen is utilized as the energy source for the migration (44).

After the 4th day of incubation, gonads develop largely. At this period, the PGCs are located not only in the cortical layer but also in the medulla. They are still round and in contact with surrounding somatic cells by means of short projections, although the two cell types are separated slightly by intercellular spaces. At such contact places, tightly associated structures such as desmosomes are often found (Fig. 7). By SEM, associations between PGCs and somatic cells are composed of lamellipodia or filopodia (Fig. 6).

In the cytoplasm of PGCs at the gonadal phase, the endoplasmic reticulum and the Golgi apparatus are well developed (Fig. 8). The Golgi apparatus often appears to be concentrated in one area of the cytoplasm. This aggregation may participate in the formation of the Balbiani body (36). The endoplasmic reticulum consists of elongated cisternae arranged more or less in parallel with the nuclear membrane. A portion of the nucleoplasm often protrudes into the cytoplasm. In this area, the nuclear membrane is continuous with the endoplasmic reticulum (Fig. 9).

4. Comparison of PGCs between chick and other vertebrates

PGCs of vertebrates, especially those of amniotes, have common characteristics. For example, nuclei of PGCs contain fine and homogeneous chromatin and exhibit low electron-density. Condensation of the chromatin near the nuclear membrane is not as great as that of the somatic cells, except in amphibians.

The shape of the nucleolus does differ, however, depending on the species. For example, the nucleolus in amphibians (24) and reptiles (4, 27), is highly electron-dense and has a condensed spherical contour. In rodents (17, 28, 29) and humans (20), it is a high electron-dense mass or a reticular structure. By contrast, the chick nucleolus is an aggregation of electron-dense, rod-like structure. In reptiles, a structure specific to PGCs is found in the vicinity of the nucleolus and is designated the 'paranucleolar mass'. This mass is composed of fine fibrils and particles (27).

The PGCs of chicks and reptiles contain many yolk platelets in the early stages of development, but they are not as prominent as compared with amphibians. Mammals, which are viviparous, rarely have yolk in their PGCs.

In PGCs of the rat, aggregations of electron-dense granules are situated near mitochondria or the nuclear membrane. These aggregations have been termed 'nuage' (45). PGCs of anuran amphibians possess a special portion of the cytoplasm that stains with basic dyes. This region is called the 'germinal plasm' (46). Ultrastructurally, it is composed of mitochondria, ribosomes, vesicles and electron-dense fine granules termed germinal granules (47). Both nuage and germinal plasm and considered to originate by equivalent substance. They are thought to have a role as germ cell determinants. However, no such structures have been identified in the chick.

References

1. Swift CH: Origin and early history of the primordial germ-cells in the chick. Am J Anat 15: 483–516, 1914.
2. Everett NB: Observational and experimental evidences relating to the origin and differentiation of the difinitive germ cells in mice. J Exp Zool 92: 49–92, 1943.
3. Witschi E: Migration of the germ cells of human embryos from the yolk sac to the primitive gonadal folds. Carnegie Contr Embryol 32: 67–80, 1948.
4. Fujimoto T, Ukeshima A, Miyayama Y, Horio F, Ninomiya E: Observations of primordial germ cells in the turtle embryo (Caretta Caretta): Light and electron microscopic studies. Develop, Growth and Differ 21: 3–10, 1979.
5. Meyer DB: Application of the periodic acid-Schiff technique to whole chick embryos. Stain Technol 35: 83–89, 1960.
6. Meyer DB: The migration of primordial germ cells in the chick embryo. Devel Biol 10: 154–190, 1964.
7. Clawson RC, Domm LV: Developmental changes in glycogen content of primordial germ cells in chick embryo. PSEBM 112: 533–537, 1963.
8. Clawson RC, Domm LV: Origin and early migration of primordial germ cells in the chick embryo: A study of the stages definitive primitive streak through eight somites. Am J Anat 125: 87–112, 1969.
9. Fujimoto T, Ukeshima A, Kiyofuji R: Light- and electron-microscopic studies on the origin and migration of the primordial germ cells in the chick. Acta Anat Nippon 50: 22–40, 1975.
10. Fujimoto T, Ukeshima A, Kiyofuji R: The origin, migration and morphology of the primordial germ cells in the chick embryo. Anat Rec 185: 139–154, 1976.
11. Ukeshima A, Fujimoto T: Observations on the migration and distribution of the chick primordial germ cells by application of the PAS reaction to whole embryos. Acta Anat Nippon 50: 15–21, 1975.
12. Hubert J: Etude cytologique et cytochemique des cellules germinales des reptiles au cours du développement embryonnaire et après la naissance. Z Zellforsch 107: 249–264, 1970.
13. Fuyuta M, Miyayama Y, Fujimoto T: Histochemical identification of primordial germ cells in human embryo by PAS reaction. Okajimas Fol anat Jpn 51: 251–262, 1974.
14. Chiquoine AD: The identification, origin, and migration of the primordial germ cells in the mouse embryo. Anat Rec 118–135–146, 1954.
15. Mintz B: Continuity of the female germ cell line from embryo to adult. Arch Anat Microsc Morphol Exp 48: 155–172, 1959.
16. Mintz B: Formation and early development of germ cells. In: Symposium on germ cells and development. Baselli A (ed) Milan, 1960, pp 1–24.
17. Clark JM, Eddy EM: Fine structural observations on the origin and associations of primordial germ cells of the mouse. Devel Biol 47: 136–155, 1975.
18. Jeon KW, Kennedy JR: The primordial germ cells in early mouse embryos: Light and electron microscopic studies. Devel Biol 31: 275–284, 1973.

19. McKay DG, Hertig AT, Adams EC, Danziger S: Histochemical observations on the germ cells of human embryos. Anat Rec 117: 201–219, 1953.
20. Fujimoto T, Miyayama Y, Fuyuta M: The origin, migration and fine morphology of human primordial germ cells. Anat Rec 188: 315–330, 1977.
21. Dubois R, Cuminge D, Smith J: Interpretation of some recent results in experimental embryology and the problem of the germ line. In: Organ culture in biomedical research. Balls M, Monnickendam MA (eds), London, Cambridge University Press, 1976, pp 61–93).
22. Hardisty MW: Primordial germ cells and the vertebrate germ line. In: The vertebrate ovary. Jones RE (ed) New York, Plenum Press, 1978, pp 1–45.
23. Nieuwkoop PD, Sutasurya L: Primordial germ cells in the chordates. Developmental and cell biology series. Abercrombie M, Newth DR, Torrey JG (eds), London, Cambridge University Press, 1979).
24. Al-Mukhtar KAK, Webb AC: An ultrastructural study of primordial germ cells, oogonia and early oocytes in Xenopus laevis. J Embryol exp Morphol 26: 195–217, 1971.
25. Ikenishi K, Nieuwkoop PD: Location and ultrastructure of primordial germ cells (PGCs) in Ambystoma Mexicanum. Devel, Growth and differ 20: 1–9, 1978.
26. Kamimura M, Kotani M, Yamagata K: The migration of presumptive primordial germ cells through the endodermal cell mass in Xenopus laevis: A light and electron microscopic study. J Embryol exp Morphol 59: 1–17, 1980.
27. Hubert J: Ultrastructure des cellules germinales au cours du développement embryonnaire du Lézard vivipare (Lacerta vivipara Jacquin). Z Zellforsch 107: 265–283, 1970.
28. Zamboni L, Merchant H: The fine morphology of mouse primordial germ cells in extragonadal locations. Am J Anat 137: 299–336, 1973.
29. Spiegelman M, Bennett D: A light- and electron-microscopic study of primordial germ cells in the early mouse embryo. J Embryol exp Morphol 30: 97–118, 1973.
30. Merchant H: Rat gonadal and ovarian organogenesis with and without germ cells. An ultrastructural study. Devel Biol 44: 1–21, 1975.
31. Fukuda T: Ultrastructure of primordial germ cells in human embryo. Virchows arch, Ser B Cell Pathol 20: 85–89, 1976.
32. Miyayama Y, Fuyuta M, Fujimoto T: Ultrastructural observations on the origin, amoeboid features and associations of human primordial germ cells. Acta Anat Nippon 52: 255–268, 1977.
33. Ukeshima A, Fujimoto T: Scanning electron microscopy of primordial germ cells in early chick embryos. J Electron Microsc 27: 19–24, 1978.
34. Lee H, Nagele RG, Goldstein MM: Scanning electron microscopy of primordial germ cells in early chick embryos. J Exp Zool 206: 457–462, 1978.
35. Heasman J, Wylie CC: Electron microscopic studies on the structure of motile primordial germ cells of Xenopus laevis in vitro. J Embryol exp Morphol 46: 119–133, 1978.
36. Yamada K, Amanuma A: Fine structure of Balbiani body in germ cells of the chick embryo. J Predental Fac Gifu Coll Dent 6: 87–103, 1980.
37. Hamburger V, Hamilton HL: A series of normal stages in the development of the chick embryo. J Morphol 88: 49–92, 1951.
38. Dubois R: Le mécanisme d'entrée des cellules germinales primordiales dans le réseau vasculaire, chez l'embryon de poulet. J Embryol exp Morphol 21: 255–270, 1969.
39. Singh RP, Meyer DB: Primordial germ cells in blood smears from chick embryos. Science 156: 1503–1504, 1967.
40. Ando Y, Kuwana T, Fujimoto T: Chick primordial germ cells escaping from the blood vessel to reach the gonadal primordium. Devel, Growth and Differ 22: 707, 1980.
41. Fujimoto T, Ninomiya T, Ukeshima A: Observations of the primordial germ cells in blood samples from the chick embryo. Devel Biol 49: 278–282, 1976.
42. Dubois R, Cuminge D: Les propriétés sécrétrices et excrétrices de l'epithlium germinatif de l'embryon de poulet: étude morphologique et dynamique par l'autoradiographie au microscope électronique. Ann Biol 9: 479–490, 1970.
43. Cuminge D, Dubois R: Etude ultrastructurale et autoradiographique de l'organogenèse sexuelle précoce chez l'embryon de poulet. Exp Cell Res 64: 243–258, 1971.
44. Dubois R, Cuminge D: Sur l'aspect ultrastructural et histochimique des cellules germinales de l'embryon de poulet. Ann Histochim 13: 35–50, 1968.
45. Eddy EM: Fine structural observations on the form and distribution of nuage in germ cells of the rat. Anat Rec 178: 731–758, 1974.
46. Beams HW, Kessel RG: The problem of germ cell determinants. Int Rev Cytol 39: 413–479, 1974.
47. Ikenishi K, Kotani M: Ultrastructure of the 'germinal plasm' in Xenopus embryos after cleavage. Devel, Growth and Differ 17: 101–110, 1975.

Authors' addresses:
A. Ukeshima
College of Medical Science
Kumamoto University
Kumamoto 862, Japan

T. Fujimoto
Department of Anatomy
Kumamoto University
Medical School
Kumamoto 860, Japan

CHAPTER 3

Germ and somatic cell interactions during gonadal morphogenesis

HORACIO MERCHANT-LARIOS

Introduction

The extragonadal origin of primordial germ cells (PGCs) is now accepted unanimously (1). From the point of view of gonadal morphogenesis, interaction between germ and somatic cell lines is viewed as the interaction between two cell types differing in the time and place of their appearance. The gonad is different from other organs which have been studied as models of morphogenesis (2). The fundamental difference between gonadal morphogenesis and that of other organs is that, although the basic morphogenetic processes (localized cell division, cell movements, programmed cell death, modifications in the extracellular matrix, and cellular differentiation) take place in the gonad, these processes can be affected by the presence of 'foreign' cells, i.e., the PGCs. A series of interactions between germ and somatic cell lines can be established for each stage of gonadal morphogenesis. This chapter will describe somatic cell interactions and the roles played by PGCs in the two principal morphogenetic stages through which the gonad passes during its development: an undifferentiated stage and sexual differentiation.

2. Non-sexual differentiation

In human embryos, gonadal ridges appear on about embryonic day 32 (when crown rump measures approximately 5 mm). These organs are longitudinal narrowings of the coelomic epithelium and are found at the sides of the dorsal mesentery root in the ventromedial region of each mesonephros. Soon these gonadal ridges shorten and become more prominent until, on about ovulation day 40 (when the human embryo crown rump measures between 17 and 20 mm), they are two cylindrical bodies with rounded extremities (3).

The first interactions which occur between PGCs and somatic cells of the gonadal ridge are evident before the PGCs colonize the gonadal ridge. This interaction implies an early differentiation of the coelomic epithelium of the gonadal ridge; apparently only gonadal ridge cells have a specific attraction for PGCs. Although the mechanism and intervening factors of this process have yet to be defined, some authors studying normal chick embryos propose that this attraction is mediated by a steroidal chemotactic agent (4); others hypothesize a mediation by glycoproteins (5). Didier et al. (6) recently compared the number of PGCs during different stages of gonadal development in chick embryos with experimental unilateral mesonephric agenesis; they found that the gonadal somatic cells clearly control germ cell attraction and proliferation. Mintz and Russell report that these two processes are separable in the mutant mouse W/W^v (7). In W/W^v embryos, PGCs migrate to the genital ridge but fail to proliferate, and their number remains very low.

2.1. The origin of the somatic cells

High-resolution techniques (e.g. light and electron microscope comparison of thin and thick sections of Epon embedded tissue, histochemical analysis, and radioautographic observation) applied to the study of gonadal morphogenesis have revived an old controversy about the embryonic origin of somatic cells. Zamboni et al. (8), studying sheep, Upadhyay et al. (9), working with the mouse, and Wartenberg (10), examining rabbit and human embryos, conclude that *most* (10) or *all* of the gonadal somatic cells are derived from the mesonephros (8, 9). These authors base their conclusion on the observation that mesonephric cells appear to leave the mesonephros and go to the gonads to participate in the formation of gonadal somatic cells. Wartenberg further concludes that gonadal and mesonephric interaction and development in small mammals is different from these processes in large mammals. Other investigators (11, 12, 13) claim that they can find no clear evidence to

Van Blerkom, J. and Motta, P.M. (eds.), Ultrastructure of reproduction. ISBN 0-89838-572-5.

support this hypothesis of a mesonephric origin of gonadal somatic cells. Although structural observations alone are unlikely to end the controversy, ultrastructural studies and experimental embryologic studies favor the idea of the non-mesonephric origin of gonadal somatic cells.

If we were to accept the mesonephric origin of these cells, we would also have to accept the de-differentiation of cells with a defined function (in this case, excretory) that form part of an organ, and their subsequent migration to another embryonic position where they acquire a totally new function. Yamada (14) produced this phenomenon in chicks from which he removed lens tissue; subsequently, the adjacent cells disjoined, migrated, and formed the lens. However, this type of event has never been reported during normal embryonic development. Even if this process occurred, representing an interesting model for the study of de-differentiation, the particular circumstances in which gonadal formation takes place do not support the origin of somatic cells from the mesonephros. Because the mesenchymal blastema, from which the mesonephros arise, is situated in the same zone as is the gonadal mesenchyme, the gonad has a great store of undifferentiated tissue from which to draw during its formation. De-differentiation of mesonephros to provide tissue would therefore be unnecessary and unlikely.

The mesonephros continue functioning for various lengths of time, depending on the taxonomic group. They function throughout the lifetime of fish and amphibians but only during the fetal stage in reptiles, birds, and mammals. In mammalian species with short gestation periods, they apparently never function. Thus, mesonephric participation in gonadal formation must be considered according to species. The first thing to establish would be if and for how long they function.

Recent studies attempting to assess the possibility of mesonephric de-differentiation to explain the origin of gonadal somatic cells have led to the consideration of various types of cells from which somatic cells may be derived. It has been hypothesized that somatic cells originate from a disjunction of mesonephric tubules in the mouse (9, 10) or, in the goat (8), initially from the epithelial and mesangial cells of a giant nephron and from the walls of the excretory tubules. Although high-resolution techniques were employed in these studies, a difinite ultrastructural distinction between mesonephric and coelomic epithelial cells was never established. Cell identification was based generally on topographical location in embryos fixed at different ages, which makes the elimination of the subjective factor difficult. This is a problem in classic studies (15), because ascertaining the direction of cell movement in two adjoining structures is practically impossible.

In morphological studies, the electron microscopic has been invaluable. With it, the basal lamina can be seen as soon as it has formed, a feat impossible with the light microscope. This is important because the basal lamina can serve as a criterion for the presence of somatic cells in different locations. It has been observed that the basal lamina on the coelomic epithelium is broken in a localized region in the gonadal ridge. In this particular region, these cells are ultrastructurally identical to the epithelium but different from the mesonephric blastema, and seem to provide the cells for the gonadal medulla. This process has been described in the gonadal development of *Xenopus laevis* and *Rana pipiens* (16).

Changes in the basal lamina are not as clear in mammals as they are in amphibians, because the ultrastructural distinction between mesothelial, mesenchymal, and mesonephric cells at the beginning of gonadal development has not been established. The names of these cells are derived exclusively from their location.

Curiously, an alternative criterion that has served to 'mark' the mesonephric cells has been the presence of large numbers of autophagic lysosomes (Figs. 1, 2). This phenomenon is probably due to intense degenerative activity resulting from programmed cell death, a common phenomenon in other developing organs. In mesonephric tubule cells of the rat and mouse, numerous mitotic figures can be observed simultaneously with the appearance of degenerating cells (Fig. 1). This finding can be understood as the disorganization and reorganization of an organ (mesonephros) that will never function in this species, but which nevertheless forms an integral part of the male genital tract and plays an important role in metanephric induction. In mammals, therefore, the mesonephros might play a crucial role in gonadal morphogenesis as well, possibly that of induction. They do not neccesarily donate their own previously de-differentiated cells to the process.

Careful observation during the start of urogenital-crest formation in mammals with short gestation periods reveals a condensation of mesenchymal urogenital cells. This condensation leads to the organization of the mesonephric tubules. The basal lamina of the mesothelium is interrupted in the region of the future genital crest, which is thickened due to mesothelial cell stratification. This is a region of great proliferative activity and cellular movement, implying a large consumption of energy. A marked

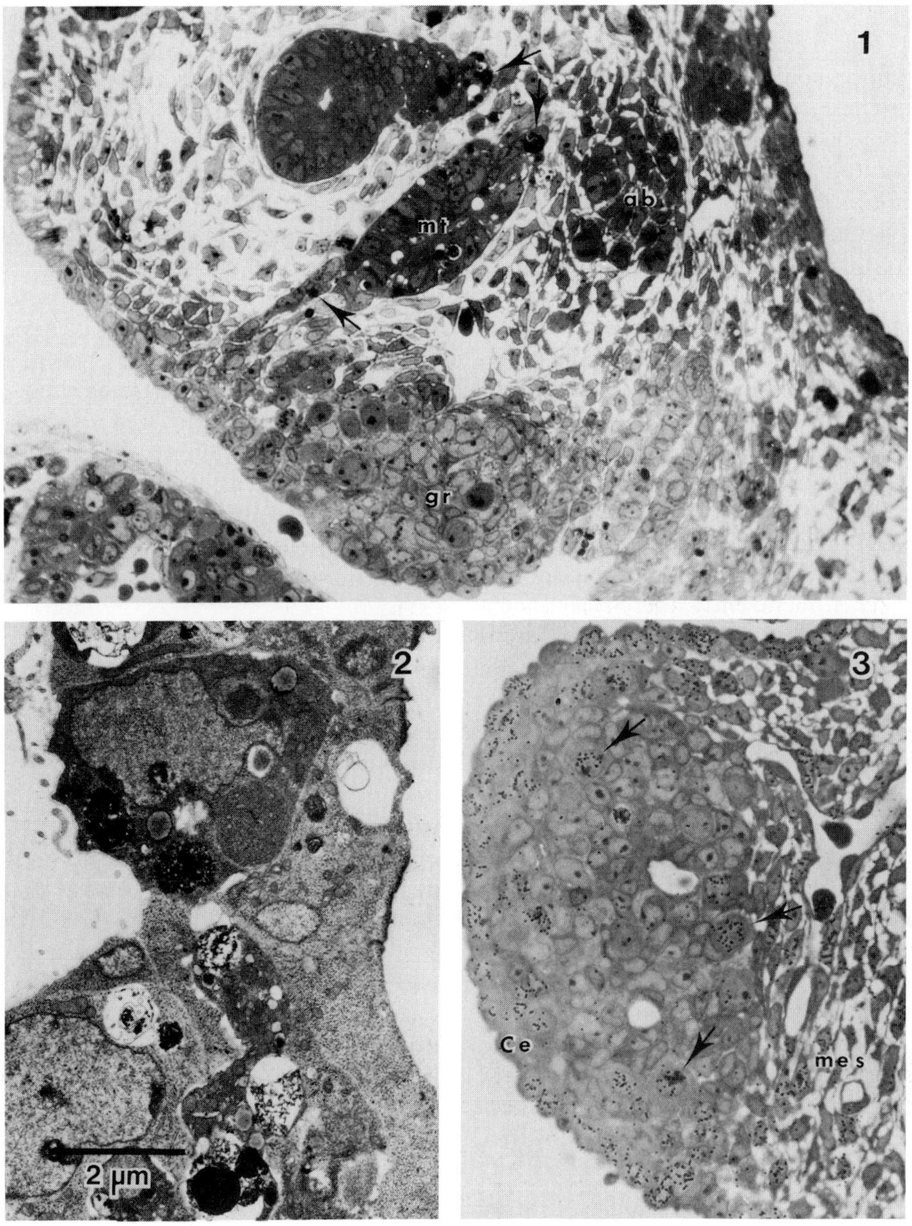

Figs. 1–3. (1) Urogenital region of a rat embryo on the twelfth day of gestation. Two mesonephric tubules (mt), condensing adrenal blastema (ab), and the gonadal ridge (gr) are shown. Mitotic figures appear in the three structures, and numerous dense autolytic bodies are present among the mesonephric cells (arrows) (× 250). (2) Part of mesonephric tubule taken from the same embryo as in Figure 1. A massive degeneration of cells can be observed (× 9,600). (3) Autoradiography of an undifferentiated gonad from a rat which had been injected i.p. with ^{3}H-thymidine twelve hours befores fixation. Coelomic epithelial cells (Ce), between the mesenchyme (mes) and mesothelium, and primordial germ cells (arrows) are labeled (× 400).

invasion of blood vessels from the mesonephric region, together with the approach of the PGCs toward the genital crest, gives the impression that growth of the genital crest occurs through the migration of mesonephric cells.

Which mesonephric cells could potentially participate in gonadal formation? To answer this question, it is necessary to consider whether the mesonephros function as an excretory organ in the species studied. In those species in which the mesonephros are not functional, as in the rat and mouse, this organ consists of the mesonephric or Wolffian duct and the mesonephric tubules; most of the tubules are still unformed when the genital crest begins to grow.

Mesonephric tubules are located close to the genital crest, and the basal lamina is absent. The cells in the ventral region of the forming mesonephric tubules are loosely arranged. All of these facts together could give the impression that the tubules are 'sending' cells toward the region of the coelomic epithelium. Here the basal lamina is interrupted and proliferative activity is very clear.

We have demonstrated, in an autoradiographic study of the urogenital region of rat embryos, extensive proliferative activity in the three histologically distinguishable compartments, i.e., the mesonephric tubules, mesenchyme, and mesothelium. Because massive cell death is found in the first compartment, we propose that the morphogenesis of the urogenital region of mammalian species with short gestation periods, such as the mouse and rat, occurs as follows. The mesonephric canal induces the organization of the mesonephric tubules from the mesenchymal cells of the urogenital region; the mesonephros never function because their destruction begins before they develop completely. Concurrently, there is a condensation of epithelial and mesenchymal cells at the level of the coelomic epithelium. Using ^{3}H-thymidine, we have found that the inner cells of this 'gonadal blastema' have a longer cell cycle than the mesothelial and loose mesenchymal cells (13). We have interpreted this behavior as an early sign of differentiation (Fig. 3).

It is well established that urogenital mesenchymal tissue forms the mesonephros and adrenal cortex. Directly, or indirectly if its prior organization as mesonephros is accepted, it also forms at least some of the gonadal somatic cells. Unfortunately, various authors have given this tissue different names, such as 'mesonephric blastema' (17), 'interrenal blastema' (18) and 'mesangial cells' (8), which further confuse the situation. In every case these terms have been used to refer to a group of mesenchymal cells that, upon aggregation, form into a definite structure with an epithelial arrangement.

Strictly speaking, the term 'mesonephric blastema' refers only to cells embryologically responsible for the formation of the mesonephros, and 'adrenal blastema' to cells which forms the adrenal cortex. Development of both structures usually precedes establishment of the genital crest. Therefore most of the urogenital mesenchyme not involved in the formation of the mesonephros and adrenal cortex is available to form the gonad. Thus, the 'de-differentiation' of cells as already-established structures to initiate gonadal construction is unlikely.

A number of mesenchymal cells from the mesonephric region may be mobilized to participate in the formation of the gonad. This fact, taken together with the invasion of blood vessels and the PGCs from the mesonephric region, suggests a mesonephric origin of the somatic cells of the gonad. Nevertheless, these cells are part of the embryological *multivalent urogenital mesenchyme*, and their differentiation occurs during or after the formation of the mesonephros. At no time does de-differentiation appear to be necessary, as there still are undifferentiated mesenchymal and mesothelial cells that can participate in the formation of the gonad.

Experimental evidence supports the above interpretation. Humphrey's studies of *Rana sylvatica* (19), corrobotated by Cambar and Mesnage in other frogs and toads (20), demonstrated that experimentally provoked mesonephric agenesis does not block the formation of the gonad. Bishop-Calamé, working with chick embryos in which pronephric cord growth was experimentally inhibited, thus preventing mesonephric differentiation, reported the formation of a gonad, although smaller than that of normal animals (21).

Combining this experimental approach with high-resolution techniques, we concluded that, apart from a retardation in growth, the establishment of an undifferentiated gonad in the chick occurs in the absence of mesonephros (Figs. 4–6) (22). It was impossible to demonstrate morphologically the contribution of the coelomic epithelium to the formation of the gonad in this study. However, it was clear that mesenchyme participating in the formation of the gonadal medulla *does not* require prior organization as mesonephros. This finding permits clarification of two important points concerning this organ's role in the formation of chick gonadal somatic cells. First, urogenital mesenchyme is a pluripotential tissue from which the adrenal cortex and gonadal medulla are organized independent of the mesonephros. Second, to designate the mesenchyme which forms part of the adrenal gland and gonad as the 'mesonephric blastema' is confusing. When no stimulus organizes the mesonephros, the same tissue can be called 'adrenal blastema' and 'gonadal blastema' as it develops into these organs.

2.2. The role of the basal lamina

The importance of the basal lamina has been shown in the morphogenesis of several organs (23, 24). Presence of the basal lamina indicates the separation of two different cell types. Laminal interruption has been considered a sign of a short-range cellular interaction between two disparate tissues (25). Systematic study of the establishment of the undifferentiated gonad has revealed that the basal lamina appears gradually. In mammals, this organ first

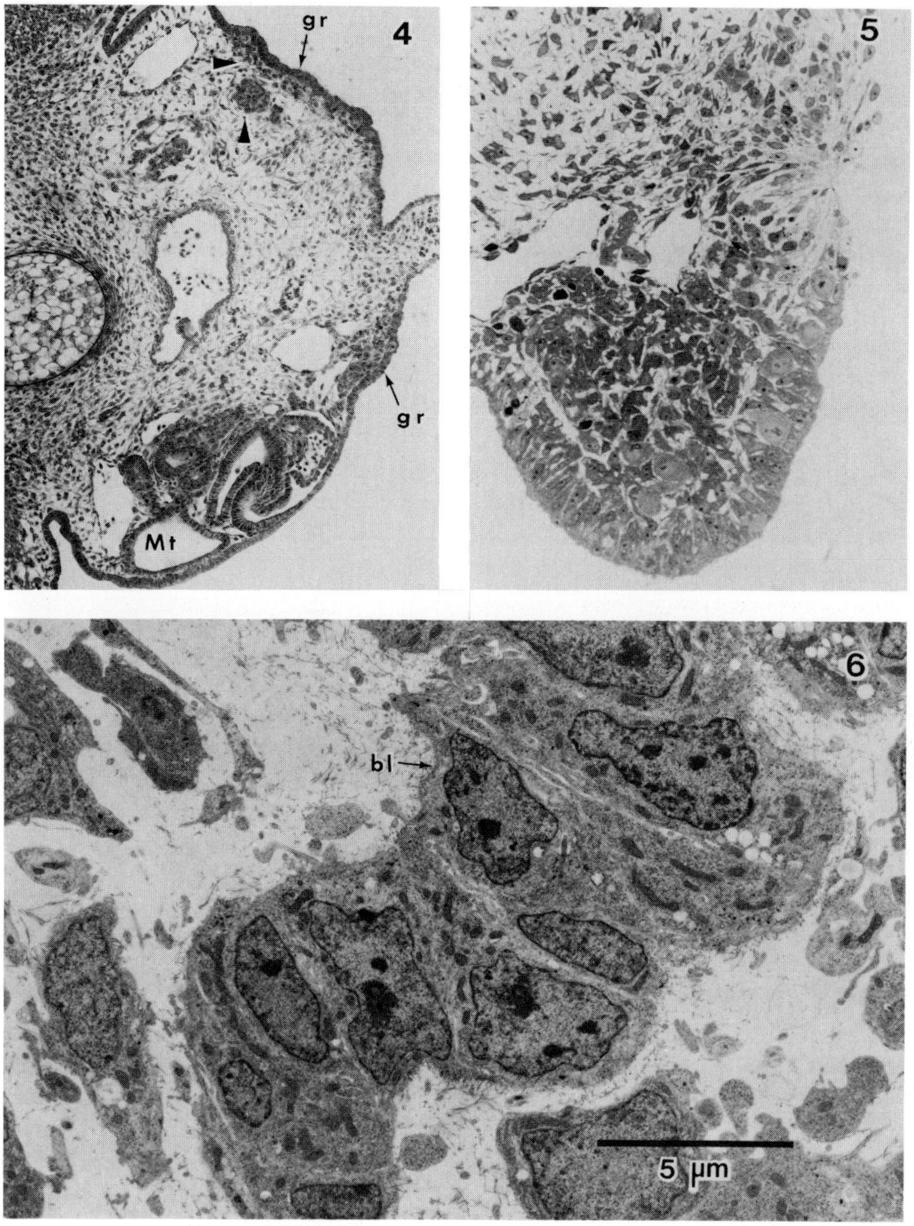

Figs. 4–6. (4) Cross section of an operated chick embryo fixed on fifth day of incubation. In the lower part of figure are a normally developing mesonephros (Mt) and a gonadal ridge (gr). The upper part of figure corresponds to the embryo's left side, where mesonephric formation was prevented. Mesenchyme is condensing normally, and the gonadal ridge is thickening normally (× 100). (5) Normally developing gonadal ridge of a chick embryo on sixth day of incubation. Mesonephric agenesis is complete (× 250). (6) Medullary sex cords of the gonadal ridge from the same embryo as in Figure 5. The basal lamina (bl) and the cellular ultrastructure of the cells of the sex cord are similar to those of chicks with mesonephros. This figure, and 4 and 5, are from Reyss-Brion, Popova, and Merchant-Larios (unpublished observations) (× 5,600).

becomes visible in two opposing regions of the genital crest. In the ventral region of the coelomic epithelium, it might be the remnants of the basal lamina that originally lined the epithelium. This lamina may have been interrupted by cellular proliferation. In contrast, this organ can be clearly distinguished in the dorsal region of the genital crest, where it completely or partially lines several mesonephric tubules. Those tubules which the basal lamina partially lines either have not yet completely formed or are disintegrating. As gonadal development advances, the lamina becomes increasingly clear, at first apparently separating morphologically identical cells of the gonadal blastema. Subsequently, epi-

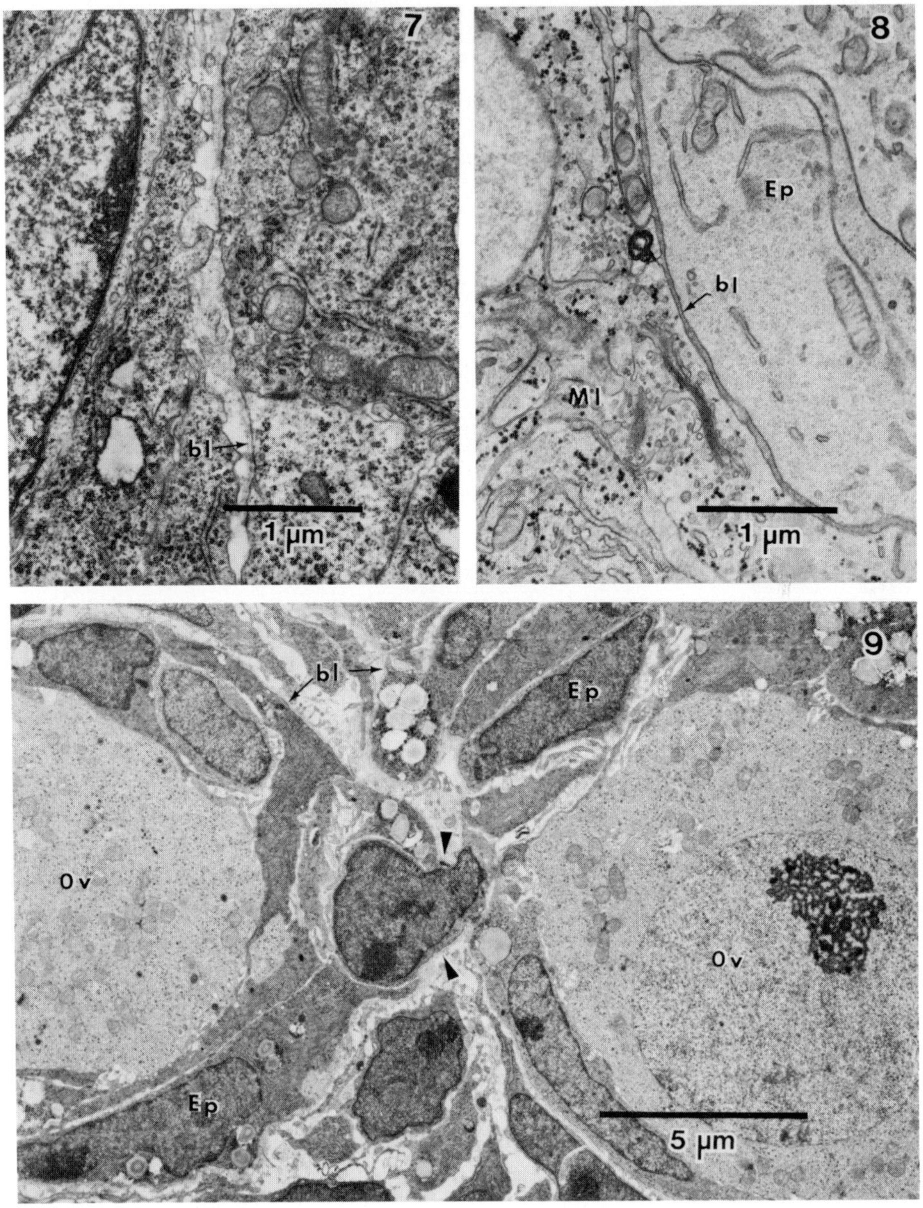

Figs. 7–9. (7) Undifferentiated rat gonad showing the early formation of a thin basal lamina (bl) between cells of the gonadal blastema (19, 200). (8) Part of the basal lamina (bl) separating tightly arranged epithelial (Ep) and mesenchymal (Ml) cells. The appearance of glycogen particles in the mesenchyme has been enhanced by fixation with O_3O_4-ferrocyanide. Mouse ovary, 14 days of gestation (× 19,200). (9) Early stage of folliculogenesis. The epithelial cords (Ep) are partially fragmented. Oocytes (ov), surrounded by epithelial cells (pregranulosa cells), are pulling away from each other. In this figure, an epithelial cell (arrows) bridges two forming follicles. Basal lamina (bl) surrounds the epithelial cells.

thelial cells separate from mesenchyme (Figs. 7, 8).

One exception to this scheme has been noted in mammals. In the horse, steroidogenic cells appear very early. They simultaneously segregate into two compartments separated by a basal lamina (26). In males, the steroidogenic cells are segregated on the outside of a poorly differentiated epithelial tissue containing PGCs. In females, however, the epithelium containing the PGCs becomes steroidogenic and the stromal cells on the outside remain undifferentiated. This precocious differentiation and segregation of gonadal somatic cells might be a model peculiar to the horse. But it may suggest an early, genetic, sex-determined differentiation of somatic cells in other mammalian species as well. Possibly such early differentiation is established at the molecular level but not manifested, as it is in the horse, at the ultrastructural level. Such a pattern of differentiation would require that the genomes of cells in different locations be selectively activated, depending upon the genetic sex of the embryo.

Contrary to most classic studies (15) and some recent reviews (3, 27), the electron microscope reveals that the *basal lamina appears at the same time in the gonad of both sexes*. Cellular mechanisms leading to basal lamina formation contribute to the establishment of an undifferentiated gonad. Thus, when this stage ends there exists an epithelial component containing most of the germ cells and a slightly differentiated stromal component. However, the complete segregation of the two tissue components by the basal lamina occurs after sexual differentiation. It is common to find areas of direct contact between epithelial and stromal cells in the ovaries as well as in the testis. This structural evidence can be interpreted as supporting the idea that all gonadal cells arise from the gonadal blastema.

3. Gonadal sex-differentiation

3.1. Somatic and germ cell interactions

Apart from their early extragonadal differentiation, female primordial germ cells, at least in mammals 'differ genetically' from their somatic neighbors in at least one important way: they have two functional X chromosomes (28). Ovocytes with only one X chromosome degenerate (29, 30). Having two X chromosomes apparently is not sufficient for survival, however, if the surrounding somatic cells are XY. Several authors (31, 32) have studied allophenic mice, derived from the fusion of blastomeres from a male and a female embryo. The ovocytes of these mice contained the usual two X chromosomes. But, because these animals also had XY chromosomes, testicular tissue surrounded the ovocytes, which soon died (31, 32).

The first clear manifestation of sexual differentiation in vertebrate germ cells is the early initiation of meiosis in females. Byskov (33) first recognized the important role that somatic cells play in meiotic control. She proposed that the mesonephros produce a meiosis inducing substance (MIS) and a meiosis preventing substance (MPS), both of which mediate somatic-germ cell interaction. Purification of these substances and establishment of their modes of action have yet to be achieved. Byskov's work indicates that germ-somatic cell interaction is mediated by relatively long-range diffusible factors. She proposes that these substances are secreted by neighboring mesonephros. However, gonadal epithelial and/or stromal cells also likely play a role in the control of meiosis. This likelihood has been suggested in artifical chimeric gonads obtained by reaggregation of germ and somatic cells *in vitro* (34), and demonstrated in postnatal ovarian follicular cells (35). Our idea is supported by differences between the sexes at the time meiosis starts. Both the genome and genesis of the somatic cells directly in contact with the germ cells within the epithelial cords differ between the sexes.

3.2. Early morphogenetic events

Structurally, the most dramatic morphogenetic event which occurs during sexual differentiation of the vertebrate gonad is *the early separation of the epithelial compartment from the coelomic epithelium* in genetic males (Fig. 10). This separation occurs *via* three mechanisms:

a) Those PGCs that originally arrived at the coelomic epithelium migrate toward the medullary epithelium. The number of cells involved in this migration varies according to species. All frog PGCs migrate (1, 19). In the human (3), rat (1), mouse (1), pig (1), and chick (5, 6), about half of the PGCs migrate. The one reptile studied, *Lacerta vivipare*, did not show migration (36). This secondary migration from the gonadal cortex to the medulla suggests an early differentiation of somatic cells in genetic males.

b) Mesenchyme and blood vessels from the mesonephric region invade the gonad and separate the medullary epithelium from the coelomic epithelium. This leads to formation of the *tunica albuginea*. A stromal invasion simultaneously takes place among the cords, which are easily distinguished under the light microscope.

c) Due to the above processes, an active deposition of basal lamina occurs on the surfaces of both the medullary and coelomic epithelia where the two zones are separated. This deposition is the completion of a process of basal lamina formation that begins when the gonad is still sexually non-differentiated. The same process subsequently occurs in the ovary.

3.3. Folliculogenesis

In mammalian ovaries, in contrast to the testis, the coelomic epithelium separates from the superficial epithelium late in development. Differentiation begins in the remnants of the mesonephric tubules, the deepest region of the ovary. The tubules are often continuous with the gonadal epithelial cores, forming the *rete ovarii*. From this zone grow primary ovocytes. They are surrounded by somatic cells organized as cubic epithelium and are lined by a basal lamina separating the epithelium from the gonadal stroma. The farther the epithelial cells are from the medullary zone, the flatter they become until they are reduced to thin plates separating clusters of germ cells still proliferating in the subcortical region. Two aspects of this process should be emphasized. First, on reaching this stage of development, 'nude' germ cells are no longer found in the stroma; they are embedded in the epithelium and later will become a pool of primordial follicles. Second, the coelomic epithelium, continuous with the superficial epithelium *via* a common basal lamina, is actively proliferating (13). These data underscore the important contribution which the ovarian coelomic epithelium makes to the prefollicular cell population.

The next stage in ovarian morphogenesis is the formation of follicles by 'fragmentation' of the ovarian epithelial cords. This process starts in the vicinity of the *rete ovarii* and gradually extends to the cortical region of the ovary. One can distinguish at least two events which appear to be directly associated with epithelial fragmentation. First, prefollicular cells proliferate. Then the ovocytes separate from each other and individualize, surrounded by a ration of somatic epithelial cells and basal lamina (Fig. 9).

In humans, folliculogenesis of the earliest-forming ovaries begins during the eighteenth week of embryonic development when crown rump measures about (50 mm). By birth (when crown rump measures 300 mm), some ovaries have fully matured and have become antral follicles. Many have not yet reached that stage (37).

3.4. The role of the H-Y antigen in gonadal sexual differentiation

It has been proposed by Watchel et al. (38) and Ohno et al. (39) that the H-Y antigen is directly related to gonadal sexual differentiation. Although genetic evidence indicates a definite relationship between the presence of this antigen and testicular differentiation, the mechanism by which the H-Y antigen leads to testicular morphogenesis remains hypothetical. *In vitro* experiments designed to produce sexual reversion in reaggregated ovarian and testicular rat cells in the presence of the H-Y antigen and/or its antiserum have only been suggestive of the antigen's role (40, 41). The observation *in vitro* of tubule-like and spherical aggregates resembling follicles still awaits confimative high-resolution studies. In an interesting experiment performed by Nagai et al. (42), the formation of a *tunica albuginea* and seminiferous cords was observed in an XX bovine gonad cultivated for five days in the presence of an 18,000-dalton polypeptide secreted by Daudi human male Burkitt lymphoma cells and identified as the H-Y antigen; in contrast to previous studies in which only the epithelial component of the gonad was considered, stromal component participation also was noted during the process of sexual reversion. If these studies prove to be reproducible in other species, we will have an excellent model of morphogenesis in which the presence or absence of a single molecule triggers an entire chain of events leading to differentiation and morphogenesis.

Regarding mechanisms responsible for sexual differentiation of the vertebrate gonad, experimental induction of sexual reversion by means of steroid hormones has been known for some time. Although the degree and duration of this reversion are species-variable, the importance of steroid hormones in the process of gonadal sexual morphogenesis is clear.

Recently, it has been established in the chick (43) and frog (44) that the appearance or disappearance of the H-Y antigen is a consequence of reversion induced by steroid hormones. This and other findings have led Ohno to conclude that, except in mammals, expression of the H-Y antigen is suppressible in heterogametes and inducible in homogametes (39). This suggests that in non-mammalian vertebrates the synthesis of certain steroidogenic aromatizing enzymes, regulation of their activity, and presence of specific receptors for certain steroids, such as testosterone and estradiol, precede the presence of the H-Y antigen. In male mammals, these characteristics of the antigen might be evolutionary adaptations designed to protect the fetal gonad from the maternal endocrine environment.

4. Gonadal development in the absence of germ cells

4.1. The undifferentiated stage

Ever since they first observed the extra-gonadal origin of the PGCs, embryologists have been trying to determine the inductive role of these cells in the establishment of the undifferentiated gonad. However, experiments carried out in amphibians and birds have shown that this process can occur in the absence of PGCs (45–49). Although experimental intervention has proved more difficult in mammals, it has been observed that, following the elimination of most PGCs with busulfan (butilen-1, 4-dimethane sulfonate, an alkylating agent used in the treatment of human leukemia, and highly specific for germ cells) all structural events occurring in a normal gonad also take place in the sterile one (50). Following treatment, at this stage of gonadal development morphogenesis depends exclusively on somatic cells differentiation and interaction.

4.2. Gonadal sexual differentiation

On the basis of morphogenetic events which occur during gonadal sexual differentiation, two processes can be considered in terms of the role of PGCs: 1) tissue rearrangements leading to the early separation of the epithelial cords from the superficial epithelium in the testis; and 2) the initiation of folliculogenesis due to 'fragmentation' of the epithelial cords in the ovary. Gonadal sex can be identified histologically after the first process is completed. Evidently, differentiation of the Leydig cells in the stroma occurs later. Studies done with sterile male embryos suggest that sexual differentiation occurs in the absence of the germ line. Here again, as in the case of the establishment of the undifferentiated gonads, morphogenesis appears to be controlled exclusively by the somatic cells.

The differentiation of the ovary, however, appears to be so seriously affected by the absence of a germ line that folliculogenesis does not occur even if somatic precursor cells, which normally participate in this process, are present (Fig. 11). The involvement of short-range inducing factors has been proposed (50). They are produced by germ cells, which would be the organizational centers of the ovary if this proposal is correct. Recently, Ohno and Matsunaga proposed that male germ cells lack the H-Y antigen and its specific receptor until puberty (51), which is associated with the initiation of meiosis (40). In males, then, definitive organization of the seminiferous cords would be independent of the germ cells. However, in females these cells would play a fundamental role in folliculogenesis. This model was originally advanced to explain folliculogenesis in female birds, which are heterogametic and H-Y positive. Female mammals, on the other hand, are H-Y negative. Therefore, it is difficult to explain female mammalian folliculogenesis on the basis of the H-Y antigen. It is clear that ovocytes are necessary for sex cords to become follicles; an interaction between ovocytes and somatic cells is crucial. We still do not know exactly what agent plays this role in mammals (Fig. 12).

4.3. Oocytes and the establishment of ovarian steroidogenic tissue

Oocytes play a determining role in fragmentation of the epithelial compartment of the ovary leading to the formation of the primary follicles. They also apparently are critical for the establishment of ovarian steroidogenic tissue.

Studies of rat ovaries sterilized with busulfan during fetal life have demonstrated that the quantity of steroidogenic tissue present at the onset of adult life is directly related to the number of follicles that are formed (52). In extreme cases in which there is a total destruction of germ cells, adult ovaries appear histologically similar to fetal ovaries. They show no sign of steroidogenic tissue at the ultrastructural nor the histochemical level, as shown by an absence of $\Delta 5$-3β-HSD reaction. Sometimes a limited amount of steroidogenic tissue differentiates or large masses of steroidogenic tissue occupy a large space within the ovary (Fig. 13). Histochemical and ultrastructural studies in the adult rat suggest a dual origin for the steroidogenic tissue in these ovaries, all of which are sterile. Some steroidogenic cells develop from the internal theca cells of the initially formed follicles but undergo atresia during the first weeks following birth. Others originate in the granulosa of those follicles reaching ovulation during puberty; some of these eventually develop *corpora lutea* and secrete progesterone. A third group arises in the granulosa of those follicles whose atresia is restricted to the degeneration of the ovocytes; what type of steroids these cells produce remains undetermined. Quantity and quality of steroidogenic cells appear to depend on how long the follicles survive, which in turn depends on the number of follicles that have formed.

These observations permit us to speculate on cellular interactions and the morphogenesis of the postnatal ovary. In the rat, the so-called interstitial gland of the normal ovary is formed both by the theca cells of healthy antral follicles and by the remnants of those that have undergone atresia. The activity of this gland appears to be important both for the normal functioning of the ovary as an endocrine gland and for the regulation of internal steroidogenesis in the ovary. With this activity in mind, it is reasonable to propose that the process of atresia that takes place in the neonatal and prepubertal rat ovary is a *necessary condition* for the formation of steroidogenic interstitial tissue in this organ.

In experiments on ovarian development in a mutant W/W^v mouse, we found a morphogenetic pattern in which only a first generation of follicles is

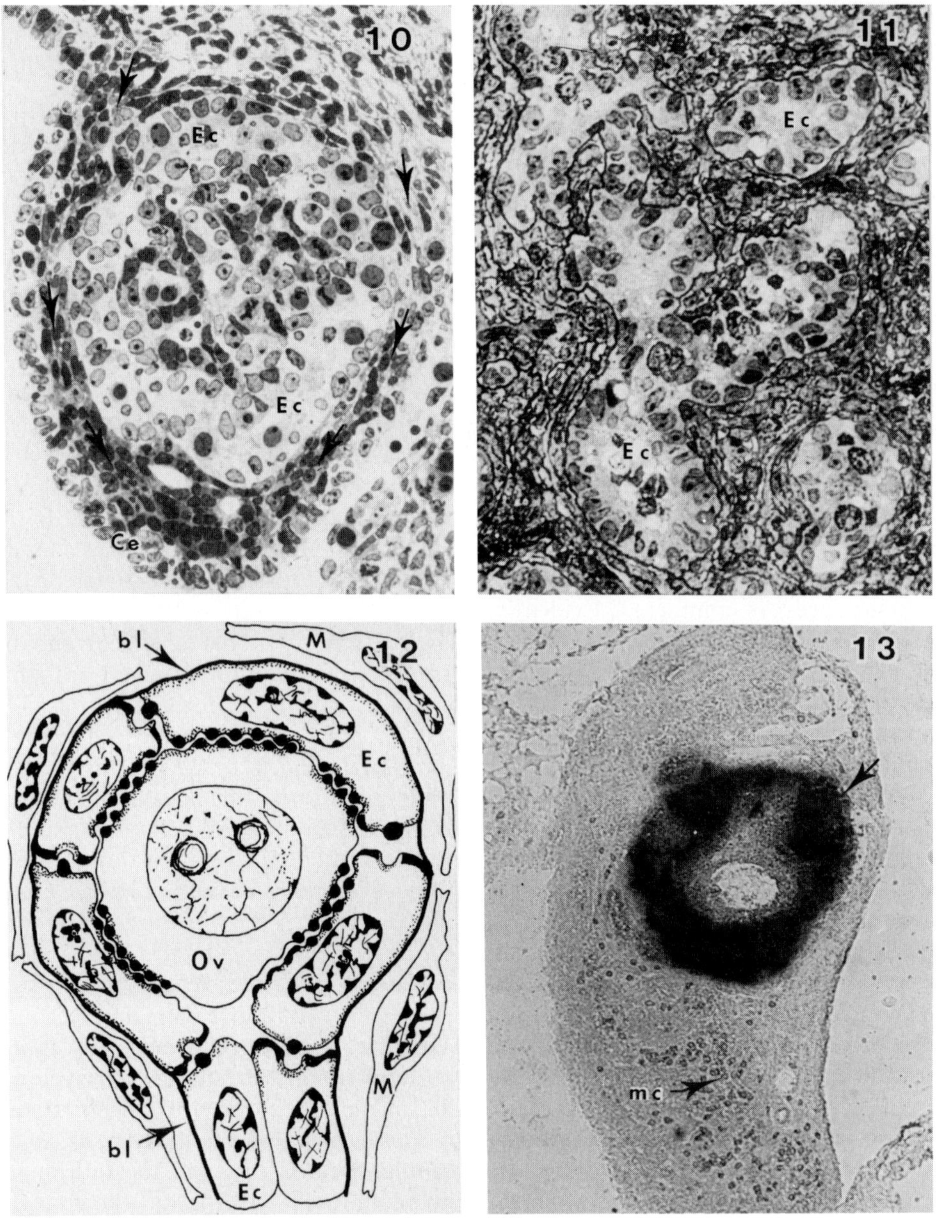

Figs. 10–13. (10) A rat testis fixed on gestation day 13. An invasion (arrows) by mesenchyme and blood vessels separated the epithelial cords (Ec) from the coelomic epithelium (Ce). (11) Epithelial cords (Ec) of a rat ovary fixed ten days after birth. Germ cells were completely eliminated by busulfan administered i.p. to the mother on gestation day 14. Pregranulosa cells and several mast cells form the cords. Semithin section stained with silver methenamine (× 400). (12) Diagram showing the possible róle of the ovocyte (Ov) in organizing the primary follicles derived from the epithelial cords (Ec). Black dots represent surface antigenic molecules, through which the cells interact. Basal lamina (bl) and mesenchymal cells (M) are also indicated. Modified from Ohno and Matsunaga (51), by permission. (13) In this busulfan-treated ovary, an oocyte has organized a follicle which, in turn, has induced the differentiation of abundant Δ^5-3βHSD-positive tissue (arrow) around it. Clumps of this tissue will remain after atresia. Several mast cells (mc) are present in, and surround, the sterile epithelial cords (× 52).

formed from surviving oocytes. These oocytes die, leaving behind scattered masses of steroidogenic tissue (primarily made up of theca cells).

This pattern is similar to that present in rats sterilized with busulfan (52). The observation of a massive death of fetal ovocytes in the mule and hinny during to prophase of the first meiotic division, apparently due to flaws in the coupling of homologous chromosomes from different species (53), also suggests a situation similar to that in the busulfan-treated rat. Finally, cases of gonadal agenesis or dysgenesis in humans could also be explained in terms of the survival of oocytes.

Several studies of Turner's Syndrome support this second idea, that oocytes survive in different stages of development. Singh and Carr (54), reported that a few germ cells may be present in some human fetuses. These ovaries have a fibrous stroma surrounding the epithelium, formed by cords and tubules of both the *rete ovarii* and medullary sex cords. Frequently, variable amounts of steroidogenic (hilus) cells are found in these streak gonads. Froland et al. (55) observed that these germ cells completely dissappear during infancy.

Jones and Scott (56) explain gonadal dysgenesis on the basis of the absence of two normal sex chromosomes. Interchange of genetic material between homologous chromosomes is disturbed during the prophase of the first meiotic division, and the oocyte dies.

This is yet another example in which the absence of oocytes delays folliculogenesis and differentiation of steroidogenic tissue. This tissue acquires its special form and function if some ovocytes survive and organize follicles that grow until their steroidogenic internal theca cells differentiate. After atresia of the follicle and granulosa, these cells remain in the ovary as the sole representatives of the interstitial gland.

Various lines of evidence suggest compartmentalization of steroidogenic activity in the ovary (57). Interstitial tissue appears to be fundamentally androgenic; the cells of the granulosa, estrogenic. Luteal bodies secrete progesterone. Sterilized ovaries offer an opportunity to study *in vivo* the activity of the first compartment in the absence of the other two during different developmental stages.

5. Conclusions

Ultrastructural and experimental evidence presented in this review support the non-mesonephric origin of gonadal somatic cells. The most plausible origin would be that of a 'gonadal blastema' formed by the condensation of mesenchymal and urogenital mesothelial cells (coelomic epithelium). The urogenital mesothelium specifically attracts PGCs. Such an attraction suggests the early differentiation of urogenital mesothelial cells as precursors of gonadal somatic tissue. This process makes highly unlikely a de-differentiation of mesonephros to form the somatic part of the gonad. Sexual differentiation of the gonad implies the programming of various mechanisms that manifest themselves in diverse morphogenetic movements and cell differentiations. Any model proposed to explain gonadal sexual differentiation must take these events into account.

The presence of germ cells does not appear to be necessary for the establishment of the undifferentiated gonad; neither is it necessary for sexual differentiation. Nevertheless, the absence of ovocytes in the ovary during the initial stage of folliculogenesis impedes ovarian development and blocks the cytodifferentiation of the steroidogenic tissue. When few oocytes are present at the start of folliculogenesis, the follicles develop normally but suffer atresia during infancy; they leave behind masses of steroidogenic tissue of the interstitial variety. It is likely that oocyte interaction with ovarian somatic cells is crucial for ovarian development and differentiation as an endocrine organ. Moreover, the process of atresia of the first generation of follicles probably is part of the developmental program of the ovary leading to the establishment of the so-called interstitial gland.

Acknowledgements

I wish to express my thanks to Mr. Felipe Olivera for excellent photography, to Mr. José G. Baltazar for technical assistance, and to M.K. Coyle for editorial assistance.

References

1. Nieuwkoop PD, Sutasurya LA: Primordial germ cells in the chordates. Embryogenesis and Phylogenesis, Cambridge, Cambridge University Press, 1979.
2. Wolpert L: Positional information and pattern formation. Curr Top Dev Biol (6): 183–224, 1971.
3. Mossman HW, Duke KL: The Mammalian Ovary, Madison, University of Wisconsin Press, 1973.
4. Swartz WJ: Effect of steroids on definitive localization of primordial germ cells in the chick embryo. Am J Anat (142): 499–513, 1975.
5. Dubois R, Cumminge D: Chimiotactisme et organization biologique. Etude de l'installation de le lingnée germinale dans les ébauches gonadiques chez l'embryon de poulet. Ann Biol (13): 241–258, 1974.
6. Didier E, Fargeix N, Bergaud Y: Age-dependent control exerted by the somatic part of the gonad upon gonocyte proliferation in the chick embryo. Dev Biol (77): 488–493, 1980.
7. Mintz B, Russell ES: Gene-induced embryological modifications of primordial germ cells in the mouse. J Exp Zool (134): 207–238, 1957.
8. Zamboni L, Bézard J, Mauléon P: The role of the mesonephros in the development of the sheep fetal ovary. Ann Biol Anim Bioch Biophys (19): 1153–1178, 1979.
9. Upadhay S, Luciano JM, Zamboni L: The role of mesonephros in the development of the mouse testis and its excurrent pathways. In: Development and function of reproductive organs- V workshop on development and function of reproductive organs. Copenhagen, July

6–9, 1981. Byskov AG, Peters H (eds), Amsterdam, Excerpta Medica, 1981, pp 18–27.
10. Wartenberg H: The influence of the mesonephric blastema on gonadal development and sexual differentiation. Copenhagen, July 6–9, 1981. Byskov AG, Peters H (eds), Amsterdam, Excerpta Medica, 1981, pp 3–12.
11. Deanesly R: Follicle formation in guinea-pigs and rabbits: a comparative study with notes on the rete ovarii. J Reprod Fert (45): 371–374, 1975.
12. Dang DC, Fouquet JP: Differentiation of the fetal gonad of Macaca Fascicularis with special reference to the testis. Ann Biol Anim Bioch Biophys (19): 1197–1209, 1979.
13. Merchant-Larios H: Origin of the somatic cells in the rat gonad: an autoradiographic approach. Ann Biol Anim Bioch Biophys (19): 1219–1229, 1979.
14. Yamada T: Cellular and subcellular events in wolffian lens regeneration. In: Current topics in developmental biology. Moscona AA, Monroy A (eds), New York, Academic Press, pp 247–283, 1967.
15. Franchi LL, Mandl Am, Zuckerman S: The development of the ovary and the process of oogenesis. In: The ovary I. Zuckerman S (ed), New York, Academic Press, 1962, pp 1–88.
16. Merchant-Larios H, Villaplando I: Ultrastructural events during early gonadal development in Rana pipiens and Xenopus Laevis. Anat Rec (199): 349–360, 1981.
17. Witschi E: Biochemistry of sex differentiation in vertebrate embryos. In: The biochemistry of animal development, Weber R (ed), New York, Academic Press, 1967, pp 193–225.
18. Vannini E: Organogénèse des gonades et déterminisme du sexe chez les amphibiens et les amniotes. Arch Anat Micr Exp (39): 295–313, 1952.
19. Humphrey RR: The development and sex differentiation of the gonad in the wood frog (Rana sylvatica) Following extirpation or orthotopic implantation of the intermediate segment and adjacent mesoderm. J Exp Zool (65): 243–269, 1933.
20. Cambar R, Mesnage J: L' agénésie expérimentale du mésonéphros n' influence pas le développement de la glande génitale chez les Amphibiens Anoures. C R Acad Sci Paris (257): 4021–4023, 1963.
21. Bishop-Calamé S: Etude expérimentale de l' organogénèse du systéme uro-génitale de l'embryon de poulet. Arch Anat Micr Morph Exp 55 (Supp. # 2): 215, 1966.
22. Reyss-Brion M, Popova L, Merchant-Larios H: Unpublished observations.
23. Bernfield MR, Cohn RH, Banerjee SB: Glucosaminoglycans and epithelial organ formation. Am Zool (13): 1067–1083, 1973.
24. Kelley RO, Bluemink JG: An ultrastructual analysis of cell and matrix differentiation during early limb development in Xenopus laevis. Dev Biol (37): 1–17, 1974.
25. Cutler LS, Chaudhry AP: Intercellular contacts at the epithelial-mesenchymal interface during the prenatal development of the rat submandibular gland. Dev Biol (33): 229–240, 1973.
26. Merchant-Larios H: Ultrastructural events in horse gonadal morphogenesis. J Reprod Fert (27 supp): 479–485, 1979.
27. Peters H, McNatty PJ: The ovary: a correlation of structure and function in mammals, London, Granada Publishing, 1980.
28. Ohno S: Life history of female germ cells in mammals. Proc. 2nd int conf on congenital malformations. New York, Int Med Congr Ltd, 1963, pp 36–42.
29. Carr DH, Haggar RA, Har AG: Germ cells in the ovaries of XO female infants. Am J Clin Pathol (49): 521–526, 1968.
30. Burgoyne PS: The role of the sex chromosomes in mammalian germ cell differentiation. Ann Biol anim Bioch Biophys (18): 317–325, 1978.
31. Mystkowska ET, Tarkowski AK: Behaviour of germ cells and sexual differentiation in late embryonic and early postnatal mouse chimaeras. J Embyol Exp Morph (23): 395–405, 1970.
32. McLaren A, Chandley AC, Kofman-Alfaro A: A study of meiotic germ cells in the gonads of foetal mouse chimaeras. J Embryol Exp Morph (27): 515–524, 1972.
33. Byskov AG, Saxen L: Induction of meiosis in fetal mouse testis in vitro. Dev Biol (52): 193–200, 1976.
34. O WS, Baker TG: Germinal and somatic cell interrelationships in gonadal sex differentiation. Ann Biol Anim Bioch Biophys (18): 351–357, 1978.
35. Tsafriri A, Pomerantz SH, Channing CP: Inhibition of ovocyte maturation by porcine follicular fluid: partial characterization of the inhibitor. Biol Reprod (14): 511–516, 1976.
36. Dufaure PH: Recherches descriptives et expérimentales sur les modalités et facteurs du développement de l' appareil génital chez le lézard vivipare (Lacerte vivipare). Arch Anat Micr Morph Exp. (55, supp): 437–537, 1966.
38. Watchel SS, Ohno S, Koo GC, Boyse EA: Possible role for H-Y antigen in the primary determination of sex. Nature (257): 235–236, 1975.
37. Gillman J: The development of the gonads in man, with a consideration of the role of fetal endocrines and the histogenesis of ovarian tumors. Contrib Embryol Carnegie Inst Washington (32): 81–83, 1948.
39. Ohno S, Nagai Y, Ciccarese S, Iwata H: Testis organizing H-Y antigen and the primary sex-determining mechanism of mammals. Rec Progr Horm Res (35): 449–476, 1979.
40. Zenzes MT, Wolf U, Gunther E, Engel W: Studies on the function of H-Y antigen: dissociation and reorganization experiments on rat gonadal tissue. Cytogenet Cell Genet (20): 365–372, 1978.
41. Ohno S, Nagai Y, Ciccarese S: Testicular cells lysostripped of H-Y antigen organize ovarian folliclelike aggregates. Cytogenet Cells Genet (20): 351–364, 1978.
42. Nagai Y, Ciccarese S, Ohno S: The identification of human H-Y antigen and testicular transformation induced by its interaction with the receptor site of bovine fetal ovarian cells. Differentiation (13): 155–164, 1979.
43. Müller U, Zenzes MT, Wolf U, Engel W, Weniger JP: Appearance of H-W (H-Y) antigen in the gonads of oestradiol sex-reversed male chicken embryos. Nature (280): 142–144, 1979.
44. Wachtel SS, Bresler PA, Koide SS: Does H-Y antigen induce the heterogametic ovary? Cell (20): 859–864, 1980.
45. Bounoure L: Sur le développement sexuel des glandes génitales de la grenouille en l'absence des gonocytes. Arch Anat Micr Morph Exp (39): 247–256, 1950.
46. Padoa E: I differenziamiento sesualle dele gonadi di Rana esculenta rese sterili dall'irradiamento con ultravioletto delle indivise. Boll Zool (31): 811–825, 1964.
47. Simon D: Organogénese et différentiation sexuelle des glandes génitales de l'embryon de poulet en l'absence totale des cellules germinales. C R Acad Sci Paris (251): 449–451, 1960.
48. Mc Carrey JR, Abbot UK: Chick gonad differentiation following excision of primordial germ cells. Dev Biol (66): 256–265, 1978.
49. Merchant H: Rat gonadal and ovarian organogenesis with and without germ cells. An ultrastructural study. Dev Biol (44): 1–21, 1975.
50. Merchant-Larios H: The roles of germ cells in the morphogenesis and cytodifferentiation of the rat ovary. In: Progress in differentiation research. Müller-Bérat N Company (ed), Amsterdam, North Holland publishing 1976, pp 453–462.
51. Ohno S, Matsunaga T: The role of H-Y plasma membrane antigen in the evolution of the chromosomal sex determining mechanism. In: Levels of genetic control in development. Subtelney S, Abbot UK (eds), New York, Alan R Liss Inc, 1981, pp 235–246.
52. Merchant-Larios H, Centeno B: Morphogenesis of the ovary from the sterile W/W^v mouse. In: Advances in the morphology of cells and tissues. Acosta Vidrio E, Galina MA (eds), New York, Alan R Liss Inc, 1981, pp 383–392.
53. Benirschke K, Brownhill LE, Beath MM: Somatic chromosomes of the horse, the donkey and their hybrids, the mule and the hinney. J Reprod Fert (4): 319–326, 1962.
54. Singh RP, Carr DH: The anatomy and histology of XO human embryos and fetuses. Anat Rec (155): 369–384, 1966.
55. Froland A, Lykke A, Zachau-Christiansen B: Ovarian dysgenesis (Turner's dyndrome) in the newborn. Acta Path Microbiol Scand (57): 21, 1963.
56. Jones WH, Scott WW: Turner's syndrome and other conditions with streak gonads. In: Hermaphroditism, genital anomalies and related endocrine disorders. Jones HW, Scott WW (eds), Baltimore, the Williams and Wilkins Co, 1971, pp 76–96.
57. Fortune JE, Armstrong DT: Androgen production by theca and granulosa isolated from proestrus rat follicles. Endrocrinology (100): 1341–1347, 1977.

Author's address:
Instituto de Investigaciones Biomédicas
Universidad Nacional Autónoma de México
Apartado Postal 70228
Ciudad Universitaria
04510 México, D.F.
México

CHAPTER 4

Germ cell differentiation and intercellular bridges

BERNARD GONDOS

1. Introduction

Germ cell differentiation proceeds in a highly ordered manner. Stages of migration, mitosis and meiosis follow a regular pattern, with a temporal sequence characteristic of the species. Such a consistent arrangement requires complex and intricate regulatory mechanisms. These mechanisms must be mediated at genetic, biochemical and structural levels enabling coordination of the differentiation process.

Structural factors involved in germ cell differentiation have been studied extensively by electron microscopy. A structural element which first became evident with the use of electron microscopy is the widespread occurrence of intercellular bridges. The formation of true intercellular bridges provides direct interconnection of germ cells. Intercellular connections also occur in other tissues, such as squamous epithelium and smooth muscle, but the connections in these tissues are of a different nature and do not allow open cytoplasmic communication of the type provided by the germ cell bridges.

The existence of intercellular bridges during germ cell differentiation in a wide variety of animal species suggests that the structures represent a key factor in the maturation of germ cells. The functions and significance of intercellular bridges are thought to be similar in the ovary and testis, although with certain obvious differences related to divergence in the patterns of gonadal differentiation. The ultrastructure of the bridges has a characteristic appearance which has been found to be identical in male and female gonads.

Emphasis in this review is placed on considerations related to the ovary. Testicular aspects are also included to enhance general understanding. The opening section reviews ultrastructural aspects of germ cell differentiation in the ovary and testis, particularly in relation to the coordination and regulation of differentiation. The main section reviews in detail information currently available on intercellular bridges, including their structure, formation and significance. Subsequent sections consider the role of the bridges in germ cell degeneration and neoplasia and, finally, suggestions are made as to possible future directions in studies of germ cell interconnections.

2. Germ cell differentiation

After arrival in the gonads, germ cells initially are individually distributed among the more numerous somatic elements (Fig. 1). The primitive germ cells have an ultrastructural appearance characterized by a large round nucleus with prominent reticular nucleolus and evenly distributed chromatin (1). Cytoplasm contains relatively few organelles resulting in an almost clear, watery appearance by electron microscopy. Ribosomes and spherical mitochondria with eccentric cristae are scattered through the cytoplasm. Glycogen, smooth-walled vesicles and a few lipid droplets may also be seen. Cell surfaces are generally smooth with occasional short projections evident at irregular intervals. Some cells retain an ameboid appearance with eccentric nuclei, while others become rounded with centrally placed nuclei.

2.1. Ovary

Germ cell differentiation in the ovary begins with the formation of oogonia from primitive germ cells. The general appearance of the two is similar. The distinction is based largely on location and degree of mitotic activity, oogonia representing those cells which become situated in the cortical cords and begin active multiplication (Fig. 2).

Among the ultrastructural features associated with the formation of oogonia, the appearance of intercellular bridges is particularly noteworthy. Bridges are generally not seen prior to the oogonial stage and it is only with the establishment of active mitotic

Van Blerkom, J. and Motta, P.M. (eds.), Ultrastructure of reproduction. ISBN 0-89838-572-5.

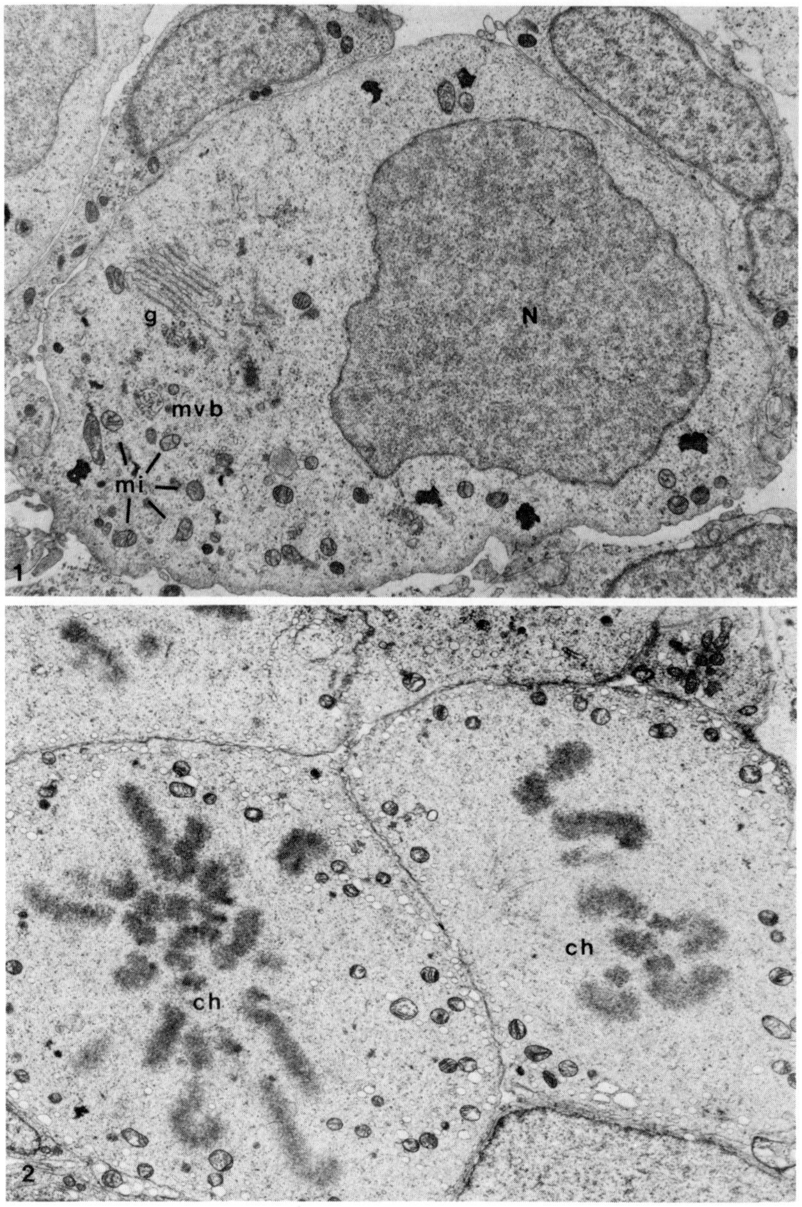

Figs 1–2. (1) Primitive germ cell, human fetal ovary. Ameboid appearance is provided by eccentric nucleus (N) and concentration of organelles at one pole with irregular projections extending from cell surface. g: Golgi; mi: mitochondria; mvb: multivesicular body (× 6,000) (2) Group of oogonia in mitosis. Activity in adjacent cells is closely synchronized as indicated by similarity of chromosomal arrangement (ch) (× 4,000).

division that the pattern of intercellular connection becomes evident (Table 1). Divisions of primitive germ cells during their migration to the genital ridges are usually not associated with bridge formation which first occurs after arrival of the germ cells in the gonadal region (2).

In human fetal ovaries studied by electron microscopy, oogonia appear more rounded than the ameboid-shaped primitive germ cells (3). Blandau (4) noted that germ cells present within the cortical cords were capable only of localized undulatory movement, differing from the ameboid type of movement utilized during the migration of primitive germ cells to the gonads. Differences in mitochondrial morphology have also been noted (5).

Oogonia are recognized by light microscopy as large, round cells with a high nucleocytoplasmic ratio. Their ultrastructural appearance is similar in the different species studied (6). Nuclei are spherical with large conspicuous nucleoli. The chromatin ar-

Table 1. Germ cells and intercellular bridges in ovary.

Stage	Activity	Type of division	Intercellular bridges
Primitive germ cells	Migration	Mitosis	Absent
Oogonia	Proliferation	Mitosis	Present
Oocytes, prefollicular	Maturation	Meiosis*	Present
Oocytes, follicular	Maturation	Meiosis**	Absent

* Meiotic prophase I
** Oocytes in meiotic arrest during most of follicular stage

rangement is uniform and finely granular, without the prominent rim of heterochromatin seen in most somatic cells. The latter enables easy distinction between oogonia and granulosa cells within the cortical cords. Nucleoli are often multiple and usually have a prominent reticular configuration. The paired nuclear membranes are well defined, with nuclear pores easily visualized in appropriate sections.

The oogonial cytoplasm has only a limited number of organelles, including scattered mitochondria, a few elongated cisternae of endoplasmic reticulum and occasionally a small Golgi complex. Ribosomes are uniformly distributed throughout the cytoplasm, mostly in a polysomal arrangement. During oogonial division, centrioles with the usual morphologic appearance can be seen. Occasionally, dense-cored vesicles of uncertain nature and function are present. Small lipid droplets may also be seen.

The oogonial surface is generally regular, but with occasional cytoplasmic ridges interdigitating with projections from the surfaces of adjacent granulosa cells. Desmosome-like structures can be found between oogonia and granulosa cells. In freeze-fracture preparations, small gap junctions of varying configuration have been noted on the surfaces of adjacent germ cells and granulosa cells.

After a series of mitotic divisions, oogonia transform into oocytes and meiosis begins. A period of DNA synthesis in the interphase following the last oogonial division precedes the initiation of meiosis. Meiosis consists of two cell divisions, each including the stages of prophase, metaphase, anaphase and telophase. It is characteristic of mammalian oogenesis that the first meiotic division commences early in development, with progression through prophase during the fetal or neonatal period. At this point, division is arrested, to be completed in the adult at the time of ovulation/fertilization. The oocytes present during the early differentiation period are in the first meiotic prophase. Connection of oocytes by intercellular bridges is seen throughout this period.

The ultrastructure of prefollicular meiotic oocytes in various mammals is similar to that of oocytes and spermatocytes of vertebrates in general, invertebrates and plants, indicating a common structural basis for early meiotic phenomena.

The nuclear changes occurring in the oocytes of different mammalian species can be described as follows (7). At preleptotene and leptotene, thin electron-dense strands appear within the matrix of the nucleus. The threads contain a dense axial core at leptotene. During zygotene, the strands become loosely paired in a 'bouquet' arrangement, and the chromosomal patterns are frequently polarized in one nuclear hemisphere. At pachytene, transformation of the paired chromosomal bivalents into tripartite synaptinemal complexes occurs. The synaptinemal complex is the ultrastructural hallmark of the meiotic stage of germ cell differentiation (Fig. 3). Diplotene is characterized by the reappearance of single threads surrounded by a complex fibrillar sheath organized into lateral projections and loops with associated granules. In the dictyate stage seen in rodents, the chromosomes appear as a reticular arrangement of fine threads associated with beads of heterochromatin.

Throughout the stages of meiotic prophase, the nucleus remains large and rounded and the nuclear envelope generally smooth. Nuclear pores similar to those found in other cell types are frequently seen. Occasionally, localized enlargements of the perinuclear space containing granular material become evident (Fig. 4). These structures may play a role in the production of organelles from the nuclear envelope. The nucleolus remains small and relatively compact until diplotene, when it becomes a large reticular or compact structure. In the human oocyte, numerous micronucleoli associated with segments of heterochromatin appear during this period.

The cytoplasm of meiotic oocytes consists of widely dispersed organelles in a relatively loose matrix. Mitochondria are generally rounded and ribosomes are arranged in diffusely scattered rosettes. Golgi elements occur in small clusters. Centrioles, which are lacking in later stages of oocyte development, are recognizable during meiotic prophase. Smooth-surfaced vesicles of varying size are seen, becoming more numerous at pachytene. They appear to be portions of endoplasmic reticulum, but elements of endoplasmic reticulum in the more usual tubular or cisternal arrangements are rarely seen until diplotene, when multivesicular bodies become prominent and the Golgi complex is also more developed. The general appearance of the cytoplasm at diplotene begins to resemble that of the follicular oocyte.

The appearance of mitochondria has been studied

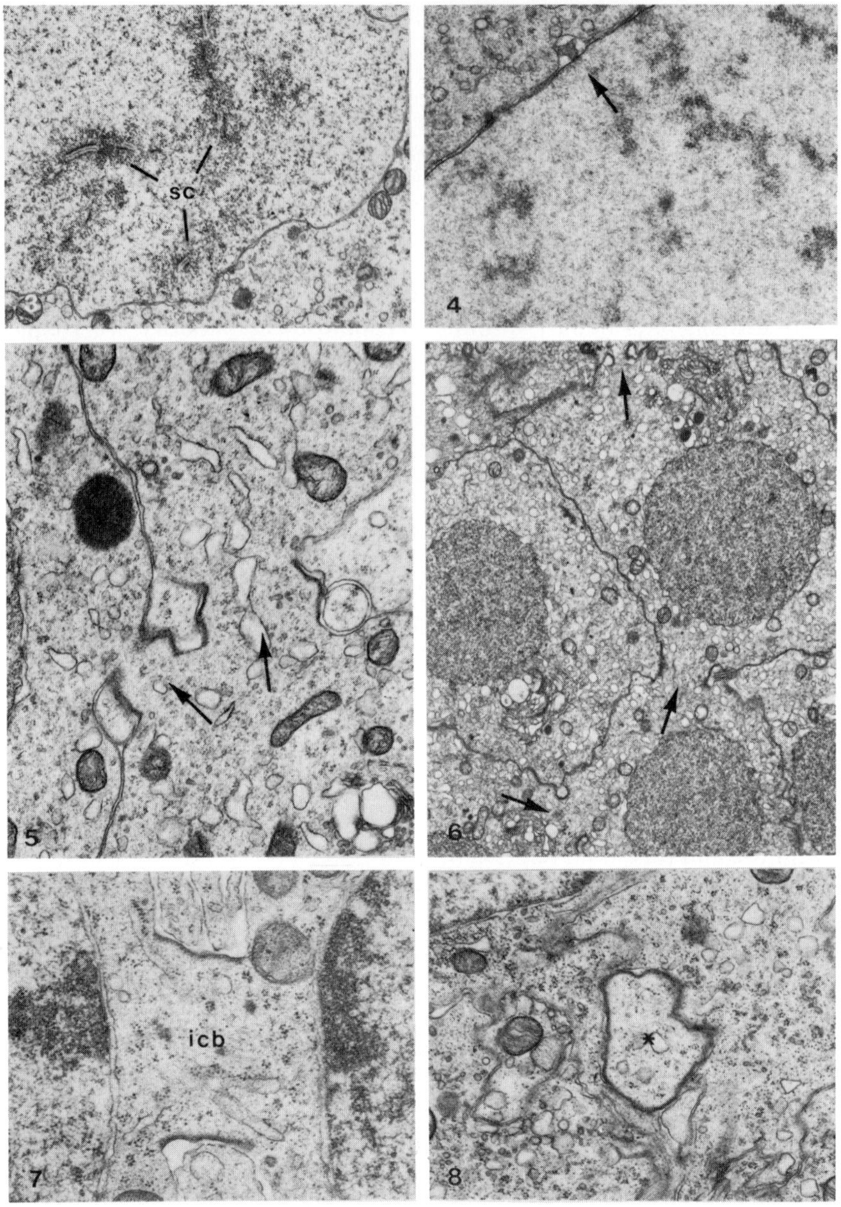

Figs. 3–8. (3) Human fetal oocyte, showing synaptinemal complexes (sc) characteristics of meiotic prophase (× 8,000). (4) Rabbit oocyte with nuclear bleb (arrow) containing dense material (× 14,240). (5) Group of three germ cells connected by two intercellular bridges (arrows), monkey testis (× 16,480). (6) Series of interconnected spermatids, monkey testis. Arrows indicate intercellular bridges (× 5,120). (7) Higher magnification of intercellular bridge (icb) reveals presence of polyribosomes, microfilaments and strands of endoplasmic reticulum (× 19,600). (8) Transverse section of intercellular bridge (*) demonstrates electron density of bridge boundary (× 12,800).

in oocytes of different species and some variations have been noted. In human prefollicular oocytes, mitochondria are preferentially arranged around the nucleus (8). This is a consistent finding regardless of the fixation technique. The perinuclear arrangement is also seen in the monkey (9), although to a lesser extent, and has not been described in other species. Intramitochondrial transformations during oocyte maturation have been observed in the mouse, the changes being attributed to increasing metabolic needs (10). Clusters of mitochondria associated with granular-fibrillar cytoplasmic bodies and intermitochondrial substance are seen in hamster oocytes during meiotic prophase, reaching a peak incidence in diplotene or early dictyate (11).

The surface of prefollicular oocytes is generally

smooth, with occasional microvillus-like projections interdigitating with similar extensions from the surfaces of adjacent granulosa cells. As oocyte maturation progresses and zona pellucida formation takes place, the number of microvilli on the oocyte surface increases greatly. Extensive development of pinocytotic vesicles also becomes evident in the maturing oocyte.

2.2. *Testis*

Table 2. Germ cells and intercellular bridges in testis.

Stage	Activity	Type of division	Intercellular bridges
Primitive germ cells	Migration	Mitosis	Absent
Prespermatogonia	Proliferation	Mitosis*	Present*
Spermatogonia	Proliferation	Mitosis	Present
Spermatocytes	Maturation	Meiosis	Present
Spermatids	Maturation	none	Present**

* Limited mitotic activity, relatively few intercellular bridges
** Bridges disappear in late spermatids

Germ cell differentiation in the testis, in contrast to the ovary, is characterized by a delay in the onset of meiosis. Spermatogenesis does not begin until the time of pubertal maturation. In spite of the difference in timing between oogenesis and spermatogenesis, fundamental similarities exist in the ultrastructural appearance of germ cells at comparable stages of maturation.

The early differentiation of testicular germ cells prior to the onset of spermatogenesis has been referred to as prespermatogenesis (12). Prespermatogenic differentiation includes an initial fetal stage of proliferation, a quiescent period during which germ cell mitosis ceases and a second mitotic stage postnatally preceding the onset of spermatogenesis.

In the indifferent gonad, primitive germ cells are recognized as large round cells distributed among the more numerous and irregularly shaped nongerminal elements (13). After testicular differentiation has occurred, as indicated by the formation of cell cords and a distinct overlying tunica albuginea, the germ cells are designated as gonocytes. Gonocytes are characterized ultrastructurally by their regular round to oval shape, large central nucleus with evenly distributed chromatin and prominent reticular nucleoli. Cytoplasm includes loosely scattered ribosomes, large spherical mitochondria, occasional lipid droplets and granular inclusions resembling lysosomes. The concentration of cytoplasmic organelles is relatively sparse in comparison to adjacent Sertoli cells.

The gonocytes gradually move to the periphery of the cell cords and become situated adjacent to the basal lamina surrounding the cords. At this stage, the germ cells are often arranged in pairs connected by intercellular bridges (Table 2). The term prespermatogonia has been used to indicate the similarity of the prespermatogenic germ cells to spermatogonia in location and pattern of association. In other respects, prespermatogonia differ from spermatogonia by their large size, more regular contour and limited mitotic activity. Also, unlike spermatogonia, prespermatogonia are not directly flattened against the basal lamina, but remain separated from it by intervening Sertoli cell cytoplasm.

The formation of spermatogonia is associated with the onset of spermatogenesis which begins with active mitotic proliferation. Spermatogonia are the direct descendants of the fetal germ cells, although they are smaller in size than either gonocytes or prespermatogonia and show other structural differences. Spermatogonia are arranged at the tubular periphery as rows of cells with slightly flattened oval shape, dense heterochromatin and prominent nucleoli. The latter appear as single compact structures attached to the inner nuclear membrane or in diffusely distributed form, depending on the stage of maturation. The rows of spermatogonia are connected by multiple intercellular bridges resulting in the formation of extensive syncytial aggregates (Fig. 5).

Spermatocyte differentiation is also associated with syncytial arrangement, the interconnection persisting through the two successive meiotic divisions characterizing the primary and secondary spermatocyte stages. During the course of meiotic prophase, the spermatocytes enlarge to several times the size of spermatogonia. Tripartite synaptinemal complexes can be readily seen in the nuclei. Cytoplasm of spermatocytes includes spherical mitochondria with a characteristic central clear zone, abundant endoplasmic reticulum and an increasingly complex Golgi apparatus.

Spermatids are the haploid cells resulting from the second meiotic division. They undergo marked structural changes during their maturation. Nuclear transformation from the spherical shape of the early spermatid to the elongated shape of the spermatozoon is associated with progressive aggregation of chromatin granules, resulting in an extremely dense homogeneous appearance. The acrosome, a glycoprotein-rich structure surrounding the sperm nucleus, is initially formed by contributions from the Golgi complex and endoplasmic reticulum. Proacrosomal granules produced by these structures aggregate into a single spherical dense acrosomal granule located adjacent to the nucleus of the early spermatid. Next, dense material radiates along the surface of the nucleus forming an acrosomal cap. Finally, with progressive elongation, the nucleus is surrounded by several layers of structures, the nuclear membrane, inner acrosomic membrane, acrosome, outer acrosomic membrane and cytoplasmic membrane. Cytoplasmic interconnection of rows of cells persists through much of the spermatid stage (Fig. 6), but the bridges eventually disappear. The spermatid cytoplasm includes a proximal centriole from which develops the locomotor apparatus consisting of an axial filament surrounding a microtubular central canal, a caudal sheath or manchette and a helical arrangement of mitochondria around the axial filament. With the completion of maturation, the cells are released into the tubular lumen as spermatozoa.

3. Intercellular bridges

Electron microscopic studies have revealed that adjacent germ cells are connected by intercellular bridges during the stages of mitosis and meiosis. The bridges are open cytoplasmic connections allowing free intercellular communication. The presence of intercellular bridges, or canals, in invertebrate gonads had been recognized for many years (14, 15), but the observation of bridges in mammalian gonads was a relatively recent finding. Initial investigations described spermatocyte and spermatid connections in several different mammalian species (16, 17). It was subsequently recognized that similar cytoplasmic interconnections could be found at earlier stages of testicular germ cell differentiation (18, 19). Bridges were also found in the ovary in a number of mammalian species throughout much of the early period of development (20–24).

Numerous reports have now appeared in which intercellular bridges between germ cells are described and it is evident that cytoplasmic interconnection of developing germ cells is a widespread, probably ubiquitous phenomenon (see 25). Many of these reports are primarily descriptive in nature, referring to the bridges only in passing. However, several investigations have considered the possible significance and implications of the extensive germ cell interconnection and have begun to place in perspective certain aspects of the role of intercellular bridges in germ cell differentiation.

3.1. Structure

The bridges described in different species have a similar structure. They are short, cylindrical channels bounded by a membrane continuous with the plasma membranes of the conjoined cells. Because of the small size of the bridges, 0.5 to 1.0 μm in diameter, their identification depends on the use of electron microscopy. It is possible under optimal conditions of resolution and magnification to identify them with the light microscope, but ordinary techniques of optical microscopy will fail to show bridges.

By transmission electron microscopy, the thick electron-opaque membranes of the bridges are easily recognized. The membranes of the bridges are thicker than the cell membranes with which they are continuous. The prominent electron density of the bridge boundaries enables them to stand out even at low magnification. Contained within the bridges are cytoplasmic elements identical with those in the interconnected cells. Mitochondria, filaments, microtubules, ribosomes and strands of endoplasmic reticulum are among the organelles that have been seen lying within the confluent cytoplasm. With both ends of the bridge open, free flow of material can occur between the connected cells.

The appearance of intercellular bridges depends on the plane of section. When sectioned parallel to the long axis, an open-ended configuration results and the bridges can be best studied in this manner (Fig. 7). Oblique sections produce an image in which the bridge appears to be closed at one end. Perpendicular sections at right angle to the long axis result in a circular area of electron density (Fig. 8). By analyzing the various configurations observed, it can be determined that the bridges must be cylindrical, allowing open intercellular communication (Fig. 9).

Scanning electron microscopy has confirmed the cylindrical nature of the bridges (26). In addition, it has been observed that they may vary somewhat in length and diameter (27). Examination of the bridge surfaces reveals the presence of microvilli in scattered arrangement.

Freeze-fracture studies have also indicated a cylindrical configuration (Fig. 10). They further demonstrate the continuity of the bridge surface with the membranes of the connected cells. Particles located on the bridge membranes appear to have a random distribution.

Further analysis of the internal structure of intercellular bridges by transmission electron microscopy has revealed some pertinent findings. At high resolution, it is evident that the thickness of the boundaries is a result of a layer of electron-dense material which is deposited on the inner aspect of the bridge membrane (Fig. 11). The nature of this material is not clear, but it has a uniform thickness and texture (Fig. 12) suggesting an intrinsic structural

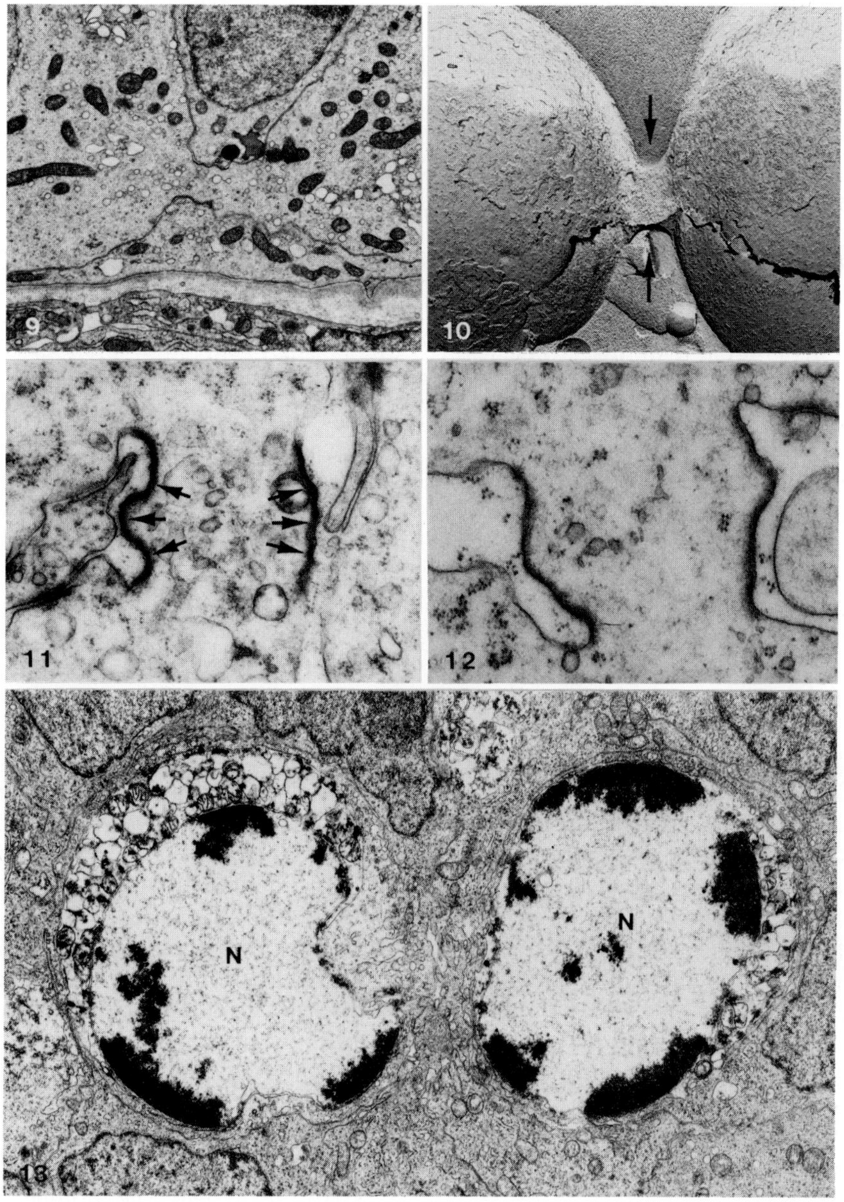

Figs. 9–13. (9) Intercellular bridge at periphery of seminiferous cord, prespermatogenic rabbit testis (× 6,480). (10) Freeze-fracture replica of germ cell interconnection (arrows) comparable to Figure 9 (× 10,960). (11) Deposition of electron-dense material along inner aspect of bridge boundary (arrows) (× 24,480). (12) Note continuity of bridge membrane with membranes of interconnected cells (× 30,480). (13) Degenerating germ cells, human fetal ovary, showing synchronous changes indicated by similarity in pattern of chromatin condensation, clearing of center of nucleus (N) and swelling of cytoplasmic organelles (× 6,480).

component rather than material that is being transported from cell to cell.

A somewhat variable feature is the presence of microtubules within the bridges. The microtubules are situated parallel to the long axis of the bridge. They can occasionally be seen extending in the direction of centrioles in the attached cells (28) and it is therefore likely that they represent remnants of the mitotic apparatus. The implications of this observation are discussed further below (Section 3.2).

It has also been observed that similar microtubular structures can be found traversing the bridge in a perpendicular manner. This is a relatively infrequent observation, but one which suggests that certain structural modifications can occur within the bridges, possibly in relation to specific circumstances affecting germ cell differentiation.

3.2. *Formation*

The origin of germ cell interconnection is by incomplete cell division resulting from failure in separation of daughter cells at telophase (29, 30). The bridges evidently represent persistent mid-bodies, as indicated by their shape, location and inclusion of microtubules consistent with residual spindle fibers. This type of incomplete cytokinesis produces true intercellular bridges associated with open cytoplasmic connection. The factors responsible for prevention of complete cytokinesis in developing germ cells remain to be determined. Genetic regulation in some form appears likely (see Section 5).

The pattern of incomplete cytokinesis begins during the phase of active mitotic proliferation of gonial cells. Primitive germ cells undergoing mitosis during their migration to the genital ridges are evidently able to complete cytokinesis in the usual manner, since intercellular bridges are not ordinarily present at this stage. In the mouse, it has been observed that the first mitotic divisions associated with incomplete cytokinesis occur shortly after the arrival of the germ cells in the genital ridges (2).

Studies in the rabbit indicate that bridges are first formed in both the ovary (31) and testis (32) several days after the male gonad has become identifiable as a testis. This is the time when gonocytes begin to develop the characteristics of prespermatogonia and when the primitive germ cells in the developing ovary take on the appearance of oogonia. Similar observations have been made in the human fetal ovary, in which bridges are seen at 10 weeks gestation but not before (3). Primitive germ cells present between 7 and 10 weeks have an irregular shape, eccentric nucleus and abundant glycogen. At 10 weeks, glycogen is no longer present, the nucleus is central and the germ cells have the characteristic regular round structure of oogonia.

These observations indicate that bridge formation depends on attainment of a state of maturation corresponding to the gonial stage. Although morphologic evidence of testicular differentiation (organization of cell cords and appearance of tunica albuginea) generally precedes evidence of ovarian differentiation, the time of appearance of intercellular bridges is the same in both sexes. This suggests that early stages of male and female germ cell maturation may be similar, with divergence occurring only during the gonial stage. Similarity in appearance of primitive germ cells in the ovary and testis supports this view.

3.3. *Arrangement*

The arrangement and number of germ cell bridges undergo changes during development related to the pattern of germ cell division. Increase in the number of bridges during the course of differentiation is a reflection of continued incomplete cell division. It appears that once the pattern of incomplete cytokinesis is initiated, all subsequent divisions result in a similar persistence of cellular connection. This produces an ever enlarging syncytium, with the interconnected cells differentiating as a unit.

Ultrastructural studies utilizing serial sections have confirmed the existence of multiple interconnections linking groups of cells (33, 34). Although the arrangement is indeed a syncytial one, the confluence of cytoplasm is limited to the narrow region of the intercellular bridges. Each cell otherwise maintains its individual structure. Consequently, the syncytial arrangement would not be appreciated at the light microscopic level.

The presence of a true syncytium is further supported by the similarity in appearance of the interconnected cells. It has been consistently observed that cells joined by intercellular bridges are at the same stage of development and have identical ultrastructural features (Figs. 5, 6). Studies of spermatogenic development have revealed that hundreds of cells may be linked together, all at the same stage of differentiation (35). These findings suggest that the extensive pattern of interconnection is responsible for establishing a common cytoplasmic milieu for large numbers of cells in a manner analogous to more obvious biological examples of syncytial organization.

The arrangement of intercellular bridges differs in the ovary and testis, related to differences in the timing and pattern of oogenesis and spermatogenesis. In the ovary, the bridges are most widely distributed during fetal and neonatal development corresponding to the time of oogonial division and oocyte proliferation. At the time when dense aggregates of germ cells are present in the prefollicular ovarian cortex, bridges are numerous. Once follicle formation begins, individual germ cells become surrounded by granulosa cells and germ cell interconnections disappear. In the testis, bridges are easily found in the adult and are also present early in fetal development when mitotic division takes place. During the subsequent quiescent period, in which many of the germ cells degenerate, bridges are only occasionally seen and generally only involving pairs or small groups of cells. With the onset of spermatogenesis, the bridges reappear in large numbers, progressively increasing with each successive spermatogonial division and reaching a peak that persists into the spermatocyte and early spermatid stages. As spermiogenesis proceeds, bridges gradually disappear and individual spermatozoa are released into the tubular lumen.

3.4. Elimination

Since mature gametes are not connected to one another, some mechanism must exist to eliminate the bridges at a certain point in differentiation. In the ovary, the separation of individual germ cells occurs in the oocyte stage prior to follicle formation. Interconnections of testicular germ cells persist into the spermatid stage and are eliminated only in later stages of spermiogenesis.

There have been no observations to demonstrate the manner of bridge elimination and no evidence bearing on the factors involved. It is possible that supporting cells (granulosa cells, Sertoli cells) disrupt the bridges. The cytoplasmic extensions of fetal granulosa cells would appear to be well adapted for this purpose. However, ultrastructural studies have thus far failed to reveal such activity.

Another possibility is that germ cell degeneration eliminates interconnected cells along with their intercellular bridges and only those cells which had succeeded in dividing normally remain to constitute the pool of surviving germ cells. This interpretation might apply in the ovary where waves of degeneration eliminate large groups of oocytes during early development. In the testis, however, persistence of cellular interconnection into the spermatid stage indicates that degeneration is not an important factor in the elimination of bridges.

That elimination of bridges may not occur in some instances is suggested by certain anomalies, such as polyovular follicles, binucleate and multinucleated oocytes, double-headed sperm and sperm with multiple tails. A likely explanation for some of these aberrant forms is persistent interconnection (36). Other possibilities, including cell fusion and abnormal division, also exist. Since there is considerable species variation in the incidence of such findings as polyovular follicles, it would be of interest to know more about the possible relationship with the occurrence of intercellular bridges.

3.5. Function

In view of the widespread occurrence of intercellular bridges, explanations have been sought for their role in germ cell differentiation. Several theories have been suggested to explain the functional significance of the germ cell interconnections. A number of functions might apply, varying with species, stage of development and local factors. Differences may also exist in the two sexes. Little direct evidence is available for any of the theories. The discussion that follows indicates several possible functions suggested by ultrastructural observation.

3.5.1. Synchronization of maturation

The most likely and widely supported function of intercellular bridges is in the coordination and synchronization of germ cell differentiation (6, 17, 19, 25, 29, 31, 36–38). Synchronous maturation is characteristic of mammalian oogenesis and spermatogenesis. Cell associations result in identical appearance of groups of cells during the stages of mitosis, meiosis and maturation until such time that bridges are no longer present.

A prominent feature of spermatogenesis is the arrangement of cells of similar type in pairs and groups. The various stages of spermatogonial development characteristically occur in local groups. With the electron microscope, it is evident that adjacent spermatogonia are joined by intercellular bridges which extend to connect large numbers of cells. The same is true for spermatocytes, with the interconnected cells consistently showing identical fine structural features. Similar observations have been made in regard to spermatids indicating that in the early stages of spermiogenesis interconnected cells continue to mature in a synchronous manner.

Studies on ovarian development reveal a similar pattern of local synchronization of maturation, also related to the presence of intercellular bridges. Oogonia in culture have been observed to undergo repeated cell divisions in synchrony (4). In tissue preparations, clusters of mitotic oogonia are frequently seen. The initial stages of oocyte meiosis also occur in a highly synchronized manner over a well-defined time interval. As in spermatogenesis, cellular interconnections during oogenesis are found only between cells at the same stage of maturation, providing further evidence for the role of intercellular bridges in the synchronization of maturation. Synchrony of development continues until the time of follicle formation, when intercellular bridges disappear. In the mature ovary, follicle growth and development are not synchronous, some follicles proceeding to ovulation while others undergo atresia. These processes are known to be regulated by gonadotropins. Prefollicular stages of differentiation are not under gonadotropic control (39, 40) and thus a key factor regulating maturation may by the synchronization established by the presence of intercellular bridges.

Possible objections to the concept of synchronization of germ cell maturation by intercellular bridges exist. In the case of spermatogenesis, the presence of spermatogenic waves has been considered to be the critical factor regulating germ cell differentiation. Accordingly, it is the series of recurring cell associations along the course of the seminiferous tubules that is the key phenomenon and not local intercellular associations. However, the two phenomena are not mutually exclusive and in fact probably have an enhancing effect. For example, the syncytial organization of local groups assures uniform populations of cells at particular stages of maturation so that the progression of cell associations required for the establishment of the spermatogenic wave can be maintained. Ultrastructural studies support this approach, particularly when correlated with serial sections performed at the light microscopic level (41).

Another objection relates to the observation that interconnected cells are not always identical, implying that the bridges do not always signify synchronization (30). This is indeed the case in invertebrates in which the bridges clearly have a nutritive function (see Section 3.5.4). In mammals, such observations are extremely rare, and when they do occur may be a result of sporadic aberrations or abnormalities in differentiation.

3.5.2. Limitation of mitosis

Intercellular bridges may be responsible for restriction of the number of divisions of gonial cells. The presence of germ cell interconnections would thereby serve to limit the number of mature germ cells derived from a single stem cell (42). By restricting the mitotic activity of germ cells, a limit would be placed on the number of cells entering meiosis. Thus, formation of intercellular bridges during the early proliferative stage of differentiation might be the mechanism which limits the population of cells in meiosis (43).

Preventing oogonia and spermatogonia from undergoing unlimited mitotic division would also be important in coordinating the onset of meiosis. Coordinated entry of germ cells into meiosis is characteristic of both the ovary and testis. In the latter, this applies both to the initial appearance of spermatocytes in the developmental period and the repeated cyclic differentiation of spermatogonia into spermatocytes in the adult. Comparative studies of spermatogenesis have shown that the number of mitotic divisions which spermatogonia undergo varies in different animals but is constant for a particular species (44).

Although it is evident that the number of germ cell mitoses is limited and that the formation of intercellular bridges is associated with the mitotic process, this does not necessarily mean that the bridges themselves are responsible for the limitation of mitosis and entry into meiosis. In the testis, it is though that hormonal factors, including testosterone and/or gonadotropins, might be involved in the onset of meiosis, although direct evidence is lacking. Other factors, particularly the influence of Sertoli cells, are the subject of active investigation. In the ovary, it has been observed that oogonia will differentiate into oocytes in the absence of steroid or gonadotropic hormones (39, 40). The possible role of granulosa cells in the onset of meiosis remains to be clarified. Of particular interest in recent years has been the suggestion that meiosis-inducing factors are produced by cells from the rete ovarii (45). This work is still in preliminary stages but the evidence that such substances regulate the onset of meiosis is quite compelling. It may well be that a combination of humoral and structural factors, such as intercellular bridges, might be required in this process.

3.5.3. Restriction of motility

There is no direct evidence that cellular interconnection inhibits motility, although it might be expected that formation of syncytial aggregates would have a limiting effect on movement. Groups of oogonia have been observed to exhibit motile activity *in vitro* (4), but pseudopodial cytoplasmic projections found in primitive germ cells are not seen in cells with bridges (3). It is of interest that intercellular bridges generally appear after completion of the migratory stage, suggesting that cellular interconnection might aid in fixing differentiating cells in certain areas.

In the testis, intercellular bridges become most conspicuous at the time that spermatogenesis begins. This is associated with the appearance of long rows of spermatogonia at the periphery of the seminiferous cords. Earlier, the germ cells are distributed both centrally and peripherally in seemingly random fashion, but with an increasing tendency to peripheral migration as the onset of spermatogenesis approaches. Since it has been proposed that the initial differentiation of spermatogonia is controlled by testosterone produced by Leydig cells in the interstitial region (46), it may be that the formation of increasing numbers of interconnections among the peripherally situated cells helps to fix them in a position closest to the source of the testosterone.

Similar considerations would be difficult to apply in the ovary, in view of the early onset of oogenesis and the lack of evidence that hormonal factors are involved in oogonial differentiation. However, restriction of germ cell motility in the developing ovary might be important in enabling establishment of interactions between granulosa cells and germ cells required for germ cell nutrition and maturation (47). Regional factors are evidently important in ovarian differentiation, since germ cells located more centrally and closer to the medulla are consistently at a more advanced stage of maturation than those situated in the outer cortex. The extent to which intercellular bridges might fix groups of germ cells in a more central location or adjacent to rete structures producing meiosis regulating factors would be intersting to determine.

3.5.4. Germ cell nutrition

There is ample evidence for a nutritive function of cellular interconnections in insects (48, 49), but such a function in mammals is unlikely. In Drosophila, each egg chamber contains a cluster of interconnected cystocytes surrounded by a layer of follicle cells. The oocyte is the most posterior cystocyte and the others serve as nurse cells, pouring their cytoplasm into the oocyte through a system of intercellular bridges. This type of arrangement is not characteristic of vertebrates, although an apparent exception is the lizard in which intercellular bridges have been described between follicle cells and oocytes (50, 51).

In mammals, germ cells are connected only to one another. In no instance has an example of intercellular bridge formation between a germ cell and a supporting cell been shown. Furthermore, while there is evidence of exchange of organelles between connected cells, preferential distribution of organelles does not occur. The evidence indicates that mammalian oocytes derive their nutritional support from adjacent granulosa cells by pinocytosis and other means of transmembranous exchange.

3.5.5. Genetic regulation

A genetic function for the bridges is also a possibility. The bridges form during the diploid phase of germ cell development and are present at the time when genetic material is redistributed. The presence of free intercellular communication during meiosis may have important implications. For example, it has been suggested that gene products might wander across the bridges (52) and the free intermingling of such materials could have a role in preserving gametic neutrality (53). As yet, there is no evidence that genetic materials are exchanged across the bridges, but this is an attractive hypothesis which provides an interesting integration of genetic and morphologic events in developing germ cells.

4. Germ cell degeneration and intercellular bridges

A possible relationship between intercellular bridges and germ cell degeneration was first suggested by Burgos and Fawcett (16). They considered that germ cell interconnection in the testis might be an early event in a sequence of regressive changes leading to the degeneration of connected cells. However, they rejected this hypothesis because of the absence of signs of degeneration in conjoined spermatids.

Subsequently, in studies on the developing ovary, it was proposed that during the period of extensive germ cell degeneration, survival might be dependent on separation of individual oocytes from the syncytial groups of degenerating cells (36). The separated cells would then constitute the pool of surviving cells. This explanation would seemingly run counter to the suggested role for the bridges in the synchronization of maturation. Furthermore, it fails to account for the manner in which individual cells are disconnected from their neighbors, although presumably granulosa cells could be involved in this process.

Whatever the relationship of germ cell degeneration and intercellular bridges, it is evident that the presence of interconnections results in synchronous degeneration. In studies on the human fetal ovary (54), connections between degenerating cells were frequently observed in early stages of degeneration and the connected cells showed similar morphologic alterations. In later stages of degeneration, intercellular bridges were infrequently found but adjacent cells continued to show similar regressive changes (Fig. 13). The final stages of the degenerative process were associated with disappearance of intercellular bridges and phagocytosis of individual degenerating cells by surrounding granulosa cells. Similarity in the appearance of groups of degenerating cells has also been observed in studies on rat spermatogonia and the findings attributed to the presence of intercellular bridges (55).

A phenomenon which links the degeneration of germ cells and intercellular bridges in a different regard is the occurrence of germ cell multinucleation in association with degenerative processes in the testis. Binucleated and multinucleated forms are often seen during early development (Fig. 14) when extensive germ cell degeneration takes place. In the adult, it is not unusual to find large numbers of multinucleated cells in the seminiferous tubules in the presence of inflammatory conditions or other conditions interfering with normal spermatogenesis. Some observers have suggested that the multinucleation is a result of fusion of individual cell membranes (56). However, it is more likely a consequence

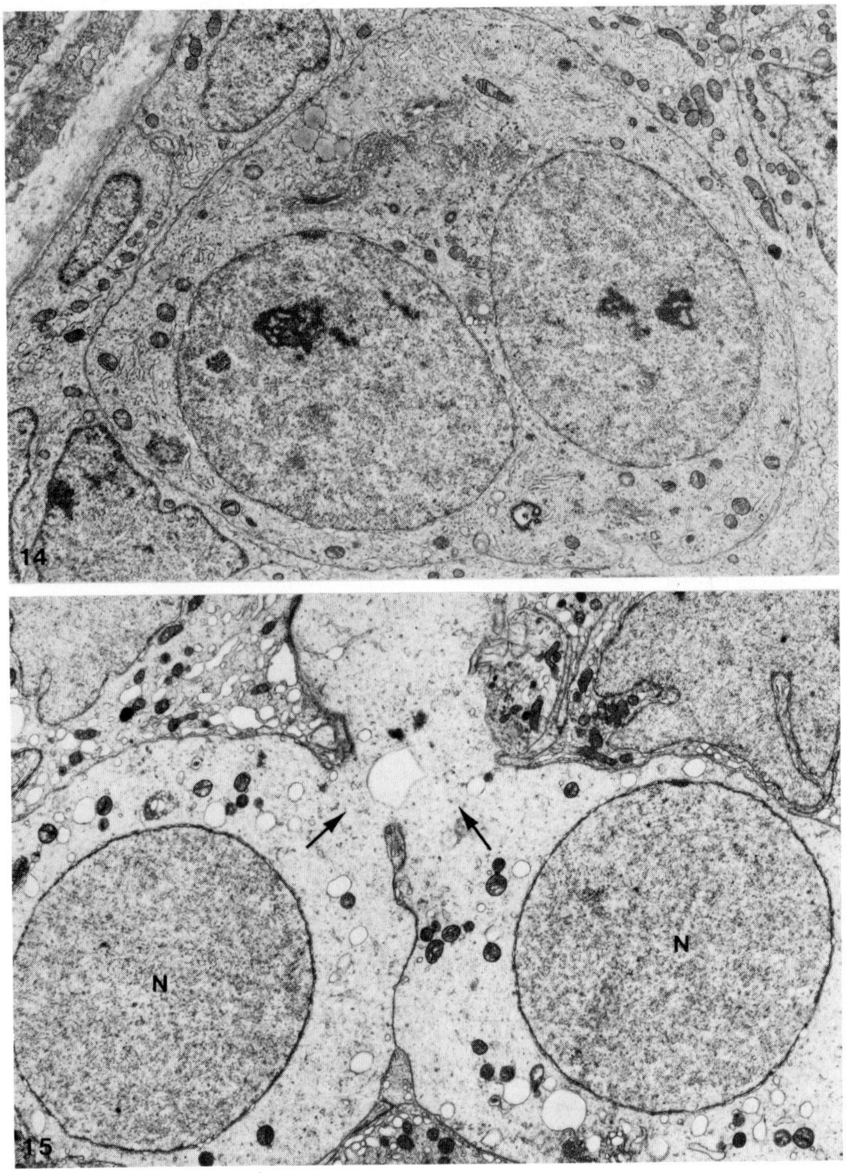

Figs. 14–15. (14) Binucleated germ cell, postnatal rabbit testis ($\times$ 6,000). (15) Broad intercellular connection (arrows), monkey spermatogonia. Note similarity in appearance of nuclei (N) and cytoplasm ($\times$ 5,760).

of widening of existing intercellular connections resulting in increasing cytoplasmic confluence and finally multinucleation (33). This explanation is supported by the observation of broad intercellular connections between adjacent germ cells (Fig. 15) in instances of impaired spermatogenesis (57).

5. Intercellular bridges and germ cell tumors

The possibility that intercellular bridges might be associated with certain forms of germ cell neoplasia relates to some interesting clinical and experimental observations.

Intercellular bridges have been found in human testicular tumors known as spermatocytic seminomas (58), an unusual variant of seminoma. Bridges have not been described in the more typical seminomas. The latter are composed of malignant cells with the light microscopic and ultrastructural appearance of primitive germ cells. It is interesting therefore and consistent with observations in normal cells that bridges are only present in those tumors in which the neoplastic cells have attained a certain level of maturation. Spermatocytic seminomas, as their name implies, are characterized by cells with the morphologic features of spermatocytes, including the presence of typical synaptinemal complexes.

Another report describes the occurrence of intercellular bridges in a mixed germ cell-sex cord stromal tumor, the connected cells showing ultrastructural features characteristic of spermatogonia (59). Bridges have thus far not been described in electron microscopic studies of dysgerminomas of the ovary or germinomas of extragonadal origin, suggesting that these tumors arise from primitive germ cells. This interpretation is consistent with the ultrastructural appearance of the neoplastic cells in such cases.

Studies in Drosophila (60) on mutations resulting in ovarian tumors have suggested that the presence of intercellular bridges may serve to prevent unrestricted germ cell proliferation, i.e. neoplasia, and that cells in which intercellular communication is lost may undergo neoplastic transformation. This would be in line with the proposed function of intercellular bridges in the limitation of mitosis (see Section 3.5.2). Several mutations that appear to affect cytokinesis and result in the production of ovarian tumors are known in Drosophila (61).

King (62) has proposed the following to explain the occurrence of such tumors. A class of genes exists that functions to control the incomplete cytokinesis of germ cells. These genes might only be activated in germ cells and might function to arrest the advance of the cleavage furrow and stabilize the cellular bridges. Mutations in these genes could lead to a high incidence of complete cytokinesis, resulting in germ cell tumors.

This intriguing hypothesis would fit with all currently available evidence regarding the formation and occurrence of intercellular bridges. Whether or not lack of bridge formation leads to tumor development is a question that offers special challenges to investigators working in the general area of germ cell neoplasia. Studies considering the role of intercellular communication in carcinogenesis have suggested that a genetic correlation exists between coupling and tumorigenicity (63). Loewenstein (64) has postulated that, assuming there are growth control molecules flowing freely from one cell to another and regulating cell division, loss or interruption of junctional communication could become an etiologic factor in some forms of cancer. While these comments were made in regard to gap junctions which are found in a great variety of cell types, they might be applied with equal relevance to intercellular bridges connecting germ cells.

6. Concluding remarks

Interconnection of germ cells by intercellular bridges represents a widespread phenomenon which results in extensive interlinking of germinal elements during certain stages of development. Chains of interconnected cells have been observed in animals as diverse as crustaceans, insects, reptiles, birds and mammals. Throughout the animal kingdom, germ cells undergo incomplete cytokinesis prior to meiosis to form clones of interconnected cells.

Genetic regulation that functions to control incomplete cytokinesis of germ cells appears likely. It is reasonable to predict that gene products arrest the advance of the cleavage furrow at telophase and facilitate the transformation of the mid-body into a persistent intercytoplasmic bridge. However, specific evidence to support this hypothesis remains to be established.

Similarly, the precise function or functions of intercellular bridges would be difficult to state in general terms at the present time. For, while their nutritive role in certain invertebrates has been well documented, a more likely function in birds and mammals is in the synchronization of maturation. In insects, there is actually a degree of asynchrony resulting from the redistribution of organelles across the bridges. This does not occur in mammals, in which similarity in appearance and stage of differentiation of interconnected cells is maintained. To what extent the bridges provide synchronization in mammals is difficult to assess, since there is considerable variation in the degree of synchrony and the extent of bridge formation in different species. Limitation of mitosis, restriction of motility and genetic regulation are additional functions suggested by morphologic observations but without direct evidence.

Electron microscopic observations have been crucial to the demonstration of intercellular bridges. It is doubtful that much would have been learned about germ cell interconnection without the application of ultrastructural techniques in the study of gonadal differentiation. With the use of transmission electron microscopy, the structure, formation and occurrence of intercellular bridges have been well characterized. Scanning electron microscopy and freeze-fracture techniques have added further information on the form and distribution of the bridges.

What still remains to be accomplished is clarification of the mechanism of bridge formation. In addition, explanation of how the bridges are eliminated at the appropriate stage of maturation is needed. Further work will also be required to establish the precise functions of the germ cell interconnections. This will require additional innovative approaches to the morphologic characterization of the bridges in combination with biochemical, physiologic and genetic studies. For, it is unlikely that ultrastructural methods which have been utilized up to this point will provide significant new information bearing on the function of the bridges.

Impetus for such studies should come from several directions. It is evident that interconnection of germ cells by intercellular bridges is an important and widespread feature of normal germ cell development. Therefore, thorough understanding of germ cell differentiation is intimately tied to knowledge regarding the formation and function of the bridges. In addition, the possibility that certain abnormalities of germ cell maturation may be related to aberrations in bridge formation suggests that insight into the pathogenesis of disorders of gonadal differentiation might be enhanced by such studies. Finally, the unique and well characterized nature of the intercellular bridges connecting differentiating germ cells should be considered. For the developmental biologist, this provides an excellent opportunity for studying a specific developmental phenomenon which is related to both cell division and differentiation.

References

1. Gondos B: Differentiation and growth of cells in the gonads. In: Differentiation and growth of cells in vertebrate tissues. Goldspink G (ed), London, Chapman and Hall, 1974, pp 169–208.
2. Zamboni L, Merchant H: The fine morphology of mouse primordial germ cells in extragonadal locations. Amer J Anat 137(3): 299–336, 1973.
3. Gondos B, Bhiraleus P, Hobel CJ: Ultrastructural observations on germ cells in human fetal ovaries. Amer J Obstet Gynec 110(5): 644–652, 1971.
4. Blandau RJ: Observations on living oogonia and oocytes from human embryonic and fetal ovaries. Amer J Obstet Gynec 104(3): 310–319, 1969.
5. Wartenberg H: Spermatogenese-Oogenese: ein cytomorphologischer Vergleich. Verh Anat Ges 68: 63–92, 1974.
6. Gondos B: Oogonia and oocytes in mammals. In: The vertebrate ovary. Jones RE (ed), New York, Plenum Press, 1978, pp 83–120.
7. Zamboni L: Comparative studies on the ultrastructure of mammalian oocytes. In: Oogenesis, Biggers JD, Schuetz AW (eds), Baltimore, University Park Press, 1972, pp 5–45.
8. Stegner HE, Wartenberg H: Elektronenmikroskopische Untersuchungen an Eizellen des Menschen in verschiedenen Stadien der Oogenese. Arch Gynäk 199(2): 151–172, 1963.
9. Baker TG, Franchi LL: The fine structure of oogonia and oocytes in the Rhesus monkey (Macaca mulatta). Z Zellforsch 126(1): 53–74, 1972.
10. Wischnitzer S: Intramitochondrial transformation during oocyte maturation in the mouse. J Morph 121(1): 29–46, 1967.
11. Weakley BS: Electron microscopy of the oocyte and granulosa cells in the developing ovarian follicles of the golden hamster (Mesocricetus auratus). J Anat 100(3): 503–534, 1966.
12. Hilscher W, Makoski HB: Histologische und autoradiographische Untersuchungen zur 'Präspermatogenese' and 'Spermatogenese' der Ratte. Z Zellforsch 86(3): 327–350, 1968.
13. Gondos B: Testicular development. In: The testis, Vol IV. Johnson AD, Gomes WR (eds), New York, Academic Press, 1977, pp 1–37.
14. Brown EH, King RC: Studies on the events resulting in the formation of an egg chamber in Drosophila melanogaster. Growth 28(1): 41–81, 1964.
15. Hannah-Alava A: The premeiotic stages of spermatogenesis. Adv Genet 13: 157–226, 1965.
16. Burgos MH, Fawcett DW: Studies on the fine structure of the mammalian testis. I. Differentiation of the spermatids in the cat (Felis domestica). J Biophys Biochem Cytol 1(4): 287–300, 1955.
17. Fawcett DW, Ito S, Slautterback D: The occurrence of intercellular bridges in groups of cells exhibiting synchronous differentiation. J Biophys Biochem Cytol 5(3): 453–460, 1959.
18. Gondos B, Zemjanis R: Fine structure of spermatogonia and intercellular bridges in Macaca nemestrina. J Morph 131(4): 431–446, 1970.
19. Togawa Y: Occurrence and structure of intercellular bridges between the human spermatogonia. Arch Histol Japon 33(4): 301–317, 1971.
20. Franchi LL, Mandl AM: The ultrastructure of oogonia and oocytes in the foetal and neonatal rat. Proc Roy Soc B 157: 99–114, 1962.
21. Stegner HE: Die elektronenmikroskopische Struktur der Eizelle. Ergeb Anat Entwickl 39(6): 1–113, 1967.
22. Zamboni L, Gondos B: Intercellular bridges and synchronization of germ cell differentiation during oogenesis in the rabbit. J Cell biol 36(1): 276–282, 1968.
23. Ruby JR, Dyer RF, Skalko RG: The occurrence of intercellular bridges during oogenesis in the mouse. J Morph 127(3): 307–340, 1969.
24. Ruby JR, Dyer RF, Gasser RF, Skalko RG: Intercellular connections between germ cells in the developing human ovary. Z Zellforsch 105(2): 252–258, 1970.
25. Gondos B: Intercellular bridges and mammalian germ cell differentiation. Differentiat 1(2): 177–182, 1973.
26. Gondos B, Connell CJ: Transmission and scanning electron microscopy study of the developing testis. In: Proc VIIIth Intl Cong Animal Reprod and Artificial Insemination, Vol. 3, Krakow, 1974, pp 50–53.
27. Eddy EM, Kahri AI: Cell associations and surface features in cultures of juvenile rat seminiferous tubules. Anat Rec 185(3): 33–358, 1976.
28. Clérot J: Les ponts intercellulaires du testicule du Gardon: organisation syncitiale et synchronie de la différentiation des cellules germinales. J Ultrastruct Res 37: 690–703, 1971.
29. Fawcett DW: Intercellular bridges. Exp Cell Res, Suppl 8: 174–187, 1961.
30. Anteunis A, Fautrez-Firlefyn N, Fautrez J: La structure de ponts intercellulaires 'obturés' et 'ouverts' entre oogonies et oocytes dans l'ovaire d'Artemia salina. Arch Biol 77(4): 645–664, 1966.
31. Gondos B: Germ cell relationships in the developing rabbit ovary. In: Gonadotropins and ovarian development. Butt WR, Crooke AC, Ryle M (eds), Edinburgh, E & S Livingstone, 1970, pp 239–246.
32. Gondos B, Conner LA: Ultrastructure of the developing germ cells in the fetal rabbit testis. Amer J Anat 136(1): 23–42, 1973.
33. Dym M, Fawcett DW: Further observations on the numbers of spermatogonia, spermatocytes, and spermatids connected by intercellular bridges in the mammalian testis. Biol Reprod 4(2): 195–215, 1971.
34. Moens PB, Go VLW: Intercellular bridges and division patterns of rat spermatogonia. Z Zellforsch 127(2): 201–208, 1972.
35. Moens PB, Hugenholtz AD: The arrangement of germ cells in the rat seminferous tubule: An electron-microscope study. J Cell Sci 19: 487–507, 1975.
36. Gondos B, Zamboni L: Ovarian development: the functional importance of germ cell interconnections. Fertil Steril 20(1): 176–189, 1969.
37. Schleiermacher E, Schmidt W: The local control of mammalian spermatogenesis. Humangenet 19: 75–98, 1973.
38. Huckins C: Spermatogonial intercellular bridges in whole-mounted seminiferous tubules from normal and irradiated rodent testes. Amer J Anat 153(1): 97–122, 1978.
39. Blandau RJ, Odor DL: Observations on the behavior of oogonia and oocytes in tissue and organ culture. In: Oogenesis. Biggers JD, Schuetz AW (eds), Baltimore, University Park Press, 1972, pp 301–320.
40. Baker TG, Neal P: Oogenesis in human fetal ovaries maintained in organ culture. J Anat 117(3): 591–604, 1974.
41. Ross MH: The organization of the seminiferous epithelium in the

mouse testis following ligation of the efferent ductules. A light microscopic study. Anat Rec 180(4): 565–580, 1974.
42. King RC, Akai H: Spermatogenesis in Bombyx mori. I. The canal system joining sister spermatocytes. J Morph 134(1): 47–56, 1971.
43. Skalko RG, Kerrigan JM, Ruby JR, Dyer RF: Intercellular bridges between oocytes in the chicken ovary. Z Zellforsch 128(1): 31–41, 1972.
44. Roosen-Runge EC: Comparative aspects of spermatogenesis. Biol Reprod, Suppl 1: 24–39, 1969.
45. Byskov AG: The role of the rete ovarii in meiosis and follicle formation in the cat, mink and ferret. J Reprod Fertil 45(2): 201–209, 1975.
46. Steinberger E: Hormonal control of mammalian spermatogenesis. Physiol Rev 51: 1–22, 1971.
47. Gondos B: Granulosa cell-germ cell relationship in the developing rabbit ovary. J Embryol Exp Morph 23(2): 419–426, 1970.
48. King RC, Mills RP: Oogenesis in adult Drosophila. XI. Studies of some organelles of the nutrient stream in egg chambers of Drosophila melanogaster and Drosophila willistoni. Growth 26(3): 235–253, 1962.
49. Koch EA, King RC: Further studies on the ring canal system of the ovarian cystocytes of Drosophila melanogaster. Z Zellforsch 102(1): 129–152, 1969.
50. Neaves WB: Intercellular bridges between follicle cells and oocyte in the lizard, Anolis carolinensis. Anat Rec 170(3): 285–302, 1971.
51. Bou-Resli M: Ultrastructural studies on the intercellular bridges between the oocyte and follicle cells in the lizard Acanthodactylus scutellatus Hardyi. Z Anat Entwickl 143(3): 239–254, 1974.
52. Beatty RA: The genetics of the mammalian gamete. Biol Rev 45: 73–119, 1970.
53. Erickson RP: Haploid gene expression versus meiotic drive: the relevance of intercellular bridges during spermatogenesis. Nature New Biol 243: 210–211, 1973.
54. Gondos B: Germ cell degeneration and intercellular bridges in the human fetal ovary. Z Zellforsch 138(1): 23–30, 1973.
55. Huckins C: The morphology and kinetics of spermatogonial degeneration in normal adult rats: An analysis using a simplified classification of the germinal epithelium. Anat Rec 190(4): 905–926, 1978.
56. Bryan JHD: Spermatogenesis revisited. I. On the presence of multinucleate spermatogenic cells in the seminiferous epithelium of the mouse. Z Zellforsch 112(3): 333–349, 1971.
57. Gondos B, Zemjanis R, Cockett ATK: Ultrastructural alterations in the seminiferous epithelium of immobilized monkeys. Amer J Path 61(3): 497–505, 1970.
58. Rosai J, Khodadoust K, Silber I: Spermatocytic seminoma: II. Ultrastructural study. Cancer 24(1): 103–116, 1969.
59. Bolen JW: Mixed germ cell-sex cord stromal tumor: A gonadal tumor distinct from gonadoblastoma. Amer J Clin Path 75(4): 565–573, 1981.
60. Gollin SM, King RC: Studies of fs(1)1621, a mutation producing ovarian tumors in Drosophila melanogaster. Dev Genet 2(2): 203–218, 1981.
61. King RC: The hereditary ovarian tumors of Drosophila melanogaster. Nat Cancer Inst Monographs 31: 323–345, 1969.
62. King RC: Ovarian development in Drosophila melanogaster, New York, Academic Press, 1970.
63. Azarnia R, Larsen WJ: Intercellular communication and cancer. In: Intercellular communication. DeMello WC (ed), New York, Plenum Press, 1977, pp 145–172.
64. Loewenstein WR: Communication through cell junctions. Implications in growth control and differentiation. Dev Biol 19(Suppl 2): 151–183, 1968.

Author's address:
Department of Pathology,
University of Connecticut Health Center,
Farmington, CT 06032, USA

CHAPTER 5

Spermiation – the sperm release process: Ultrastructural observations and unresolved problems

LONNIE D. RUSSELL

1. Introduction

The release of germ cells from the epithelium of the seminiferous tubule signals the end of spermatogenesis and the beginning of the passage of these highly differentiated cells through the excurrent duct system. At the time of release germ cells no longer should be termed *spermatids* but spermatozoa or, simply, sperm. Sperm have attained virtually their final form at the time they are liberated; however, they are functionally immature and still lack the capacity to fertilize.

The release of sperm from the seminiferous epithelium is termed *spermiation*. Two types of spermiation have been described – *spontaneous spermiation* and that induced by *mating* (1). Mating-induced or postcoital spermiation has been described in amphibians (2), but its occurrence in mammals is not recognized by most researchers. Spermiation has never been rigorously defined; consequently, the limits of its meaning have remained vague, and its usage often has reflected this lack of clarity. In a narrow sense, spermiation may be defined as the time when germ cells separate or disengage from the seminiferous epithelium. In a much broader sense, the meaning includes the entire process which leads to the eventual separation of cells from the seminiferous epithelium (3). The latter definition is used here, as the narrower definition does not encompass the mechanisms which underlie sperm release or their control. The process by which sperm are released is a major event in the reproductive process, and it appears to be sensitive to interruption. Consequently, it is worthy of study with respect to correction of human infertility problems as well as to development of suitable contraceptives.

This review will emphasize morphological approaches to understanding spermiation, especially those involving electron microscopy, since it is this tool which, to date, has provided the bulk of our information on the topic. The appearance of descriptive information relative to spermiation has been slow. This lag might be attributable to difficulties in preservation of the seminiferous tubules especially their lumina in which there are no blood vessels.

Ultrastructural techniques utilized to study spermiation are varied, but always yield static images which must be interpreted in the context of an ongoing dynamic process. Consequently, there have arisen numerous theories relating to the mechanisms of spermiation. This chapter will focus on those areas in which our understanding of spermiation is less than adequate in order to provide direction for future research. The emphasis will be mammalian, because the spermatogenic process and spermiation seem to follow a 'similar plan' and are more completely studied, although only a handful of mammalian species have been studied to date. Human spermiation has not been studied adequately. Similarities in testis morphology of humans and other mammalian species indicates that spermiation does not occur differently in the human.

This chapter utilizes specific terminology (italicized in text) to describe cellular parts, relationships or processes. In general, designations for structures vary in the literature, and here the attempt is made to define clearly terms for future reference. In most cases the terminology is intentionally selected as not intended to imply a functional connotation where there is inadequate justification for doing so.

Having defined spermiation as a process rather than an event, it is necessary also to define a point of departure from which a meaningful description of the process might ensue. If spermiation involves the detachment of germ cells from the seminiferous epithelium, it then is appropriate to describe first the morphological relationships existing immediately before the initiation of spermiation.

2. Sertoli-germ cell relationships immediately prior to the beginning of spermiation

Nearly all investigators studying the spermiation process suggest an active role for the Sertoli cell in facilitating sperm release (2–11) and thus both Sertoli and germ cells must be considered together because the two cells are intimately related throughout the spermatogenic process (10–12). Germ cells begin their development at the base of the Sertoli cell and eventually are released into the tubular lumen at the apical aspect of the Sertoli cell. In the course of luminal transit, germ cells occupy a multitude of positions within the seminiferous tubule; consequently, their relationship to the Sertoli cells is continually changing. The morphological relationship between germ cells and Sertoli cells throughout spermatogenesis is maintained by cell-to-cell junctions, surface specializations, and specific configurational relationships (10).

For a long period in spermiogenesis the apical aspect of the Sertoli cell is markedly indented by the late spermatids, and they may be said to lie within deep Sertoli crypts or recesses formed from the apex of the cell (Figs. 1a, 2, 7). From this position, the movement of the germ cells to the surface of the seminiferous epithelium places spermatids at or near the tubular lumen in a position for release. In the mouse and rat, for example, the spermatid head lies only a short distance (8–12 μm) from the basal aspect of the seminiferous tubule and very close to Sertoli-Sertoli junctions (12) and the nucleus. In the rabbit, degu, dog, opossum, guinea pig, hamster, monkey and human, the crypts are of a variable depth, but generally are somewhat shallower than those seen in the mouse and rat. The late spermatids are inserted

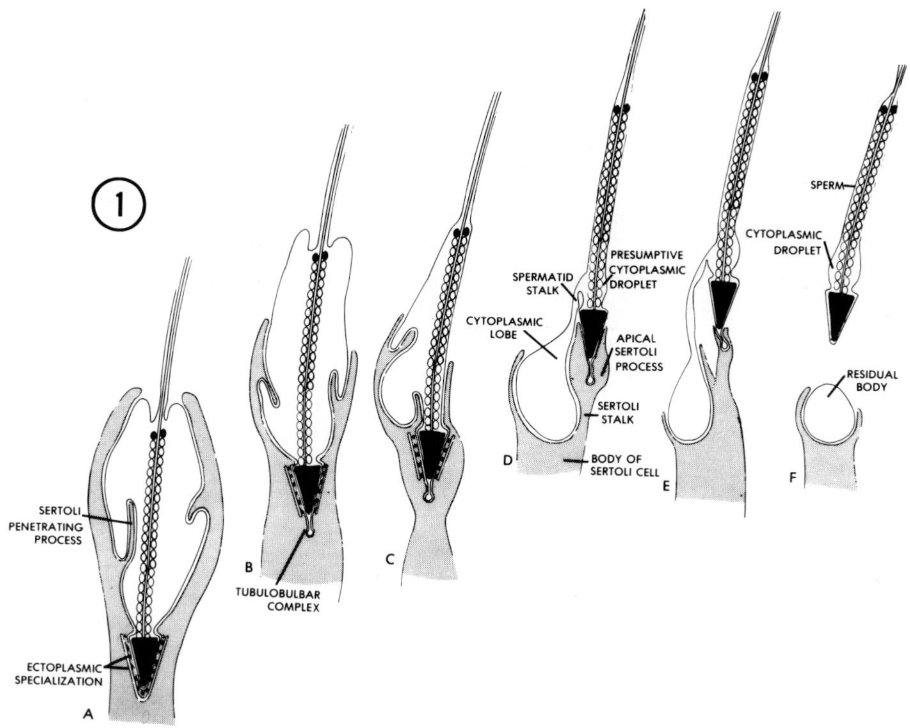

Fig. 1. Summary drawing depicting the general relationship of a spermatid to the Sertoli cells before, during, and after spermiation. (A) Prior to the beginning of spermiation the spermatid is embedded in the deep crypts of a Sertoli cell. The Sertoli cell shows *ectoplasmic specializations* (EcSs) facing the spermatid head and Sertoli *penetrating processes* indenting the cytoplasm around the flagellum. (B) After rapid ascent to the lumen the spermatids retain their configuration and are embedded in a shallower crypt of the Sertoli cell. (C & D) As the spermatid moves toward the lumen the cytoplasm along the flagellum remains stationary with respect to the tubule and gathers into a *cytoplasmic lobe* which is attached to the *presumptive cytoplasmic droplet* by a *spermatid stalk*. The apical Sertoli cell constricts to form a *Sertoli stalk* which connects to an expanded extremity surrounding the heads termed the *apical Sertoli process*. Sertoli EcSs and penetrating processes are no longer present, although the timing of their disappearance is species-dependent. (E) The apical Sertoli process has withdrawn from most of the head, and the spermatid stalk has lengthened. (F) *Disengagement* occurs with the separation of the apical Sertoli process and the spermatid head and the severence of the spermatid stalk. The liberated spermatid cytoplasm is termed the *residual body*. The meager cytoplasm around the flagellum is termed the *cytoplasmic droplet*.

within the crypts to only one-half to one-third the length of the tall columnar Sertoli cell. The plasma membrane at the spermatid head is uniformly close (6–8 nm; refs. 6, 13) to the Sertoli plasma membrane, but this tight relationship between the two membranes does not extend to the flagellar region where the two cells are further apart (about 20 nm) and their spacing is less regular (Figs. 8 and 9). Lanthanum infused through the lumina of seminiferous tubules enters this space but does not reveal the fine detail of the linking structures which might pass from one plasma membrane to another (13), probably because section thickness is much greater than that of the linking structures, and/or these structures are obscured by overlapping lanthanum. Occasionally small bridges can be observed joining the plasma membranes of the two cell types, but these bridges are infrequently seen and certainly are not regular in their distribution along the membranes (Fig. 9). This region of the spermatid head at its interface with the Sertoli cell is a functionally tight region, because the two cells resist certain forces which tend to separate other cells (14, 15). To free the spermatids from the Sertoli crypts, pre-incubation of testicular fragments with trypsin is necessary (14, 16). From the results of such experiments it may be inferred that spermatids and Sertoli cells are linked by proteinaceous substances, and that their desengagement requires enzymatic digestion.

Overlying the head region of the late spermatid, the Sertoli cell displays subsurface bundles of filaments and more deeply placed saccules of endoplasmic reticulum (Figs. 7–9) (9, 17–21). Presumedly, these structures are associated with adhesive or gripping properties referred to above (3, 5, 6, 10, 11, 17–20, 22) and their original designation, junctional specialization (19), implies this functional capability. The term Sertoli *ectoplasmic specialization* (EcS) is used to describe the filament bundles and subsurface endoplasmic reticulum which line the surface of the cell. This term distinguishes the morphological entities, filament bundles and associated endoplasmic reticulum, as being distinct from the functional property of adhesiveness as demonstrated by the adjoining plasma membranes of two cells at EcS sites. The tight membrane associations between the Sertoli and germ cells are thought to be important in the prevention of premature sperm release, and loss of this property must be seriously considered in subsequent descriptions of spermiation.

Caudal to the spermatid head, the apical Sertoli cell surrounds the bulk of the spermatid cytoplasm lying around the midpiece of the flagellum (Figs. 1a and 2). The principal piece of the flagellum generally projects into the tubular lumen unsheathed by the Sertoli cell. Numerous processes of the Sertoli cell invaginate the cytoplasm of the spermatid around the flagellar midpiece (Figs. 1a, 7, 16, 31, 32) (5, 18, 24, 25). The terminology for these processes in the literature is diverse, and includes the terms cytoplasmic extension (4), 'crampons' (4) and invading branches (5). The term *penetrating processes* of the Sertoli cell is used in this chapter (see also 18). Sertoli penetrating processes contain various organelles, but display no obvious surface contacts with the spermatid. The general shape and extensive penetration of the Sertoli cell into the spermatid has suggested a mechanical anchoring function by which these structures would prevent premature release of elongate spermatids lodged within deep crypts of the Sertoli cell (5). The significance of the Sertoli penetrating processes which indent the spermatid has drawn considerable attention. Their fate, as well as various theories related to their function, will be examined in this chapter.

3. Spermiation

3.1. Movement of the spermatids to the tubular lumen

The positioning of the spermatid at the tubular lumen is a phenomenon which signals the beginning of the release process. In most species studied, the distance to be traversed to position the spermatid at the tubular lumen is not great (about 15–25 μm), because the majority of spermatids have not become inserted into the Sertoli crypts to any great depth. In some species such as the rat and mouse there is a considerable distance to be traveled to afford a luminal position for all spermatids. The time involved in transit is a matter of only a few hours, as evidenced by the small number of tubules showing the process in an ongoing state (personal observations). The movement to the lumen is impressive in that the spermatid head traverses a distance of about 50 μm or more in a relatively short period of time (compare Figs. 2 and 3). In these species and in others where a few or all spermatids are deeply embedded, this movement is referred to as the rapid phase of movement (compare Figs. 1a and 2 with 1b and 3), as opposed to the continued migration at a slower pace which appears to take place in all species examined.

Intuitively, one would expect that the spermatid has little part in the translocation process because it is not well equipped with the cytoplasmic elements (free microtubules and microfilaments) which usually are regarded as important for motile processes and

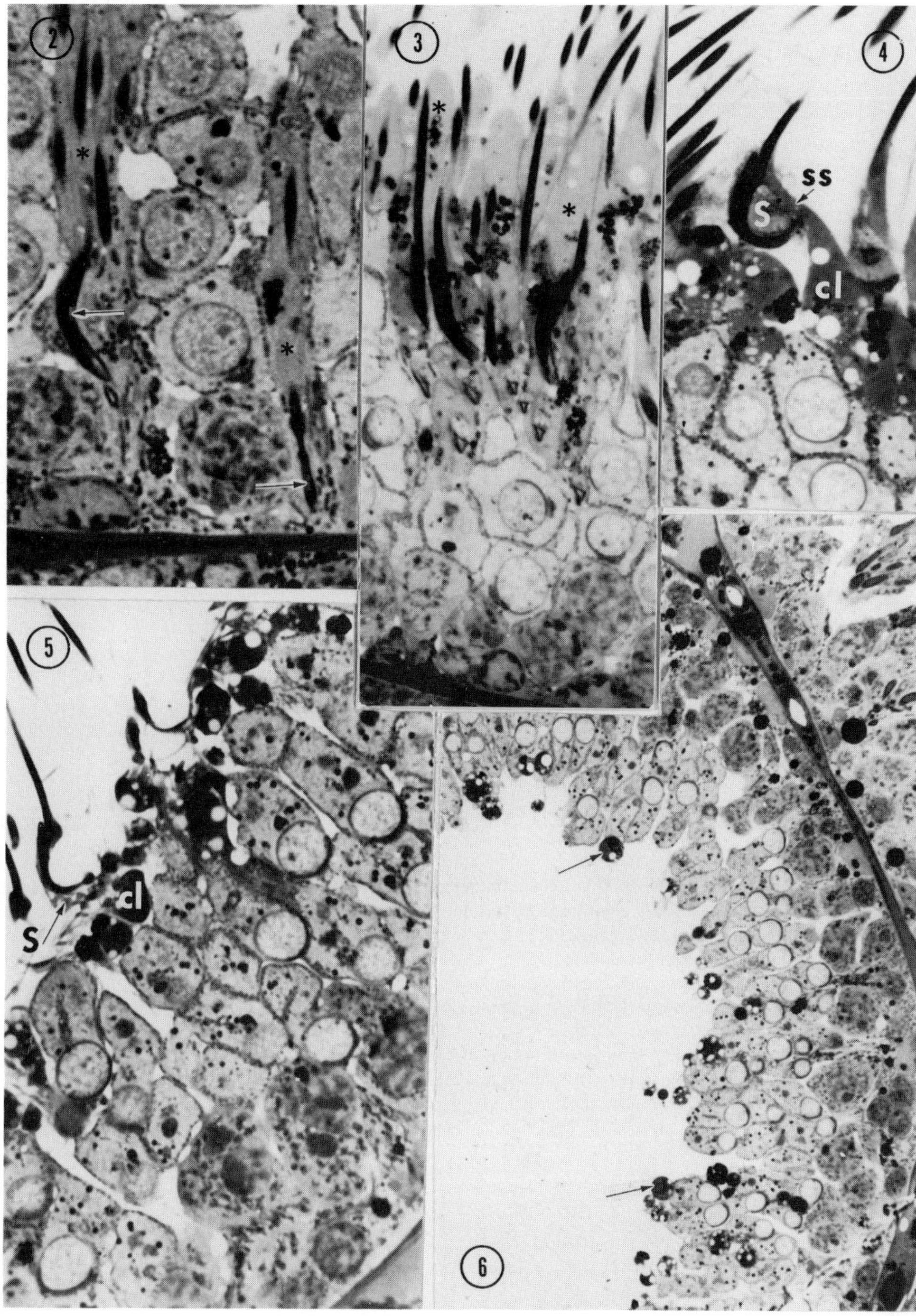

Figs. 2–6. Light micrographs showing the sequence of events prior to (Fig. 2) and during (Figs. 3–6) spermiation in the rat. (2) Several spermatids (arrows) are deeply inserted in Sertoli crypts and their cytoplasm surrounds the flagellum (asterisk) (× 1,000). (3) After the rapid phase of movement to the lumen the spermatid cytoplasm remains along the flagellum (asterisk). The spermatid cytoplasm in subsequent figures of this plate stains increasingly more intensely and is progressively decreased in volume (× 460). (4) The spermatid head and cytoplasm have reversed positions and a cytoplasmic lobe (cl) and spermatid stalk (ss) have formed. The Sertoli cell (S) occupies the concavity of the sickle-shaped spermatid head (× 1,000). (5) As spermiation progresses, the head of the late spermatid shows less of a relationship to the Sertoli cell (S) and appears to extend into the lumen. The spermatid stalk is not seen, although the cytoplasmic lobe is apparent (cl) (× 680). (6) Disengagement has occurred and, for the most part, residual bodies (arrows) line the lumen, although a few have been resorbed. Some residual bodies are larger than expected and may represent residual bodies which have joined (× 300).

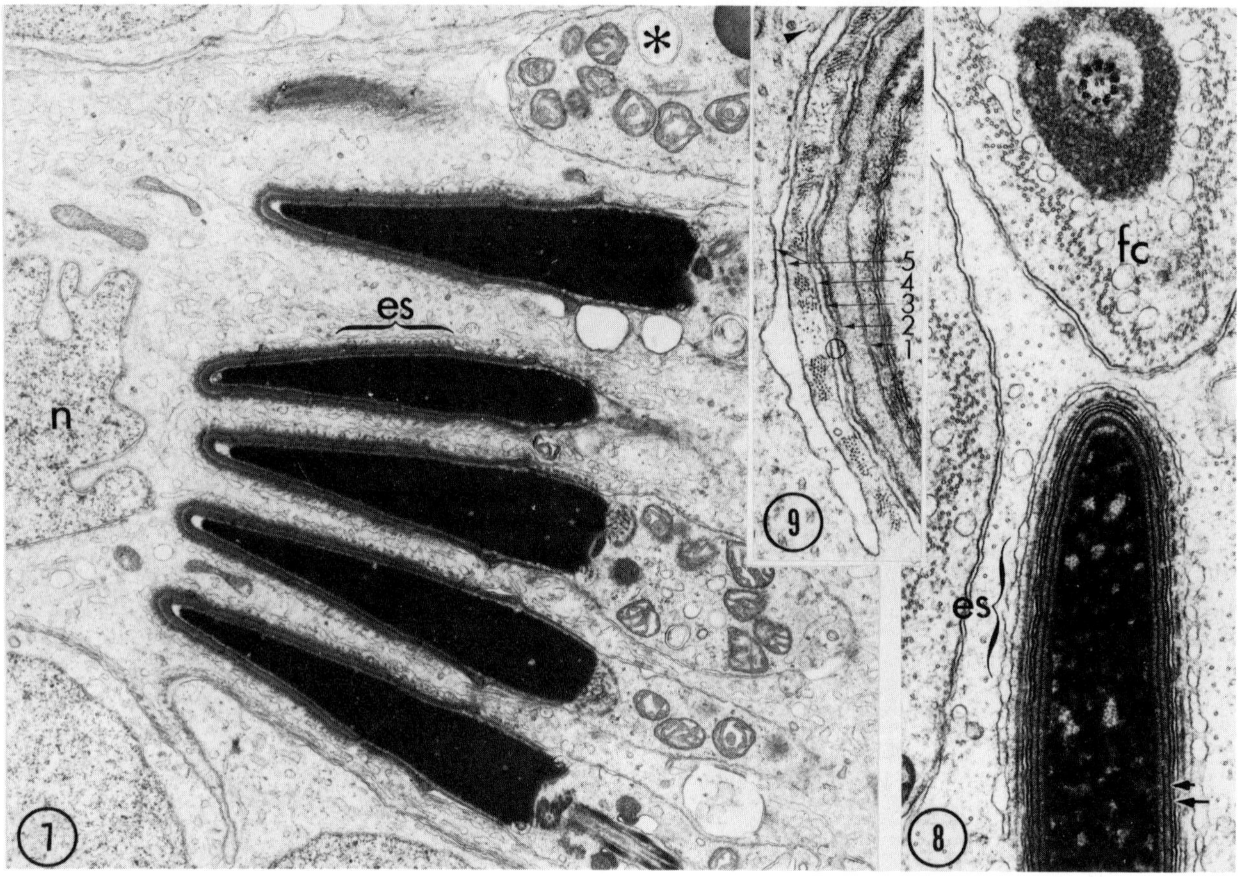

Figs. 7–9. The physical relationship of elongate spermatids within Sertoli (S) crypts prior to the onset of spermiation is depicted. (7) The heads of late spermatids nearly reach the basally positioned Sertoli nucleus (n). A mantle of Sertoli filaments and more deeply positioned endoplasmic reticulum, the EcS (es) apposes the spermatid head. A Sertoli penetrating process is shown (asterisk). Monkey (× 1,125). (8) Higher magnification micrograph of spermatids inserted within Sertoli crypts. The Sertoli EcS (es) faces only the head region of the spermatid, whereas none is evident apposing the flagellar cytoplasm (fc). The two plasma membranes are closer in the region of the spermatid head than along the flagellum, and in the former note the close relationship of the outer acrosomal membrane (short arrow) to the spermatid plasma membrane (long arrow). Monkey (× 2,250). (9) Elongating spermatid showing the EcS (es) Membranes are indicated by numbered arrows; (1) inner acrosomal membrane, (2) outer acrosomal membrane. (3) spermatid plasma membrane. (4) Sertoli plasma membrane, (5) Sertoli endoplasmic reticulum. Note a microtubule within the EcS (es) and another microtubule showing a link to the endoplasmic reticulum (large arrowhead). Occasional filamentous structures (encircled) bridge the spermatid and Sertoli plasma membranes. Note the close relationship of the spermatid plasma membrane to its outer acrosomal membrane. Rat (× 56,250).

changes in cell shape. Cell proliferation at the base of the tubule may be responsible for this luminal transit (6), as well as active movements undertaken by the Sertoli cell. Regarding the latter hypothesis, Sertoli microtubules are linked directly to the cisternae of the Sertoli ectoplasmic reticulum which surrounds the spermatid head. Similarly, filaments of EcS appear to be linked to both the subsurface cisternae and Sertoli plasma membrane (9, 26). Thus, by this mantle of EcS lying over the spermatid head, and its link to widely distributed cytoskeletal agents, a structural basis has been proposed (9) by which these elements may influence the contour of the surface zone of the Sertoli cell (i.e., maintenance of a crypt). Through the bonding of the two plasma membranes at EcS sites, the late spermatids may be moved to the vicinity of the tubular lumen (9, 11, 14). Additional data are needed to establish the role of the Sertoli cell in luminal positioning of the spermatid.

3.2. General relationship of the spermatid and Sertoli cell during spermiation

Depending on the species, the spermatid spends a variable time (in the rat, about three days) in a luminal position prior to release. Near the lumen the spermatid head is related to a much shallower crypt within the Sertoli cell than previously. The body of

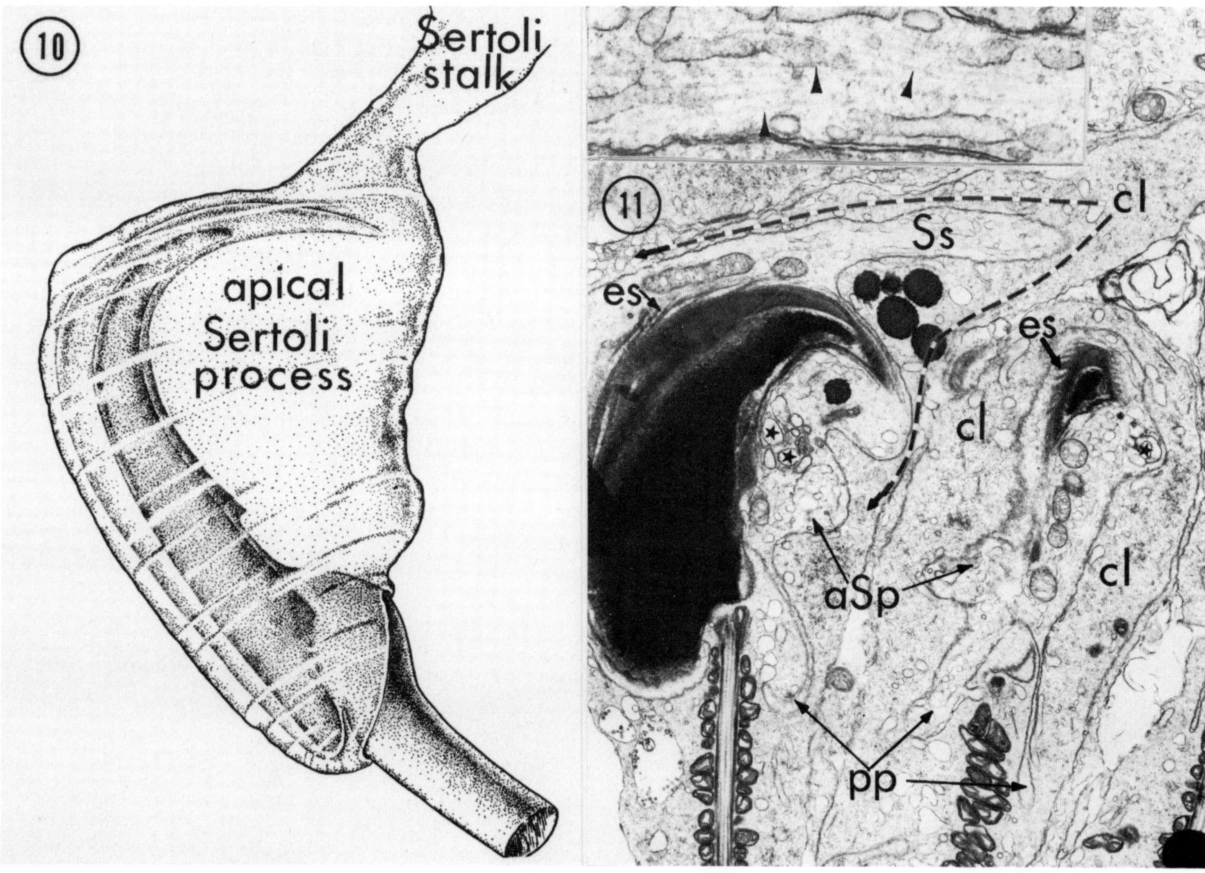

Figs. 10–11. Typical relationship of the spermatid to the apical Sertoli cell during spermiation (comparable to Fig. 1–d). (10) In this drawing, the rat apical Sertoli cell is shown forming a transparent enlargement. the *apical Sertoli process*, which surrounds the spermatid head and is connected to the remainder of the Sertoli cell by a narrow *Sertoli stalk*. For clarity the cytoplasmic lobe and spermatid stalk are shown. (11) Electron micrograph depicting the relationship shown in Figure 10. The *cytoplasmic lobe* (cl) forms a hood over the spermatid head and apical Sertoli process (aSp) and allows for entry of the *Sertoli stalk* (Ss). The two spermatids shown in Figure 11 are sectioned at right angles to each other. Tubulobulbar complexes (stars), EcS (es), penetrating processes of the Sertoli cell (pp), and microtubules in the Sertoli stalk (arrows in Fig. 11 inset) are indicated. Mouse (× 7,500) From Russell and Malone (1980) used with permission.

the Sertoli cell usually terminates at the neck region of the spermatid and gives rise to finer terminal sheetlike processes which extend to a variable degree along the cytoplasm surrounding the proximal flagellum (Figs. 1b and 34). After the rapid phase of movement, the head and flagellum move toward the lumen at a much slower pace. When this phase is completed, the spermatid head is no longer at a circumferential level even with the young generation of spermatids, but is extended into the tubular lumen (Figs. 4 and 5). The bulk of the spermatid cytoplasm appears not to take part in the slow phase of movement, but remains along the surface of the seminiferous epithelium. The net effect of this is the reversal in position of the cytoplasm of the spermatid relative to the remainder of the cell (3, 27). While remaining essentially stationary, the spermatid cytoplasm, which was spread out along the flagellum, gathers into a lobular mass. This cytoplasmic mass received various designations, such as protoplasmic lobe (27), caudal tag, and cytoplasmic tag or lobe (4, 28), residual cytoplasmic lobe (8), and cytoplasmic lobule (29). In this chapter the term *cytoplasmic lobe* is used (Figs. 1d, 3, 11, 12 and 16–18) (30). Many investigators have assumed that the cytoplasmic lobe is the body of cytoplasm eliminated at a later time known as the residual body. Thus, the terms residual body, residual cytoplasm, residual cytoplasmic bodies or residual cytoplasmic lobule are frequently used by one or more of cited works referring to the attached spermatid cytoplasm (1, 3–8, 11, 14, 18, 31). Such designations should be avoided because they infer by their use of the adjective 'residual' that this same cytoplasm will be later discarded as a single mass known as the residual body. There is evidence to suggest that much of the cytoplasm from the cytoplasmic lobe flows back into the spermatid head (21, 32, 33) and is not eliminated as the residual body. In this review the term *residual body* refers only to detached spermatid cytoplasm (as used by 5 and 17) which is separated at the time of liberation of the sperm.

Relative to the spermatid, the cytoplasmic lobe is now much closer to the base of the tubule (Figs. 4, 5; compare Fig. 16 with 17 and 18). It may be lobular in form which is usually the case in the rat (Fig. 4), guinea pig (Fig. 16–18), and opossum (Fig. 12); or, in the mouse and monkey, it may assume a more flattened configuration which covers the head of the late spermatid like a hood or umbrella (Fig. 11). The Sertoli cell is insinuated between the spermatid head and this umbrella-like hood. A gap in the hood allows passage of an *apical Sertoli stalk* which connects the body of the Sertoli cell to the Sertoli cytoplasm around the head (Figs. 10, 11 and 27). This relationship has led to the suggestion (5) that in traveling toward the lumen the spermatid had been 'pushed through its own cytoplasm'.

The cytoplasmic lobe is connected to the scanty cytoplasm remaining around the flagellum by a narrow cytoplasmic stalk (11, 18, 34). The term *spermatid stalk* (Figs. 1d, 3, 12, 37 and 38) is used here to describe this connection. Scanning electron microscopy has provided the best views of the spermatid stalk (11). The cytoplasm around the flagellum, which creates a slight bulge at the midpiece of the spermatid flagellum, is designated the *presumptive cytoplasmic droplet* (Figs. 27 and 38). This structure remains with the sperm after spermiation is completed.

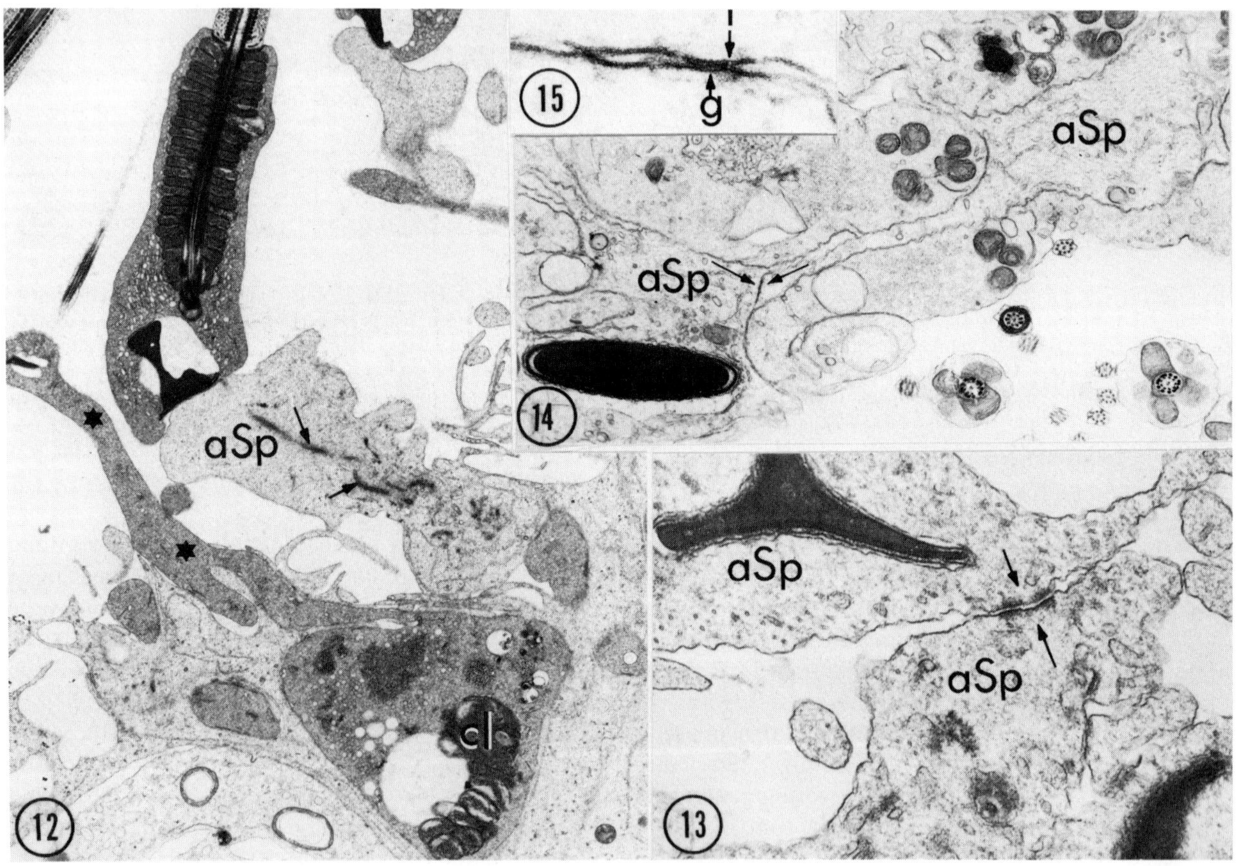

Figs. 12–15. Apical Sertoli processes (aSp) and their relationships to spermatids are depicted. (12) The apical Sertoli process in the opossum extends to impact on the more or less flat head of the spermatid. EcSs have been lost (compare with Fig. 15) and only tubulobulbar complexes (arrows) are present. The spermatid stalk (*) is fortuitously sectioned and connects to the cytoplasmic lobe (cl) (× 5,700) From Russell and Malone (1980). (13–15) Apical Sertoli processes in the rat (Fig. 13) and monkey (Fig. 14–15) frequently are joined by junctions. In the rat they often appear desmosome-like (arrows). In the monkey (Fig. 15) they show features characteristic of gap (g) or tight (t) junctions (× 3,150; × 7,500). Figure 15 is a consecutive serial section of Figure 14 (× 97,500).

Clermont et al. (35) have classified stage VII* for the rat into three substages which defines the general position of the spermatid cytoplasm (1) along the flagellum (Figs. 3 and 24), (2) alongside the head, and (3) associated closer to the base or epithelial side of the tubule than the head (Fig. 4). Such a classification, along with changes in the staining affinity of the cytoplasmic lobe (Figs. 3–5 and 16–18) (23, 30), has proved a useful tool in distinguishing a sequence of events in spermiation. Similar tools for timing related events in spermiation may be devised for other species.

Electron micrographs show that spermatids are related at first to the apical aspect of the large columnar portion of the Sertoli cell. As time progresses, a spermatid individually becomes associated with a much smaller apical Sertoli extension, the *Sertoli stalk* or 'tongue of Sertoli cytoplasm' (Figs. 10, 11, 27) (5, 8, 23, 30) which itself is an apical branch of the main body of the cell separated from another Sertoli stalk by a cytoplasmic lobe (Fig. 27). The Sertoli stalk and trunk contain numerous microtubules in their long axis (Fig. 11 inset and Fig. 7), and these organelles are apparently responsible for maintenance of its narrow, elongate form extending in the direction of the tubular lumen (36). In most species studied, the end of the stalk is expanded and provides a crypt for the spermatid head and, as such, is termed here the *apical Sertoli process*. These processes also have been termed the Sertoli sleeves (11). Such apical processes may be viewed as encompassing as many late spermatids (one or two spermatids per apical process) as there are spermatids associated with a single Sertoli cell (Figs. 1, 1d, 10 and 11). In the rat and monkey, and possibly other species, desmosome-like and tight junctions are encountered which apparently join adjacent apical process and provide some stability to these delicate structures as they extend into the lumen (Figs. 13–15). An exception to the general configuration is seen with respect to the apical processes of the Sertoli cell in the opossum (30) and possibly other marsupials (20). The apical Sertoli process impacts directly on the somewhat flat head of the spermatid without surrounding it (i.e., there is little or no Sertoli crypt) (Fig. 12).

The dynamic relationships between the late spermatids and Sertoli cells are characterized by withdrawal of the apical Sertoli process and gradual obliteration of the Sertoli crypt from the spermatid head (8, 17). The more caudal portions of the sperm head first are unveiled from their Sertoli cover, and this unveiling continues until the crypts within the apical Sertoli process either are no longer evident or are very shallow. Only the leading edge of the spermatid head may be related to the apical Sertoli process (Figs. 1b-e, 5, and 16–18).

During the period of withdrawal of the Sertoli cell from the sperm head, a flow of Sertoli cell cytoplasm toward the base of the tubule has been postulated. At the same time a flow of cytoplasm has also been postulated to facilitate the upward movement of spermatocytes (15). Peripherally directed cytoplasmic flow also is indicated by a marked reduction in size of the apical Sertoli process. Flow of Sertoli cytoplasm probably is associated with microtubules of the Sertoli stalk and the body of the cell, because in other systems microtubules often are associated with cytoplasmic streaming.

The mode of displacement of the Sertoli cell from the spermatid head is variable and apparently more dependent on species than on sperm head morphology. In sperm with symmetrical head morphology (e.g., monkey sperm), the Sertoli cell appears to withdraw uniformly from both surfaces (dorsal and ventral) until only the tip of the head faces the apical Sertoli cell (Fig. 27). In the rat, where the spermatid head is roughly sickle-shaped, the Sertoli cell withdraws first from the ventral concave surface and then later the dorsal convex surface (23). In the musk shrew, where the sperm are especially large and asymmetrical, the Sertoli cell surface apposing the ventral portion of the sperm head first loses its close relationship with the spermatid, followed by the portion facing the dorsal aspect of the sperm head, leading to disengagement of the two cells. This pattern has been correlated with diminished acquiescence of negative surface charge, visualized by binding of charged colloidal iron on the sperm surface (31).

3.3. Fate of the Sertoli ectoplasmic specializations (EcSs)

Having arrived at the tubular lumen with the Sertoli EcS intact, it is presumed that the spermatid remains bound to the Sertoli cell at EcS sites (7, 9, 26, 31, 34). In most species studied, there appears to be a gradual loss of the EcS from the head of the spermatid (18, 23, 30, 31). Precisely when this occurs is apparently more dependent on species than on other events taking place at the lumen. In the mouse, EcS remain

* A stage is defined as a particular association of germ cells seen in a cross section of a seminiferous tubule. Placed side-by-side in logical order the cell associations reveal the cycle of the seminiferous epithelium and are a useful roadmap in studying spermatogenesis (35).

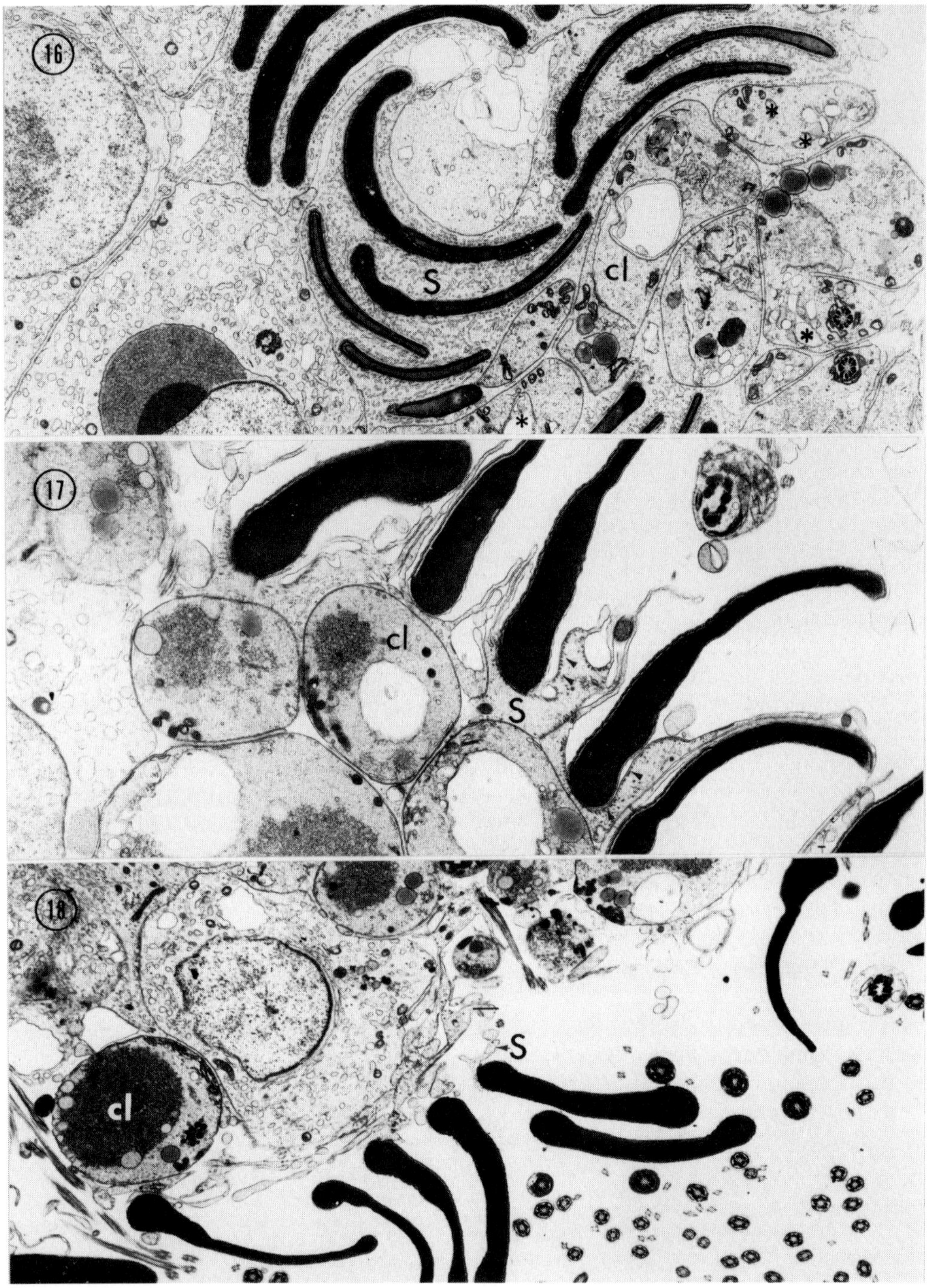

Figs. 16–18. The general relationship of the late spermatids to the Sertoli cell(s) in successive phases of spermiation in the guinea pig is depicted. Cytoplasmic reversal has occurred as shown in Figure 17 and 18. Progressively less contact with the apical Sertoli process and crypt is depicted as the time spermiation is approached as shown in these micrographs of guinea pig testes. Cytoplasmic lobes (cl) show increasing electron density with increasing time (Fig. 16 → Fig. 17 → Fig. 18). EcSs in Figure 17 have lost their endoplasmic reticulum component and filaments line both closely apposed and widely separated parts of the two cells (arrowheads). Sertoli penetrating processes (*) are evident only on Figure 16, but not in the cytoplasmic lobes of more advanced phases of spermiation (× 3,400; × 6,800, × 3,400).

after the time of sperm release (7). In the guinea pig and chinchilla (3), mouse (18, 19, 37), stallion, bull, sheep, pig, dog, cat, rabbit, and guinea pig (18), and rat (17) they depart concomitant to sperm disengagement or slightly before. In the rat (30) and hamster (25), they dissolve as spermiation proceeds; whereas in the opossum (30) and bandicoot (5), they dissolve before spermiation begins. Examination of guinea pig, rat and rabbit tissues shows that the endoplasmic reticulum frequently departs (or dissolves) and leaves a trace of filamentous material where the bundles of filaments of the EcS were originally found. This material is often seen until near the time of disengagement (Fig. 17). In species such as the opossum, early loss of the EcS (compare Figs. 29 and 30) is a clear indication that this subsurface modification plays no role in spermiation once spermatids lose their relationships to the deep crypts of the Sertoli cell.

Exactly how EcSs depart or dissociate from the apical surface of the Sertoli cell is not known, but it is also probably somewhat variable from species to species. Three theories have been brought forward to explain this dissociation. First, it has been suggested that the endoplasmic reticulum and filaments may dissipate into the cytoplasm in the hamster (34), rat (see Fig. 15 of 29), dog (37), opossum (30), guinea pig and rabbit (personal observations) and become internalized by the Sertoli cell. Some electron micrographs support this means of dissolution of EcS (Fig. 20), and because intact EcS cannot be found elsewhere in some species, there is reason to believe that they have dissociated (37, 30). It is possible that depolymerized filaments add to an actin pool, the individual components of which are later mobilized to form new EcSs (37). A second theory relates to the removal of EcSs from the surface as an intact structure (Fig. 19). In some species such as hamster (21) and mouse (37), concurrent with the departure of EcS from the spermatid head are found profiles of EcSs in surface regions of the Sertoli cell which do not oppose the head. They may be encountered at the free surfaces of the Sertoli cell, apposing the cytoplasmic lobe, or in contact with younger germ cells (Fig. 22). From electron micrographic images of such profiles, the concept of reutilization of EcSs has been suggested (9). According to this theory, EcSs may be disengaged from the surface of the spermatid head and move along the surface of the Sertoli cell eventually to face younger generations of germ cells. In the mouse, morphometric studies indicate that shortly after the late spermatids lose their EcS, these structures appear in abundance in developing round germinal cells, which are just beginning to enter the crypts of the Sertoli cell (37). From observations in several species, it has been suggested that a combination of these first two theories more adequately explains the fate of EcSs. Apparently both dissociation and reutilization of EcSs may take place (37).

Gravis (38) also has noted EcSs in an intermediate position, lying opposite the cytoplasmic lobe. He believes that by virtue of the cohesive forces between the plasma membranes usually associated with EcS sites that they are responsible for the selective retention of the residual bodies. This idea would imply that plasma membranes first lose adhesiveness from the spermatid head at EcS sites and then regain this property when their transit brings EcSs into apposition along the cytoplasmic lobe. To provide substance to Gravis' theory, adhesion of EcSs to the cytoplasmic lobe must be demonstrated as well as their continued presence during and after spermiation.

It has yet to be demonstrated that the physical separation of EcSs signals the loss of adhesiveness between the Sertoli cell and spermatid plasma membranes. In general, the loss of EcS from the caudal regions of the sperm head occurs first, and it is also these areas which show a wider intercellular space (31) and often a loss of their typical relationship to the Sertoli cell and the Sertoli crypt (9). In some species where EcSs have been lost, the membranes of the two cells stay in very close apposition. This feature is illustrated best in the opossum (although it is not a feature exclusive to the opossum), where EcSs are removed early (shortly after transit to the lumen) and a column of Sertoli cell cytoplasm, about 10–12 μm long, extends into the lumen and is related to the flattened head of the spermatid (Fig. 12). The intercellular space remains about the same width and it is presumed that the two cells remain bound for some time in spite of the loss of the prominent EcSs (Fig. 30). Occasional subsurface densities of the Sertoli cell, which may be remnants of the original EcS, may persist for some time. At these sites the Sertoli and germ cell are usually spaced quite close together. This relationship is reminiscent of a modified desmosome (Fig. 21). Thus, answers to the questions relating to the temporal relationship of loss of adhesiveness at the level of the two plasma membranes and removal of the EcSs are fundamental to understanding the detachment of the spermatid from the hold by the Sertoli cell and should be pursued in future research.

In addition to their association with cell-to-cell adhesion, EcSs have been implicated in regulating the surface contour of the Sertoli cell in regions where they are associated with elongate(ing) germ cells (9).

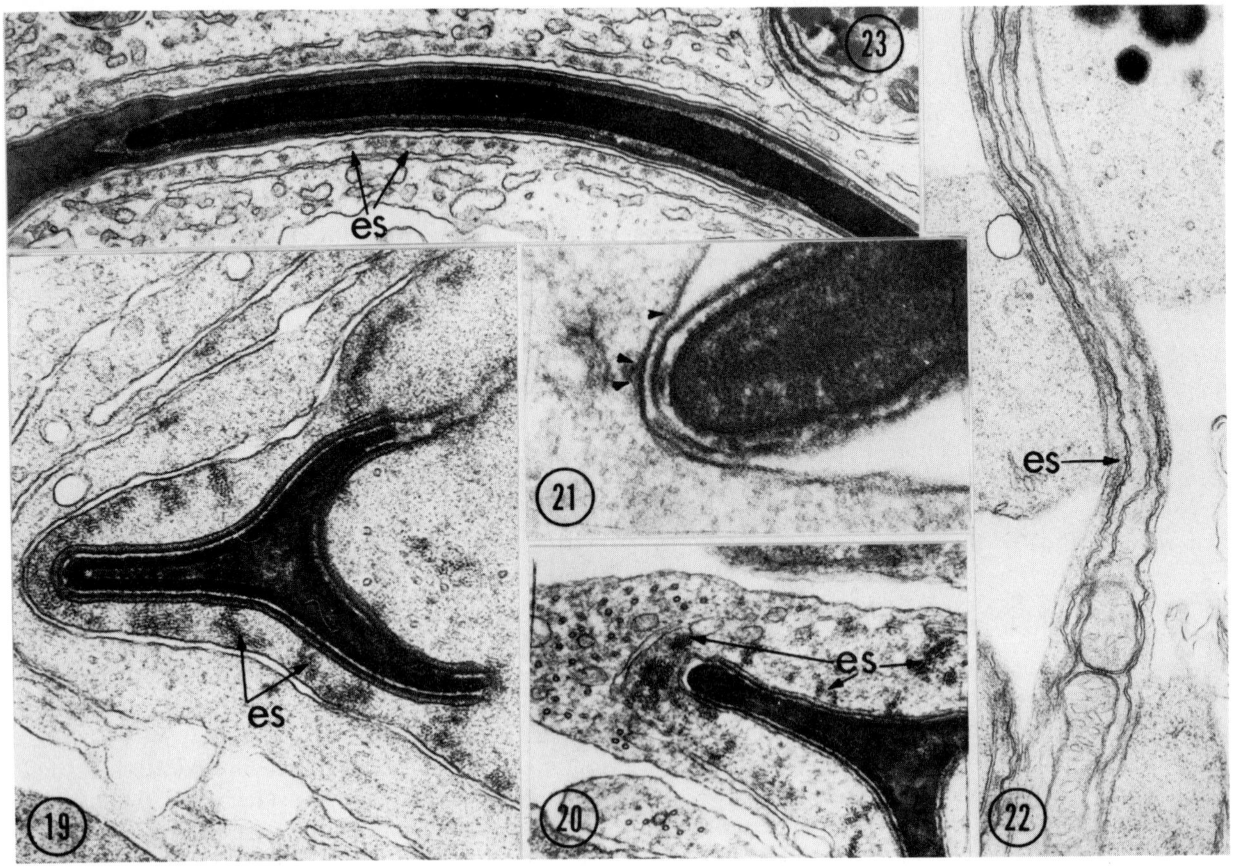

Figs. 19–23. Micrographs showing the relationship of Sertoli EcS (es) as spermiation proceeds. (19–22) In the rat, EcSs (es) may appear either along the spermatid head as an intact unit (Fig. 19) or along the free surface of the Sertoli cell (Fig. 22). It may dissociate and leave filament bundles scattered (Fig 20; arrow), only to disappear later (Fig. 21). Figure 21 shows the only region of close association between the Sertoli cell and germ cell is at the convex surface of the head. Here a small subsurface density of the Sertoli cell is seen associated with fine fibrils (arrow heads) which traverse the intercellular space. Rat (×48,000; ×41,250; ×108,750; ×18,750). (23) Sertoli EcS (es) faces the spermatid head, but the intercellular space is widened more than usual. Guinea pig (×18,000).

The development and loss of the Sertoli crypt in which the germ cell is lodged is related temporally to the appearance and loss of EcSs respectively, over the spermatid head (31, 35). The abolishment of the crypt in which the spermatid head is held would seem to be a necessary prerequisite for completion of spermiation.

3.4. The development and fate of tubulobulbar complexes

After spermatids move to a position near the tubular lumen, they develop minute projections at the cell surface of the spermatid head in regions which have lost EcSs (Figs. 24–28). As the head is located within a Sertoli crypt, these projections invaginate the Sertoli cytoplasm forming the apical Sertoli process (Fig. 27). They generally take the form of long (1–9 μm) and narrow (50 nm) cytoplasmic tubes which are dilated (1–3 μm across) near their termination, but narrow down again into a short, terminally ending tube (Fig. 25). This structure is called the *tubulobulbar complex* (8, 23). The membranes involved in the spermatid protrusion and corresponding invagination of the Sertoli cell are very close together (~5 nm apart). Freeze-fracture and thin section electron microscopy indicate that the membranes forming the tubulobulbar complex are unlike those forming the cytoplasmic lobe in that they appear to be relatively free of intramembranous particles. The Sertoli cell has fine filaments which radiate and surround the tubular portion of the complex and a saccule of smooth endoplasmic reticulum which apposes the bulbous portion (Fig. 25) (8, 23). This endoplasmic reticulum is continuous with an extensive network of the endoplasmic reticulum located elsewhere in the

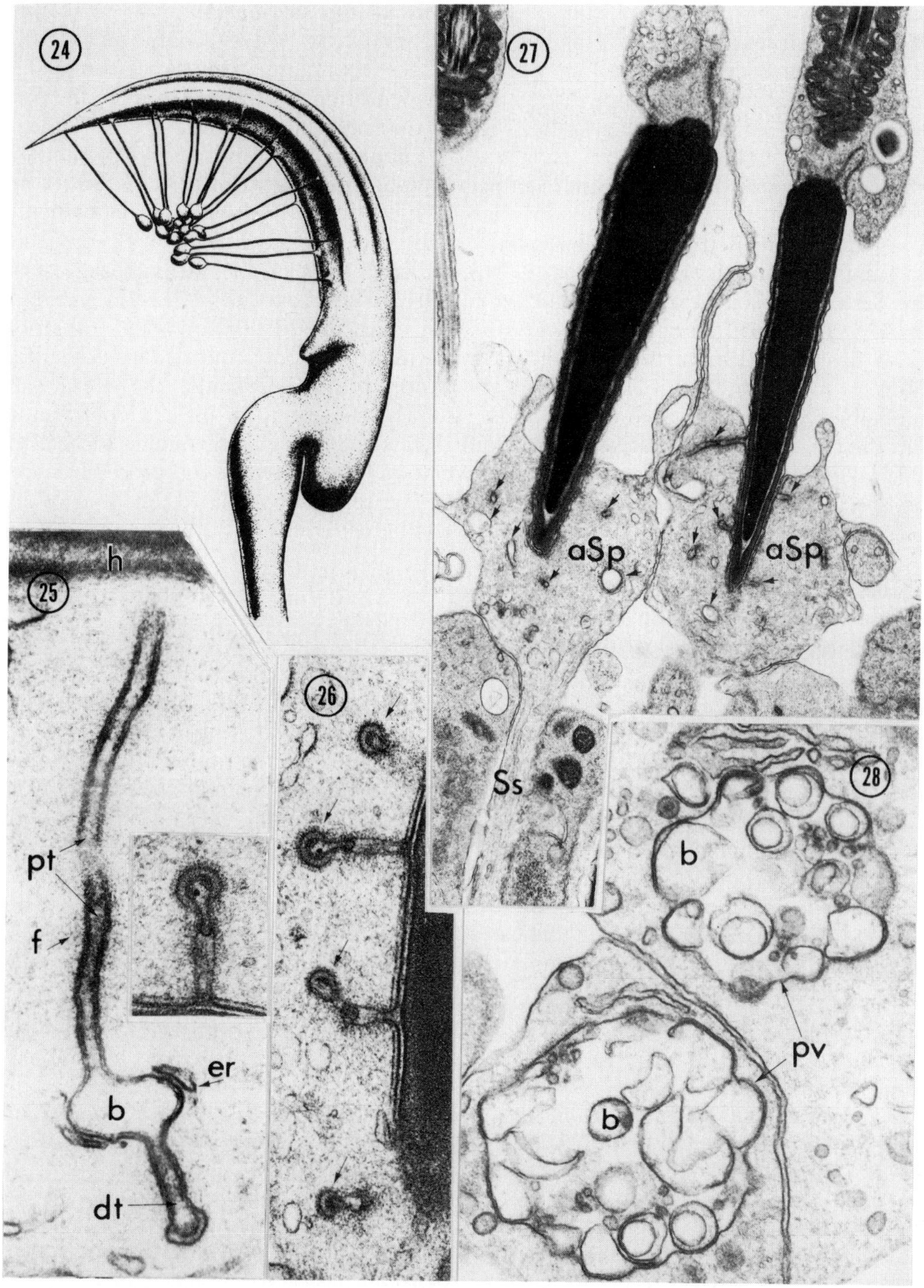

Fig. 24–28. Tubulobulbar complexes are depicted. (24) Drawing of the spermatid head from the rat showing tubulobulbar complexes extending from the ventral concave surface of the head. (25) Micrograph of a rat tubulobulbar complex showing the spermatid head (h), the proximal tube (pt), bulb (b), and distal tube (dt). Filaments (f) and endoplasmic reticulum (er) are also depicted. Rat (× 18,700) From Russell (23) used with permission. (26) Bristle-coated pits (arrows) associated with the tubulobulbar complex demonstrate fine bridges which span the spermatid and Sertoli plasma membranes (more clearly shown in the inset). Rabbit (× 27,200; × 45,900). (27) General relationship of the spermatid head to the Sertoli cell in the monkey. The *Sertoli stalk* (Ss), apical Sertoli process (aSp) and tubulobulbar complexes (arrows) are indicated. Note that the crypt containing the spermatid head is nearly abolished (× 8,500). (28) Degradation of the bulbous portion (b) of tubulobulbar complexes occurs in large phagocytic vacuoles (pv) of the apical Sertoli process. Monkey (× 34,000).

apical Sertoli process (29). The distal end of the tubulobulbar complex faces a coated pit (Fig. 26) of the type frequently seen at cell surfaces. Fine fibrils traverse the intercellular space extending from the spermatid and Sertoli plasma membranes (Fig. 26, inset). The presence of linking structures suggests complementary receptors on the sperm and Sertoli cell which may participate in the formation of the complex in a manner similar to receptor mediated-endocytosis. Tubulobulbar complexes are eventually degraded by Sertoli lysosomes or phagocytic vacuoles (Fig. 28). Transport of coated vesicles to lysosomes also has been demonstrated in the vas deferens (39).

The tubulobulbar complex has been studied most extensively in the rat, but a comparative study also has been undertaken to demonstrate its presence and fate in other species (30). Since tubulobulbar complexes were demonstrated only recently, the possible function of such a cellular relationship in spermiation may alter earlier views of the mechanisms of spermiation (2, 3, 6). Tubulobulbar complexes were first described as anchoring structures which hold the spermatids in place. This function was assigned by virtue of their shape and deep insertion into the apical Sertoli process (8), although the relative role of tubulobulbar complexes as compared to other adhesive devices has yet to be ascertained. In addition, the suggestion has been made that some or most of the excess cytoplasm is eliminated from the spermatid head through tubulobulbar complexes (32, 33). Evidence to support this theory comes from (1) electron micrographic observations which indicate that several generations of tubulobulbar complexes form in the period prior to sperm disengagement, and these are sequentially degraded by lysosomes within the Sertoli cell (32); (2) morphometric data which indicates that in the rat, up to 70% of the cytoplasm is eliminated in this manner as opposed to the more traditional view of cytoplasmic elimination as exclusively taking part by the elimination of the residual body at the time of sperm disengagement; (3) electron micrographs which show that the cytoplasm eliminated by tubulobulbar complexes appears 'watery' and organelle-free, whereas cytoplasm remaining as part of the residual body is packed with organelles with few areas which are organelle-free (32); and (4) data from experimental studies which indicate that, if tubulobulbar complexes are prevented from developing, an enlargement of the spermatid head occurs, apparently the result of congestion of the watery component of the cytoplasm in the head region of the spermatid (21, 32). Collectively, these observations indicate that the final maturation steps of the spermatid appear to require the elimination of the 'watery' component of the cytoplasm of the spermatid. This process may facilitate the condensation of material forming the perforatorium of the sperm head.

Tubulobulbar complexes are present until near the time of disengagement (23). In the rat and monkey (Fig. 27), their presence is evident when the contact with the spermatid head with the Sertoli cell is minimal. Such profiles strongly suggest that the two cells are held together, at least in part, by tubulobulbar complexes. In the opossum, tubulobulbar complexes are the only conspicuous cell surface specialization present, although the spermatid is at the lumen, suggesting at least a contributory role for these structures in maintaining this cell-to-cell association (30). Experimental procedures are needed to further test the postulated functions of tubulobulbar complexes.

3.5. The fate of the penetrating processes of the Sertoli cell

There is no general agreement concerning the time of disappearance of Sertoli penetrating processes from the spermatid cytoplasm. They have been reported to be present (1) in the rabbit, bull and guinea pig until the formation of the mid-piece of the flagellum (18), (2) in the opossum, rat and monkey until the completion of the rapid phase of ascent (personal observations), (3) in the rabbit, bull and guinea pig until shortly before spermiation (18), or (4) remain as in residual bodies in hamster (26), bandicoot and ram (5). Micrographs of Sertoli penetrating processes from nine mammalian species indicates that these processes are found infrequently within the cytoplasmic lobes at times near spermiation (compare Figs. 16 and 17) (unpublished observations).

There is no concensus of opinion as to the function(s) of Sertoli penetrating processes. It has been suggested that they act as clamps which retain the residual cytoplasm and facilitate breakage of the spermatid stalk (24), or they facilitate metabolic exchange, or both (24). In the bandicoot, a sequence of events has been described by which the penetrating processes of the Sertoli cell are responsible for retaining the spermatid cytoplasm during its rapid phase of movement to the lumen (5).

In the ram, the reversal in position of cytoplasm relative to the head in the rapid phase of ascent has been attributed to the grasp of the penetrating processes. If penetrating processes have any function in regard to cytoplasmic retention, it seems that they would be active in the slow phase of ascent.

Spermatids which ascend rapidly to the lumen after being embedded deeply in the Sertoli crypts arrive at the lumen in basically the same configuration as before they started their ascent (Fig. 3, early Stage VII) (29, 35) indicating that the penetrating processes have not retarded this movement. However, the cytoplasm remains stationary in the slow phase of ascent, and therefore this change in cytoplasmic position could theoretically be due to the retarding influence of Sertoli penetrating processes.

Detailed studies of Sertoli penetrating processes in the bandicoot and ram (5, 20) demonstrate that penetrating Sertoli processes take a preferential path within the spermatid cytoplasm. The processes are thought to dissect the bulk of the spermatid cytoplasm away from the flagellum and thus facilitate the formation of the cytoplasmic lobe early in spermiogenesis (see Fig. 11 for profiles of penetrating processes which suggest this activity.)

With time Sertoli penetrating processes in the rat and most other species display degenerative features (compare Fig. 31 with Fig. 32). Profiles of this type are numerous and indicate internalization and eventual degradation of Sertoli penetrating processes within the spermatid cytoplasmic lobe, although only serial sections would demonstrate clearly a loss of the connection to the Sertoli cell proper. A recent report indicates that there are two types of Sertoli penetrating processes (40). One of the types is associated with smooth endoplasmic reticulum of the adjoining spermatid and is phagocytosed by the spermatid prior to the initiation of spermiation.

Because some species show Sertoli penetrating processes only in early phases of spermiation, it is assumed that they function only at this time. The original formation of penetrating processes of Sertoli cells is approximately one cycle previous to their disappearance, i.e., they form as spermatids com-

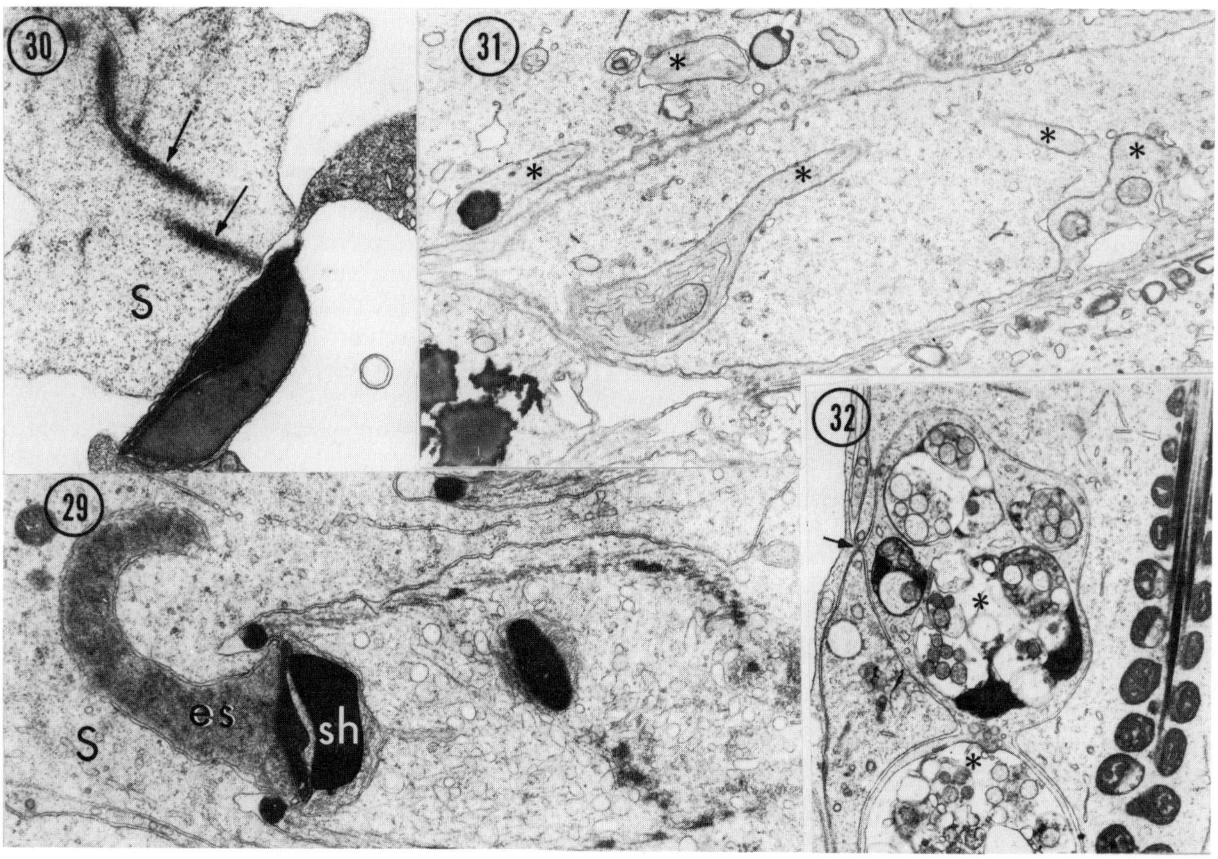

Figs. 29–32. (29–30) Micrographs showing the general relationship of the Sertoli cell and spermatid in the opossum. In Figure 29, the spermatid head (sh) is deep within the Sertoli (S) crypt and shows a prominent EcS (es). After ascent to the lumen (Fig. 16) this specialization is absent (see also Fig. 12), although the two cells remain closely related. Note the tubulobulbar complexes (arrows) (× 7,500; × 14,250) Micrographs courtesy of J. Malone. (31–32) Sertoli penetrating processes (*) in the rat, prior to ascent of the spermatid, appear normal early in spermiation (Fig. 31). Various organelles are present within these processes. In Figure 32 they appear in early stages of degradation. Only a narrow connection between the penetrating process of Figure 32 and the Sertoli cell is evident (arrowhead) and only in rare instances is the connection to the remainder of the Sertoli cell found (× 19,500; × 11,250).

mence their initial elongation phase of development. They are present at the tubular lumen for a relatively short time compared to their total lifespan. If such structures functioned only in the spermiation process, their appearance and disappearance seemingly would be correlated with this activity. Thus, any possible function of these structures may commence earlier in spermiogenesis and terminate with the ascent of the spermatid to the lumen at the beginning of spermiation. Perhaps alternative functions such as those associated with metabolic exchange from the Sertoli cell to the spermatid should be investigated (40).

3.6. Formation of the residual body from the cytoplasmic lobe

The term *residual body* is defined as that portion of the spermatid cytoplasm which is rich in organelles and remains after the spermatid has detached (4, 27). As mentioned above, this body of cytoplasm is distinguished clearly from that of the cytoplasmic lobe, since only a portion of the latter may be released as the residual body (23, 33).

After the rapid ascent phase, the cytoplasm of at least some of the individual spermatids of a clone remain joined by cytoplasmic bridges (Figs. 33 and 34). These bridges are similar in form to those known to exist joining less mature germ cells. The cells remain joined by bridges as the cytoplasmic lobes come to lie more basally than the spermatid heads (Fig. 35). Near the completion of spermiation and after spermiation has taken place, bridges are only rarely found. The electron density associated with the internal aspect of the membranes at the bridge is greatly diminished (compare Figs. 33 and 34 with Fig. 35; compare Figs. 11 with 12 of ref. 4). It is emphasized that the bridges which join germ cells do so in the region of the cytoplasmic lobe. When sperm disengagement takes place, the result is the liberation of individual spermatozoa, although residual bodies may remain joined (3).

After spermatid detachment, the excess spermatid cytoplasm, termed the residual body, remains at the surface of the seminiferous epithelium (Figs. 1f, 6 and 39, Fig. 22 of ref. 18; Fawcett's illustration in ref. 2) (36). Light microscopy sections taken of individual residual bodies immediately after spermiation indicate that the shape of residual bodies is rounded and that their sizes are highly variable (Fig. 6 and 39). Fewer than expected residual bodies appear to rim the tubule, considering that there should be an equal number of residual bodies to sperm liberated (Fig. 6). Individual residual bodies appear larger than expected, considering a loss of up to 70% of their cytoplasmic area has occurred (Figs. 6 and 39) (10, 32). These observations suggest that one or more residual bodies have coalesced as the result of a breakdown of intercellular bridges, leaving fewer and larger residual bodies.

After the formation of the cytoplasmic lobe, it, like the head of the spermatid, becomes associated with recesses in the apical aspect of the Sertoli cell (Figs. 12, 17 and 18). After sperm disengagement, the cytoplasmic lobe remains lodged in a recess of the Sertoli cell and thin apical Sertoli cell sheets encompass it. In this position the spermatid often is described as 'clasped' (18) or 'grasped' (3) by the Sertoli cell. The interface between Sertoli cells and spermatids has received considerable attention in our laboratory, because the selective retention of the residual body after sperm release is a phenomenon which has not been explained adequately (6, 7). The presence of surface specializations which usually provide cell-to-cell adherence might be important in the retention of the residual body. Like Nicander (18) we have not found desmosomes linking the two cells. Delicate rod-like structures occasionally are seen traversing the space between the two cells and connecting their plasma membranes (Fig. 36), but there are no data to suggest functional adhesion at these sites. The surface of the residual body appears to contain specific molecules not found elsewhere on the spermatozoon. These may be recognized in some way by the apical Sertoli processes to ensure retention of the residual body (41). Special techniques are needed to demonstrate regions of cell-to-cell adhesion in areas other than those which display classical types of junctions.

Other theories relating to selective retention of the residual body by EcSs (38) and/or by penetrating processes of the Sertoli cell (5, 24, 25) have been discussed in previous sections of this chapter. Thus, the means by which selective retention of the residual body occurs is not known presently, although the hypotheses mentioned above remain to be experimentally tested.

3.7. Disengagement of the spermatid from the cytoplasmic lobe and release of the sperm into the tubular lumen

In all mammalian species studied, the release of the spermatid into the tubular lumen takes place simultaneously with its detachment from the cytoplasmic lobe. This phenomenon is termed *disengagement*. There is only one report which describes severence of the connection of the spermatid stalk prior to release

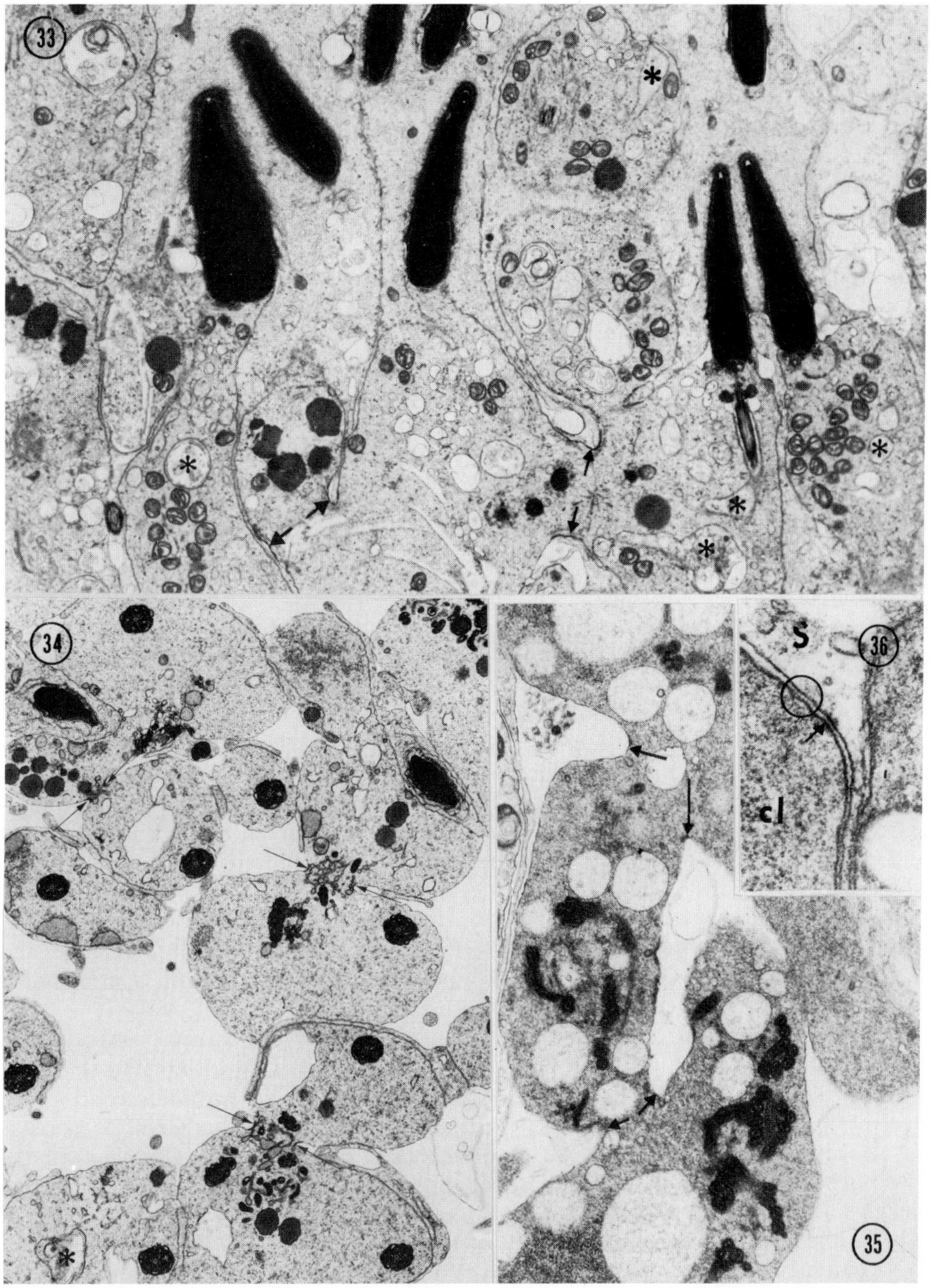

Figs. 33–36. Cytoplasmic lobes of spermatids showing intercellular bridges. (33–35) At the beginning of the slow phase of ascent monkey spermatids are joined by intercellular bridges (arrows) at the caudal region of the cell (i.e., in the flagellar region). Numerous penetrating processes are evident (asterisks). In Figure 34 spermatids have reached the lumen and demonstrate prominent intercellular bridges (arrows) in the cytoplasm around the flagellum. Near the completion of spermiogenesis in the rat, intercellular bridges are difficult to find. However, when they are encountered (arrow heads), they do not show conspicuous densities associated with the plasma membrane. (× 6,800; × 3,400; × 14,450). (36) Thin rod-like structures are seen connecting this cytoplasmic lobe (cl) and Sertoli cell (S) of the rat (× 37,400).

(5). This interpretation of disengagement results from the difficulty of obtaining sections which traverse the entire stalk, since the thin stalk takes a sinuous course and is rarely sectioned in a longitudinal dimension (Figs. 3 and 12). We have sectioned serially spermatids of the rat in which disengagement appears to be eminent, yet we find the spermatid stalk intact. Regauds' (27) original account of disengagement is in agreement with these findings, although he is often quoted (3) as saying that the residual body and sperm are cast off into the tubular lumen, with the residual body being retrieved and phagocytosed later.

Several theories have been proposed to explain the mechanism of disengagement of the spermatid. The first of these, initially suggested by Fawcett and Phillips (3) and later receiving support from other laboratories (11, 25), is that as the head of the spermatid is extruded by the Sertoli cell, the cytoplasmic lobe is simultaneously grasped by the Sertoli cell. The final separation occurs as a phenomenon related to increased tension placed on the spermatid stalk. Details relating to how 'grasping and extrusion' might be accomplished were not provided, but the spermatid stalk does appear to get longer and thinner with increasing time and this configuration does suggest that tension might be a factor in severence of the spermatid stalk (Fig. 37). Stretching, increasing tension and severence of the spermatid stalk as an explanation for disengagement must be examined more carefully. Our observations suggest that sperm release takes place in a relatively narrow time span. When longitudinal sections of seminiferous tubules are obtained, it is apparent that groups of sperm are released simultaneously at specific sites along the tubule (23, 27). The mechanism appears not to involve a slow process of 'thinning out' of sperm but a relatively rapid disengagement of sperm in a particular segment of the tubule – the 'sperm release zone'. Perhaps this represents the activity of one 'unit segment' as visualized by Perey et al. (42). It is difficult to imagine how sperm would all be released simultaneously as the result of a slowly evolving phenomenon of stretching caused by gradual luminalward extention of the spermatid head.

It seems probable that severence of the spermatid from the cytoplasmic droplet must occur at the spermatid stalk in the region at which the stalk joins the cytoplasm around the flagellum (presumptive cytoplasmic droplet). Often, in profiles showing the stalk and its connection with the cytoplasmic droplet, we have noted clear membrane-bound vacuoles which span one-half or more of the diameter of the stalk in this region (in monkey, Fig. 37: rat, Fig. 38; Figs. 2 and 3 of ref. 3). Their strategic position where the stalk would seem to be severed in the disengagement process suggests that this particular site is not reinforced strongly by cytoskeletal elements. In sperm about to be disengaged, we have observed continuity of the vacuolar wall with the sperm plasma membrane (Fig. 38), suggesting that this area is fragile and that final severence occurs here.

Burgos et al. (2) have summarized their views of spermiation in the toad, hamster, and rat. The mechanism proposed involves periodic water imbibition by and consequent swelling of the Sertoli cell. The result of Sertoli swelling is the obliteration of the apical recesses of the Sertoli cell and subsequent sperm expulsion. The micrographs presented by these investigators reveal less than optimal cell preservation, especially in animals undergoing spermiation. This alone may account for some of their observations (see Fawcett's discussion of ref. 2). Other authors specifically have noted that Sertoli vacuolizations and swelling associated with spermiation were not observed in their studies (11, 18). The interpretations provided by Burgos and collaborators in the hamster (1, 2) do not take into account the role of structures such as Sertoli EcSs and tubulobulbar complexes in the hamster (30, 38), nor do they account for the selectivity of the process i.e., how sperm are released while at the same time the residual bodies are retained. Finally, it is suggested (2) that spermiation induced by hormones prior to that normally described would yield morphologically normal sperm. Clearly, this can not be the case because significant changes in the spermatid and its cytoplasm occur rather late in spermiation (Fawcett's discussion of ref. 2; 23, 33).

There are several other possible mechanisms by which sperm that lie within very shallow recesses of the Sertoli cell might be prompted to enter the tubular lumen. One mechanism involves peristaltic movements of the seminiferous tubule. The periphery of the tubule wall contains contractile elements usually termed 'myoid cells', which are probably responsible for movements of seminiferous tubules and, by producing peristaltic waves, influence the transport of fluid in the tubular lumen (43). The contractility of myoid cells and resulting peristalsis may be sufficient to cause dislodgement of the spermatid (7). Directly related to myoid cell contractility are 'flushes' which have been observed in rats to occur on average, at 13 minute intervals, whereby the seminiferous tubule fluid is transported rapidly (44). Rapid fluid movements caused by peristalsis could be important in dislodging very tenuously held spermatids, while at the same time the fluid flow is instrumental in

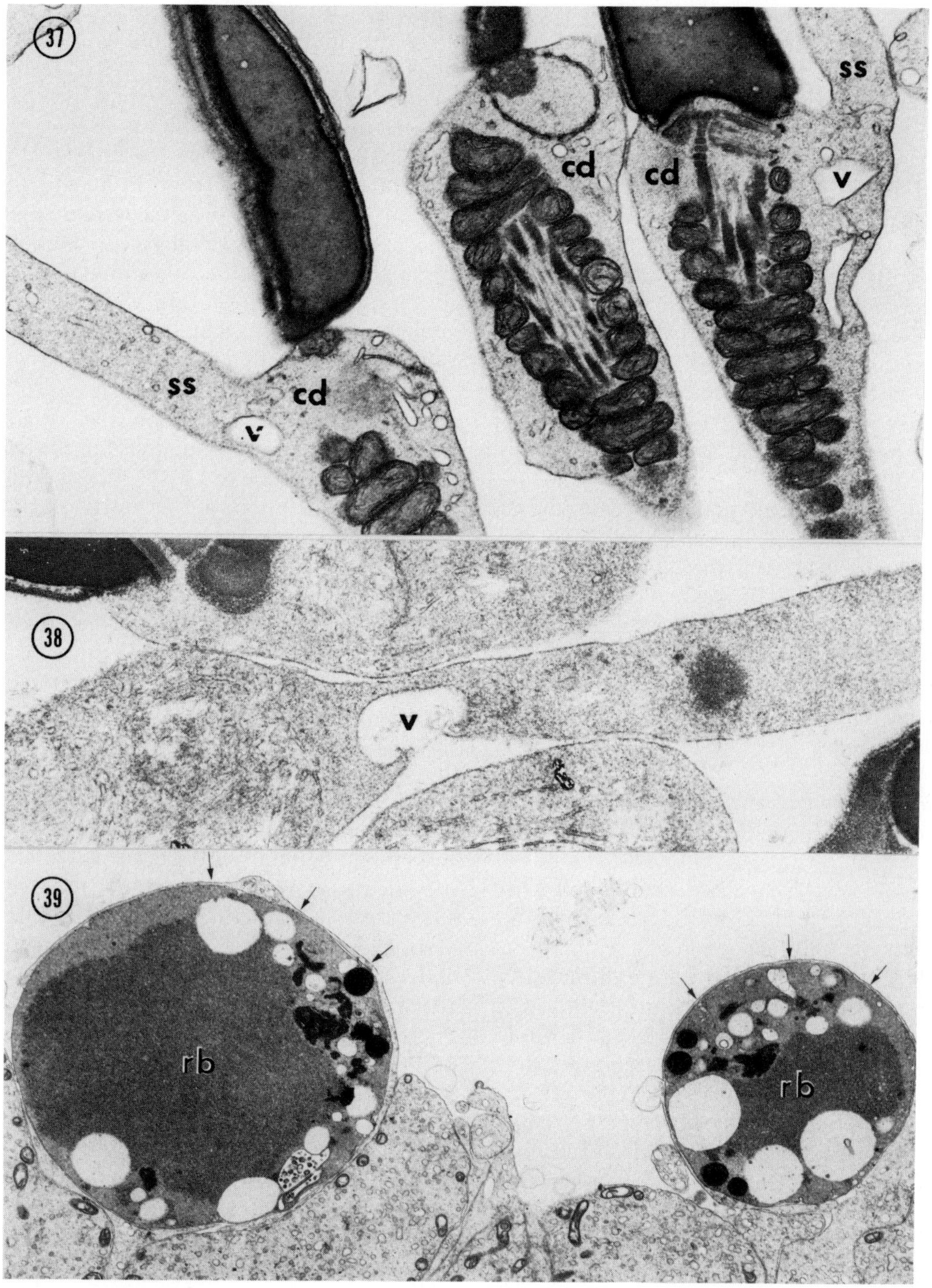

Figs. 37–39. Spermatids near disengagement showing the spermatid stalk (ss) and its connection to the presumptive cytoplasmic droplet (cd). Clear vacuoles (v) are seen at this juncture and in Figure 38 one of these has ruptured into the extracellular space. Monkey (× 20,400); Rat (× 31,450). (39) After disengagement residual bodies (rb) showing a very electron dense and packed cytoplasm remain for a short time at the surface of the seminiferous epithelium. They are almost completely surrounded by thin apical sheets of the Sertoli cell (arrows). Rat (× 6,800).

carrying them down the lumen of the tubule toward the rete testis. In amphibia, secretion of fluid from the testis has been reported to increase after LH stimulation (2). Oxytocin increased the frequency of contractions of seminiferous tubules, and the output of seminiferous tubule fluid and the sperm concentrations in rete testis fluid (46).

Our lab has examined many seminiferous tubules, primarily from the rat and, to a lesser degree, other species. We rarely see released sperm in their tubular lumina. The sparcity of sperm in the tubular lumina also has been noted in the mouse (43). The rapidity with which sperm are transported must account for their scarcity in the lumina of seminiferous tubules. According to Voglmayr (46), they reach the rete of the ram, after having traversed the ductuli efferentes, within about 40 minutes.

It is possible that sperm aid in their own disengagement. Sperm in the tubular lumen show slight motions which can best be described as 'vibratory or twitching'. Perhaps a slight degree of motility is initiated while spermatids are at the surface of the seminiferous epithelium and this activity is sufficient to dislodge them. The observations mentioned relating to fluid movement and sperm motility warrant further investigation.

3.8. Spermiation throughout the seminiferous tubule and testis as a whole

Thus far, this review has focused on that phase of spermiation leading to the disengagement of the spermatid from one specific portion of a seminiferous tubule which is at the appropriate state of maturity. There has been almost no information dealing with sperm release in the seminiferous tubule as a whole, and little information relevant to release in the entire testis. Perey et al. (42) have shown for the rat that many cell associations which show sperm release (Stage VIII) are present in one seminiferous tubule. It appears that rat Stage VIII regions are not synchronized within a tubule or even other parts of the same or opposite testis to produce simultaneous sperm release. The work of Voglmayr (46) indicates that the number of sperm reaching the rete testis in the ram remains relatively constant with time. The influence of contractions of the testicular capsule on the movement of fluid in the seminiferous tubules and the consequent dislodgement of sperm should also be investigated (47) and should not be discounted.

3.9. Control of spermiation

There is a tendency to compare spermiation with its apparent counterpart in the female, ovulation. Clearly, ovulation is associated with luteinizing hormone (LH) discharge, but in a strict sense the two events are not comparable since the gametes are released at very different phases in their maturation. In amphibia, the evidence for hormonal control of spermiation is stronger than in mammals. Pituitary extracts appear to increase the weight and volume of the testis and cause swelling of the Sertoli cells. The active substance for Sertoli stimulated spermiation appears to be LH. Mating with concomitant LH discharge are the possible natural stimuli for sperm discharge in amphibia (2).

The responsiveness of the Sertoli cell to LH and/or testosterone in the manner described by Burgos et al. (2) for mammals would suggest that the Sertoli cell is cyclically susceptible to LH (5). The spermatid would be released whether or not its morphological maturation was complete. In mammals, it has been shown clearly that gonadotrophic hormones or testosterone do not influence the duration of spermatogenesis (48). No changes in the seminiferous epithelium have been noted with exogenous LH (unpublished).

Gravis (34) claimed to delay spermiation in the hamster by administering dibutryl cyclic-AMP. For spermiation to be inhibited, however, requires that the process occur at a later time, providing the inhibiting substance(s) are withdrawn or inactiviated.

One suggestion regarding the mechanism by which EcSs might be removed from the spermatid has been provided (34, 49). The protease, plasminogen activator, secreted from cultured Sertoli cells (49), converts plasminogen to plasmin. Plasmin has been widely implicated in cell migratory phenomena and tissue shape change. The control of plasma membrane-plasma membrane adherence at EcS site might be influenced by this Sertoli cell protease. Sertoli-Sertoli EcS and surface contacts between these cells at the level of the blood-testis barrier also dissipate at the same approximate time in the cycle (23) to allow upward movement of spermatocytes. Plasminogen activator has been localized at these critical time periods (50).

In a recent report (51), spermiation was reported to occur prematurely usually after injection of the microtubular depolymerizers colchicine and viblastine directly into the lumina of the seminiferous tubules. In another report (52), microtubular inhibitors injected intratesticularly (into the intertubular lymphatic space) caused shedding of the apical Sertoli cell processes associated with late spermatids. This class of compounds may act on the delicate Sertoli stalks containing microtubules which extend the spermatid head into the tubular lumen. These compounds either cause them to sever (52, 51) or to retract (51).

As mentioned, contractions of the wall of the seminiferous tubule could directly affect sperm release (7) or produce periodic 'flushes' which dislodge spermatids. Similarly, contractions of the testicular capsule might affect release and/or transport of sperm (47). Studies have shown that an intact pituitary is necessary for the development of contractility of tubules *in vivo* (53) and testosterone for their development in organ culture (54). Both oxytocin (55) and PGE_2 (56) increase the rate of tubular contractions.

To date our knowledge about the physiological stimulation of spermiation is hampered by our inability to observe the process. It is suggested that new techniques be developed to directly observe or even measure sperm release in particular segments of seminiferous tubules under natural and experimental conditions.

4. Summary

Spermiation in mammals is a complex series of events transpiring in the final days before sperm traverse the male duct system. In a broad sense, spermiation involves positioning of spermatids at the tubular lumen and progressive disengagement of spermatids from attachments to the Sertoli cell and from their excess cytoplasm. Spermiation has been investigated most actively through the use of a variety of electron microscopic techniques where it has been shown that the Sertoli cell appears to be actively involved in (1) positioning the spermatid for release, (2) retaining the spermatid in a position for release, and (3) in the eliminating of excess spermatid cytoplasm. Numerous theories have arisen to explain the roles of potential anchoring structures such as Sertoli ectoplasmic specializations, Sertoli penetrating processes and tubulobulbar complexes and, by and large, these theories remain to be explored methodically and experimentally tested. To explain spermiation adequately, interpretations must account for the selective retention of the residual body and the relative roles of various attachment devices, and the mechanism by which each is lost. There are little data relating to the control and/or coordination of spermiation or how spermiation is accomplished in the testis as a whole. Because there are differences in individual species, comparative taxonomic studies are valuable in determining the validity of various theories relating to spermiation and their applicability to human spermiation. The need for standardized terminology to describe events and morphological entities relating to the spermiation process is emphasized.

Acknowledgements

The use of the Springfield Campus School of Medicine electron microscope is acknowledged. The artistry of Ms. Brenda Kester and photographic help of Mr. John Vercillo is greatly appreciated.

References

1. Vitale-Calpe R, Burgos MH: The mechanism of spermiation in the hamster I. Ultrastructure of spontaneous spermiation. J Ultrastruc Res (31): 381–393, 1970.
2. Burgos MH, Sacerdote FL, Russo J: Mechanism of sperm release. In: Regulation of Mammalian Reproduction. Segal SJ, Grozier R, Corfman PA, Condliffe PG (eds), Charles C. thomas, Publishers, Springfield, Ill, 1970, pp 166–182.
3. Fawcett DW, Phillips DM: Observations on the release of spermatozoa and on changes in the head during passage through the epididymis. J Reprod Fert Suppl (6): 405–418, 1969.
4. Dietert SE: Fine structure of the formation and fate of the residual bodies of mouse spermatozoa with evidence for the participation of lysosomes. J Morph (120): 317–346, 1966.
5. Sapsford CS, Rae CA: Ultrastructural studies on Sertoli cells and spermatids in the bandicoot and ram during the movement of mature spermatids into the lumen of the seminiferous tubule. Aust J Zool (17): 415–445, 1969.
6. Fawcett DW: Ultrastructure and function of the Sertoli cell. In: Handbook of Physiology. Male Reproductive System. Vol. V. Section 7: Endocrinology. Hamilton DW, Greep RO (eds), American Physiological Society, Washington, D.C. Waverly Press Inc., Baltimore, Md, 1975, pp 21–53.
7. Ross MH: The Sertoli cell junctional specialization during spermiogenesis and at spermiation. Anat Rec (186): 79–103, 1976.
8. Russell LD, Clermont Y: Anchoring device between Sertoli cells and late spermatids in rat seminiferous tubules. Anat Rec (185): 259–278, 1976.
9. Russell LD: Observations on rat Sertoli ectoplasmic ('junctional') specializations in their association with germ cells of the rat testis. Tissue and Cell (9): 475–498, 1977b.
10. Russell LD: Sertoli-germ cell interrelations: A review. Gamete Research. (3): 179–202, 1980b.
11. Gravis CJ: A scanning electron microscopic study of the Sertoli cell and spermiation in the Syrian hamster. Am J Anat (151): 21–38, 1978a.
12. Fawcett DW: Interrelation of cell types within the seminiferous epithelium and their implications for control of spermatogenesis. In: The Regulation of Mammalian Reproduction. Segal S, Grozier R, Corfman P, Condliffe, P (eds), C.C. Thomas Pub., Springfield, Ill, 1970, pp 91–99.
13. Ross MH: Sertoli-Sertoli junctions and Sertoli-spermatid junctions after efferent ductule ligation and lanthanum treatment. Am J Anat (148): 49–56, 1977.
14. Romrell LJ, Ross MH: Characterization of Sertoli cell-germ cell junctional specializations in dissociated testicular cells. Anat Rec (193): 23–42, 1979.
15. Russell LD: Movement of spermatocytes from the basal to the adluminal compartment of the rat testis. Amer J Anat (148): 313–328, 1977a.
16. Sapsford CS: The development of the Sertoli cell of the rat and mouse; Its existence as a mononucleate unit. J Anat (97): 225–238, 1963.
17. Brökelmann J: Fine structure of germ cells and Sertoli cells during the cycle of the seminiferous epithelium in the rat. Z Zellforsch Mikrosk Anat (59): 820–850, 1963.

18. Nicander L: An electron microscopical study of cell contacts in the seminiferous tubules of some mammals. Z Zellforsch Mikrosk Anat (83): 375–397, 1967.
19. Flickinger C, Fawcett DW: The junctional specializations of Sertoli cells in the seminiferous epithelium. Anat Rec (158): 207–221, 1967.
20. Sapsford CS, Rae CA, Cleland KW: Ultrastructural studies on maturing spermatids and on Sertoli cells in the bandicoot (Perameles nasuta). Aust J Zool (17): 195–292, 1969.
21. Gravis CJ: Ultrastructural observations on spermatozoa retained within the seminiferous epithelium after treatment with dibutyryl cyclic AMP. Tissue and Cell (12): 309–322, 1980.
22. Ross MH, Dobler J: The Sertoli cell junctional specializations and their relationship to the germinal epithelium as observed after efferent ductule ligation. Anat Rec (183): 267–292, 1975.
23. Russell LD: Further observations on tubulobulbar complexes formed by late spermatids and Sertoli cells in the rat testis. Anat Rec (194): 213–232, 1979.
24. Fouquet JP: Le mecanisme de la spermiation chez le hamster: Signification des relations entre cellules de Sertoli et spermatides. C R Acad Sci D (275): 2025–2028, 1972.
25. Fouquet JP: La spermiation et la formation dus corps résiduals chez le hamster: Rôle des cellules de Sertoli. J Microscopie (19): 161–168, 1974.
26. Franke WD, Grund C, Fink A, Weber K, Jockusch BM, Zentgraf H, Osborn M: Location of actin in the microfilament bundles associated with the junctional specializations between Sertoli cells and spermatids. Biol Cell (31): 7–14, 1978.
27. Regaud C: Etude sur la structure des tubes séminiféres et sur la spermatogénése chez les mammiferes. Arch Anat Micr (4): 101–156, 1901.
28. Firlit CF, Davis JR: Morphogenesis of the residual body of the mouse testis. Quart J Microsc Sci (106): 93–98, 1965.
29. Clermont Y, McCoshen J, Hermo L: Evolution of the endoplasmic reticulum in the Sertoli cell cytoplasm encapsulating the heads of late spermatids in the rat. Anat Rec (196): 83–99, 1980.
30. Russell LD, Malone J: A study of Sertoli-spermatid tubulobulbar complexes in selected mammals. Tissue and Cell (12): 263–285, 1981.
31. Cooper GW, Bedford JM: Asymetry of spermiation and sperm surface charge patterns over the giant acrosome in the musk shrew Suncus murinus. J Cell Biol. (69): 415–428, 1976.
32. Russell LD: Spermatid-Sertoli tubulobulbar complexes as devices for elimination of cytoplasm from the head region of late spermatids of the rat. Anat Rec (194): 233–246, 1979b.
33. Russell LD: Deformities in the head region of late spermatids of hypophysectomized-hormone-treated rats. Anat Rec (197): 21–31, 1980a.
34. Gravis CJ: Inhibition of spermiation in the Syrian hamster using dibutryl cyclic-AMP. Cell Tiss Res (192): 241–248, 1978b.
35. Clermont Y, Leblond CP, Messier B: Duré du cycle de l'épithélium séminal du rat. Archs Anat Microsc Morp Exp (48): 37–55, 1959.
36. Malone J: On the sperm release mechanism and the final separation of the spermatid from its attachments. Anat Rec (196): 118a, 1980.
37. Russell LD, Myers P, Ostenburg J, Malone J: Sertoli ectoplasmic specializations during spermatogenesis. In: Testicular Development, Structure, and Function. Steinberger A, Steinberger E (eds), Raven Press, New York, 1980b, pp 55–63.
38. Gravis CJ: Interrelationships between Sertoli cells and germ cells in the Syrian hamster. Z mikrosk Anat Forsch Leipzig (93): 321–342, 1979.
39. Friend DS, Farquhar MG: Functions of coated vesicles during protein absorption in the rat vas deferens. J Cell Biol (35): 357–376, 1967.
40. Clermont Y, Morales C: Transformation of Sertoli cell processes invading the cytoplasm of elongating spermatids of the rat. Anat Rec (202): 32–33A, 1982.
41. Millette CF, Bellve AR: Selective partitioning of plasma membrane antigens during mouse spermatogenesis. Develop Biol (79): 309–324, 1980.
42. Perey B, Clermont Y, Leblond CP: The wave of the seminiferous epithelium in the rat. Am J Anat (108): 47–78: 1961.
43. Barack B: Transport of spermatozoa from seminiferous tubules to epididymis in the mouse: A histological and quantitative study. J Reprod Fert (16): 35–48, 1968.
44. Setchell BP, Scott TW, Voglmayr JK, Waites GMH: Characteristics of testicular spermatozoa and the fluid which transports them into the epididymis. Biol Reprod (1): 40–66, 1969.
45. Niemi N, Kormano M: Contractility of the seminiferous tubule of the postnatal rat testis and its response to oxytocin. Ann Med Exp Biol Fenn. (43): 40–42, 1965.
46. Volgmayr JK: Output of spermatozoa and fluid by the testis of the ram and its response to oxytocin. J Reprod Fert (43): 119–122, 1975.
47. Davis JR, Langford GA, Kirby PJ: The testicular capsule. In: The Testis. Johnson AD, Gomes WR, Vandemark NL (eds), Academic Press, New York and London, 1970, pp 282–337.
48. Clermont Y, Harvey SC: Duration of the cycle of the seminiferous epithelium of normal, hypophysectomized and hypophysectomized-hormone treated albino rats. Endocrinology (76): 80–89, 1965.
49. Lacroix M, Smith FE, Fritz IB: Secretion of plasminogen activator by Sertoli cell enriched cultures. Molec and Cell Endocrinol (9): 227–236, 1977.
50. Lacroix M, Parvinen M, Fritz IB: Localization of testicular plasminogen activator in discrete portions (Stage VII and VIII) of the seminiferous tubule. Biol Reprod (25): 143–146, 1981.
51. Aoki A: Induction of sperm release by microtubule inhibitors in rat testis. European J Cell Biol (22): 467, 1980.
52. Russell LD, Malone JP, McCurdy DS: Effect of microtubule disrupting agents, colchicine and vinblastine, on seminiferous tubule structure in the rat. Tissue and Cell (13): 349–367, 1981.
53. Bressler RS, Ross MH: Differentation of peritubular myoid cells of the testis: Effects of intratesticular implantation of newborn mouse testis into normal and hypophysectomized adults. Biol Reprod (6): 148–159, 1972.
54. Hovatta O: Effect of androgens and antiandrogens on the development of the myoid cells of the rat seminiferous tubules (organ culture). Z Zellforsch Mikrosk Anat (131): 299–308, 1972.
55. Suvanto O, Kormano M: The relationship between *in vitro* contractions of the rat seminiferous tubules and the cyclic stage of the seminiferous epithelium. J Reprod Fert (21): 227–232, 1970.
56. Ellis LC, Buhrley LE: Inhibitory effects of melatonin, prostaglandin El, cyclic-AMP, dibutyrylcyclic AMP and theophylline on rat seminiferous tubular contractility *in vitro*. Biol Reprod (19): 217–222, 1978.

Author's address:
Department of Physiology, School of Medicine
Southern Illinois University
Carbondale, IL 62901, USA

CHAPTER 6

Fine structural features of Sertoli cells

LEIF PLÖEN and E. MARTIN RITZÉN

1. Introduction

The Sertoli cell may be defined as, 'a somatic cell associated individually and simultaneously with several generations of germ cells' (30). In 1865, Sertoli (37) described these cells as 'celluli ramificato' and since 1888 (6) they have borne his name. Attempts have been made to introduce other names – sustentacular cells or sustenocytes, supporting cells, nurse cells – but the name Sertoli cells is generally accepted today.

During the last 15 years, the Sertoli cells have been extensively studied, and their vital importance for normal testicular function is generally accepted. Detailed morphological investigations have elucidated the complex structure of the Sertoli cells and also shed some light on their function (4, 5, 7, 8, 9, 12, 21, 22, 23, 31, 32, 33, 34, 35, 36 and 41). Biochemical and physiological studies *in vivo* and *in vitro* have contributed to the understanding of their function, and endocrinological approaches have established their role as 'key cells' in the regulation of spermatogenesis (25 and 27). More recently, their involvement in the prevention of auto-immune reactions towards meiotic and post-meiotic cells has begun to be recognized (44 and 45).

Many functions have been ascribed to the Sertoli cells (40). Some of these functions are obvious and have been more or less convincingly demonstrated while others are more hypothetical.

2. Morphology of the Sertoli cell

2.1. General morphology

The Sertoli cells of sexually mature animals comprise a sessile, nonproliferating cell population. Proliferation takes place during the fetal and postnatal period but ceases at or before the onset of spermatogenesis at puberty (10). During puberty the final differentiation of the Sertoli cells takes place and the cells attain their mature shape and functions which persist throughout life (15, 19 and 46). In seasonal breeders, Sertoli cell morphology varies with the season (1).

In adult, sexually active mammals the Sertoli cells are tall and extend from the lamina propria to the lumen of the seminiferous tubules. Their outline is extremely irregular and the germ cells are located in dilations between or in invaginations of the Sertoli cell cytoplasm, but in sections through the central portion a main trunk can be distinguished (Fig. 1). From this trunk thin, sheet-like, often branched projections extend (Figs. 1 and 6). Thus spermatocytes and all except late spermatids are almost completely surrounded by Sertoli cell projections. The spermatogonia are situated along the lamina propria and completely separated from the other germ cell types by Sertoli cell cytoplasm. The Sertoli cells change shape as the cycle of the seminiferous epithelium proceeds, but otherwise there are few cyclical changes in Sertoli cell morphology. The elaborate and continuously changing shape of the Sertoli cells make interpretations of sections of Sertoli cells difficult since it is often impossible to determine to which cell the projections belong. Individual Sertoli cells form special occluding junctions with neighbouring cells and thus the Sertoli cells form a continuous layer inside the seminiferous tubules. This continuous layer comprises the blood-testis barrier, which will be discussed further below.

2.2. Nucleus

The Sertoli cell nucleus is located in the basal part of the cell, often close to the basal plasmalemma (Fig. 1) although more apical location is common in some species, e.g. man (Fig. 11). The nucleus is basically ovoid but very irregular in outline, often with deep invaginations (Fig. 1). The long axis of the nucleus is

Van Blerkom, J. and Motta, P.M. (eds.), Ultrastructure of reproduction. ISBN 0-89838-572-5.

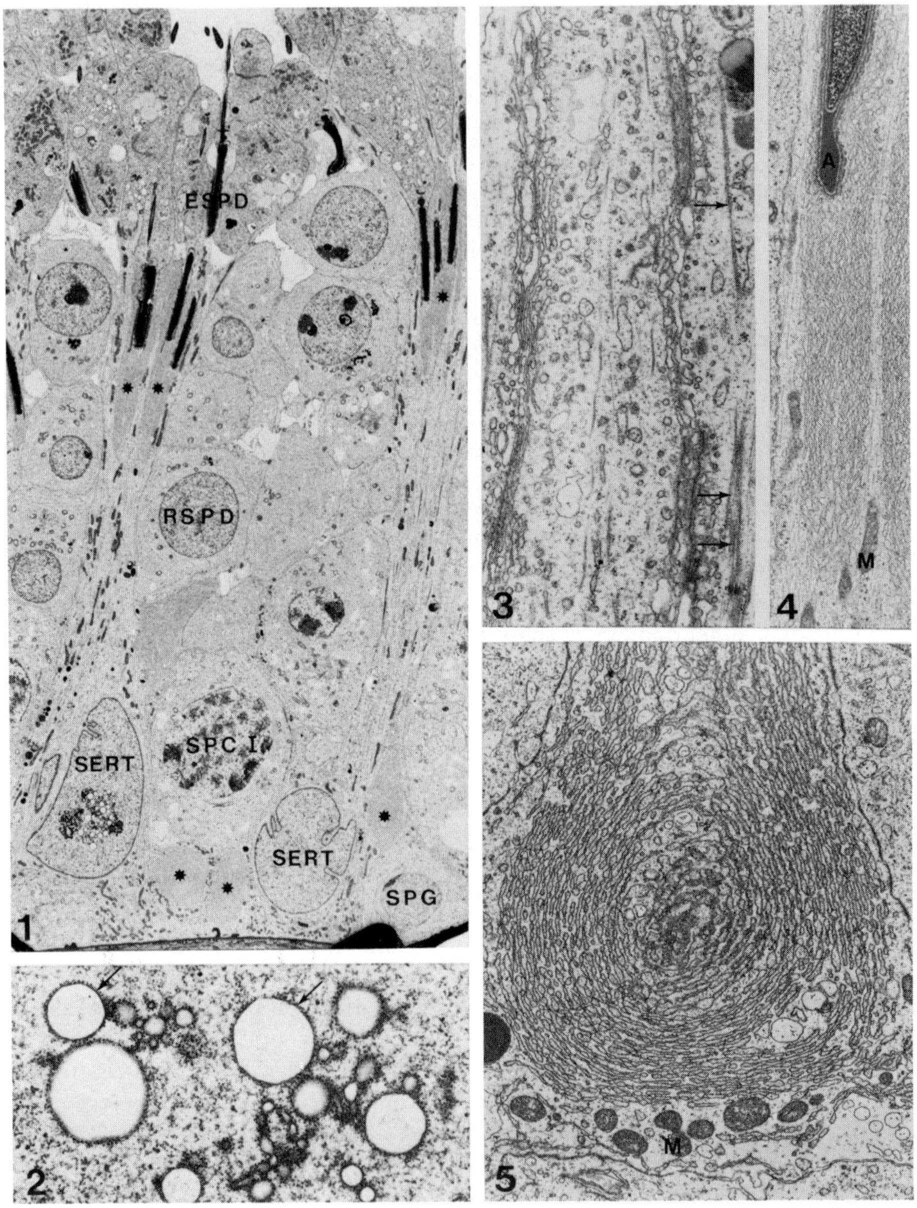

Figs. 1–5. Bull Testis. (1) Section along the main trunk of two Sertoli cells. Note the nucleolus and the vacuoles in the left Sertoli cell nucleus and the accumulations of agranular endoplasmic reticulum (*). SPG = spermatogonium; SPCI = primary spermatocyte; RSPD = round spermatid; ESPD = elongated spermatid (× 1,100). (2) Higher magnification of the vacuoles and tubules in a Sertoli cell nucleus. Note the granules towards the nucleoplasm (arrows) (× 16,000). (3) Golgi apparatus in the main trunk. Note also the bundles of filaments (arrows) (× 19,000). (4) Accumulations of agranular endoplasmic reticulum in the area surrounding the developing acrosome (A) of an acrosome phase spermatid. M = mitochondrion (× 7,000). (5) A basal accumulation of concentrically arranged cisternae of agranular endoplasmic reticulum. M = mitochondrion (× 10,000).

either parallel or perpendicular to the cell axis and it is surrounded by a zone of filamentous material. The nucleus has very little heterochromatin. The nucleoulus is prominent (Fig. 11), and in most species consists of a distinct nucleolonema which is flanked by two heterochromatic satellite karyosomes or perinucleolar spheres that are Feulgen-positive and do not incorporate uridine (7). In several ruminant species there is no satellite heterochromatin but instead membrane-limited vacuoles and tubules of various sizes (Figs. 1 and 2). On the surface towards the nucleoplasm they show small granules, not unlike ribosomes (Fig. 2). The lumen of the vacuoles and tubules appears empty and no function has been ascribed to them yet.

2.3. Cytoplasm

The *Golgi apparatus* is found in the main trunk (Fig. 3). In thin sections it is often seen as several separate smaller units, scattered throughout the cytoplasm. High voltage electron microscopy of thick sections from the mouse testis reveals that the Golgi apparatus consists of a continuous cylindrical structure located in the main trunk (26), and a similar appearance probably occurs in many other species. Compared to cells with extensive merocrine secretion, the Golgi apparatus of the Sertoli cell does not appear very active but occasional vacuoles, resembling condensation vacuoles, can be encountered. Vacuoles, sometimes with dense contents, are often found in the apical cytoplasm.

Mitochondria are numerous and found in all parts of the Sertoli cell except in the thinnest cytoplasmic projections between germ cells. In some species there are basal accumulations of mitochondria. The mitochondria in the basal part of the cell are generally rounded (Figs. 1 and 5) whereas in the main trunk and in cytoplasmic projections they are long and slender and oriented longitudinally (Figs. 1 and 4).

2.3.1. Endoplasmic reticulum is a conspicuous feature of Sertoli cells. Most of the endoplasmic reticulum is agranular, but cisternae of granular reticulum occur in the basal portion of the cell and in the main trunk. Agranular reticulum occurs throughout the cell, but accumulations of concentrically arranged agranular cisternae are often encountered in certain locations which vary between species. Basally located accumulations are common in many species (Figs. 1 and 5), sometimes as annulate lamellae (36). Large accumulations of regularly arranged cisternae occur in a number of species in the part of the Sertoli cell cytoplasm that immediately surrounds the developing acrosome of spermatids from early acrosome phase up to the stage when the spermatids move towards the lumen of the seminiferous tubule (Figs. 1 and 4). Such accumulations are prominent in a number of Artiodactyla and also in some rodent species (7). Cisternae of endoplasmic reticulum are also found in association with inter-Sertoli cell junctions and 'half-Sertoli cell junctions' (see below).

2.3.2. Lipid inclusions are common. The amount, the size and the density of the lipid droplets vary considerably among species. In some species they constitute a prominent feature, whereas in others they are fairly inconspicuous. In the domestic boar for example they are up to 25 μm in diameter, but in most other species they are rarely larger than 5 μm. The appearance of the lipid droplets varies; to some extent this may be due to different procedures utilized during fixation and dehydration. On the other hand, differences in the chemical composition of the lipids will result in differences in morphology. In the rat, cyclic changes in the lipid contents of the Sertoli cells have been described (16). Lipid droplets are sometimes surrounded by aggregations of concentrically arranged cisternae of agranular endoplasmic reticulum (23).

2.3.3. Microfilaments are abundant in the Sertoli cells. Most of them are found in association with cell contacts (see below), but bundles of microfilaments can be found in the main trunk of the Sertoli cell (Fig. 3). In the domestic boar such bundles sometimes form crystalloids (41). Filamentous material is also found in a narrow, organelle-free zone around the nucleus and in association with the tubulobulbar complexes (see below). In human Sertoli cells, densely packed parallel filaments form crystalloids of Charcot-Böttcher in the basal cytoplasm (36).

2.3.4. Microtubules also comprise a prominent feature of Sertoli cells. They are mostly located in the main trunk and oriented parallel to the cell axis.

Sertoli cells are very active phagocytotic cells, and this activity is greatest at the time of spermiation when the residual bodies are shed from the mature spermatids. In the normal testis there is also a certain amount of germ cell degeneration (29) and the degenerated cells are phagocytosed by the Sertoli cells. The digestion of the ingested material seems to be very rapid and is not well understood. It is assumed that much of the material is reutilized. In the basal cytoplasm numerous membrane-bound dense bodies are seen (Fig. 11). These probably represent primary and secondary lysosomes, many of which may be involved in the digestion of ingested material. Dense, pleomorphic deposits of lipochrome pigment occur, but considering the high phagocytotic activity, they are comparatively few (7). In some species accumulations of microbodies are seen in the basal cytoplasm in close relation to the endoplasmic reticulum (23).

3. Cell contacts

Contacts between neighbouring Sertoli cells are of two types: occluding junctions and nexuses or gap junctions. The occluding inter-Sertoli cell junctions are extensive junctional complexes, continuously interconnecting adjacent Sertoli cells at a level 'above'

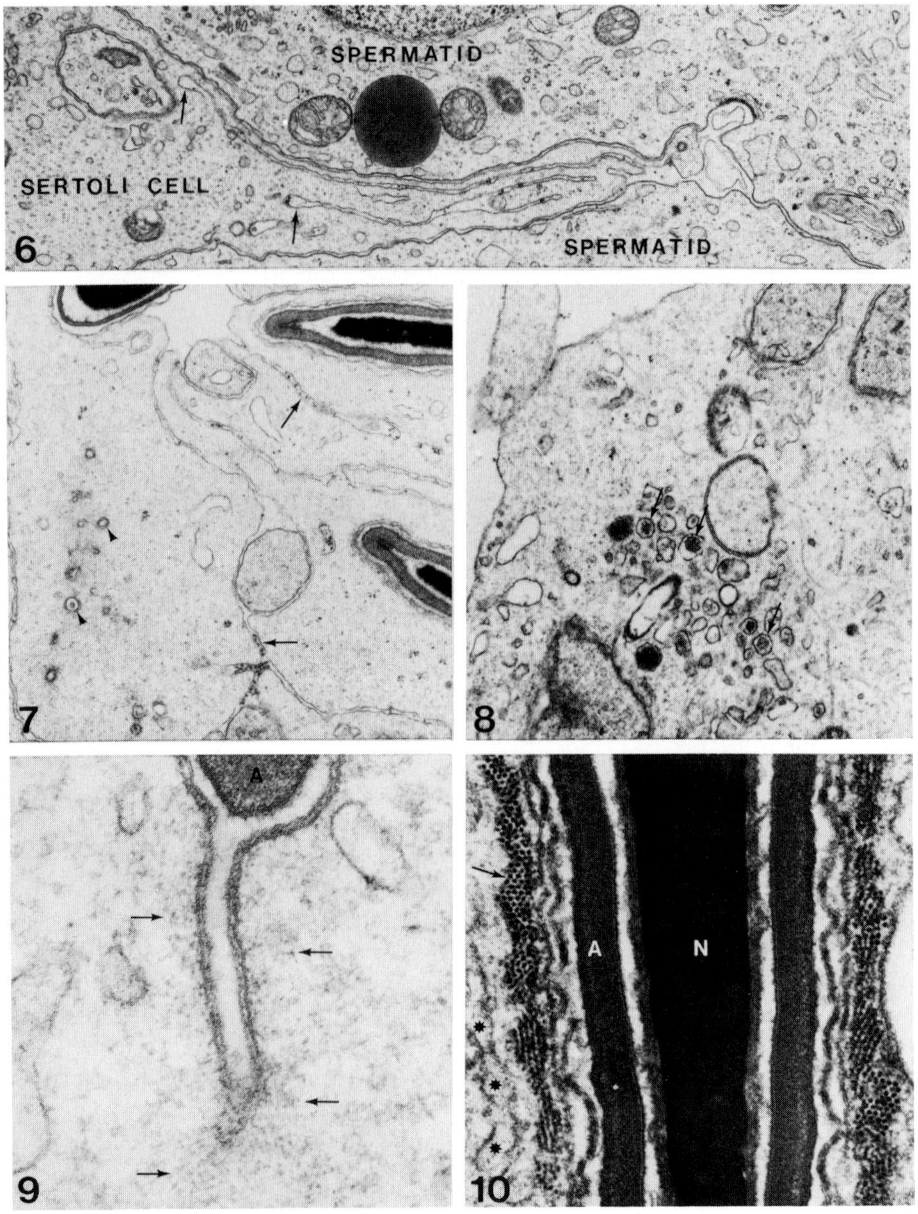

Figs. 6–10. Baboon testis. (6) Section of a thin Sertoli cell projection between two spermatids. Note the deep invaginations of Sertoli cell plasmalemmae into the cytoplasm (arrows) (× 14,000). (7) Section through the apical region of Sertoli cells to show cross-sectioned tubulo-bulbar complexes (arrowheads) and dense granules (arrows) in deep invaginations into the cytoplasm (× 16,000). (8) Another Sertoli cell from the same tubule as Figure 7 to show vacuoles with dense contents (arrows) (× 16,000). (9) Tubulo-bulbar complex from the rostral end of a late maturation phase spermatid. The two plasmalemmae run very close together. Note the filamentous material in the Sertoli cell (arrows). A = acrosome (× 72,000). (10) Boar testis. Longitudinal section of a late maturation phase spermatid surrounded by Sertoli cell ectoplasmic specialization. Note the regular arrangement of the filaments (arrow) and subsurface cisternae of endoplasmic reticulum (*). A = acrosome and N = nucleus of the spermatid (× 72,000).

the spermatogonia but 'below' the meiotic and post-meiotic germ cells. By means of these junctions the Sertoli cells form a continuous layer inside the seminiferous tubules, thus subdividing the interior of the tubules into two compartments: basal and adluminal (5). This continuous layer comprises the ultimate part of the blood-testis barrier (the name is somewhat misleading since it is an intratubular barrier). In the cytoplasm of both cells subjacent to the occluding juctions there are bundles of actin-like microfilaments (42) running parallel to the plasmalemma. Outside these bundles are subsurface cisternae

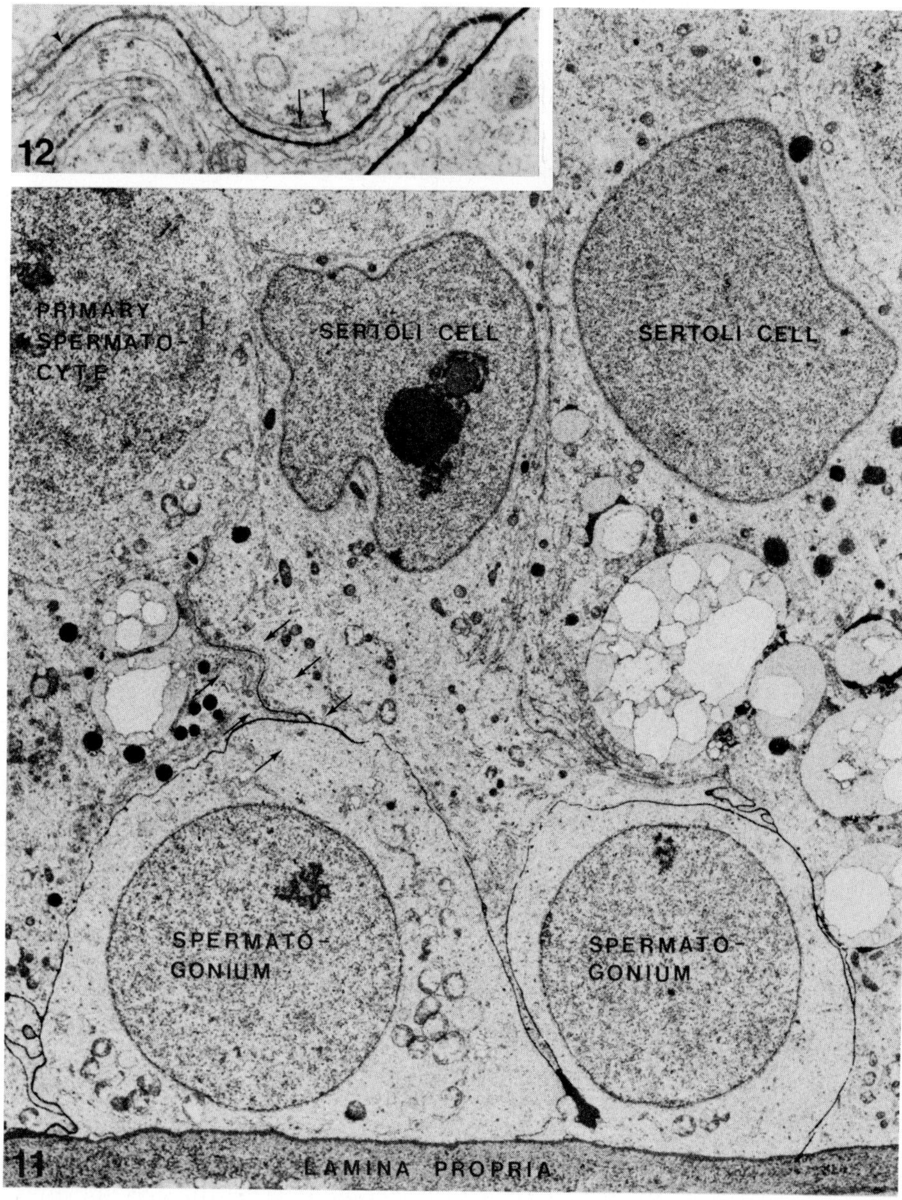

Figs. 11–12. (11) Section of human testis where lanthanum was included in the fixative as an extracellular tracer. Note that lanthanum is confined to the basal compartment and penetrates only a short distance into the inter-Sertoli cell junctions. Area between arrows is shown in higher magnification in Figure 12. Note the prominent nucleolus in the left Sertoli cell and also the pleomorphic lipid droplets and dense bodies (× 4,000). (12) The inter-Sertoli cell junction indicated with arrows in Figure 11. Lanthanum penetrates into the junction but is stopped (arrowhead) after a short distance. Note also the ribosomes (arrows) on the subsurface cisternae of endoplasmic reticulum (× 19.000).

of endoplasmic reticulum (for further details see 12). Occasional ribosomes occur on the cytoplasmic surface of these cisternae (Fig. 12).

Different types of contacts between Sertoli cells and germ cells occur. Desmosome-like devices have been observed between Sertoli cells and all types of germ cells (23, 32). The most prominent Sertoli-germ cell contact is found towards the acrosome region of acrosome- and maturation-phase spermatids, but similar, less conspicuous specializations occur toward germ cells from pachytene spermatocytes (34) through cap phase spermatids. These specializations have been called Sertoli ectoplasmic specializations (34), 'half-Sertoli junctions' (23) or – those towards acrosome-maturation phase spermatids – the mantle (47). In these specializations the Sertoli cells have the same components as in the inter-Sertoli cell junctions but with no corresponding specializations in the germ cells (Fig. 10).

Special devices, called tubulo-bulbar complexes (23, 34) are present in the region of the Sertoli cell facing the acrosome of late maturation phase spermatids (Fig. 9). These complexes consist of tubular evaginations of spermatid plasmalemma into corresponding invaginations in the Sertoli cell. The plasmalemmae of the two cells run parallel and very close together for a considerable distance and end with a bulbous swelling. Filamentous material in the Sertoli cell surrounds these complexes (Fig. 9).

4. Structure-function relationships

4.1. Steroid metabolism

For a long time various functions in steroid synthesis and/or metabolism have been ascribed to the Sertoli cells. These assumptions have partly been based on the analogies with the granulosa cells of the ovary and partly on some of the morphological features of the Sertoli cell. Sertoli cells contain large amounts of agranular endoplasmic reticulum resembling that found in steroid synthesizing cells such as Leydig cells or cells of the adrenal cortex, although the reticulum is not as extensively developed in Sertoli cells. In Sertoli cells of immature rats there is a marked steroid aromatase activity that vanishes during early maturation (2), but several other androgen metabolizing activities persist (cf 25). Probably, there is litte *de novo* synthesis of steroid hormones in the mature Sertoli cell (20) although the abundance of agranular endoplasmic reticulum suggests that hitherto unknown functions may be present.

4.2. Protein synthesis

The Sertoli cell has been shown to be the source of several proteins found in the rete testis fluid. Androgen binding protein (ABP) was first shown to be a specific protein of the Sertoli cell (11 and 14). Later *in vitro* studies have indicated that a great number of proteins are secreted (plasminogen activator, albumin-like proteins, peptides etc), ABP constituting less than 1% of the total. Morphological signs of protein secretion are inconspicuous in the Sertoli cells. The nucleolus is certainly prominent and has a high level of RNA synthesis (43). Autoradiographic studies show the incorporation of ^{3}H-lysine into vacuolated Sertoli cell projections (17). Vacuoles resembling condensation vacuoles in the Golgi region and the apical vacuoles with small dense granules may be interpreted as signs of merocrine secretion. This together with the biochemical evidence, indicates that the Sertoli cell secretes protein by merocrine secretion although in amounts quantitatively inconspicuous compared to typical serous cells such as exocrine pancreatic cells. Hypothetical functions in the regulation of spermatogenesis have been ascribed to some of the factors secreted by the Sertoli cells.

4.3. Blood-testis barrier

The morphological basis for the blood-testis barrier is, as already mentioned, the Sertoli cells and the occluding inter-Sertoli cell junctions. One of the main functions of this barrier is to establish and maintain a microenvironment in the adluminal compartment that differs in chemical composition from that outside the barrier (38, 39). Disturbances in the barrier function may lead to disruption of spermatogenesis due to alterations in the microenvironment of the adluminal compartment. Such disturbances may be due to alterations in the occluding junctions that can be demonstrated by the use of extracellular tracers such as lanthanum. They may also be due to other changes in Sertoli cell functions, which are difficult or, at present, impossible to demonstrate. During the last ten years much work has been devoted to clarify Sertoli cell functions, and considerable information has accumulated on specific Sertoli cell markers (25). The second main – perhaps the primary – function of the blood-testis barrier is to prevent contact between more advanced stages of germ cells and the immune system. Spermatocytes in the pachytene stage of the meiotic prophase and in all successive developmental stages express specific surface antigens (44) that would evoke an immune response if exposed to the immune system. It also appears that the Sertoli cell plasmalemma facing the adluminal compartment expresses specific antigenic determinants (45). The occluding inter-Sertoli cell junctions are very stable (12), but nevertheless, they have the ability to open temporarily and to transport newly formed spermatocytes from the basal to the adluminal compartment. During this process the junctions are never open along their entire length; the basal part becomes closed before the spermatocytes reach the adluminal compartment. Thus for a brief period during each cycle of the seminiferous epithelium there is a transient, intermediate compartment (33). It seems reasonable to assume that the actin-like filaments (42) that are always associated with the junctions are active in the transportation of spermatocytes.

4.4. *Sertoli-germ cell relations*

Ectoplasmic specializations (35) (see Chapter 5) or 'half-Sertoli junctions' first appear in relation to primary spermatocytes and persist throughout spermatogenesis. The amount of these specializations in various developmental stages varies, and quantitative data indicate a cyclical reutilization of the actin-like filaments in their formation (35). During the acrosome-phase, spermatids become located in deep recesses in the Sertoli cell and, as spermiation approaches, they 'back out' towards the lumen. The spermatids lack the ability to move and thus it seems probable that the actin-like filaments of the ectoplasmic specializations of the Sertoli cells are involved in the movement of the spermatids. The tubulobulbar complexes (see above) were first thought to anchor the germ cells to Sertoli cells, but later they have been interpreted as devices for elimination of redundant plasmalemma (23) and cytoplasm (34).

The process of spermiation is not fully understood and different opinions have been held on the mechanisms involved (7, 9, 23 and 31). It is, however, generally accepted that an active role is played by the Sertoli cells, which are also responsible for the demarcation of the residual bodies (7). Recent studies of the occurrence of plasminogen activator at different stages of the cycle of the seminiferous epithelium show that this enzyme has maximum levels during the stages when spermiation and the concomitant transport of spermatocytes from the basal to the adluminal compartment take place (18). A possible role of plasminogen activator for these processes has been suggested (18).

4.5. *Sertoli-germ cell communication*

As noted above, the postmeiotic germ cells are totally dependent on the Sertoli cells. The means of communication between these cell types may be manyfold. Direct cell-to-cell contacts through specialized structures are not found. Thus, diffusible messengers may be more important for the Sertoli-germ cell communication. Such messengers may be *stimulatory* (energy substrates, growth promoting factors, hormones?) or *inhibitory*. The latter phenomenon is indicated in recent *in vitro* studies (3). Such paracrine secretions of the Sertoli cells (and germ cells?) are just beginning to be uncovered.

4.6. *Sertoli-Sertoli cell communication*

The existence of nexuses or gap junctions between adjoining Sertoli cells provides a basis for direct cell-to-cell communication. Thus it is possible that Sertoli cells can coordinate their activity and thereby effect the synchronization of the differentiation process that occurs in most species.

4.7. *Cyclic changes in Sertoli cell function*

Through the cycle of the seminiferous epithelium the Sertoli cells are exposed to different combinations of germ cells of different developmental stages. Recently a number of Sertoli cell functions have been shown to vary in different stages of the cycle of the seminiferous epithelium. Binding of FSH, FSH dependent cyclic AMP formation (24), adenylyl cyclase activity (13) and secretion of ABP (28) and plasminogen activator (18) vary along the spermatogenic wave in rats. There are, however, few corresponding changes in Sertoli cell morphology. Cyclical changes in the lipid content of Sertoli cells have been reported in rats (16), but in other species there is little or no evidence for similar cyclical changes in Sertoli cell lipids.

There is now much evidence that the Sertoli cells are 'key cells' in the regulation of spermatogenesis, but many of the postulated functions remain to be confirmed. The morphological basis for some Sertoli cell functions seems obvious, such as the blood-testis barrier, whereas for other functions new methods must be employed to further elucidate the relation between morphology and function.

Acknowledgements

Dr. R.E. Johnsonbaugh, Naval Med. Center, Bethesda, is gratefully acknowledged for letting us use unpublished data on baboon testicular morphology. Our thanks are due to Hans Ekwall and Åsa Jansson for excellent technical assistance.

This work was supported by grants from the Swedish Council for Forestry and Agricultural Research (A4989/B3601) and the Swedish Medical Research Council (3168).

References

1. Andersen K, Sundby A, Hansson V: Fine structure and FSH binding of Sertoli cells in the blue fox (Alopex lagopus) in different stages of reproductive activity. Int J Androl (4): 570–581, 1981.
2. Armstrong DT, Dorrington JH: Estrogen biosynthesis in the ovaries and testes. In: Advances in sex hormone research. Regulatory mechanisms affecting gonadal hormone action. Thomas JA, Singhal RL (eds). Baltimore. University Park Press, 1977, pp 217–258.
3. Boitani C, Ritzen EM, Parvinen M: Inhibition of rat Sertoli aromatase by factor(s) secreted specifically at spermatogenic stages VII and VIII. Mol Cell Endocrinol (23): 11–22, 1981.
4. Dym M: The fine structure of the monkey (Macacca) Sertoli cell and its role in maintaining the blood-testis barrier. Anat Rec (175): 639–656, 1973.
5. Dym M, Fawcett DW: The blood-testis barrier in the rat and the physiological compartmentation of the seminiferous epithelium. Biol Reprod (3): 308–326, 1970.
6. von Ebner V: Zur Spermatogenese bei den Säugetieren. Arch. mikrosk. Anat (31): 236–292, 1888.
7. Fawcett DW: Ultrastructure and function of the Sertoli cell. In: Handbook of Physiology. Hamilton DW, Greep RO (eds). Washington. American Physiological Society, 1975, pp 21–55.
8. Fawcett DW, Leak LV, Heidger PM: Electron microscopic observations on the structural components of the blood-testis barrier. J Reprod Fert (suppl 10): 105–122, 1970.
9. Fawcett DW, Phillips DM: Observations on the release of spermatozoa and on changes in the head during passage through the epididymis. J Reprod Fert (suppl 6): 405–418, 1969.
10. Flickinger, CJ: The postnatal development of the Sertoli cells of the mouse. Z Zellforsch (78): 92–113, 1967.
11. Fritz IB, Rommerts FG, Louis BG, Dorrington JH: Regulation by FSH and dibutyryl cyclic AMP of the formation of adrogen-binding protein in Sertoli cellenriched cultures. J Reprod Fertil (46): 17–24, 1976.
12. Gilula NB, Fawcett DW, Aoki A: The Sertoli cell occluding junctions and gap junctions in mature and developing mammalian testis. Develop Biol (50): 142–168, 1976.
13. Gordeladze JO, Parvinen M, Clausen OPF, Hansson V: Stage dependent variation in Mn2+-sensitive adenylyl cyclase (AC) activity in spermatids and FSH-sensitive AC in Sertoli cells. Arch Androl (8): 43–51, 1982.
14. Hagenäs L, Ritzen EM, Plöen L, Hansson V, French FS, Nayfeh SN: Sertoli cell origin of testicular androgen-binding protein (ABP). Mol Cell Endocrinol (2): 339–350, 1975.
15. Hagenäs L, Plöen L, Ekwall H, Osman DI, Ritzen EM: Differentiation of the rat seminiferous tubules between 13 and 19 days of age. Int J Androl (4): 257–264, 1981.
16. Kerr JB, de Kretser DM: Cyclic variations in Sertoli cell lipid content throughout the spermatogenic cycle in the rat. J Reprod Fert (43): 1–8, 1975.
17. Kierszenbaum AL: Distribution of newly synthesized proteins in Sertoli cells as traced by electron microscope autoradiography. In: The testis in normal and infertile men. Troen P, Nankin HR (eds), New York. Raven Press, 1977, pp 125–136.
18. Lacroix M, Parvinen M, Fritz IB: Localization of testicular plasminogen activator in descrete portions (Stages VII and VIII) of the seminiferous tubule. Biol Reprod (25): 143–146, 1981.
19. Meyer R, Posalaky Z, McGinley D: Intercellular junction development in maturing rat seminiferous tubules. J Ultrastruct Res (61): 271–283, 1977.
20. Van der Molen HJ, Grootegoed JA, de Greef-Biljeveld MJ, Rommerts FFG, van der Vusse GJ: Distribution of steroids, steroid production and steroid metabolizing enzymes in rat testis. In: Hormonal regulation of spermatogenesis. French FS, Hansson V, Ritzen EM, Nayfeh SN (eds), New York. Plenum Press, 1975, pp 3–23.
21. Nagano T, Suzuki F: Freeze-fracture observations on the intercellular junctions of Sertoli cells and of Leydig cells in the human testis. Cell Tiss Res (166): 37–48, 1976.
22. Nicander L: An electron microscopical study of cell contacts in the seminiferous tubules of some mammals, Z Zellforsch, (83): 375–397, 1967.
23. Osman DI, Plöen L: The ultrastructure of Sertoli cells in the boar. Int J Androl (1): 162–179, 1978.
24. Parvinen M, Marana R, Robertson DM, Hansson V, Ritzen EM: Functional cycle of rat Sertoli cells: Differential binding and action of FSH at various stages of the spermatogenic cycle. In: Testicular Development Structure, and Function. Steinberger A, Steinberger E (eds). New York. Raven Press, 1980, pp 423–432.
25. Purvis K, Hansson V: Hormonal regulation of spermatogenesis: Regulation of target cell response. Int J Androl (suppl 3): 81–143, 1981.
26. Rambourg A, Clermont Y, Marraud A: Three-dimensional structure of the osmium-impregnated Golgi Apparatus as seen in the high voltage electron microscope, Am J Anat (140): 27–46, 1974.
27. Ritzen EM, Hansson V, French FS: The Sertoli cell. In: The Testis. Burger H, de Kretser D (eds) New York. Raven Press, 1981, pp 171–194.
28. Ritzen EM, Boitani C, Parvinen M, French FC, Feldman M: Stage-dependent secretion of ABP by rat seminiferous tubules. Mol Cell Endocrinol (25): 25–33, 1982.
29. Roosen-Runge EC: The process of spermatogenesis in mammals. Biological reviews (37): 343–377, 1962.
30. Roosen-Runge E: The process of spermatogenesis in animals. Developmental and cell biology 5. Cambridge University Press, 1977.
31. Ross MH: The Sertoli cell junctional specialization during spermiogenesis and at spermiation. Anat Rec (186): 79–104, 1976.
32. Russel L: Desmosome-like junctions between Sertoli and germ cells in the rat testis. Am J Anat (148): 301–312, 1977.
33. Russel L: Movement of spermatocytes from the basal to the adluminal compartment of the rat testis. Am J Anat (148): 313–328, 1977.
34. Russel L: Spermatid-Sertoli tubulobulbar complexes as devices for elimination of cytoplasm from the head region of late spermatids of the rat. Anat Rec (194): 233–246, 1979.
35. Russel L, Myers P, Ostenburg J, Malone J: Sertoli ectoplasmic specializations during spermatogenesis. In: Testicular Development, Structure, and Function. Steinberger A, Steinberger E (eds), New York. Raven Press, 1980, pp 55–63.
36. Schulze C: On the morphology of the human Sertoli cell. Cell Tiss Res (153): 339–355, 1974.
37. Sertoli E: Dell'esistenza de particolari cellule ramificati nei canalicoli seminiferi del testiculo humano. Morgagni (7): 31–39, 1865.
38. Setchell BP, Voglmayr JK, Waites GMH: A blood-testis barrier restricting passage from blood into rete testis fluid but not into lymph. J Physiol (200): 73–85, 1969.
39. Setchell BP, Waites GMH: The blood-testis barrier. In: Handbook of physiology. Hamilton DW, Greep RO (eds), Washington. American Physiological Society, 1975, pp 143–172.
40. Steinberger A, Steinberger E: The Sertoli cells. In: The testis IV. Johnson AD, Gomes WR, (eds), London. Academic Press, Inc, 1977, pp 371–399.
41. Toyama Y: Ultrastructural study of crystalloids in Sertoli cells of the normal, intersex and experimental cryptorchid swine. Cell Tiss Res (158): 205–213, 1975.
42. Toyama Y: Actin-like filaments in the Sertoli cell junctional specializations in the swine and mouse testis. Anat Rec (186): 477–492, 1976.
43. Tres LL, Kierszenbaum AL: Transcription during mammalian spermatogenesis with special reference to Sertoli cells. In: Hormonal regulation of spermatogenesis. French FS, Hansson V, Ritzen EM, Nayfeh SN (eds), New York London. Plenum Press, 1975, pp 455–478.
44. Tung PS, Fritz IB: Specific surface antigens on rat pachytene spermatocytes and successive classes of germinal cells. Develop Biol (64): 297–315, 1978.
45. Tung PS, Fritz IB: Histopathological changes in testes of adult inbred rats immunized against pachytene spermatocytes or Sertoli cells. Int J Androl (suppl 2): 459–477, 1978.
46. Vitale R, Fawcett DW, Dym M: The normal development of the blood-testis barrier and the effects of clomiphene and estrogen treatment. Anat Rec (176): 333–344, 1973.
47. Yasuzumi G, Tanaka H, Tezuka O: Spermatogenesis in animals as revealed by electron microscopy. VIII. Relations between the nutritive cells and the developing spermatids in a pond snail. J Biophys Biochem Cytol (7): 499–504, 1960.

Authors' addresses:
Department of Anatomy and Histology
Faculty of Veterinary Medicine
Swedish University of Agricultural Sciences
S-75007 Uppsala, Sweden
and
Pediatric Endocrinology Unit
Karolinska Sjukhuset
S-10401 Stockholm, Sweden

CHAPTER 7

Membrane organization and differentiation in the guinea-pig spermatozoon

DANIEL S. FRIEND

1. Introduction

Studied in extensive *morphological* detail, the plasma membrane of the guinea-pig spermatozoon (1–11) – like those of all other mammalian sperm – is divided into structural regions (*macrodomians*) closely correlating with the *functional* segments of the cell (2, 12–15). In each macrodomian, plasmalemmal, cytoplasmic, and functional diversities exist. Moreover, the membrane macrodomains themselves contain *microdomains*, thus contributing to further heterogeneity in the major vicinities of the cell. The parameters in which the primary sectors differ – that is, the acrosomal cap of the head, the equatorial and postacrosomal segments, and the midpiece, annulus, and principal-piece of the tail – involve many factors. These include the surface coat (morphology, antibody-specificity, lectin binding, surface-charge); the plasma membrane (intramembranous particle numbers and configurations, and the lining of the membrane); and possibly membrane sterol content, anionic-lipid asymmetry, and lipid-phase state. Similar forms of heterogeneity (6, 9) are provided in several internal organelles, particularly the nuclear envelope and the mitochondrion.

In this chapter, we review the predominant structural features of each principal functional area of the sperm membrane, identify their maturational sites in the genital tract, and relate them (whenever possible) to specific cellular activity. Of paramount focus here are plasmalemmal domains such as the acrosomal cap, which fuses with the underlying acrosome to release enzymes for sperm penetration to the oocyte surface. Others of importance are the equatorial and postacrosomal segments, which adhere to and fuse with the oolemma. The midpiece, including the retained cytoplasmic droplet, houses mitochondria for producing cellular energy, while the annulus separates the tail's mid- and principal-pieces. And, not least, the principal-piece propagates the flagellar wave (Fig. 1).

2. The acrosomal cap

Observed in freeze-fractures and surface-replicas, the most noteworthy characteristic of the acrosomal cap in guinea-pig sperm (derived from the tail of the epididymis or vas deferens) is its patchwork-quilt pattern. This pattern mirrors the superficial portion of the thick, periodic glycocalyx seen in thin sections (Figs. 1–3) (2–5). First emerging at the end of the epididymal mid-portion, the quilt design vanishes after short (0.5-h) incubation in physiologic salt solutions to permit capacitation – the several-hour-long maturational process required before calcium can induce fusion between the acrosomal and plasma membranes (14–16). Alternately, capacitation can be accelerated by brief treatment with crude trypsin (17). After removal of the quilt, a thin, downy glycocalyx, clearly visible only in surface-replicas, still coats the membrane. Judging from results of the procedures used to assess plant-lectin-binding and antibody reactivity of the sperm head, we think it probable that this second layer of glycocalyx (in conjunction with the plasma membrane proper) is responsible for the specific reactivity of the head with wheat-germ agglutinin, which recognizes β-galactose (18), as well as with certain monoclonal antibodies directed only against the acrosomal cap of the plasma membrane (19). Labeling with other lectins is generally more uniform (18, 20), indicating that most sugar residues, unlike β-galactose, are not confined to specific areas of the spermatozoan surface. The regionality of surface-charge, which becomes increasingly negative during the passage of sperm through the male genital ducts, with the cells also increasing negative charge starting from the tip of the head to the end of the tail, has been fully considered by other investigators (21, 22).

Van Blerkom, J. and Motta, P.M. (eds.), Ultrastructure of reproduction. ISBN 0-89838-572-5.

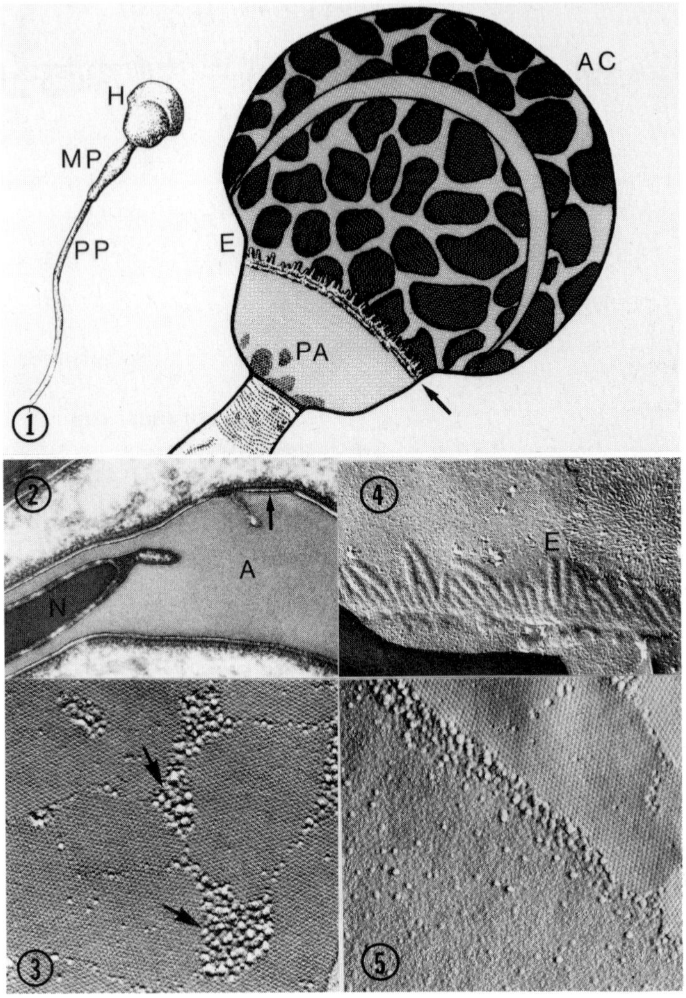

Figs. 1–5. (1) Diagram of the guinea-pig sperm head incorporating features observed in freeze-fracture and surface replicas. Slender projection (arrow) separate the stippled acrosomal (AC) portion of the membrane from the postacrosomal segment (PA). The whole sperm appears on the left. H – head; MP – midpiece; PP – principal piece; E – equatorial segment.* (2) In thin-section, tannic-acid fixation enhances the periodicity of the glycocalyx. Observe the close apposition of the acrosomal and plasma membranes in front of the nuclear tip. This region corresponds to the plaque-free band in the diagram above. N – nucleus; A – acrosome. The arrow indicates a region of dense nap (× 30,000). (3) Detail of the quilt-patterned plaques in freeze-fractured, filipin-treated sperm before incubation in a capacitating medium. Filipin/sterol complexes form 250-Å protrusions (arrows) in the aisles between the plaques (× 27,600). (4) Freeze-fracture image of the slender projections commonly seen in surface replicas at the acrosomal: postacrosomal juncture. E – equatorial segment (× 32,400). (5) Filipin-treated, freeze-fractured sperm. The usual image of the acrosomal (quilted): postacrosomal juncture (× 24,000).

2.1. Filipin/sterol complexing in the sperm plasma membrane

Applied to the guinea-pig sperm membrane, filipin, a polyene that binds β-hydroxysterols and demonstrates them by complexing with sterol, reveals complexes only in the aisles between plaques (23, 24) (Figs. 3 and 5). Viewed in freeze-fracture replicas, the complexes appear as $\sim$250-Å protuberances and depressions. In thin sections, the sterol-rich areas of membrane seem scalloped. Dissolved in membrane, the reagent also fluoresces. However, filipin fluorescence indicates that sterols may exist *in toto* throughout this region – that is, the entire acrosomal cap, not only between the plaques. Such high concentrations of sterols (cholesterol and desmosterol) contrast with their amount and disposition in the postacrosomal segment, which is nonfusigenic under mutual conditions (Fig. 5). The findings obtained from preliminary studies in which we applied filipin to liposomes composed of phosphatidylcholine and cholesterol and examined the protrusions and indentations

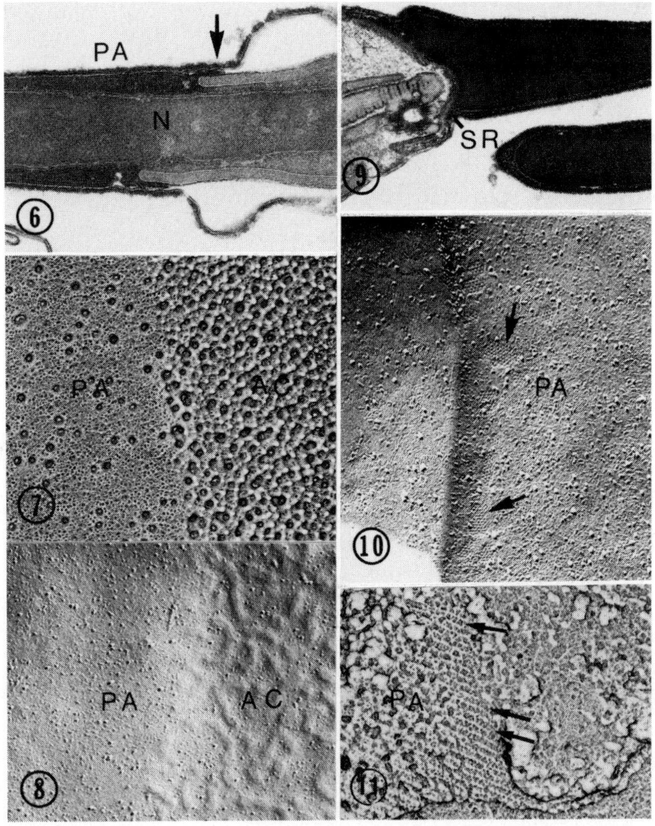

Figs. 6–11. (6) Thin section of a tannic-acid-fixed sperm showing the close apposition of the plasma membrane to a dense postacrosomal substructure originating near the end of the equatorial segment (arrow). The deep-etched appearance of this dense material is depicted in Figure 11. N – nucleus; PA – postacrosomal segment (× 26,400). (7) The acrosomal (AC): postacrosomal (PA) juncture in a rotary shadowed freeze-fracture replica of a filipin-treated sperm after incubation in capacitating medium. The acrosomal-portion quilt-pattern has disappeared. More than five times the number of filipin/sterol complexes emerge in the acrosomal region (right side) compared to those of the postacrosomal sector (left side), reflecting the maintenance of a major difference in sterol concentration in the two regions of this continuous membrane (× 22,800). (8) A similar compositional disparity exists in the membrane's outerleaflet anionic-lipid concentration as reflected by the polymyxin-B-induced crenulations confined to the acrosomal portion (AC) of the plasmalemma (right side). PA – postacrosomal segment (× 27,600). (9) Thin section of the postacrosomal tail juncture. N – nucleus; SR – striated ring (× 18,000). (10) The E-fracture face of the plasma membrane in the most distal region of the head contains ordered clusters of pits (arrows). Complementary patches of particles would appear on the P-face. PA – postacrosomal segment (× 75,000). (11) Parallel periodic rods (arrows), 150-Å apart, composed of 50–70-Å subunits, underlie the entire postacrosomal portion of the plasma membrane. PA – postacrosomal segment (× 72,000).

of the liposomal membrane are interpreted as proving the reagent's ability to detect gradients of cholesterol across the bilayer (9). In circumstances where the majority of sterol is presumably in the outer leaflet, the membrane indents. But when the sterol is assumed to be equal in both membrane leaflets, the membrane becomes deformed into both protrusions and indentations. In the acrosomal sector of sperm, *filipin/sterol complexes* protrude, as would be compatible with a model consigning the high sterol concentration to the inner half of the membrane bilayer (9, 25). Yet in the postacrosomal segment (Fig. 6), which contains less that one-fourth the number of complexes observed in the acrosomal area (Fig. 7), such complexes often indent the membrane, implying that the sterols inhabit the outer half of the bilayer. We suspect that this distribution characterizes stable, nonfusigenic regions of the membrane.

2.2. Anionic lipids in the cap

High anionic-lipid concentration, as determined by *polymyxin-B membrane perturbation* (26, 27), typifies the acrosomal-cap portion of the sperm plasma membrane (Fig. 8) (17, 27). Before incubating the cells in capacitating medium to prepare them for plasmalemmal: acrosomal membrane fusion, we noted anionic lipids only at the tip of the cell. During capacitation, however, they are present throughout the entire cap, except in a transcellular band fronting the tip of the nucleus. Even under capacitating conditions, cytochemically-detectable high anionic-lipid concentration halts at the end of the equatorial segment (Fig. 8) (9, 17, 27), the remotest line of fusion between the acrosomal and plasma membranes during calcium-triggered discharging (exocytosis) of the acrosomal contents. The antibiotic polymyxin B binds anionic phospholipids in equimolar ratio. In freeze-fracture replicas, the binding is reflected by broad (500–1000 Å) crenulations of anionic-phospholipid-rich membrane (17, 28).

2.3. The role of calcium in the acrosome reaction

Physiologically, sperm undergo capacitation in the uterus and fallopian tubes. They can also be capacitated *in vitro* in a variety of physiologic salt solutions. The end-point of the capacitating process is fusion between the acrosomal and plasma membranes – the *acrosome reaction* (15). Without calcium in the capacitating medium, the acrosome reaction does not occur (16). Therefore, by the strategy of using calcium-free solutions, the processes of capacitation and the acrosome reaction can be studied separately.

2.4. Merocyanin as a tester of membrane fluidity

Roughly parallel with the increasingly decipherable sterol and anionic lipid concentrations during incubation of the cells in *calcium-free Tyrode's solution*, the degree of membrane fluidity consistent with the pattern of merocyanin S-540 (Eastman Kodak, Rochester, N.Y.) incorporation and fluorescence (29, 30) is also enhanced all through the acrosomal cap (31). Merocyanin S-540, a fluorescing lipophilic reagent, differentially stains various regions of the membrane. One interpretation is that it retains its fluorescence after intercalating into highly fluid lipid areas of biological membranes but quenches it and/or does not enter gel-state lipid portions of membrane. Although this interpretation is not universally accepted, we favor it as a working hypothesis. Therefore, we consider merocyanin incorporation and fluorescence as mapping extremely fluid parts of cell membranes. From this viewpoint, a high degree of membrane fluidity is initially apparent at the tip of the cell and then proceeds onward into the equatorial segment. Thus during capacitation, the total acrosomal and equatorial segments of the plasma membrane resolvable by fluorescence microscopy seem to be in a state of considerable fluidity – a state which may be generally characteristic of imminently fusigenic membrane.

2.5. The anterior band and initiation of fusion

Both before and during capacitation, a band of membrane in front of the nucleus (anterior band) is usually devoid of reagent/sterol and reagent/anionic-lipid membrane perturbation. And it is this microdomain which often contains circles of barren membrane, cleared of sterols and intramembranous particles preceding the induction of membrane fusion by calcium ions (4, 16, 17, 27, 32). Less pervasive after preparation of the cells by rapid-freezing without fixation or glycerol cryoprotection (6), these denuded patches may still be demonstrable in the guinea-pig acrosomal cap. The anterior band also corresponds in topography to the disposition of an intracellular reservoir of pyroantimonate-trapping cations, including calcium (3). On the bases of 1) the probable instability of membrane in areas of sharp interfaces between sterol-rich and sterol-poor as well as anionic-lipid-rich and anionic-lipid-poor portions; 2) the possibility for very intimate apposition of the acrosomal and plasma membranes provided by the intramembranous particle-free patches; and 3) the presence of an intracellular cation pool – we are inclined to consider this the primary microdomain in which fusion originates during the acrosome reaction.

A cytoplasmic tuft exclusively underlies this particle-bare, filipin/sterol-complex-sparse, anionic-lipid-deficient band of membrane in the foreground of the nuclear tip (17). Otherwise, no cytoskeletal structures can be observed until we arrive at the end of the *finger-like projections* stretching from the posterior margin of the equatorial segment (Figs. 1 and 4) (3, 33).

The tuft beneath the anterior band, closely adhering to the acrosomal membrane, is visible in thin sections as a tannic-acid-stained density (Fig. 2) (17), and in surface-replicas as a pebbled region on that membrane. Since it demarcates the pre-capacitation anterior anionic-lipid domain, plausibly it helps to restrain the backward flow of anionic lipids within the plane of the membrane. And because this site also possesses a high affinity for cations, it may reasonably represent a calmodulin-type calcium-binding protein.

3. The equatorial segment

In the posterior part of the equatorial segment, the acrosomal membrane grasps the plasma membrane in a tight sheath (Fig. 6). Thin sections expose a generally dense matrix, punctuated by several, more intensely staining globules, in the 300-Å interspace between the two membranes. These globes appear in fractures and surface-replicas as a picket-fence arrangement of slender, attenuated projections pointing toward the anterior tip of the nucleus (Fig. 4) (3, 5). Our ability to observe them in surface-replicas is probably a consequence of cell shrinkage, with the projections then indenting the plasma membrane. Occasionally, they imprint the nuclear envelope as well. Perceived early in acrosomal development, during spermatid formation, the projections reach their ultimate destination in the mature cell's entrance to the epididymis. Their function is still undetermined. The composition of these digitations, so prominent in surface-replicas, remains an enigma, too. They precisely segregate the plasma membrane into two domains – an anterior one which fuses with the acrosomal membrane, and the postacrosomal segment, which commonly fuses with the egg. So situated, they also mark the region where the final splicing of acrosomal and plasma membranes occurs at the end of the acrosome reaction. (Do they perhaps stow away a cache of molecular solder for that event?)

3.1. The postacrosomal region

The transition between the acrosomal and postacrosomal regions is very sharp (Figs. 1, 5–8). Most often, it is seen as a termination of the acrosomal-area glycocalyx and quilt patterns (Figs. 1 and 5); and less frequently, also as the base of the equatorial-segment projections discussed above (Fig. 4). This sector of membrane (Fig. 9), non-fusigenic under capacitating conditions, contains low concentrations of sterols (Fig. 6), no discernable anionic lipids in its outer leaflet (Fig. 8) as determined by polymyxin-B-binding, and some inner-leaflet anionic lipids demonstrable by labeling with adriamycin (27, 31, 32) (note 2.3). Here the membrane clings to long, rod-like structures (Fig. 11) extending horizontally from the acrosomal: postacrosomal juncture to the striated ring between the nuclear envelope and the plasma membrane. Before capacitation, this region does not absorb merocyanin S-540 (see 1.4), suggesting that its outer-leaflet membrane lipids exist in a gel state (29).

In ordinary and rapidly-frozen freeze-fracture preparations, the postacrosomal part of the membrane exposes scattered intramembranous particles, with distinctive clusters at the facade of the striated ring (Fig. 10) (5). Though such clusters are augmented in capacitated sperm (7), the overlap in their population with their numbers in non-capacitated sperm makes this facet an unreliable guide to the functional status of the sperm. Many mammalian species – including man, rabbit, and hamster – have individual markings in this area. While the subunits within the species mentioned are identical, their arrangement differs in each sperm. Therefore, these markings truly act as fingerprints, emphasizing the uniqueness of every sperm cell within the population. At any given stage of maturation, all other surface differentiations – the quilt pattern, the out-stretching projections, the striated ring, the midpiece strands, the annulus, and zipper – appear indistinguishable from one cell to the next.

3.2. Membrane changes during capacitation

Capacitation is a complex process involving removal of material adsorbed to the sperm head and changes in the sperm membrane's glycocalyx, proteins, and lipids (14, 15). These, in concert, render the plasma membrane permeable to calcium ions and capable of fusing with the acrosomal membrane. The process generally takes place in the female genital tract, most effectively in the uterus and fallopian tubes (15). *In vitro*, it is simulated with considerable success: Consistent alterations occur in the plasma membrane covering the sperm head as it increases in adaptability for multiple sites of acrosomal: plasma membrane fusion after the addition of 5–15 mmol $CaCl_2$ (16). The quilt disappears, filipin/sterol complexes fill the membrane (4), *polymyxin-B-induced* perturbation extends all the way from the anterior cap to the end of the equatorial segment (17), and membrane fluorescence with merocyanin proceeds as far as the middle of the postacrosomal segment (31). Above the tip of the nucleus, a band of membrane remains unaffected by polymyxin B (17). This band discloses naked circles of membrane, cleared of filipin/sterol complexes and intramembranous particles. Less often, when prepared with conventional freeze-fracture techniques, patches devoid of particles emerge here after rapid-freezing without fixation or cryoprotection. Comparable circular clearances persist in the postacrosomal segment immediately behind the equatorial-region boundary (4, 7, 17).

The following interpretations are compatible with freeze-fracture: cytochemical observations: Usually, the fusigenic portion of plasma membrane covering the acrosome 1) is rich in free sterols within the inner leaflet of the bilayer; 2) is abundant in outer-leaflet anionic lipids; 3) is fluid; and 4) contains concisely demarcated lipid microdomains where fusion may originate (the acrosome reaction). The band of membrane with these attributes reveals interfaces of sterol-rich and -poor membrane, collared by anionic lipids and anchored by a cytoplasmic feltwork. Less is known about the non-fusigenic postacrosomal segment. Although it also contains a row of circles free of intramembranous particles, it still lacks outer-leaflet anionic lipids and plentiful inner-leaflet sterols at this time. After incubation in calcium-free medium for 6 hours, the acrosomal region becomes adaptable to fusion following the addition of $CaCl_2$; whereas the stable postacrosomal segment remains resistant. By comparing the cytochemical properties of the acrosomal and postacrosomal segments of plasma membrane, we can speculate that the non-fusigenicity of the postacrosomal domain is due to the anchoring of anionic phospholipids in its inner leaflet, its meager sterol content, dan its low fluidity. While this view is consistent with the rest of our data, please bear in mind that the methodologies by which the observations were obtained (filipin/sterol complexing, polymyxin-B perturbation, and adriamycin (34) and merocyanin incorporation) are all too new to have been tested for degrees of specificity *in situ*. Our hypotheses are by no means proven.

3.3. Membrane changes after the acrosome reaction

The acrosome reaction consists of multiple fusions between the outer acrosomal membrane and the overlying sperm plasma membrane, which allows the release of acrosomal enzymes probably necessary for sperm passage through the egg's zona pellucida (15). In the process, vesicles composed of both acrosomal and plasma membrane frequently cluster around the sperm head. A hybrid, continuous membrane is also created by merger of the remaining acrosomal membrane with the retained membrane of the equatorial and postacrosomal regions (3, 13, 14). We know little about the properties of this new encasement, but its very formation, obligatory for subsequent sperm: egg fusion, suggests that the splicing conveys fusigenic features to the remaining equatorial and postacrosomal segments. The unmasked inner acrosomal membrane is the site which usually binds to the zona pellucida of the egg (15).

Currently, scant information is available concerning the transmuted postacrosomal membrane following the acrosome reaction. We do know, however, that it achieves the capacity to fuse with the oolemma, and in the presence of fusigens such as monolyoglycerol, can unite with the plasmalemmae of other cell-types (35). Its intramembranous-particle density decreases to that of the acrosomal membrane; and round, particle-free patches girdle the cell directly behind the acrosomal: postacrosomal juncture (7). Further, we know that membrane reactivity to filipin is intensified to an extent approaching that of the acrosomal cap in capacitated sperm. Filipin/sterol bulges as well as indentations emerge, attesting to the presence of non-esterified sterols in both leaflets of the bilayer. Although the amount and transbilayer location of anionic phopholipids is uncertain, *adriamycin* (34) incorporation does indicate that some anionic lipids inhabit this portion of the membrane (31).

Adriamycin, an anthracyclic sugar, binds to acidic phospholipids (34). Since it crosses the plasma membrane, it binds to lipids in both leaflets of the bilayer. Taking advantage of the molecule's multiple double bonds, E. Bearer developed an osmium-reduction procedure for detecting the presence of the compound in thin-sectioned tissues (27). With the use of this procedure, lighter osmium reduction is associated with the plasma membrane in the postacrosomal segment than in the acrosomal cap – in agreement with the theory that there is a lesser amount, or a less accessible pool, of anionic phospholipids in the postacrosomal sector.

4. The sperm tail

Understanding of the sperm tail's meld of structure and function lags behind our comprehension of the head – but so much greater the challenge! The regular, morphologically diverse arrays in the membranes of the tail are truly remarkable (Fig. 12). During freeze-fracture, the plasma membrane splits within its hydrophobic interior, exposing two novel membrane 'faces' – the P-face, which presents the internal aspect of the half-membrane coating the cytoplasm; and the E-face, which presents the internal aspect of the half-membrane opposing the extracellular environment. Viewed in freeze-fracture, starting at the striated ring (Figs. 12, 13), necklaces of 80-Å particles encircle the midpiece P-face, save for the retained cytoplasmic droplet, which harbors only a few such strands in nondescript orientation (Fig. 14) (6). At the annulus, with its smooth external surface, rows of large E-face particles correspond to complementary depressions on the P-face. Orthagonal patches of sizable intramembranous particles adorn the E-face of the proximal principal-piece, from whence the staggered 150-Å particles of the zipper course opposite *fiber number one* to the end-piece (Fig. 12) (2, 3).

4.1. The midpiece

In sperm freshly removed from the epididymis, the

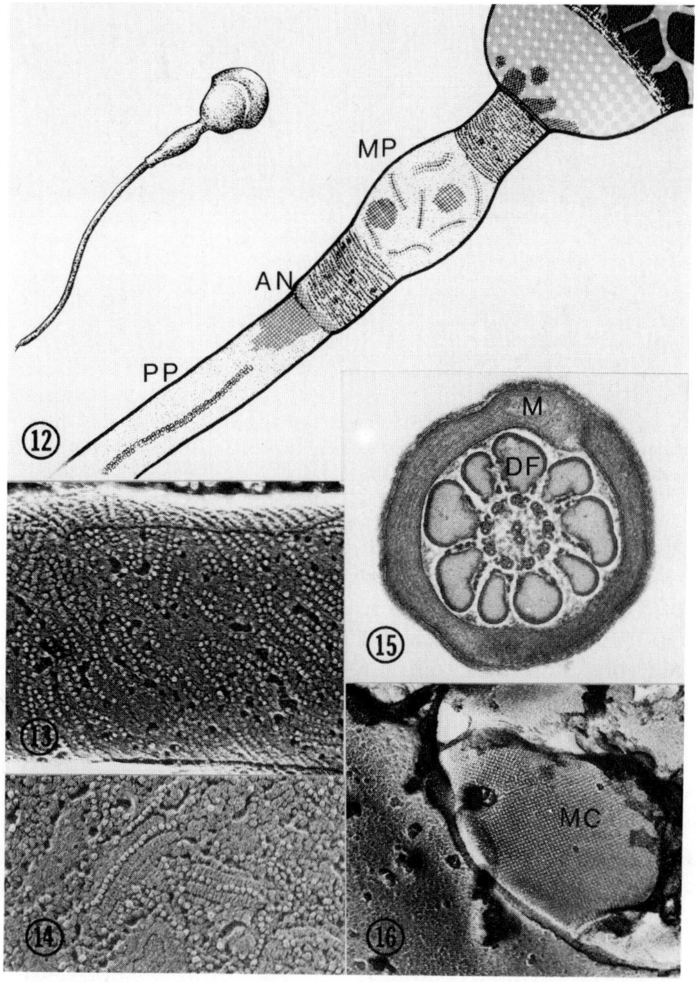

Figs. 12–16. (12) Diagram of the midpiece and proximal principal-piece of the sperm tail, illustrating microdomains exposed in conventionally- and rapidly-frozen cell preparations. A whole guinea-pig sperm is shown on the left. MP – midpiece; AN – annulus; PP – principal piece. (13) Rapidly-frozen, deep-etched preparations of the midpiece reveal the presence of particle-strands on the surface (upper portion) as well as on the P-fracture face of the membrane (× 72,000). (14) Similar strands, appearing in the cytoplasmic droplet during capacitation, often develop striated collars coplanar with the lipid matrix of the membrane (× 78,000). (15) Cross-section of the midpiece portraying the disposition of a mitochondrion (M) around the dense fibers (DF) and axonemal complex (× 39,000). (16) In rapidly-frozen preparations of sperm maintained in a calciumpoor medium, both the flattened cisternae (MC) (shown here) and the membrane comprising the retained cytoplasmic droplet occasionally disclose focal 'window-screen' patterns, which we interpret as gel states of the membrane lipids (× 75,000).

midpiece strands, tightly apposed, parallel the pitch of mitochondria (Figs. 13 and 15). Following capacitation, most of the strands disperse their particles in random distribution (7); but some of the strands remain intact and settle in the portion of membrane encasing the retained cytoplasmic droplet (Fig. 14). Here they develop streaked lipid collars, as observed in unfixed, non-glycerinated, rapidly-frozen preparations (6). In both leaflets of the flattened microcisternal vacuoles (dictyosome-like cisternae) (36, 37) of the droplet (Fig. 16), as well as in the overlying plasmalemma, window-screen-patterned circles of membrane emerge (6, 9). As in the fracturefaces, such orderly-disposed lipids can be demonstrated on the surface, too.

The droplet itself displays all the membrane insignia (similar to those of the acrosomal cap) of a secretory or fusigenic organelle – that is, it fluoresces brightly with filipin, stains with adriamycin, and avidly incorporates merocyanin S-540. Though this part of the spermatozoon is seldom discussed or considered to persist in fully mature, ejaculated sperm (1, 37), our experience dictates that, at the very least, several midpiece-droplet microcisternae remain in this region. We feel that its possible importance in the physiology and function(s) of fertile sperm should

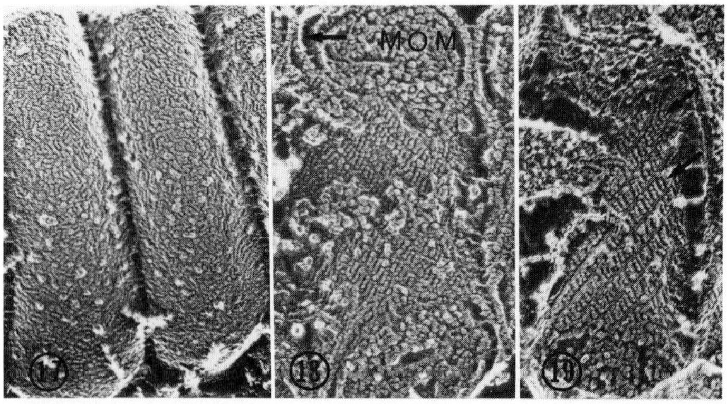

Figs. 17–19. In rapidly-frozen, deep-etched sperm preparations, particle-triplets are randomly dispersed over the mitochondrial (M) convex surface facing the plasma membrane (Fig. 17); while identical triplets (arrows) are observed in orderly arrays on the concave surface facing the axonemal complex (Figs. 18 and 19). MOM – mitochondrial outer membrane. (17: × 54,000; 18: × 60,000; 19: × 62,400).

not be discounted. Solving the riddle of this overlooked area may constitute a significant problem in reproductive biology. Does it act as an osmometer? Is it utilized in the passage of sperm through the cervical mucus or uterotubal junction? Or could it be a recognizable remnant of immaturity, used to mark cells for elimination from the race to the egg? These are the questions we pose and would like to have answered.

The surface-structure of the mitochondria which circle (Fig. 15) the axoneme throughout the midpiece is also intriguing. Examined in deep-etched samples, triplets of particles haphazardly coat the surface facing the plasma membrane (Fig. 17), while identical triplets in highly-ordered arrays (the mitochondrial ladder) engrave the surface which faces the axonemal complex (Figs. 18 and 19) (6, 9). Also associated with the mitochondrial outer membrane (6) are seven enzyme systems. If the series of particle triplets (rungs of the ladder) can be identified as one of them, there will be the opportunity for direct morphological: biochemical correlation of one aspect of mitochondrial function. For example, pharmacological manipulation of the enzyme could result in an alteration in the structure of the ladder – perhaps increased spacing between the rungs, with a gross dynamic incident attending a change in sperm metabolism. The rungs may even be an assayable target for modifying the fertilizing function of sperm.

4.2. *The principal-piece*

Just before the sperm enter the epididymis, the mitochondrial sheath ceases its retrograde growth. Concomitant with this event, the annulus reaches its final mature-cell position (Fig. 20). Only here in its fully differentiated state does it acquire characteristic freeze-fracture patterns (Fig. 21). During migration, the annular membrane is indistinguishable from the neighboring plasmalemmal fracture-faces. At its resting site – the juncture of the mid- and principal-pieces – the annulus fastens to a dense, wedge-shaped, submembranous anchor (Fig. 20). Feasibly, this adherent, detergent (digitonin, filipin, tomatin) -resistant portion of membrane (38) acts as a barrier to diffusion within the plane of the membrane, confining specialized transmembrane proteins to either the midpiece or principal-piece – whichever their appropriate foci of function.

Beyond the annulus (Figs. 22–24), micron-long lines of 150-Å E-face intramembranous particles (Fig. 25) are consistently encountered. Frequently, 20 to 30 or even more such arrays embellish the area. Whereas the P-face reveals 80-Å particles in register with the larger E-face particles, the membrane-surface shows elongated, grooved particle doublets – each occupying a little greater surface-area than one E-face particle. It is as if the outer leaflet of the bilayer sported a line of molars, with their occlusive surfaces protruding into the world outside and their bases embedded in the hydrophilic tips of the fatty-acid tails (Fig. 27). The complex-particle morphology, similar in configuration and dimensions to the acetylcholine receptor of neuromuscular junctions (39), implicates the particles as participants in some transporting channel function. Which transport function they subserve, we don't know. Dansylated (fluorescent) α-bungarotoxin, an acetylcholine antagonist, does not bind appreciably to guinea-pig sperm (Friend and Hall, unpublished observations).

Where the diagonal strands terminate (Fig. 25), the *zipper* (5, 40) begins (Figs. 22, 25, and 26). This

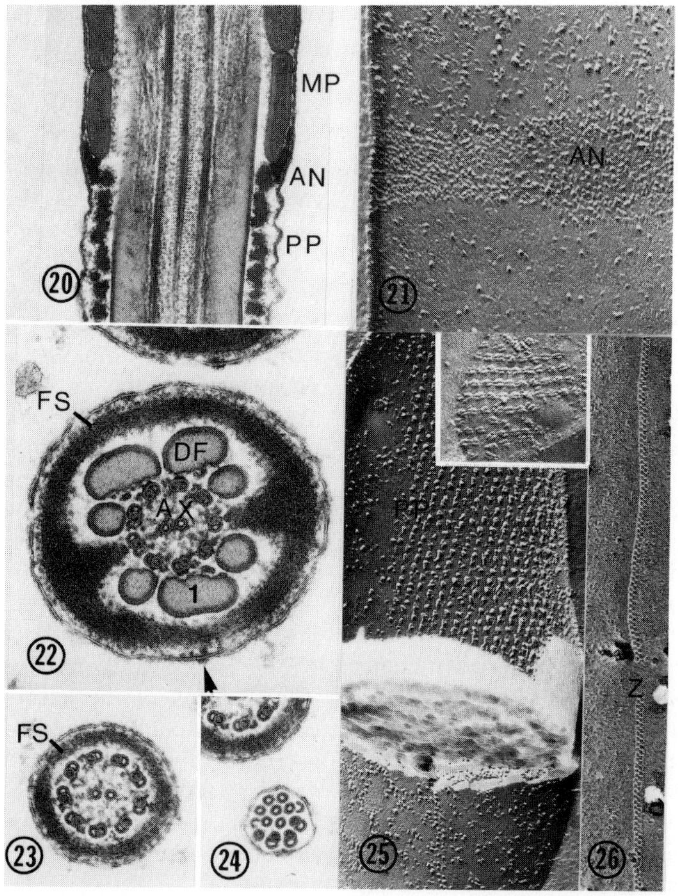

Figs. 20–26. (20) The annulus (AN) separates the mid- (MP) and principal-pieces (PP) of the tail (× 27,600). (21) The surface of the membrane at the annulus (AN) is smooth, but its E-fracture face (shown here) contains circumferential rows of large particles (× 48,000). (22–24) Cross-sections through the principal-piece of the sperm tail at the level of the dense fibers (DF) (Fig. 22), beyond the dense fibers (Fig. 23), and through the end-piece (Fig. 24). In Figure 22), the arrow points to an increase in density both above and beneath the membrane, corresponding to the position of the zipper opposite fiber 1. FS – fibrous sheath; AX – axonemal complex; 1 – dense fiber number one. (22: × 64,800; 23: × 54,600; 24: × 54,600). (25) Diagonal rows of large particles decorate the principal-piece (PP) E-fracture face beyond the annulus. Inset: The membrane surfaces of the molar-like particles. The staggered particles of the zipper are also visible in the P-fracture face in the lower half of the micrograph (× 38,400; inset: × 39,000). (26) The surface of the principal-piece zipper (Z) exposed in a rapidly-frozen, deep-etched, rotary-shadowed replica (× 36,000).

membrane structure consists of a staggered doublet of large particles, identified in the spermatozoan principal-piece in many species which propagate by internal fertilization. It was first described in the guinea pig (5). What can be stated with certainty about the zipper is that its particles clasp the plasma membrane to the fibrous ribs and remain after the cell membrane has been removed with digitonin (38, 40). Nor are they displaced by filipin/sterol protrusions (23), and only grudgingly detach from the fibrous *ribs* after treatment with Triton X-100 (37).

Other potential qualities of this unique membrane-specialization are conjectural. In the realm of the zipper, the increase in electron density both above and below the phospholipid membrane bilayer (Fig. 22), the presence of distinctive intramembranous particles (Fig. 25), and the surface-replica imaging of rectangular units (Fig. 26) all serve to establish that these zipper particles contain transmembranous integral proteins. Their surfaces expose mannose, D-galactose, and N-acetyl-D-glycosamine residues, as indicated by the binding of concanavalin A, *Ricinis communis 1*, and wheat-germ agglutinin/plant-lectins (39). The Triton X-100-soluble fraction, which has the same lectin-binding properties, includes four Coomassie-blue-stainable polypeptides with molecular weights ranging from 110,000 to 24,000 daltons (40). The foregoing facts are in concord with zipper particles containing membrane integral glycoprotein(s) with a complex-type asparagine-linked oligosaccharide moiety.

5. Concluding comments

Every primary feature of guinea-pig sperm morphology and the processes of capacitation and the acrosome reaction pertains to the sperm of all eutherian mammals. To be sure, marked variations do exist among the different species in the details of acrosome configuration, the ease of defining certain particle-clusters, and the precise patterns of the distinctive arrays. As an example, while the guinea pig has a single zipper, the mouse has several of them, and man often has two. But the basic structural: functional correlations, thoroughly examined in any one species, are probably applicable to all mammals. And thus far, the greatest degree of membrane specialization has been charted in the guinea pig, which possesses one of the more uniform sperm populations and readily discernable topographies. By way of contrast, human sperm is a heterogeneous population, difficult to preserve well, and subtle in membrane-designs. Particularly advantageous for studying, the guinea-pig sperm-cell offers a singular opportunity for helping unravel the many remaining mysteries of this indispensible cell-type.

This spermatozoon is obviously capable of sequestering its membrane components into functionally diverse groupings definable by contemporary morphological: cytochemical techniques. Although the extent and variety of patterns in the guinea-pig gamete are unusual, their 'raison d'être' probably holds true for all mammalian sperm. It is simply more convenient to explore the function(s) of each mosaic in guinea-pig sperm, where the 'parquets' are distinct, than, for example, in human sperm, where they are subtle.

Observations gleaned from studying guinea-pig sperm may apply not only to the sperm of other species, but to all secretory cells. By comparing the gamete's acrosomal cap before and after capacitation, and further comparing it to the non-fusigenic postacrosomal segment, we conclude that 1) the inner leaflet of the plasma membrane, readied for fusion, is rich in free sterols; 2) its outer leaflet contains a large proportion of anionic lipids; and 3) the membrane itself is highly fluid. In addition, sharp interfaces between sterol and phospholipid rich and poor areas exist where fusions presumably originate after the addition of calcium ions to the sperm preparation. Regions of membrane resistant to fusion sustain low levels of non-esterified sterols, hold anionic lipids in the inner leaflet, are less fluid than fusigenic regions, and evade sharp lipid-domain interfaces. It is possible that a protein submembranous lining maintains and regulates these differences in membrane composition and function.

Since many domains of the guinea-pig sperm plasma membrane are minute and sometimes variable in location, their detection and examination require the use of morphological and other *in-situ* methods. Whereas the presence of multiple, well-demarcated domains also permits the cell to operate with great efficiency because of this mosaicism, the disturbance of small regions can profoundly alter the function of the cell as a whole – and therein lies the promise for novel approaches to fertility control.

The future beckons with exciting challenges — identification of the nature and functional roles of the particle-clusters fronting the striated ring, as well as those of the midpiece strands, the cytoplasmic-droplet saccule and membrane window-screen patterns, the orthagonal arrays at the beginning of the principal-piece, and the zipper.

Acknowledgements

The author acknowledges the republication of Figures 1, 5, 7, and 12 from reference 9 and Figure 8 from reference 17, which appeared in The Journal of Cell Biology, published by the Rockefeller University Press, and the republication of Figures 13–15 and 17–19 from a paper (6) which appeared in The Anatomical Record, published by Alan R. Liss, Inc.

I thank Elaine L. Bearer, Peter M. Elias, and John E. Heuser for their substantial collaborative contributions, and Rosamond Michael for her excellent editorial help.

This work was supported by National Institutes of Health Grant HD 10445.

References

1. Fawcett DW: A comparative view of sperm ultrastructure. Biol Reprod Suppl 2: 90–127, 1970.
2. Fawcett DW: The mammalian spermatozoon. Develop Biol 44: 394–436, 1975.
3. Friend DS: Organization of the spermatozoal membrane. In: Immunobiology of the gametes. Edidin M, Johnson MH (eds). Alden Press, Cambridge, pp 5–30, 1977.
4. Friend DS: Freeze-fracture alterations in guinea-pig sperm membranes preceding gamete fusion. In: Membrane-Membrane Interactions. Gilula NB (ed) Raven Press, New York. pp 153–165, 1980.
5. Friend DS, Fawcett DW: Membrane differentiations in freeze-fractured mammalian sperm. J Cell Biol 63: 641–664, 1974.
6. Friend DS, Heuser JE: Orderly particle arrays on the mitochondrial outer membrane in rapidly-frozen sperm. Anat Rec 159: 198–199, 1981.
7. Friend DS, Orci L, Perrelet A, Yanagimachi R: Membrane particle changes attending the acrosome reaction in guinea-pig spermatozoa. J Cell Biol 74: 561–577, 1977.
8. Koehler JK: The mammalian sperm surface: studies with specific labeling techniques. Int Rev Cytol 54: 73–107, 1978.

9. Friend DS: Plasma-membrane diversity in a highly polarized cell. J Cell Biol 93: 243–249, 1982.
10. Koehler JK: Changes in the fine structure of the guinea-pig sperm head following experimental treatment. In: The Functional Anatomy of the Spermatozoon. Afzelius BA (ed) Pergamon Press. Oxford and New York. pp 105–114, 1974.
11. Koehler JK, Gaddum-Rosse P: Media induced alterations of the membrane associated particles of the guinea-pig sperm tail. J Ultrastruct Res 51: 106–118, 1975.
12. Austin CR: Membrane fusion events in fertilization. J Reprod Fertil 44: 155–156, 1975.
13. Bedford JM, Cooper GW: Membrane fusion events in the fertilization of vertebrate eggs. In: Membrane Fusion. Poste G, Nicolson GL (eds). Elsevier North-Holland Biochemical Press. pp 65–125, 1978.
14. Meizel S: The mammalian sperm acrosome reaction. In: Development in Mammals, Vol. 3. Johnson MH (ed). North-Holland, Amsterdam, pp 1–64, 1978.
15. Yanagimachi R: Mechanisms of fertilization in mammals. In: Fertilization and Embryonic Development *in vitro*. Mastroianni L Jr, Biggers JD (eds). Plenum Publishing Corp., New York. pp 81–182, 1981.
16. Yanagimachi R, Usui N: Calcium dependence of the acrosome reaction and activation of guinea-pig spermatozoa. Exp Cell Res 89: 161–174, 1974.
17. Bearer EL, Friend DS: Modifications of anionic lipid domains preceding membrane fusion in guinea-pig sperm. J Cell Biol 92: 604–615, 1982.
18. Koehler JK: Lectins as probes of the spermatozoan surface. Arch Androl 6: 197–217, 1981.
19. Myles DG, Primakoff P, Bellvé AR: Surface domains of the guinea-pig sperm defined with monoclonal antibodies. Cell 23: 433–439, 1981.
20. Millette DF: Distribution and mobility of lectin binding sites on mammalian spermatozoa. In: Immunobiology of the Gametes. M. Edidin M, Johnson MH (eds). Cambridge University Press, Cambridge. pp 51–71, 1977.
21. Flechon J-E: Ultrastructural and cytochemical analysis of the plasma membrane of mammalian sperm during epididymal maturation. Prog Reprod Biol 8: 90–99, 1981.
22. Yanagimachi R, Noda YD, Fujimoto M, Nicolson GL: The distribution of negative surface charges on mammalian spermatozoa. Am J Anat 135: 497–520, 1972.
23. Elias PM, Friend DS, Goerke J: Membrane sterol heterogeneity. Freeze-fracture detection with saponins and filipin. J Histochem Cytochem 27: 1247–1260, 1979.
24. Bradley MP, Ryans DG, Forrester IT: Effects of filipin, digitonin, and polymyxin B on plasma membrane of ram spermatozoa – an EM study. Arch Androl 4: 195–204, 1980.
25. Orci L, Miller RG, Montesano R, Perrelet A, Amherdt M, Vassalli P: Opposite polarity of filipin-induced deformations in the membrane of condensing vacuoles and zymogen granules. Science 210: 1019–1021, 1980.
26. Teuber M, Miller IR: Selective binding of polymyxin B to negatively charged lipid monolayers. Biochim Biophys Acta 467: 280–289, 1977.
27. Bearer EL, Friend DS: Anionic lipid domains: correlation and functional topography in a mammalian cell membrane. Proc Natl Acad Sci USA 77: 6601–6605, 1980.
28. Friend DS, Bearer EL: β-hydroxysterol distribution as determined by freeze-fracture cytochemistry. Histochem J 13: 535–546, 1981.
29. Schlegel RA, Phelps BM, Waggoner A, Terada L, Williamson P: Binding of merocyanin S-540 to normal and leukemic erythroid cells. Cell 20: 321–328, 1980.
30. Mercado E, Rosado A: Structural properties of the membrane of intact human spermatozoa. Biochim Biophys Acta 298: 639–652, 1972.
31. Bearer EL, Friend DS: Maintenance of lipid domains in the guinea-pig sperm membrane. J Cell Biol 91: 266a, 1981.
32. Papahadjopoulos D: Calcium-induced phase changes and fusion in natural and model membranes. In: Membrane Fusion. Poste G, Nicolson GL (eds). Elsevier/North-Holland Biomedical Press. pp 765–790, 1978.
33. Phillips DM: Surface of the equatorial segment of the mammalian acrosome. Biol Reprod 16: 128–137, 1977.
34. Karim M, Duarte J, Ruysschaert M, Hildebrand J: Affinity of adriamycin to phospholipids – a possible explanation for cardiac mitochondrial lesions. Biochem Biophys Res Commun 71: 658–663, 1976.
35. Holt WV, Bott HM: Chemically induced fusion between ram spermatozoa and avian erythrocytes: an ultrastructural study. J Ultrastruct Res 71: 311–320, 1980.
36. Mollenhauer HH, Morré DJ: Dictysome-like structures with cylindrical intersaccular connections (microtubules?) in guinea-pig spermatocytes. Am J Anat 150: 381–394, 1977.
37. Fawcett DW: Unsolved problems in morphogenesis of the mammalian spermatozoon. In: International Cell Biology. Brinkley RB, Porter KR (eds). The Rockefeller University Press, New York, 1977.
38. Friend DS, Elias PM, Rudolf I: Disassembly of the guinea-pig sperm tail. In: The Spermatozoon. Fawcett DW, Bedford JM (eds). Urban and Schwarzenberg. Baltimore-Munisch. pp 157–168, 1979.
39. Kistler J, Stroud RM, Klymkowsky MW, Lalancette RA, Fairclough RH: Structure and function of an acetylcholine receptor. Biophys J 37: 371–383, 1982.
40. Enders G, Werb Z, Friend DS: Lectin binding to sperm zipper particles. J Cell Biol 91: 116a, 1981.

Author's address:
Department of Pathology,
University of California
School of Medicine,
San Francisco, CA 94143, USA

*Abbreviations used in figures: A – acrosome; AC – acrosomal cap; AN – annulus; AX – axonemal complex; DF – dense fiber; E – equatorial segment; FS – fibrous sheath; H – head; M – mitochondrion; MC – microcisterna; MOM – mitochondrial outer membrane; MP – midpiece; N – nucleus; PA – postacrosomal segment; PP – principal piece; SR – striated ring; Z – zipper.

CHAPTER 8

Comparative aspects of acrosomal formation

JEROME S. KAYE

1. Introduction

The acrosome is an organelle which is attached to the anterior part of the nucleus of animal sperm and forms the apical body of the sperm. It contains enzymes which lyse the oocyte membranes to permit entry of the sperm for fertilization. In some marine sperm the acrosome also ejects a filament which makes the initial contact of the sperm with the egg. Acrosome structure (1), and evolution (2), have been carefully reviewed recently, as well as aspects of acrosome formation and physiology (3).

The acrosome appears after the second meiotic division of spermatogenesis. It occurs only in the spermatid, with no counterpart in earlier spermatocytes. The acrosome was shown definitively to be derived from the Golgi apparatus by Bowen (4). As he described it in insects, the Golgi apparatus of primary spermatocytes occurs as scattered dictyosomal elements which are distributed among the daughter cells during meiotic divisions. In the early spermatid the dictyosomes fuse to form a single Golgi body, the acroblast. A granule termed the proacrosomal granule, appears within the acroblast apparently as a secretion product of the acroblast. This event Bowen likened to the formation of zymogen granules. The granule is deposited on the posterior region of the nuclear membrane near the centrioles and the newly forming flagellum. The acroblast migrates away from the granule, taking no further part in the acrosome formation, and is eventually cast out of the cell. The proacrosomal granule moves to the anterior part of the nucleus maintaining close contact with the nucleus throughout, and undergoes a series of structural and biochemical changes leading to the mature acrosome. The essential steps in acrosome formation in orthopteran insects are shown in Figures 1–5.

Electron microscopy has clarified many details of acrosomal formation left unclear by the earlier workers, and raised new questions. These clarifications and questions will be discussed against a background of acrosomal formation as it occurs in the cricket, an insect. The cricket is used here because acrosomal formation in it has been described in enough detail to present a whole view of the process, from the initial events of proacrosomal granule formation to the final maturation of the acrosome (5, 6). The events of acrosomal formation in other species are similar enough to the cricket that the cricket can be used as a basis for comparison with other species.

The dictyosomes of primary spermatocytes of the cricket have the typical structure of Golgi elements, stacked membranes and associated vesicles. None of the vesicles can be distinguished morphologically as a precursor of the acrosomal vesicle which will form later. In the early spermatid a single acrosomal vesicle completely filled by the proacrosomal granule appears within the acroblast. The vesicle migrates to the nucleus with the leading boundary flattened and having a sheet-like element attached, originally termed the interstitital membrane (5), but renamed the acrosomal band (6) (Fig. 6). The vesicle becomes attached to the nucleus with the acrosomal band interposed between the two, and closely attached to the nucleus over its full extent. The vesicle assumes a cone shape, and its contents become segregated into a basal portion of high electron opacity and a distal portion of low opacity (Fig. 7). The base of the vesicle then invaginates progressively, ultimately forming a hollow cone of nearly constant wall thickness. In the space formed by the invagination, material accumulates which has a fluffy appearance initially, but which rapidly becomes organized into another cone-shaped structure called the inner cone (Fig. 8).

Acrosomal formation in the cricket and in other organisms appears as a series of interrelated but independent events: i.e. formation of the acrosomal vesicle and granule, deposition on the nucleus, migration to the final position, remodeling, the development of a subacrosomal structure, and each of these events shows significant variation among different organisms.

Van Blerkom, J. and Motta, P.M. (eds.), Ultrastructure of reproduction. ISBN 0-89838-572-5.

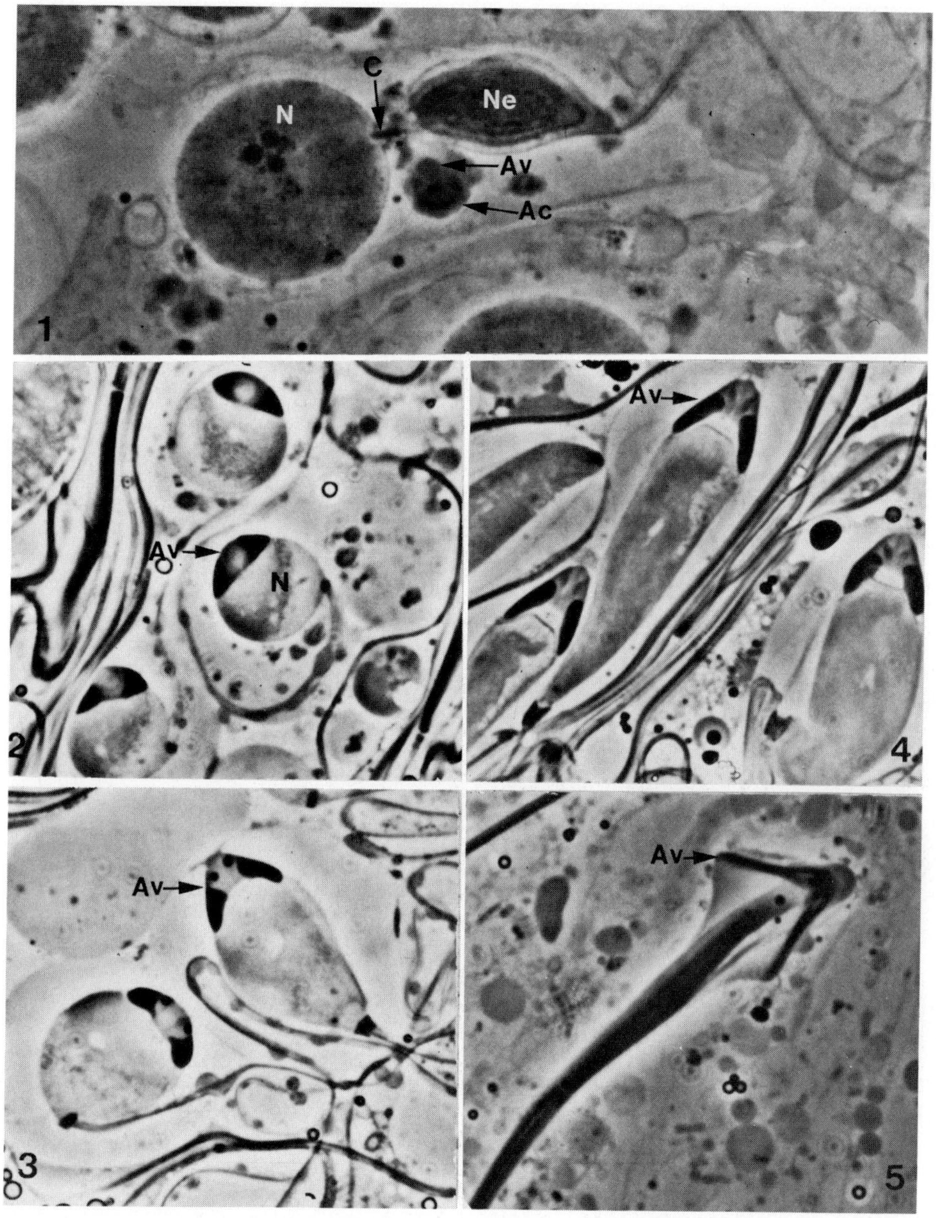

Figs. 1–5. (1) Phase-contrast photomicrograph of a squash preparation of an early spermatid of the cricket Nemobius. The acroblast (Ac) and acrosomal vesicle (Av) have migrated close to posterior part of the nucleus (N), near the centriole (C) and the nebenkern (Ne). Fixed in OsO_4, post-fixed in formalin prior to squashing (× 1,760). (2) Phase contrast photomicrograph of living cells of the Decticid grasshopper *Metrioptera roseli*. Middle spermiogenesis. Acrosomal vesicle (Av) attached to nucleus (N) showing the onset of segregation of phase-dense from less dense material (1,760). (3) Phase contrast photomicrograph of living cells of the Decticid grasshopper *Metrioptera roseli*. The onset of late spermiogenesis. Acrosomal vesicle (Av) showing well defined segregation of materials of high and low phase densities. The characteristic pyramidal shape of the mature acrosome is already evident in one cell (1,760). (4) Phase contrast photomicrograph of living cells of the Decticid grasshopper *Metrioptera roseli*. Later stages showing well developed acrosomal vesicles (Av) (× 1,760). (5) Phase contrast photomicrograph of living cells of the Decticid grasshopper *Metrioptera roseli*. Very late spermatid. The acrosomal vesicle (Av) is a flattened cone with thin walls perched on the end of the nucleus (× 1,760).

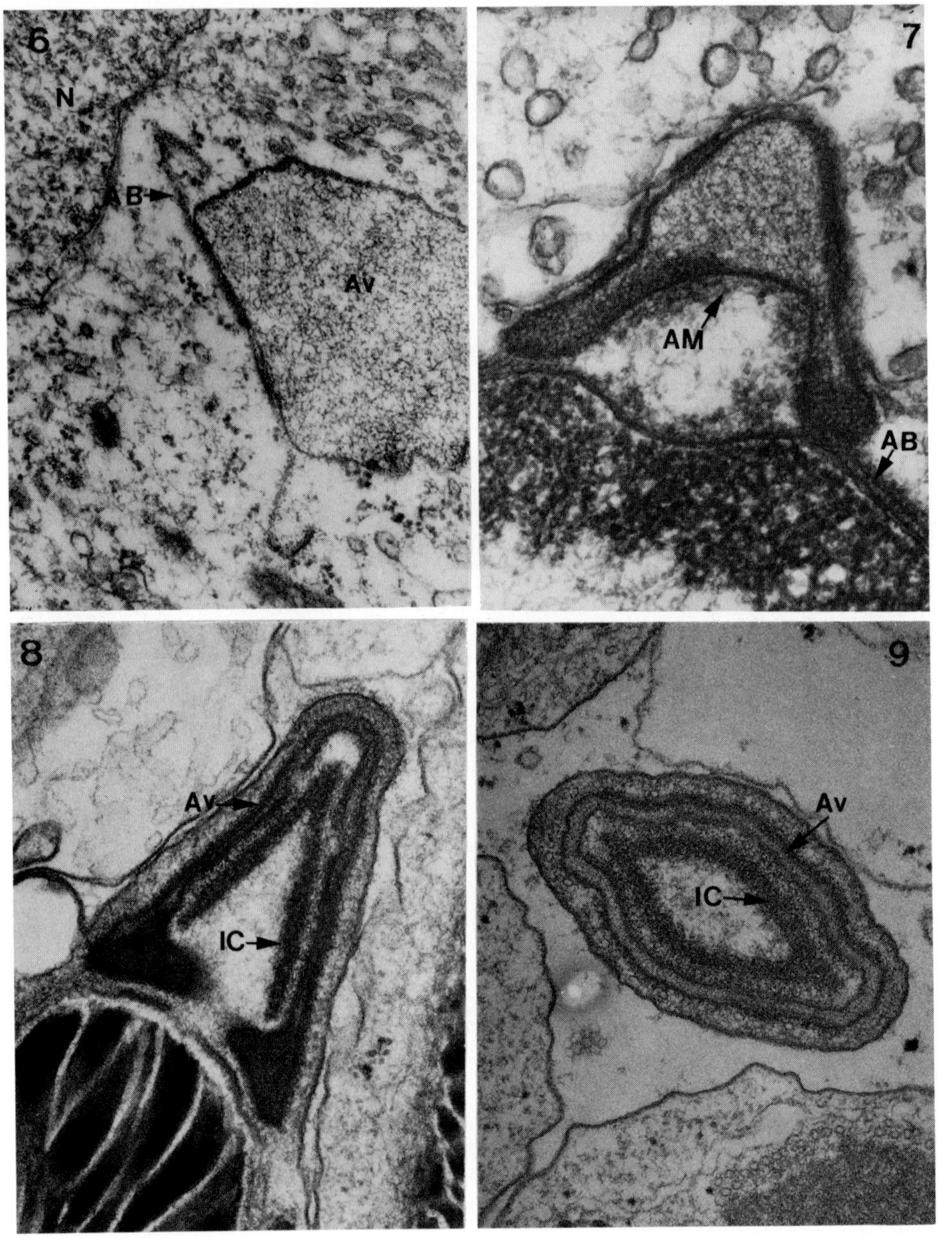

Figs. 6–9. (6) Transmission electron micrograph of the house cricket *Acheta domesticus*. An acrosomal vesicle (Av) midway in its migration to the nucleus. The acrosomal band (AB) is closely associated with the vesicle at its flattened surface. A part of the nucleus (N) is apparent (From ref. 5) (× 64,000). (7) Transmission electron micrograph of a spermatid from middle spermiogenesis of the house cricket showing initial invagination of the acrosomal vesicle (Av). Electron opaque material at the base of the vesicle, electron-lucent material at the tip. Amphorous material (Am) is present in the subacrosomal space created by the invagination. The acrosomal band (AB) stops short of the subacrosomal space (From ref. 5) (× 35,200). (8) Transmission electron micrograph of a late spermatid of the house cricket showing the deeply invaginated acrosomal vesicle (Av), now a thin walled conical shell. The inner cone (Ic) is well developed and occupies the subacrosomal space (× 28,000). (9) Transmission electron micrograph of a transverse section through a developing acrosome of about the same stage as that in Fig 8, showing the flattened shape of the acrosomal vesicle (Av) and of the inner cone (IC) (× 40,800).

2. Morphology of the acrosomal vesicle

In the cricket, one acrosomal vesicle appears to be present from the onset of acrosomal formation (5). The presence of one vesicle also is true for the cat, though two occasionally are found (7). Multiple vesicles at the outset occur in a range of organisms, including the starfish (8), and clam (9) among invertebrates, and the lemming (10) among the vertebrates. Workers who report multiple acrosomal vesicles and granules claim that the single vesicle and the formation of the granule occurs by fusion of smaller vesicles, but definitive observations of fusing vesicles are lacking. Alternatively, the selective loss of all but one vesicle and granule may occur.

In cricket spermatids, the acrosomal vesicle is filled completely by the proacrosomal granule. In contrast, cat spermatids have a large acrosomal vesicle containing a small dense granule surrounded by material of low density. (7). This arrangement occurs in other organisms as well, for example, in the ram (11).

A proacrosomal granule is usually present in the acrosomal vesicle, but exceptions have been reported. In the dogfish (12), starfish (8) and a marsupial (13), the acrosomal vesicle lacks a proacrosomal granule. The contents of the vesicle in these cases have low electron opacity which indicates either that the substance within is very dilute or that it does not bind stain as well as the substance of the typical electron opaque vesicle. The mature acrosomes of both the dogfish and marsupial spermatozoa have high electron opactities which presumably reflects material that is derived from the vesicle. However, the possibility that material is added to the vesicle during the later stages of differentiation cannot be ruled out by present evidence.

Variations in the modes of formation of the acrosomal vesicle and the presence or absence of a proacrosomal granule indicate the indispensable element in acrosome formation is a vesicle from the acroblast. The significance of variations in the concentration of material within the vesicle and in the presence or absence of a granule is obscure and has no obvious correlation with either the size or shape of the resulting acrosome. However, so little is really known about the different modes, that a firm judgement on this point is not warranted yet.

2.1. Migration of the acrosomal vesicle

In many species, as illustrated by the cricket, the acrosomal vesicle is formed within the acroblast at a distance from the nucleus. The vesicle migrates through the cytoplasm to the nucleus to which it becomes attached. Structural elements which might mediate this migration have not been observed. Involvement of microtubules would appear to be ruled out by the absence of any reports on their presence in the region of granule migration. A role for microfilaments or actin sol-gel transformations is conceivable, but vesicle migrations have not been studied with techniques considered capable of revealing such elements.

The initial site of attachment of the vesicle on the nuclear membrane is constant for each organism. In crickets (5, 14) and other insects (4) the initial site of attachment is posterior, near the centrioles; in mammals such as the cat (7) it is anterior. There obviously must be precise factors which determine the attachment site and there are hints it is an intranuclear factor. In Thyone sperm (15) the nuclear membrane is modified with the two leaflets fused at the future attachment site prior to attachment. Presumably there is some physical interaction between the granule and the nuclear membrane to cause the granule to migrate to the correct site. Schmid and Krone (16) observed in salamander spermatids that a particular heterochromatic chromosomal element migrates to the future anterior end of the nucleus. The nucleus begins elongation at this point and becomes the attachment site for the acrosome. The salamander is particularly interesting because in animals heterozygous for the heterochromatic segment, half the spermatids undergo only partial nuclear elongation and lack acrosomes, which is consistent with heterochromatin determining the site of granule attachment. However, the details of sperm development have not been worked out sufficiently to know when the block occurs to acrosome formation in spermatids lacking the heterochromatic segment, and whether it precedes the attachment step.

2.2. Mechanism of acrosomal vesicle attachment

In crickets, the attachment of the acrosomal vesicle to the nuclear membrane involves the acrosomal band. The acrosomal band was first observed in edge view in sections and appeared to be a membrane of some sort, because it had a thin, linear profile of constant thickness, though it did not exhibit a trilaminer-type structure (5). In whole mounts of isolated nuclei spread on specimen grids, the acrosomal band appears in face view as an oblong sheet about the width of the acrosomal vesicle and some five to six times longer than its width (6) (Figs. 10, 11). The acrosomal band makes contact with the nuclear surface and becomes attached to it, with the acrosomal vesicle firmly attached to the band. Under the

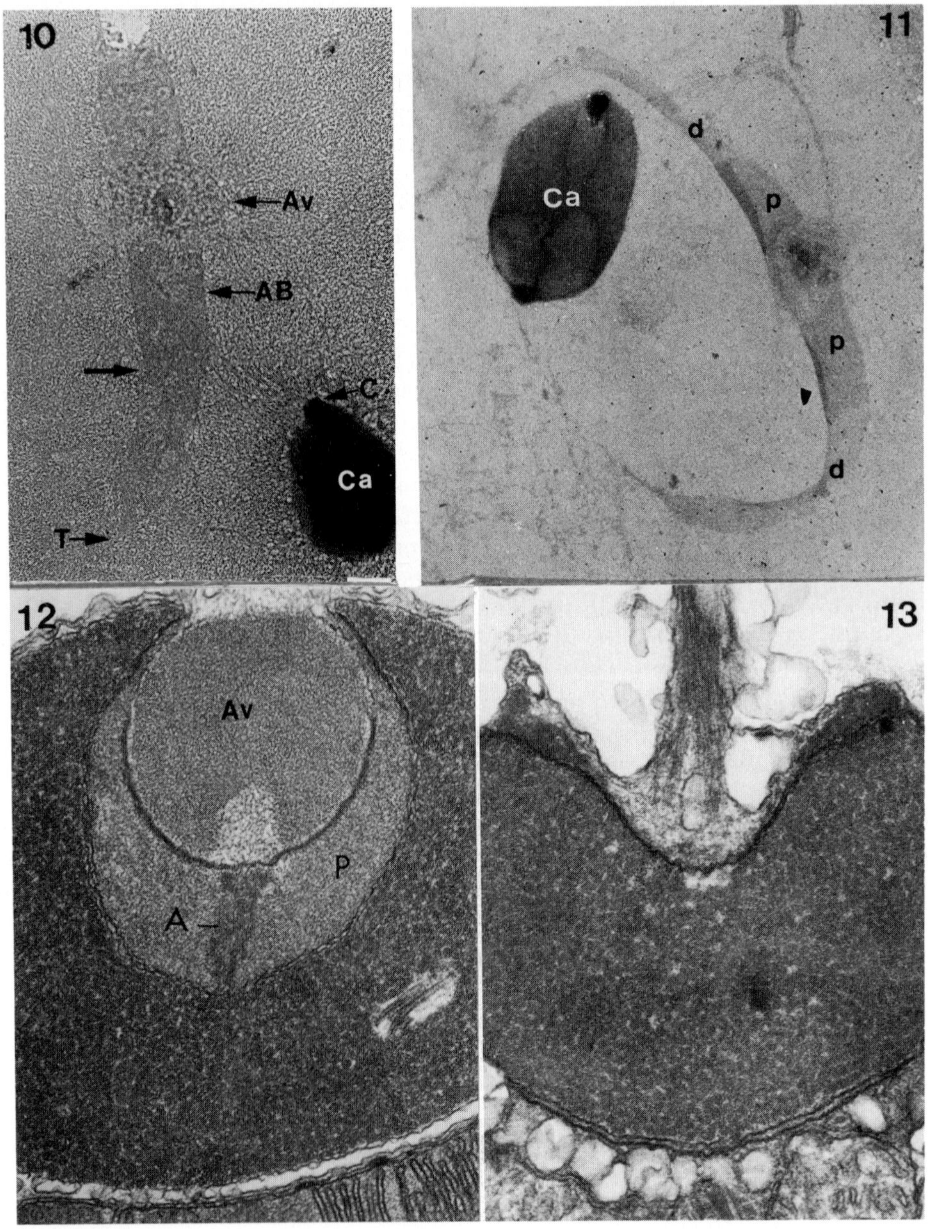

Figs. 10–13. (10) Transmission electron micrograph of the acrosomal band (AB) from middle spermatogenesis of the house cricket. The band appears as a narrow sheet with a curved end and pointed tip (T) at its lower arm. The end of the upper arm has been torn off. The acrosomal vesicle (Av) is collapsed and has a clear central area where invagination has occured. Centriole (C) and centriole adjunct (Ca) at lower right. An isolated nucleus, spread and Pt-Pd shadowed (From ref. 6) (× 8,000). (11) Transmission electron micrograph of a cell from middle spermiogenesis of the house cricket. The acrosomal band is long and curved. The proximal (p) and distal (d) segments are clear. Centriole adjunct (Ca). Isolated nucleus, formalin fixed, spread and stained with phosphotungstic acid (From ref. 6) (× 6,000). (12) Transmission electron micrograph of the acrosome of Thyone sperm. The periacrosomal space (P) is filled with profilactin. The actomere (A) spans the distance between the acrosomal vesicle (Av) and the nucleus (From ref. 22) (× 56,000). (13) An Asterias sperm which has undergone the acrosome reaction. The acrosomal process projects upward and a bundle of actin filaments can be seen within it (From ref. 30) (× 49,600).

band the two leaflets of the nuclear membrane appear to fuse. The acrosomal band apparently acts as an element which cements the vesicle to the nucleus.

It is difficult to assess whether an acrosomal band-like structure is present in acrosomal formation of other species because only in the cricket has acrosomal formation been studied by the spread-type preparations which permit definitive observations of the band to be made. It is clear, however, that in other organisms a structure lies between the granule and the nuclear membrane. This structure has a sheet-like appearance in a marsupial (13), and in boar, bull, ram and rabbit spermatids (17, 18). There is a material of high electron opacity, but of indeterminite structure in dogfish spermatids (12). The material seen in the mammalian and dogfish spermatids may be homologous to the acrosomal band of the cricket, but the certainty of this conclusion cannot be assessed because the composition and function of this material remain undetermined for any case.

In the cricket, the acrosomal vesicle moves from a posterior to an anterior position on the nucleus. Sliding of the vesicle on the nuclear surface has been suggested by Bowen (4) as the mechanism for its change of position in insect spermatids. Discovery of the acrosomal band in the cricket spermatids and its association with the acrosomal vesicle during the time of its change in position suggests that the band may mediate sliding, with the granule attached to the band while the band slides over the nuclear surface (6). An alternative explanation to sliding is that the nucleus rotates to bring about the change in position. However the centrioles and tail accessory structures remain attached to the posterior part of the nucleus at this time. The absence of rotation of the centrioles and tail accessory structures does not support the idea of nuclear rotation.

Extensions to the acrosomal band appear during the time of acrosomal vesicle movement (Fig. 11). These extensions nearly encircle the nucleus and conceivably serve as anchoring elements once vesicle migration is completed. Similar structures have not been reported for other organisms, but again they may be more common than supposed; there have been no other studies which have utilized spreading and staining techniques considered capable of revealing the presence of such structures. The two leaflets of the nuclear membrane are fused under the entire extent of the acrosomal band in the cricket, and the acrosomal band can be detected in fixed and sectioned material from this feature alone. The two leaflets of the nuclear membrane of Thyone spermatids are fused under the acrosome, and also at a point which is diametrically opposite the acrosome (15), which strongly resembles the situation in cricket spermatids and suggests an acrosomal band-type structure might be present in Thyone also.

2.3. *Acrosomal vesicle and granule differentiation*

Segregation of the contents of the acrosomal vesicle into distinct regions occurs in many species and is detected by refractive index differences within the vesicle by phase-contrast microscopy (Figs. 1–5), cytochemical staining patterns, or differences in electron opacity as seen by electron microscopy (Figs. 7, 14, 15). The granule may be heterogeneous from the outset as in the guinea pig (19) and the lemming (10), in which the material of the central region has a higher electron opacity than that at the periphery. When the acrosomal vesicle is initially homogeneous, heterogeneity may appear during the initial phase of vesicle differentiation. Thus, when the vesicle of the cricket (5), and the surf clam (9), change from a roundish to a conical profile and begin to invaginate at the base, the base of the vesicle has material of high opacity. In the Hemipteran insect Gerris, (20), the dense material of the vesicle has a polysacharide component detected in light microscope preparations by PAS staining. In ram spermatids, segregation of a polysaccharide component has been demonstrated at the level of the electron microscope (11) with the periodic acid thiocarbohydrazide silver stain. The segregation persists through the latest stages of acrosome differentiation in the ram, with the polysaccharide component forming a thin rim of material just under the outer margin of the acrosome. Holt (21) detected similar heterogeneous distribution of material in several mammalian species using phosphotungstic or silicotungstic staining over a wide pH range. This investigator concluded the material at the acrosomal periphery was a basic protein because the material incorporates stain even when the staining reaction is carried out at the higher pH ranges.

The appearance of two components in acrosomal vesicles that are initially homogeneous indicates there is some internal rearrangement of material within the vesicle. The mechanism by which this rearrangement occurs, however, remains to be determined. Addition of a new component to a restricted region of the vesicle after it is already attached to the nucleus, as reported for the guinea pig (19) and the starfish (8) is not ruled out.

In some sperm of the primitive type as in Thyone (22), the vesicle undergoes only superficial remodeling of shape to form the mature acrosome (Fig. 12).

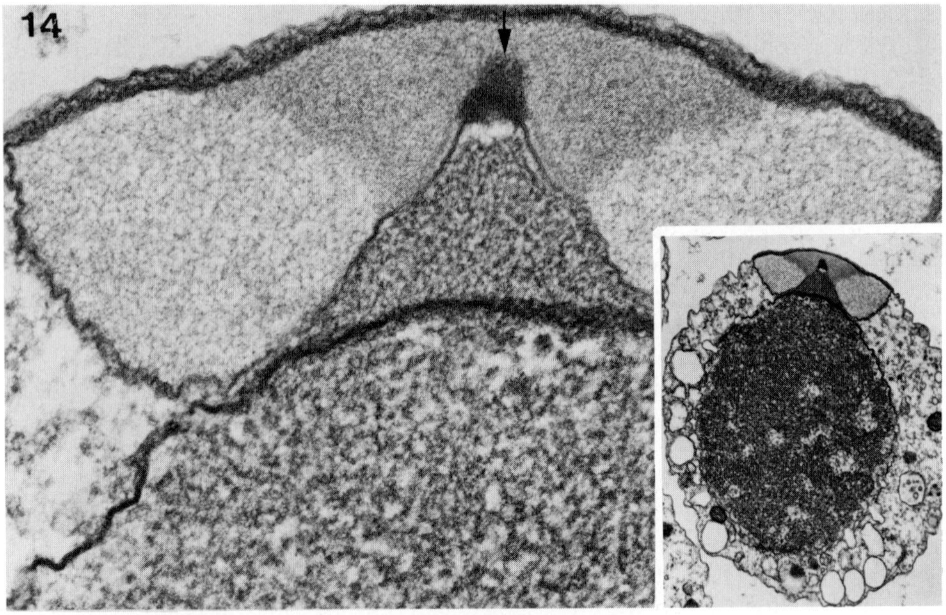

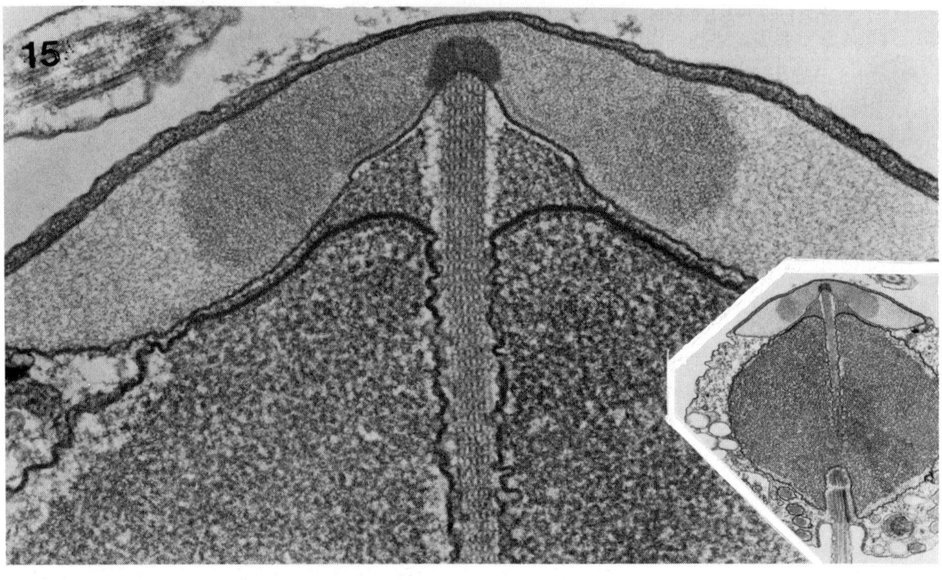

Figs. 14–15. (14) Transmission electron micrograph of a Limulus spermatid. The acrosomal vesicle has flattened and invaginated. The subacrosomal space has unorganized material, presumably actin. A dense organizing center, the acrosomal button, is marked with an arrow (From ref. 25) (× 84,000; inset × 11,200). (15) Transmission electron micrgrph of a Limulus spermatid showing a partially polymerized filament bundle. The individual filaments are clearly seen. The inset shows four coils of the filament bundle at the base of the nucleus (From ref. 25) (× 64,800; inset × 13,600).

In more advanced sperm the remodeling may be very extensive. In the cricket there is a deep invagination of the acrosomal vesicle resulting in an acrosome that is little more than a hollow shell (5) (Figs. 8, 9). Less extreme degrees of invagination are common as in the bird (23), the mussel (24) and Limulus (25). In each of these species the acrosome is a cone, the base of which is narrow in the bird and mussel and broad in the Limulus (Fig. 15).

In mammals the differentiation of the acrosomal vesicle results in its extension posteriorly to cover a large portion of the anterior end of the nucleus. The acrosome of the ram is typical of this process, forming a thin shell of nearly uniform thickness which covers over half of the anterior end of the nucleus (11). The guinea pig has an atypical acrosome for the mammal; its anterior end has a large bulbous extension (19).

Factors involved in change in shape of the acrosomal vesicle to form the mature acrosome are unknown. Control of acrosomal shape by supporting cells in intimate contact with spermatids while in the testis would appear to be ruled out by observations that the acrosome of the guinea pig continues to differentiate while the spermatozoa are in the epididymus and free of contact with other cells (19). The acrosome becomes thinner and takes on a pronounced bend. Similarly the acrosomes of epididymal spermatozoa of a marsupial undergo large changes in shape (13). The nucleus would seem to be ruled out as a contributing factor by the extraordinary circumstance that the acrosomes of annelids undergo nearly their entire differentiation at some distance from the nucleus, and only in the terminal stages do they become attached to the nucleus (26, 27). The signals for the changes in the acrosomal vesicle, and the underlying molecular mechanisms have been elucidated only partially. The ultimate control presumably is hormonal. The acrosome fails to differentiate normally in hypophysectomized rams (18). However, there would seem to be a number of secondary factors intrinsic to the acrosomal vesicle and granule (28), and perhaps the surrounding cytoplasm. Differential dehydration of portions of the vesicle and granule might be a factor; i.e., the degree of hydration could control local shrinkage or swelling. At the molecular level, it is conceivable that a precise spatial ordering of the proteins of the vesicle and granule, with accompanying conformational changes in the protein molecules themselves is reflected by larger conformational changes in the acrosome seen with the microscope. Certainly some chemical changes in the acrosome would seem to accompany the shape changes; in the guinea pig, events in capacitation appear to be temporally correlated with the shape changes that occur in the passage of the spermatozoa through the epididymus (19). It is also conceivable that shape changes are generated outside the acrosomal vesicle through an association of the vesicle membrane with actin filaments in the surrounding cytoplasm exerting pulling or pushing forces on different parts of the vesicle. However, beyond speculation on possible factors, the mechanism of acrosomal differentiation is still as baffling as it was in Bowen's time.

2.4. Actin elements of the acrosome

One of the most important discoveries on acrosome function in recent years is that the filament called the acrosomal process, which emerges from the acrosome of echinoderm and Limulus sperm during the acrosome reaction, is composed of actin (25, 29). Localization of the actin in the unreacted acrosome, its accumulation during acrosomal development, and its organization into the acrosomal process have been addressed in a series of elegant papers by Tilney et al. (15, 22, 25, 29, 30, 31, 32, 34).

In Thyone the acrosomal vesicle lies in a space formed from an invagination of the nucleus called the periacrosomal cup (Fig. 12). The vesicle is close to, but not attached to the nuclear membrane. The resulting space defined by the vesicle and nuclear membrane becomes filled with material of moderately high electron opacity, but with no ordered structure apparent. Actin in an inactive form, profilactin, has been identified from SDS elecrophoretic analysis of the proteins from isolated periacrosomal cups (29) and is a major component of this material.

In the initial stages of its accumulation in the periacrosomal space of Thyone sperm (15), profilactin appears as a fluffy material in the electron microscope. As the maturation of the sperm proceeds, the profilactin becomes more concentrated, but never is visibly ordered (Fig. 12).

The acrosomal vesicle appears to be ruled out as the source of the profilactin because the volume of the vesicle remains constant during its accumulation (15). Because the acrosomal vesicle is not the source of profilactin, it must come from the spermatozoan cytoplasm. Synthesis of the profilactin at the site of the periacrosomal cup is unlikely because there are no polysomes in that region of the cell. Tilney (15) has suggested that a profilactin pool is synthesized in the general cell cytoplasm and profilactin, being freely diffusible, diffuses to the periacrosomal space where it is trapped by some sort of as yet undetermined specializations of the acrosomal vesicle and subacrosomal nuclear membranes.

2.5. Organization of the acrosomal process

The acrosome reaction is one of the more dramatic events discovered in cell biology. In Thyone sperm an acrosomal process is extended 90 μm in about 10 seconds. The reaction can be initiated in the presence of Ca^{++} by egg water, by an increase in pH of the water, or by an ionophore that transports H^+ out of the cell (22). The first event in the reaction is the fusion of the acrosomal vesicle membrane with the cell membrane at the anterior tip of the sperm; this fusion is Ca^{++} dependent. When the pH rises above 6.5, the formation and extension of the acrosomal process proceeds by the release of actin from the proteins with which it is complexed and the actin polymerizes rapidly to form the acrosomal process (Fig. 13).

Just as the acrosomal vesicle is not a source of the profilactin that accumulates in the periacrosomal cup, neither does it contribute to the growth of the acrosomal process. During the acrosome reaction in echinoderm sperm treated with H^+ ionophore in Ca^{++}-free water, only a partial acrosome reaction occurs. Actin polymerizes, but the acrosomal vesicle remains intact, not breaking down as it does in the complete reaction. A partial acrosomal process forms, but clearly with no contribution possible from the acrosomal vesicle (22). Electrophoretic analysis of isolated acrosomal vesicles from sea urchin sperm shows directly the absence of actin (33).

Actin is added at the base of the growing acrosomal process where an organelle, termed the actomere, appears to be the organizing center. The kinetics of the extension of the acrosomal process in Thyone has been established by Tilney and Inoue (31), who showed the kinetics are consistent with the model that actin molecules are freely diffusible within the periacrosomal cup and are transported to the organizing site by diffusion.

In Limulus, the actin polymerizes during sperm development. It forms a fiber that grows progressively down through a tunnel in the nucleus, ultimately becoming so long that it coils several times around the base of the nucleus (Fig. 15). The actin is organized into parallel filaments which can be seen in both fixed and sectioned material and negative stain preparations. During the growth of the fiber, actin molecules are added to the proximal end just under the acrosomal cap at the organizing center, a small region of high electron opacity called the acrosomal button (25) (Figs. 14, 15).

During the acrosome reaction in Limulus, the extension of the acrosomal process proceeds by quite a different mechanism from that in echinoderms. In Limulus the actin is polymerized before the reaction. The molecules of the unreacted process are twisted; a change in the twist provides the force for the extension of the process (32, 34).

An actin component of the acrosome has been positively identified only in sperm of echinoderms and of Limulus. However, the sperm of a diverse variety of organisms have well defined structures which lie under the acrosomal vesicle; certainly some, and conceivably all contain actin. Cricket spermatozoa are one such example. The fluffy material that collects in the subacrosomal space formed by the invagination of the acrosomal vesicle, (Fig. 7), resembles in appearance the actin that collects in Limulus sperm prior to the formation of the actin fiber. This observation suggests, of course, that the fluffy material in the cricket is also actin. In the cricket, this material becomes organized into a hollow cone, a type of structure which has not been described for any other organism. More typically, when the sub-acrosomal structure has a definitive form, it is rod-shaped. The rod may be very short, as in sea urchin sperm (35), or slightly longer with a length nearly equal to the height of the acrosomal vesicle, as in the surf clam (9), or still longer, extending part way into a nuclear invagination, as in the earthworm (26) and the bird (23), or very long, going completely through the nucleus as in the mussel, (24) and Limulus (25).

The role of subacrosomal elements in the acrosome reaction has been studied primarily in the sperm of marine invertebrates, and the role of such elements in the acrosome reaction in other organisms is largely unknown. Thus, it might be supposed that the subacrosomal rod of bird sperm functions to form an acrosomal process, but the role of the inner cone of the cricket sperm in this formation is obscure. Subacrosomal elements of the actin type, or putative actin type, appear to be absent from mammalian sperm (2). Presumably, the functions associated with these elements are also absent from mammalian acrosomes.

In rodent sperm, a specialized rod-shaped structure called the perforatorium, lies under the ventral part of the sperm head. The perforatorium appears to be a specialization of membranes in that region of the sperm and not a subacrosomal element of the putative actin type. Some of the latter subacrosomal elements have also been called perforatoriums (2). The use of the term perforatorium for a variety of subacrosomal elements is confusing and is a reflection of the lack of information on the homologies of different parts of the acrosome in various organisms. At this time the term should be restricted to the rod-shaped structure of rodent sperm.

2.6. *Organization of the acrosomal vesicle*

A number of lytic enzymes have been found in the mammalian acrosome, the most prominent being hyaluronidase and a protease called acrosin. The existence and properties of these enzymes, as well as proteases of invertebrate sperm have been reviewed recently (2).

Little is known about the origins and behavior of these enzymes during acrosomal formation, but it is surely reasonable to suspect that the rearrangment of the substance of the acrosomal vesicle, with the segregation of materials of different electron opacities and staining properties seen in the electron microscope, reflects a compartmentalization of at least some of the enzymes of the vesicle. Acrosin is a basic protein with an affinity for membranes (36). Conceivably, it is the basic protein localized on the inner surface of the ram acrosome that stains with silicotungstic acid at high pH (21).

3. Comparative aspects of acrosomal formation

One great similarity in the general course of acrosomal formation throughout the animal kingdom is that in every species studied, visible acrosomal development does not begin until the completion of the second meiotic division. The purpose for this delay is unclear, because acrosomal development does not seem to depend on the differentiation of any other cell structure. Perhaps the delay of acrosomal formation to the spermatid stage evolved as a mechanism to assure that each spermatid has an acrosome. If acrosomes were made prior to the meiotic divisions, some additional mechanism would have had to evolve to effect their distribution to the daughter cells. The absence of any case where acrosomal differentiation begins earlier than the second meiotic division, implies that the controls for acrosomal formation are invariable among species. The search for those controls should yield important insights into general questions of the control of sperm differentiation. Other aspects of acrosomal formation such as formation of a single acroblast and formation of an acrosomal vesicle are also universal.

The myriad differences in the details of acrosomal formation offer several lifetimes of work to spermatologists to catalogue and explain. There is the fascination of working out evolutionary relationships among different groups of species with regard to differences in acrosomal formation, and the details of correlating variations in acrosomal structure with its function according to the species. However, the greatest insights into the controls and mechanisms involved in acrosomal formation probably will result from a thorough understanding of acrosomal formation in a single species, rather than cataloguing superficial differences among different species. The latter ultimately may prove to be most important as a source of entertainment for spermatologists but that, of course, is no small thing.

The underlying biochemical mechanisms are as yet not known for any single step in acrosome formation, neither for the formation of the acrosomal vesicle, nor for its morphogenetic movements and final differentiation. Crucial information on these processes is completely lacking. What are the signals for activating the genes for the vesicle proteins? When are the RNA messages transcribed? Are they translated immediately, or is there a translational control step? How are the proteins organized into a formed vesicle? How does the vesicle move, differentiate? It is clear the answers to these questions will derive mainly from studies of isolated cell fractions using the techniques of biochemistry and molecular biology. The electron microscope will not be a primary tool in such studies but rather an adjunctive one to aid in identifying isolated organelles and to assess the purity of cell fractions.

However, spermiogenesis does present some unusual problems for biochemical studies and the electron microscope can make unusual contributions to their solution. The testis is an extremely heterogeneous tissue with a very large number of cell types, considering each stage of spermiogenesis an individual cell type. Thus, cells with developing acrosomes in a particular stage will comprise only a very small fraction of the mass of the tissue. This limited availability of cell types creates considerable difficulty in isolating pure organelles in quantities sufficient for analysis. With electron microscopic cytochemical techniques, particularly immunochemical techniques, it should be possible to determine the protein compositions and the sequence, both structurally and temporally, of changes in the composition of the acrosomal vesicle and associated elements. Such knowledge is interesting because it should lead to definitive answers on the structural and functional homologies of acrosomal elements among different organisms. More importantly the knowledge would serve to outline the limits of the central problem of determining the controls and mechanisms involved in acrosomal formation. It would pinpoint the developmental stages of greatest interest for biochemical analysis.

Acknowledgement

I am indebted to professor Lewis Tilney for his conversations about acrosomes, for providing me with preprints of his recent work, and for the micrographs which appear in this work.

References

1. Baccetti B, Afzelius BA: The Biology of the Sperm Cell. Monographs in Developmental Biology, Vol. 10. S Karger, AG Basel, 1976.
2. Baccetti B: The evolution on the acrosomal complex. In: The Spermatozoon. Fawcett DW, Bedford JM (eds). Urban & Schwartzenberg, Inc, Baltimore-Munich, 1979.
3. Dan JC: Morphogenetic aspects of acrosome formation and reaction. In: Advances in Morphogenesis. Abercrombie M, Brachet J, King TJ (eds). Academic Press, New York-London, 1970.
4. Bowen R: Studies on Spermatogenesis. I. Biol Bull 39: 316–362, 1920.
5. Kaye JS: Acrosome formation in the house cricket. J Cell Biol 12: 411–431, 1962.
6. McMaster-Kaye R, Kaye JS: Acrosomal bands: Specialized structures on the nuclear surface for holding the acrosomal granule. J Ultrastruct Res 71: 233–248, 1980.
7. Burgos MH, Fawcett DW: Studies on the fine structure of the mammalian tesis. I. Differentiation of the spermatids of the cat (Felis domestica). J Biophys Biochem Cytol 1: 287–300, 1955.
8. Dan JC, Sirikami A: Studies on the acrosome. X. Differentiation of the starfish acrosome. Devel Growth and Differentiation 13: 37–52, 1971.
9. Longo FJ, Anderson E: Spermiogenesis in the surf clam Spisula solidissima with special reference to the formation of the acrosomal vesicle. J Ultrastruct Res 27: 435–443, 1969a.
10. Hopsu VK, Arstila AV: Development of the acrosomic system of the spermotozoon in the Norwegian lemming (Lemmus lemmus). Zeit f Zellforsch 65: 562–572, 1965.
11. Courtens JL: Cytochemical localization of glycoproteins in the developing acrosome of ram spermatids. J Ultrastruct Res 65: 173–181, 1978.
12. Stanley HP: Fine structure of spermiogenesis in the elasmobranch fish Squalus suckleyi. 1. Acrosome formation, nuclear elongation and differentiation of the midpiece axis. J Ultrastruct Res 36: 87–102, 1971.
13. Harding HR, Carrick FN, Shorey CD: Spermiogenesis in the Bush-tailed possum, Trichosarus vulpeculas (Marsupalia). Cell Tiss Res 171: 75–90, 1976.
14. Kaye JS, MacMaster-Kaye R: The fine structure and chemical composition of nuclei during spermiogenesis in the house cricket. I. Initial stages of differentiation and the loss of nonhistone protein. J Cell Biol 31: 159–179, 1966.
15. Tilney LG: The polymerization of actin. II. How nonfilamentous actin becomes nonrandomly distributed in sperm: Evidence for the association of this actin with membranes. J Cell Biol 69: 51–72, 1976.
16. Schmid M, Krone W: The relationship of a specific chromosomal region to the development of the acrosome. Chromosoma 56: 327–347, 1976.
17. Courtens JL, M Courot, Flechon JE: The perinuclear substance of boar, bull, ram and rabbit spermatozoa. J Ultrastruc Res 57: 54–64, 1976.
18. Courtens JL, Courot M: Acrosomal and nuclear morphogenesis in ram spermatids. An experimental study of hypophysectomized and testosterone-supplemented animals. Anat Rec 197: 143–152, 1980.
19. Fawcett DW, Hollenberg RS: Changes in the acrosome of guinea pig spermatozoa during passage through the epididymis. Zeit f Zellforsch 60: 276–292, 1963.
20. Moriber LC: A cytochemical study of hemipteran spermatogenesis. J Morph 99: 271–327, 1956.
21. Holt WV: Development and maturation of the mammalian acrosome. A study using phosphotungstic acid staining. J Ultrastruct Res 68: 58–71, 1979.
22. Tilney LG: Polymerization of actin. V. A new organelle, the actomere, that initiates the assembly of actin filaments in thyone sperm. J Cell Biol 77: 551–564, 1978.
23. Humpheys PN: The differentiation of the acrosome in the spermatid of the Budgerigar (Melopsittacus undulatus). Cell Tiss Res 56: 411–416, 1975.
24. Longo FJ, Dornfeld EJ: The fine structure of spermatid differentiation in the mussel, Mytilus edulis. J Ultrastruct Res 20: 462–480, 1967.
25. Tilney LG, Bonder EM, DeRosier DJ: Actin filaments elongate from their membrane-associated ends. J Cell Biol 90: 485–494, 1981.
26. Anderson WA, Ellis RA: Acrosome morphogenesis in Lumbricus terrestris. Zeit f Zellforsch 85: 398–407, 1968.
27. Potswald HE: An electron microscope study of spermiogenesis in Spirobis (Laeospira) morchi Levensen (Polychaeta). Zeit f Zellforsch 83: 231–248, 1967.
28. Fawcett DW, Anderson WA, Phillips DM: Morphogenetic factors influencing the shape of the sperm head. Devel Biol 26: 220–251, 1971.
29. Tilney LG: The polymerization of actin: III. Aggregates of non-filamentous actin and its associated proteins: a storage form of actin. J Cell Biol 69: 73–89, 1976.
30. Tilney LG, Hatano S, Ishikawa H, Mooseker MS: The Polymerization of Actin: Its role in the generation of the acrosomal process of certain echinoderm sperm. J Cell Biol 59: 109–126, 1973.
31. Tilney LG, Inoue S: The acrosomal reaction of Thyone sperm: II. The kinetics and possible mechanism of acrosomal process elongation. J Cell Biol 93: 820–827, 1982.
32. De Rosier D, Tilney L, Flicker P: A change in the twist of the actin-containing filaments occurs during the extension of the acrosomal process in Limulus sperm. J Mol Biol 137: 375–389, 1980.
33. Vacquier VD: The isolation of gamete surface componenets involved in the adhesion of sperm to eggs during sea urchin fertilization. In: Advances in invertebrate reproduction. Clark WH, Adams, TS (eds). Elsevier/North Holland, New York, Amsterdam, Oxford, 1981.
34. De Rosier DJ, Tilney LG, Bonder EM, Frankel P: A change in the twist of actin provides the force for the extension of the acrosomal process in Limulus sperm. J Cell Bio 93: 324–337, 1982.
35. Longo FJ, Anderson E: Sperm differentiation in the sea urchins Arbacia punctulata and Strongylocentratus purpuratus. J Ultrastruc Res 27: 486–509, 1969.
36. Strauss JW, Parrish RF, Polakoski KL: Boar acrosin. Association of an endogenous membrane proteinase with phospholipid membranes. J Biol Chem 256: 5662–5668, 1981.

Author's address:
Department of Biology
University of Rochester
Rochester, NY 14627, USA

CHAPTER 9

Changes in the sperm surface during maturation in the epididymis

GARY E. OLSON

1. Introduction

Mammalian spermatozoa are released into the lumen of the seminiferous tubule after completing a prolonged and complex process of differentiation. Although the resultant cells are highly specialized in structure and function, they are immature, being neither motile nor capable of fertilizing an oocyte. After leaving the testis, sperm rapidly move into the epididymis where they remain for several days. During passage through the epididymis, sperm achieve functional maturity, developing both the capacity for forward motility and for effecting fertilization of an ovum (1–4).

The epididymis can be subdivided into regions based on gross anatomical, histological and biochemical criteria (5). The three major subdivisions, beginning proximally, are in the caput, corpus and cauda. In species examined, spermatozoa achieve maturity in a specific region of the epididymis. In the rabbit, this region is the distal corpus and in the rat it is the proximal cauda (3). While present in the epididymis, spermatozoa may interact with the epithelial lining or with the luminal fluid. This fluid is a complex mixture of inorganic ions, small molecular weight organic compounds, proteins and glycoproteins. The composition of this fluid is modified progressively along the length of the epididymis by the secretory and absorptive activity of the epithelial cells lining the tubule (5, 6, 7). The interaction of the spermatozoa with the fluid environment produced by the epididymis plays an important role in the maturation process.

The normal activity of the epididymis and its differentiated morphology are dependent upon androgens produced by the testis. Some proteins normally synthesized and secreted by the epididymis are androgen-dependent, in that their synthesis ceases during androgen withdrawal (8–10). Moreover, studies employing organ culture of the rabbit epididymal tubule have demonstrated that androgens must be present in the culture medium for the sperm to develop fertilizing capacity (11). These studies have demonstrated further that if epididymal protein synthesis is blocked by inhibitors, the spermatozoa do not develop fertilizing capacity (12). This emphasizes the likely importance of the interaction between spermatozoa and epididymal secretions.

During passage through the epididymis the spermatozoon undergoes progressive structural and biochemical modification (1, 3, 4). The variety of changes identified to date are wide-ranging and include remodeling of the shape of the acrosome; increased disulfide bonding of some sperm structures such as the chromatin, perinuclear matrix, connecting piece, outer dense fibers, outer mitochondrial membranes and fibrous sheath; changes in concentration of cyclic nucleotides; and changes in the structural and biochemical properties of the plasma membrane (1–5). It is probable that the development of motility and fertilizing capacity by the spermatozoa results from the cumulative effects of these multiple changes. The discussion which follows will emphasize changes in the sperm plasma membrane occurring during post-testicular development.

The plasma membrane of the spermatozoon is subdivided into a number of biochemically and morphologically distinct domains (14). The mosaic character of the plasma membrane reflects the polarized structure of the sperm cell where its constituent organelles are segregated to specific locations within the cell. Generally, a chemically and structurally distinct region of the plasma membrane parallels the extent of the underlying structures; distinct membrane domains frequently include the periacrosomal and postacrosomal segments of the head and the midpiece and principal piece regions of the flagellum. The mosaic character of the sperm surface arises during spermiogenesis. However, several different types of plasma membrane changes occur as sperm mature in the epididymis. These changes can alter a discrete region(s) or even the entire plasma mem-

Van Blerkom, J. and Motta, P.M. (eds.), Ultrastructure of reproduction. ISBN 0-89838-572-5.

brane. These changes are thought to be of importance in maturation, because specific domains of the sperm plasma membrane participate in fertilization events (15, 16). Functions of the sperm membrane include participation in the regulation of flagellar movement, binding to the zona pellucida, participation in membrane fusion during the acrosome reaction and participation in fusion to the oocyte plasma membrane. Thus, an understanding of the modifications of the sperm plasma membrane during post-testicular maturation may further the understanding of fertilization events.

The sequence in which the various types of sperm membrane changes are presented in this chapter was chosen to emphasize both the mosaic character of the sperm surface and the asymmetric construction of the unit membrane. The fluid mosaic model of membrane structure defines two major classes of membrane proteins (17). Integral membrane proteins possess hydrophobic domains and are embedded at least partially in the hydrophobic regions of the bilayer. Nonetheless, hydrophilic domains of these proteins may be exposed to the inner cytoplasmic compartment, the external environment or, in the case of penetrating proteins, both surfaces. In contrast, peripheral membrane proteins are present exclusively at the inner or outer face of the membrane and are bound to the membrane by interactions either with the polar head groups of phospholipids or the exposed segments of integral membrane proteins. In accordance with this asymmetric organization of the unit membrane, the following discussion will treat sequentially the post-testicular maturation of the sperm plasma membrane with respect to changes at its inner cytoplasmic face, in its internal organization and finally at its external face.

2. Changes at the cytoplasmic surface of the sperm plasma membrane

In mammalian spermatozoa, specific regions of the plasma membrane are associated structurally with underlying cytoplasmic elements (14). The regions include the postacrosomal segment, where the plasma membrane overlies an electron-dense lamina; the posterior ring, where the plasma membrane fuses with the underlying nuclear membrane; the junction of the midpiece and principal piece, where the plasma membrane adheres to the underlying dense matrix composing the annulus; and the principal piece, where the plasma membrane attaches to the underlying fibrous sheath. Moreover, ultrastructural studies of a few species have revealed additional examples where filamentous assemblies are associated with the cytoplasmic face of specific regions of the plasma membrane of mature spermatoza. For example, in vole sperm a regular pattern of 'tubule-like' structures lies beneath the plasma membrane of the postacrosomal segment, and at the anterior end of the spermatozoon an array of filaments is associated with the periacrosomal plasma membrane (18). Microfilaments also are found associated with the plasma membrane of the postacrosomal and equatorial segments of boar spermatozoa examined after an ionophore-induced acrosomal reaction (19). In guinea pig spermatozoa, grazing sections of the midpiece plasma membrane reveal a filamentous substructure (20), and in marsupial spermatozoa structural specializations also are noted at the cytoplasmic face of the midpiece plasma membrane (21–28).

Several studies have indicated the presence of 'actin-like' proteins in the postacrosomal segment of mammalian spermatozoa (29, 30). Whether this actin-like material is in a polymerized state, is membrane-associated or forms a component of the dense lamina is not known. Both the intramembranous particles (31, 32) and the externally disposed surface residues (33) in this region can undergo a redistribution in the plane of the membrane, and it has been suggested that these rearrangements may be influenced by cytoskeletal proteins bound to the membrane.

Neither the functional role nor the chemical composition of these submembranous assemblies of mammalian spermatozoa are well understood. On the basis of their location and structure they may be operationally analogous to peripheral membrane proteins defined by the fluid mosaic model of membrane structure (17). Peripheral proteins can form submembranous assemblies which affect the properties of the membrane by regulating the mobility and/or distribution of integral membrane proteins and their glycosylated segments which are exposed at the external face of the membrane (34). In spermatozoa these components or assemblies associated with the cytoplasmic face of the plasma membrane could function in establishing and maintaining its mosaic character and in regulating the capacity for lateral motion of integral membrane proteins.

The best examples of structural alterations at the cytoplasmic face of the spermatozoan plasma membrane during epididymal transit are the result of ultrastructural studies on marsupial spermatozoa. The plasma membrane of caput epididymal marsupial spermatozoa has a typical trilaminar appearance and thin section analysis reveals no unusual differentiations over the sperm surface. However, in the species examined, a modification of the midpiece plasma membrane occurs while sperm are present in the epididymis. For example, in the brush-tailed opossum, *Trichosurus vulpecula*, a helical array of parallel 20–30 nm diameter fibers form in close apposition to the cytoplasmic face of the midpiece plasma membrane (21, 25, 26, 27). In sperm of the American opossum, *Didelphis virginiana*, or the wooly opossum, *Caluromys philander*, the cytoplasmic face of the midpiece plasma membrane progressively thickens during epididymal transit due to an accumulation of a 10–15 nm thick layer of amorphous material (Figs. 4, 7, 8) (22, 24, 28). The precursor of this membrane coat apparently is an electron-dense, amorphous material present in the cytoplasm of the immature spermatozoon (Fig. 4). Coincident with the deposition of the membrane coating material the topography of the membrane changes so that it assumes a scalloped configuration (Figs. 7, 8) and, at least for Didelphis, there is a coincident rearrangement in its internal structure (24, 28). The fact that parallel alterations occur at the cytoplasmic face of the plasma membrane and in the interior of the membrane suggests the possibility of a

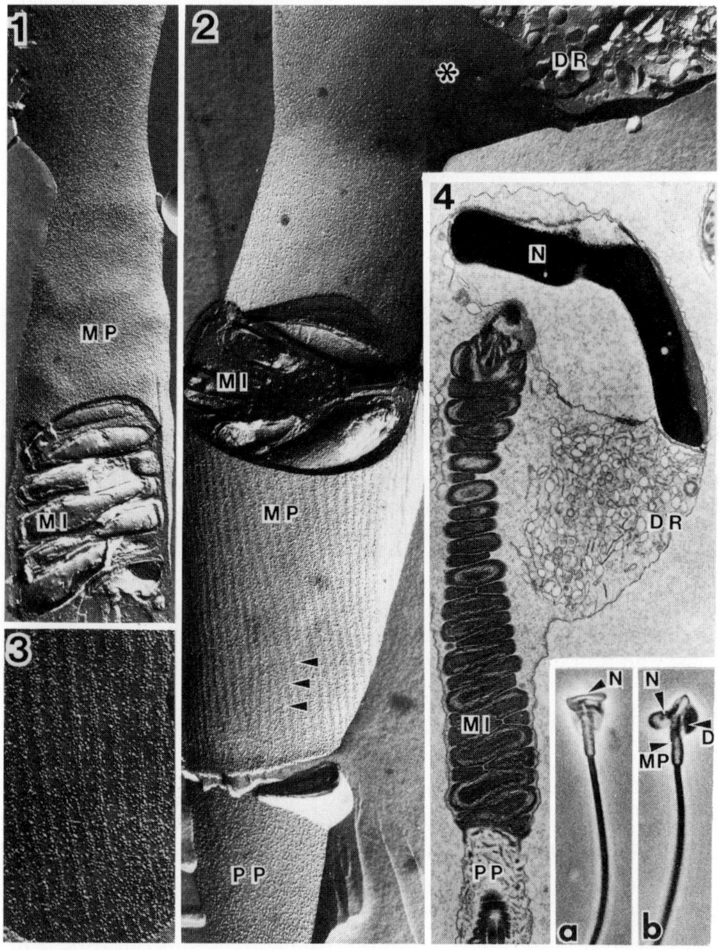

Figs. 1–4. (1) Freeze-fracture replica showing the middle piece region of caput epididymal opossum (D. virginiana) spermatozoon. A random distribution of intramembranous particles occurs on the plasma membrane (MP) overlying the mitochondrial sheath (MI). (2) Freeze-fracture replica showing the cytoplasmic droplet, middle piece and principal piece region of opossum spermatozoon from the corpus epididymis. In the midpiece (MP) plasma membrane the intramenbranous particles are redistributed into rows which parallel the flagellar long axis (arrowheads). The plasma membrane (*) over the droplet (DR) retains a random distribution of intramembranous particles. In the principal piece (PP) a random distribution of particles is seen. MI = mitochondria. (3) Freeze-fracture replica showing a segment of the middle piece plasma membrane of opossum sperm from the corpus epididymis. Note that although most particles are segregated into rows, some are found in the inter-row region. (4) Thin selection electron micrograph showing a profile of opossum spermatozoa from the corpus epididymis. The droplet (DR) is shifted to one side of the middle piece. In the distal segment of the middle piece the plasma membrane is closely applied to the underlying mitochondria (MI). N = nucleus; PP = principal piece. Inset: phase contrast photomicrograph of caput (a) and corpus (b) opossum spermatozoa. In corpus sperm the droplet (D) has shifted to one side of the middle piece (MP) and the nucleus (N) has begun to rotate with respect to the flagellar long axis.

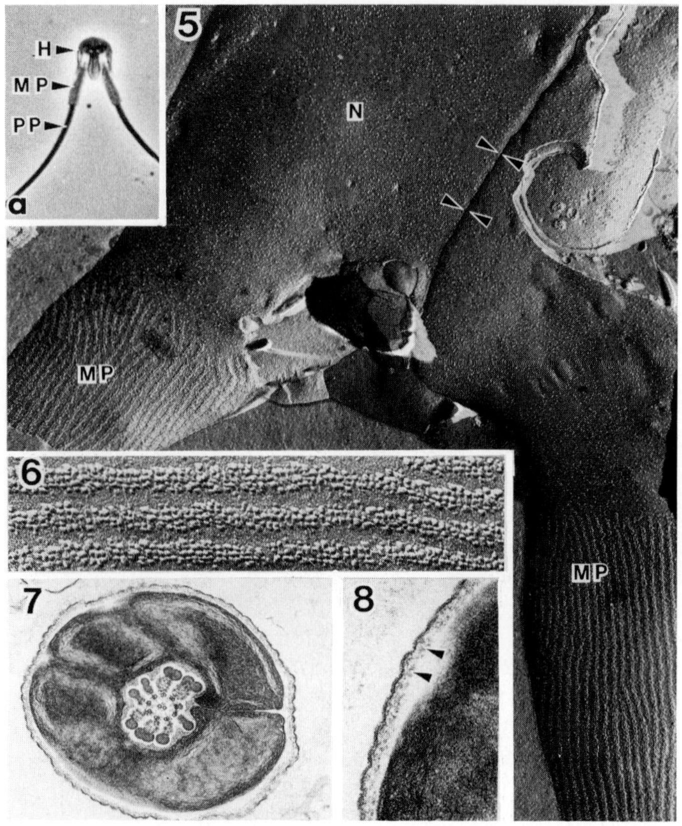

Figs. 5–8. (5) Freeze-fracture replica showing paired opossum spermatozoa obtained from the cauda epididymis. The sperm are adjoined along their acrosomal surfaces and the boundary of the fusion zone appears as a groove in freeze-fracture replicas (arrowheads). The perinuclear plasma membrane (N) possesses a random particle distribution. This arrangement is replaced distally by particle rows characteristic of the midpiece (MP). Inset is a phase contrast photomicrograph showing paired spermatozoa. (H = head; MP = midpiece; PP = principal piece). (6) High magnification view of midpiece particle rows of cauda epididymal opossum spermatozoa. The particles are packed in a lattice-like arrangement. The membrane between rows is free of particles. (7) Cross sectional view of the midpiece of cauda epididymal opossum spermatozoon. The plasma membrane has a scalloped configuration. A mat of amophous material is associated with the cytoplasmic face of the membrane. This configuration is not seen in sperm from the caput epididymis. (8) High power view of the midpiece plasma membrane of cauda epididymal spermatozoon showing the scalloped configuration and prominent undercoat of material (arrowheads). This undercoat is absent in sperm from the caput epididymis.

structural interaction between the membrane coating material and integral membrane constituents.

Whether changes at the cytoplasmic face of the plasma membrane occur in other regions of the sperm surface and whether these types of maturational changes are widespread in other mammalian species is not known. Based on their restricted distribution, it appears that these submembranous assemblies may be involved in maintaining the mosaic character of the sperm surface and in regulating the fluidity of specific membrane domains. Changes in their state of assembly or association with the plasma membrane could be an important component of the maturational process and ultimately contribute to the ability of the spermatozoon to effect fertilization.

3. Changes in the internal structure of the sperm plasma membrane during maturation

3.1. Freeze fracture data

In electron micrographs obtained from conventional thin sections, the central region of the trilaminar unit membrane appears as an electron-lucent zone with no obvious substructure. However, this zone is the hydrophobic interior of the membrane which contains hydrocarbon chains of membrane lipids and the hydrophobic domains of integral membrane proteins. This region is revealed by the freeze-fracture technique which cleaves the unit membrane along the center of the bilayer. Replicas of the exposed surfaces generally reveal an array of intramembranous particles, representing the protein component, distributed on a smooth surface, the hydrophobic surface of the fractured bilayer. In a freeze-fracture analysis of guinea pig and rat spermatozoa, Friend and Fawcett (35) elegantly demonstrated that the plasma membrane overlying the acrosome, postacrosomal segment, midpiece and principal piece had a different internal anatomy. This analysis provided a visual confirmation of the mosaic character of the sperm surface. The regional differences were reflected by the size, distribution and number of intramembranous particles per unit area. Subsequent analyses of sperm of other species by freeze-fracture techniques have confirmed regional differences in the internal anatomy of the sperm plasma membrane. However, in comparing sperm of different species, the same regions of the plasma membrane, e.g., periacrosomal, postacrosomal and midpiece, often have quite different arrangements of intramembranous particles (14). The functional significance of these arrangements, as well as the molecular mechanisms generating and maintaining the particle patterns, are poorly understood. The distribution of the particles may reflect the functional state of the membrane, because during capacitation and immediately prior to the acrosome reaction changes in their distribution have been noted (31, 32, 36, 37).

The freeze-fracture appearance of the plasma membrane of the rat sperm and opossum sperm at successive stages of maturation in the epididymis has been studied (38, 24). In rat spermatozoa, the freeze-fracture appearance of the periacrosomal plasma membrane changes. Spermatozoa from the proximal caput epididymis have a scattered distribution of 6 to 12 nm diameter particles in the periacrosomal plasma membrane. In sperm from the distal caput/proximal corpus region the particles in the periacrosomal plasma membrane segregate into plaque-like areas. Within each plaque the intramembranous particles form an ordered, two-dimensional lattice with the 9 nm diameter particles spaced about 17 nm center-to-center. In sperm from the cauda epididymis the hexagonal lattice of particles in the periacrosomal plasma membrane is absent and replaced by a random distribution of smaller particles. The mechanism accounting for this change in particle size has not been determined. A redistribution of intramembranous particles was not noted in other regions of the rat sperm surface. Analysis of thin sections revealed that structural changes at the external face of the plasma membrane of rat spermatozoa occur concurrently with the redistribution of the intramembranous particles (38). When ordered arrays of intramembranous particles are present, aggregates of amorphous material are bound to the outer surface of the periacrosomal plasma membrane. This material, termed 'variable glycocalyx material', is distributed in a patchy fashion reminiscent of the plaque-like arrangements of particles seen in freeze-fracture replicas. In distal segments of the epididymis this glycocalyx-like material is no longer associated with the membrane and its loss correlates with the return to the random distribution of the particles noted in freeze-fracture replicas. This result raises the possibility that the binding of components to the membrane exterior might affect the structural organization of the membrane interior. A noteworthy parallel is that the completion of these membrane changes correlates temporally with the development of fertilizing capacity by the spermatozoa (39).

In the American opposum, *Didelphis virginiana*, the freeze-fracture appearance of the midpiece changes during sperm maturation in the epididymis (24). In freeze-fracture replicas of caput epididymal spermatozoa, the plasma membrane over the acrosome, midpiece, and princpal piece has a scattered distribution and different density of intramembranous particles (Fig. 1). As Didelphis spermatozoa pass through the corpus epididymis, a rearrangement of intramembranous particles is initiated in the plasma membrane of the midpiece region. Short linear aggregates of particles appear and, as they elongate, orient parallel to the flagellar long axis (Figs. 2 and 3). A lateral aggregation of the strands occurs to form a row 3 to 5 particles wide. Neighboring rows are parallel to one another and, as development progresses, the inter-row region becomes devoid of particles (Fig. 2).

In sperm from the cauda epididymis the parallel particle rows extend from the annulus anteriorly to the termination of the underlying midpiece mitochondria. At this point the particles abruptly assume a random distribution (Fig. 5). The particles within each row are packed in an ordered arrangement suggestive of some type of particle-particle interaction (Fig. 6). No rearrangements of intramembranous particles in other segments of the opossum sperm surface have been noted during the course of epididymal transit (24). As with the example provided by the rat spermatozoa (38), the intramembranous particle rearrangement occurring in the midpiece plasma membrane of opossum spermatozoa correlates temporally with visible morphological changes at the membrane surface. At the cytoplasmic face of the membrane there is a progressive deposition of electron-dense material which eventually achieves a thickness of 10–15 nm (Figs. 7 and 8). Interestingly, the completion of this sequence of membrane changes correlates with the development of the capacity for forward motility by the sperm (24). The ordered distribution of intramembranous particles in the midpiece plasma membrane of cauda epididymal spermatozoa suggests that the particles have restricted lateral motion. Although a number of mechanisms could be involved, the ultrastructural evidence suggests that both particle-particle interactions and particle anchoring to the membrane coating material may act to generate the nonrandom intramembranous particle arrangements.

The extent to which rearrangements of intramembranous particles in the plasma membrane occurs in sperm of other species during their transit through the epididymis is not resolved. The middle piece plasma membrane of mature guinea pig spermatozoa is characterized by strands of particles which wrap circumferentially about the flagellum (35). In sperm recovered from more proximal segments of the epididymis, the midpiece strands tend to be fewer and a random intramembranous particle distribution frequently is encountered over much of the midpiece (Figs. 9–11) (40). This evidence suggests that in guinea pig sperm a redistribution of intramembranous particles occurs during epididymal transit. However, in boar and bovine spermatozoa, differences in the freeze-fracture appearance of the plasma membrane of caput and cauda spermatozoa have not been seen (41). Even though the formation of recognizable intramembranous particle patterns is not seen in all species or all regions of the sperm surface, it is possible that during maturation there is an insertion of new intramembranous particles, a redistribution of specific intramembranous particles, or a change in the interaction of intramembranous particles with one another or with ligands at the cytoplasmic or exterior face of the membrane. These types of changes would not necessarily be revealed by conventional analysis of freeze-fracture replicas. They could, however, have important functional effects on the membrane and be of critical importance in the maturational process.

3.2. Changes in lipid composition

Lipids are a major component of the unit membrane and affect its physiochemical properties. The lipid composition of plasma membrane fractions isolated from sperm at successive stages of maturation has not been determined and therefore lipids will not be emphasized in this chapter. However, several studies have analyzed the lipid content and composition of whole spermatozoa obtained from different epididymal regions. These studies showed significant differences among the different cell populations. In spermatozoa of the boar (42–44), ram (45, 46), and bull (47, 48), there is a decline in phospholipid content as sperm pass through the epididymis. In these three species, the degree of unsaturation of the phospholipid-associated fatty acids also increases as the sperm pass through the epididymis. In contrast, in rat sperm a decline in total lipid content (49) but a slight increase in phospholipid content occurs as a result of a large increase in choline plasmalogen during sperm maturation in the epididymis (50).

The observed changes in lipid composition of spermatozoa during maturation in the epididymis have been interpreted as reflecting a modification of the sperm plasma membrane. However, because the observed changes were obtained from measurements of bulk lipid, it cannot be ascertained to what extent they reflect similar changes in the lipid composition of the plasma membrane. If lipid changes do occur in the plasma membrane, they could have important implications for its structure and function. As noted by Johnson (51), the pattern of lipid changes observed suggests that the lipid domains of sperm membranes become less stable or more fluid during the course of sperm maturation in the epididymis. This increased fluidity of the lipid bilayer could have important effects on the lateral mobility of the protein constituents of the membrane. This idea correlates well to freeze-fracture data which demonstrates that during capacitation and fertilization a lateral redistribution of intramembranous particles occurs in the sperm plasma membrane (31, 32, 36, 37).

4. Changes at the membrane exterior during maturation

4.1. Morphological changes

Changes occurring at the exterior surface of the sperm plasma membrane of sperm during their passage through the epididymis have received considerable attention. This surface is the one on which carbohydrate constituents of membrane glycoproteins and glycolipids are exposed and it is the surface readily accessible to modification by the actions of the epididymis.

By conventional thin section electron microscopy, it occasionally has been possible to identify surface modifications of sperm during their maturation in the epididymis. In electron microscopic studies of sperm of the rabbit (52) and chincilla (53), a loosening of the plasma membrane over the acrosome has been noted during sperm transit through the epididymis. It was suggested that this loosening reflected a change in the biochemical or physical properties of the membrane. However, Jones (54) demonstrated in boar spermatozoa that the apparent loosening could be related to fixation conditions. Nonetheless, he noted a difference in the incidence of breakage of the plasma membrane of caput epididymal and cauda epididymal boar sperm when the samples were fixed identically, and suggested that this may reflect a change in the physical properties of the membrane.

Another feature observed by conventional thin section electron microscopy is the deposition of material, of presumed epididymal origin, upon the outer surface of the plasma membrane of sperm maturing in the epididymis. In sperm of the Echidna, a monotreme mammal, a specific deposition of electron-dense material upon the periacrosomal plasma membrane occurs as sperm traverse the middle level of the epididymis (55). In *Caluromys philander*, a metatherian mammal, the middle piece plasma membrane becomes coated by an amorphous, electron-dense material as spermatozoa traverse the corpus region of the epididymis (22). In the corpus epididymis of the boar (54) and rat (38), electron-dense material of apparent epithelial origin transiently binds to the periacrosomal plasma membrane of the spermatozoa. Thus, in sperm from a wide variety of mammals, electron microscopic analysis of conventionally fixed material reveals a modification of specific regions of the surface coat of the spermatozoaoon as it passes through the epididymis. Although these changes have not been demonstrated in a large number of species, closer examination may reveal their widespread occurrence.

Direct electron microscopic observation gives little insight into the chemical nature of these surface changes. However, using specific probes which can be visualized by microscopic analysis or by direct biochemical analysis, some of these changes at the external face of the spermatozoan plasma membrane during maturation have been defined in more detail. In the following paragraphs these changes will be considered with respect to alterations of surface charge, in lectin receptors and macromolecular composition.

4.2. Surface charge changes

By comparing the mobility of caput and cauda epididymal rabbit spermatozoa in an electric field, Bedford (56) provided convincing evidence for a changing surface charge of spermatozoa during their passage through the epididymis. He found that cauda spermatozoa migrated more rapidly to the anode than did caput spermatozoa, and that the flagellum of cauda sperm appears to be more electronegative than the sperm head. Thus, sperm appear to become more negatively charged during their maturation in the epididymis and specific regions of the sperm surface may accumulate more charge than other regions. Recently, the net surface charge of immature (testicular or caput epididymal) and cauda epididymal sperm of several species including rat, hamster, mouse, rabbit and ram have been determined directly by isoelectric focusing of intact cells (57, 58). In each species, spermatozoa from the caput epididymis had a higher isoelectric point than sperm from the cauda epididymis. The mean isoelectric point for caput and cauda sperm respectively was: rat – 4.4, 4.1; hamster – 5.25, 4.0; mouse – 4.9, 4.2; and rabbit – 4.95, 4.3 (57). In the ram, a pI of 5.2 was obtained for testicular sperm and 4.7 for cauda sperm (58). Although a net increase in surface negativity was found in all species, between species the magnitude of the changes differed as did the final pI values obtained by cauda epididymal spermatozoa.

These physiological data correlate well with electron microscopic studies. By observing the binding at low pH of colloidal iron hydroxide particles to the sperm surface, several groups of investigators have demonstrated a net increase in anionic sites at the sperm surface during passage of the sperm through the epididymis (59–62). In most species studied, the flagellum develops a higher density of anionic sites than does the sperm head. For example, in the rabbit the head of both caput and cauda epididymal spermatozoa does not bind colloidal iron hydroxide particles, but the flagellum acquires a substantial binding capacity during sperm maturation. The binding of colloidal iron particles along the sperm surface is discontinuous and the density of bound particles may change abruptly at the major mophological subdivisions of the sperm, verifying the mosaic character of the sperm surface. This mosaic character is further emphasized during maturation in which specific segments of the sperm surface appeared to express increased density of anionic surface sites. It should be noted that a changed pattern of colloidal iron binding has not been discovered in sperm of all species studied. In spermatozoa of the shrew, the surface distribution of anionic sites in testicular and cauda epididymal sperm are similar (63).

The chemical nature of these colloidal iron binding sites is not fully understood. In some cases the sites are resistant to neuraminidase treatment, which removes accessible sialic acid residues (60, 61). However, where examined, esterification of carboxyl groups by methylation does appear to reduce colloidal iron hydroxide binding (60). It has been suggested that the colloidal iron hydroxide binding sites on the sperm surface could represent sulfate residues, neuraminidase-resistant derivatives of sialic acid or sialic acid residues. The source of the changes in surface charge during passage of sperm through the epididymis probably results in part from an epididymal contribution. Some of the possible molecular mechanisms accounting for these changes will be discussed in a following section.

4.3. Lectin binding changes

Lectins with different saccharide binding specificities have been employed to map the distribution of specific sugar residues in different domains of the sperm plasma membrane. Approaches have included 1) radiolabeled lectins, to quantify the accessible binding sites on the sperm surface; 2) ferritin or peroxidase-conjugated lectins, for electron microscopic localization of binding sites; 3) fluorochrome-conjugated lectins, for light microscopic localization of binding sites, and 4) sperm-lectin mixtures, to assess agglutinability of the cells by the multivalent lectin (13).

With radiolabeled lectins, the number of accessible concanavalin A (Con A-binds -D-mannose-like residues), Ricinus communis I agglutinin (RCA_1 binds B-D-galactose-like residues) and wheat germ agglutinin (WGA binds N-acetyl glucosamine-like residues) binding sites have been determined for intact sperm of several species. Mouse cauda epididymal spermatozoa bind about 4.9×10^7 Con A molecules/cell (64). Rabbit cauda epididymal spermatozoa bind about 1.0×10^7 RCA_1 and 3.3×10^7 WGA molecules per cell (65). With respect to maturational changes, a 15% to 25% decrease in Con A binding by rat spermatozoa occurs as they pass through the epididymis, with caput sperm binding about 3×10^7 Con A molecules/cell (66).

In an electron microscopic study using ferritin conjugate lectins, Nicolson et al. (67) found differences between caput and cauda epididymal rabbit spermatozoa in the pattern of binding of RCA_1 and WGA but not Con A. The results suggested to these authors a decease in RCA_1 and WGA receptors occurs during sperm maturation in the epididymis. They noted that the head region of caput sperm bound more RCA_1 than did the heads of cauda sperm, but both sperm populations bound approximately equal amounts to the midpiece and principal piece. Caput sperm bound WGA to the flagellum, whereas cauda sperm did not. In this same study it was noted that Con A binding occurred over the entire cell surface of both caput and cauda spermatozoa and no dramatic changes were noted. These results paralleled agglutination experiments showing caput spermatozoa are more agglutinated by WGA and RCA_1 than are cauda spermatozoa. In contrast to the above results, Gordon et al. (68) reported significant changes in Con A binding of rabbit spermatozoa during maturation. They found that caput sperm bound little Con A, while cauda spermatozoa displayed binding to the plasma membrane over the head and flagellum. The basis for the opposing results of these two studies is not clear, but could have resulted from different protocols of sperm treatment and labeling.

Maturationally associated changes in lectin binding to discrete regions of the sperm surface have been reported in several other species. A decreased binding of Con A by the flagellar plasma membrane and an increased binding by the periacrosomal plasma membrane occurs during maturation of rat spermatozoa in the epididymis (Figs. 12 and 13) (66). In ram spermatozoa, a slight reduction in flagellar Con A binding has been noted during sperm passage from the corpus to cauda epididymis (62). In the immature sperm of the hyrax, only the periacrosomal plasma membrane and the plasma membrane overlying the cytoplasmic droplet bind WGA; the periacrosomal binding does not occur in sperm present at sites further down the epididymis. Also in the hyrax spermatozoa, there is a loss of flagellar Con A binding sites as sperm pass down the epididymis (69). The changes in lectin binding may relate to specific epididymal regions. For example a periacrosomal binding of WGA occurs in armadillo sperm obtained from the corpus epididymis but not in sperm obtained from the caput or cauda epididymis (69). This transient binding may relate to the temporary binding of epididymal secretions.

Only a few of the available examples have been cited above, but they serve to demonstrate sperm surface changes in a wide variety of species. The studies show that discrete regions of the sperm surface may be modified independent of other regions, and that the changes range from a loss to an appearance of lectin binding sites. The studies do not reveal if the observed changes result from a loss, masking, binding, or modification of components. The possible contribution of these different processes will be considered in the next section.

4.4. Macromolecular changes

Several recent studies of different species have defined specific macromolecular alterations in the plas-

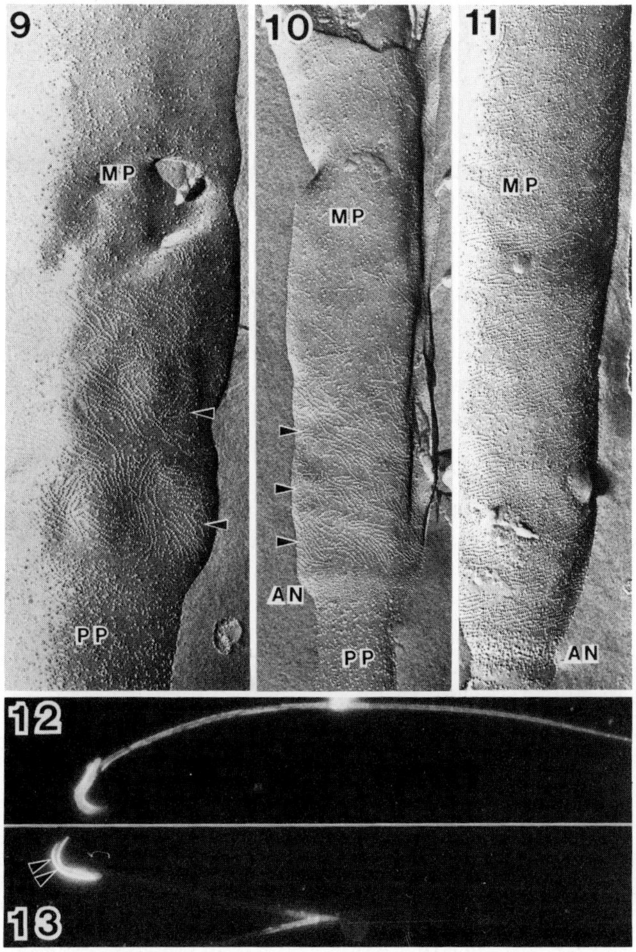

Figs. 9–13. (9) Freeze-fracture replica showing the midpiece and principal piece of guinea pig sperm recovered from the proximal region of the epididymal tubule. The principal piece (PP) plasma membrane has a random particle distribution. In the distal region of the midpiece (arrowheads) the particles are associated into chains but anteriorly (MP) the particles are distributed randomly. (10) Freeze-fracture replica showing the midpiece and principal piece of guinea pig sperm recovered from the proximal region of the epididymal tubule. Compared to Figure 9, note the more extensive development of particle chains (arrowheads). Anterior to this region a random particle arrangement is still present (MP). At the annulus (AN), a dense band of particles is present, and in the principal piece plasma membrane (PP) a random distribution of particles is noted. (11) Freeze-fracture replica showing the midpiece of guinea pig sperm from the distal cauda epididymis. Not the more extensive development of the particle chains compared to sperm from the more proximal segment of the tubule. (12) Fluorescence photomicrograph showing the pattern of FITC-ConA binding to caput epididymal rat sperm. (13) Fluorescence photomicrograph showing the pattern of FITC-ConA binding to cauda epididymal rat sperm. Compared to the previous figure, note the increased fluorescence of the acrosomal region (arrowheads) and decreased flagellar fluorescence.

ma membrane of sperm during the course of their maturation in the epididymis. One approach has been to incubate intact spermatozoa with non-penetrating reagents which either covalently bind or promote the covalent binding of radiolabeled marker molecules to externally located plasma membrane proteins of glycoproteins. Some of the probes applied to radiolabel the sperm surface include lactoperoxidase-catalyzed iodination of exposed tyrosine residues (66, 83–85); galactose oxidase-[^{3}H] sodium borohydride treatment, which radiolabels exposed galactose-like residues (70, 71); sodium metaperiodate-[^{3}H] sodium borohydride treatment to label sialic acid residues (70); and [^{125}]diazodiiodosulfanilic acid treatment to label exposed polypeptide chains (83). After labeling, the plasma membrane is solubilized and the individually labeled components are identified by electrophoretic separation. Although a number of species have been examined, only a few will be discussed in order to illustrate both the extent of the macromolecular changes which occur and the mechanisms generating these changes.

In the rat, both in the galactose oxidase and sodium metaperiodate protocol, a component of

37,000 daltons (37kd) is radiolabeled on cauda but not caput epididymal spermatozoa (70, 71 – MW of 32 kd reported in 71). This sperm surface component is found first on sperm from the proximal cauda epididymis (40, 71) which, significantly, is the zone where rat spermatozoa acquire fertilizing capacity. This component binds both Con A and RCA_1 (60, 70). On the basis of its lectin binding and labeling for sialic acid, this particular macromolecule could contribute to changes in lectin binding and surface charge of the spermatozoa which occurs during epididymal maturation.

The mechanism underlying the difference in the labeling pattern of caput and cauda rat spermatozoa is not fully resolved, and two different possibilities have been suggested. One possibility is that the 37kd component is produced by the epididymal epithelium and secreted into the tubule lumen where it binds to the sperm surface to become a type of coating protein. Evidence supporting this notion includes a demonstration of the production of an acidic epididymal glycoprotein (AEG) by principal cells of the caput and corpus epididymis and its secretion into the epididymal lumen (72–74). Immunohistochemical observations have revealed that AEG appears to coat the spermatozoa in the epididymal lumen (72–77). This glycoprotein has an apparent molecular weight of 37kd and a pI value of 4.7 (76, 78), and its binding could increase the net negative surface charge of the spermatozoa. Radiolabeling of spermatozoa and epididymal plasma by the galactose oxidase technique reveals a radiolabeled component with an apparent molecular weight of 32kd in each (71). Comparison of lactoperoxidase-labeled proteins of caput and cauda spermatozoa by SDS polyacrylamide gel electrophoresis reveals increased labeling in the 37kd region of cauda spermatozoa (79). These results suggest the binding of a 37kd glycoprotein of epididymal origin to the sperm surface occurs during maturation.

The second possibility for explaining the differences in galactose oxidase-promoted radiolabeling of caput and cauda sperm is that a preexistent 37kd membrane constituent is modified, possibly by glycosylation, while the sperm are present in the epididymis. Evidence consistent with this idea includes the fact that autoradiograms of SDS-polyacrylamide gels of lactoperoxidase-[^{125}I] NaI-labeled surface proteins of caput and cauda spermatozoa show similar banding patterns in the 37kd regions of the gels (40, 66). The galactose oxidase-labeled 37kd component of cauda spermatozoa is resistant to extraction and is not removed by high salt, chaotropic ions, chelators or disulfide reducing agents, but is removed by detergents (40). Thus, it has the solubility properties of an integral membrane protein, as opposed to a component absorbed to the outer surface. Recently, it has been demonstrated that both a soluble galactosyl transferase (80) and – lactalbumin-like activity (81) are present in rat epididymal fluid. It has been suggested that these activities could regulate glycosylation of sperm surface proteins (82). This idea is supported by data demonstrating that caput and cauda rat spermatozoa can incorporate UDP-galactose into a macromolecular fraction, suggesting that the appearance of a galactose-containing glycoprotein could be regulated in part by a galactosyl transferase complex (82). One possible resolution for the two ideas presented above is that the secreted AEG and the galactose oxidase-labeled sperm surface component may not be the same molecular species as often has been presumed.

Radiolabeling of caput and cauda epididymal rabbit spermatozoa by lactoperoxidase-catalyzed iodination or by treatment with a non-penetrating [^{125}I] labeled diazonium reagent has revealed significant differences between the sperm populations (83). A decreased labeling on cauda sperm of a 86kd component and increased labeling of components of 78kd and 50kd as compared to caput spermatozoa was found; in addition, components of 35kd and 39kd were detected on cauda epididymal spermatozoa but not caput epididymal spermatozoa. Similar patterns were obtained for lactoperoxidase-[^{125}I]NaI-labeled caput and cauda epididymal bull spermatozoa. In this species a decreased labeling of a higher molecular weight component (90kd–100kd), increased labeling of a 42kd–47kd component and disappearance of labeling of a 15kd–18kd component was noted during maturation (84). These results suggest a change in the relative exposure of some components during maturation of sperm in the epididymis, whereas the components identified exclusively on cauda spermatozoa could result from modifications such as proteolysis of existing components, unmasking of preexistent components, or the absorption of components from epididymal fluid.

Lactoperoxidase-catalyzed iodination of testicular ram spermatozoa labels seven polypeptides that migrate in the molecular weight range of 78kd to 132kd, but none of these polypeptides is noted on cauda epididymal spermatozoa (85). Cauda epididymal ram spermatozoa do, however, demonstrate unique polypeptides of 71kd, 116kd, and 24kd. The 24kd component is a soluble constituent of cauda epididymal fluid (85). Incubation of this fluid with testicular spermatozoa results in the selective binding of the 24kd component to the sperm surface (85). A similar pattern has been demonstrated for rat spermatozoa. In this species lactoperoxidase-catalyzed iodination reveals an intensely labeled component of 26kd–28kd on cauda but not caput spermatozoa (49, 66). The results of two-dimensional gel electrophoresis indicate this surface component has a molecular weight and pI identical to a major component of epididymal plasma (40). This result provides evidence for the specific binding of epididymal proteins to spermatozoa.

The functional role of most epididymal proteins which bind to the sperm surface remains undetermined. The binding of forward motility protein, which is of epididymal origin, to spermatozoa stimulates motility (86). By immunohistochemical techniques it has been demonstrated that some of the epididymal secretions are produced by specific regions of the epididymis and bind to restricted segments of the sperm surface (87–89). Moreover, exposure of spermatozoa to antibodies against these bound surface components can impair fertility. Thus these surface components likely make a functional contribution to the specific activities of restricted segments of the plasma membrane.

5. Conclusions

Recently, we have begun to understand the extent to which the sperm surface is modified during the maturation of sperm in the epididymis. With respect to discrete regions of the sperm plasma membrane, it appears that alterations entail changes at its cytoplasmic surface, in its internal anatomy and in the macromolecular composition of its exterior face. Some of the alterations at the outer face of the membrane result from the androgen-dependent secretory activities of the epididymis, so that the overall modification is the result of the combined activities of the spermatozoon and epididymis. Although many types of changes have been identified, many of the mechanisms underlying the changes are poorly understood. Further investigation of the constituents of individual plasma membrane domains at successive phases of sperm maturation in the epididymis should clarify some of these processes. The functional role of most membrane modifications occurring during sperm maturation presently is unknown. The changes occurring during passage through the epididymis are only one phase of post-testicular maturation, and additional surface alterations occur during capacitation. It is believed that these modifications are necessary to permit normal interaction of the sperm and egg during fertilization, and future studies no doubt will define the significance of individual changes with respect to the highly specialized functions of the individual plasma membrane domains of the spermatozoon.

Acknowledgements

I express my appreciation to Vera Henley and Brenda Lair for aid in preparing the manuscript and to Virgina Winfrey for assistance with the electron microscopy. Portions of the work discussed herein were supported by PHS Grant HD-11816.

References

1. Bedford, JM: Maturation, transport and fate of spermatozoa in the epididymis. In: Handbook of Physiology, Vol 5, Endocrinology, Section 7, Male Reproductive System. Greep RO, Astwood EB (eds), American Physiological Society, Washington, 1975, pp 303–317.
2. Bedford JM: Evolution of the sperm maturation and sperm storage functions of the epididymis. In: The Spermatozoon. Maturation, motility, surface properties and comparative aspects. Fawcett DW, Bedford JM (eds), Urban and Schwarzenberg, 1979, pp 7–21.
3. Orgebin-Crist M-C, Danzo BJ, Davies J: Endocrine control of the Development and maintenance of sperm fertilizing ability in the epididymis. In: Handbook of Physiology. Vol 5, Endocrinology, Section 7, Male reproductive system. Greep RO, Astwood EB (eds), American Physiological Society, Washington, 1975, pp 319–338.
4. Orgebin-Crist M-C, Olson GE, Danzo BJ: Factors influencing maturation of spermatozoa in the epididymis. In: Intragonadal regulation of reproduction. Francimont P, Channing CP (eds), Academic Press (London), 1981, pp 393–418.
5. Hamilton DW: Structure and function of the epithelium lining the ductuli efferentes, ductus epididymis, and ductus deferens in the rat. In: Handbook of Physiology, Vol V, Endocrinology. Section 7, Male reproductive system. Greep RO and Astwood EB (eds), American Physiological Society, Washington, 1975, pp 259–301.
6. Brooks DE: Biochemical environment of sperm maturation. In: The Spermatozoon. Maturation, motility, surface properties and comparative aspects. Fawcett DW, Bedford JM (eds), Urban and Schwarzenberg, 1979, pp 23–34.
7. Levine N, Marsh DJ: Micropuncture studies and the electrochemical aspects of fluid and electrolyte transport in individual seminiferous tubules, the epididymis and the vas deferens in rats. J Physiol 213: 557–570, 1971.
8. Cameo MS, Blaquier JA: Androgen-controlled specific proteins in rat epididymis. J Endocr 69: 47–55, 1976.
9. Brooks DE: Secretion of proteins and glycoproteins by the rat epididymis: Regional differences, androgen-dependence, and effects of protease inhibitors, procaine, and tunicamycin. Biol Reprod 25: 1099–1117, 1981.
10. Jones R, Brown CR, Von Glos K, Parker MG: Hormonal regulation of protein synthesis in the rat epididymis. Characterization of androgen-dependent and testicular fluid-dependent proteins. Biochem J 188: 667–676, 1980.
11. Orgebin-Crist M-C, Jahad N, Hoffman LH: The effects of testosterone, 5 α-dihydrotestosterone, 3 α-androstanediol, and 3 β-androstanediol on the maturation of rabbit epididymal spermatozoa in organ culture. Cell Tiss Res 167: 515–525, 1976.
12. Orgebin-Crist M-C, Jahad N: The maturation of rabbit epididymal

spermatozoa in organ culture: Inhibition by antiandrogens and inhibitors of ribonucleic acid and protein synthesis. Endocrinol 103: 46–53, 1978.
13. Bedford JM and Cooper GW: Membrane fusion events in the fertilization of vertebrate eggs. In: Membrane fusion, Poste G and Nicolson GL (eds), Elsevier/North-Holland Biomedical Press, 1978, pp 65–125.
14. Fawcett, DW: The mammalian spermatozoon. Dev Biol 44: 394–436, 1975.
15. Yanagimachi R: Specificity of sperm-egg interaction. In: Immunobiology of gametes. Edidin M, Johnson MH (eds) Cambridge University Press, 1977, pp 255–295.
16. Yanagimachi R 1981: Mechanism of fertilization in mammals. In: Fertilization and embryonic development *in vitro*. Mastroianni L, Biggers JD (eds), 1981, pp 81–182.
17. Singer SJ, Nicolson GL: The fluid mosaic model of the structure of cell membranes. Science 175: 720–731, 1972.
18. Koehler JK: Observations on the fine structure of vole spermatozoa with particular reference to cytoskeletal elements in the mature sperm head. Gamete Res. 1: 247–257, 1978.
19. Peterson R, Russell L, Bondman D, Freund M: Presence of microfilaments and tubular structures in boar spermatozoa after chemically inducing the acrosome reaction. Biol reprod 19: 459–466, 1978.
20. Friend DS: The organization of the spermatozoal membrane. In: Immunobiology of gametes. Edidin M, Johnson MH (eds), Cambridge University Press, 1977, pp 5–30.
21. Olson G: Observations on the ultrastructure of a fiber network in the flagellum of sperm of the brush tailed phalanger, Trichosurus volpecula. J Ultrastruct Res, 50: 193–198, 1975.
22. Olson GE, Hamilton DW: Morphological changes in the midpiece of wooly opossum spermatozoa during epididymal transit. Anat Rec 186, 387–404, 1976.
23. Olson GE, Lifsics M, Fawcett DW, Hamilton DW: Structural specializations in the flagellar plasma membrane of opossum spermatozoa. J Ultrastruct Res 59: 207–221, 1977.
24. Olson GE: Changes in intramembranous particle distribution in the plasma membrane of Didelphis virginiana spermatozoa during maturation in the epididymis. Anat Rec 197: 471–488, 1980.
25. Harding HR, Carrick FN, Shorey CD: Special features of sperm structure and function in marsupials. In: The spermatozoon. Maturation, motility, surface properties and comparative aspects. Fawcett DW, Bedford JM (eds), Urban and Schwarzenberg, 1979, pp 289–303.
26. Harding HR, Carrick FN, Shorey CD: Ultrastructural changes in spermatozoa of the brush-tailed possum, trichosurus volpecula (Marsupialia) during epididymal transit. I. The Flagellum. Cell Tiss Res 164: 121–132, 1975.
27. Temple-Smith PD, Bedford JM: The features of sperm maturation in the epididymis of a marsupial, the brush-tailed possum trichosurus volpecula. Am J Anat 147: 471–500, 1976.
28. Temple-Smith PD, Bedford JM: Sperm maturation and the formation of sperm pairs in the epididymis of the opossum, didelphis virginiana. J Exp Zool 214: 161–171, 1980.
29. Clarke GN, Yanagimachi R: Actin in mammalian sperm heads. J Exp Zool. 205: 125–132, 1978.
30. Tamblyn TM: Identification of actin in boar epididymal spermatozoa. Biol Reprod 22: 727–734, 1980.
31. Friend DS, Orci L, Perrelet A, Yanagimachi R: Membrane particle changes attending the acrosome reaction in guinea pig spermatozoa. J Cell Biol 74: 561–577, 1977.
32. Friend DS: Freeze-fracture alterations in guinea pig sperm membranes preceeding gemete fusion. In: Membrane-membrane interactions. Gilula NB (ed), Raven Press, 1980, pp 153–165.
33. Nicolson GL, Yanagimachi R: Mobility and the restriction of mobility of plasma membrane lectin-binding components. Science 184: 1294–1296, 1974.
34. Nicolson GL: Transmembrane control of the receptors on normal and tumor cells. I. Cytoplasmic influence over cell surface components. Biochem Biophys Acta 457: 57–108, 1976.
35. Friend DS, Fawcett DW: Membrane differentiations in freeze-fractured mammalian spermatozoa. J Cell Biol 63: 641–664, 1974.
36. Friend DS, Rudolf I: Acrosomal disruption in sperm. Freeze-fracture of altered membranes. J Cell Biol 63: 466–479, 1974.
37. Koehler JK, Gaddum-Rosse P: Media induced alterations of the membrane associated particles of the guinea pig sperm tail. J Ultrastruct Res 51: 106–118, 1975.
38. Suzuki F, Nagano T: Epididymal maturation of rat spermatozoa studied by thin sectioning and freeze-fracture. Biol Reprod 22: 1219–1231, 1980.
39. Dyson ALMB, Orgebin-Crist M-C: Effect of hypophysectomy, castration and androgen replacement upon the fertilizing ability of rat epididymal spermatozoa. Endocrinol 93: 391–402, 1973.
40. Olson GE, Orgebin-Crist M-C: Sperm surface changes during epididymal maturation. Ann NY Acad Sci, in press.
41. Fléchon J-E: Ultrastructural and cytochemical analysis of the plasma membrane of mammalian sperm during epididymal maturation: In: Progress in reproductive biology, V 8. Epididymis and fertility: Biology and pathology, Hubinot PO (ed), S. Karger, 1981, pp 90–99.
42. Scott TW, Voglmayr JK, Setchell BP: Lipid composition and metabolism in testicular and ejaculated ram spermatozoa. Biochem J 102: 445–461, 1967.
43. Grogan DE, Mayer DT, Sikes JD: Quantitative differences in phospholipids of ejacualted spermatozoa and spermatozoa from three levels of the epididymis of the boar. J Reprod Fert 12: 431–436, 1966.
44. Evans RW, Setchell BP: Lipid changes in boar spermatozoa during epididymal maturation with some observations on the flow and composition of boar rete testis fluid. J Reprod Fert 57: 189–196, 1979.
45. Evans RW, Setchell BP: Lipid changes during epididymal maturation in ram spermatozoa-collected at different times of the year. J Reprod Fert 57: 197–203, 1979.
46. Poulous A, Brown-Woodman, PDC, White IG, Cox RI: Changes in phospholipids of ram spermatozoa during migration through the epididymis and possible origin of prostaglandin $F_{2\alpha}$ in testicular and epididymal fluid. Biochem Biophys Acta 388: 12–18, 1975.
47. Poulus A, Voglmayr JK, White IG: Phospholipid changes in spermatozoa during passage through the genital tract of the bull. Biochem Biophys Acta 306: 194–202, 1973.
48. Lavon V, Volcani R, Danon D: The lipid content of bovine spermatozoa during maturation and ageing. J Reprod Fert 23: 215–222, 1970.
49. Terner C, MacLaughin J, Smith BR: Changes in lipase and phospholipase activities of rat spermatozoa in transit from the caput to the cauda epididymis. J Reprod Fert 45: 1–8, 1975.
50. Dawson RMC, Scott TW: Phospholipid composition of epididymal spermatozoa prepared by density gradient centrifugation. Nature 202: 292–293, 1964.
51. Johnson MH: The macromolecular organization of membranes and its bearing on events leading up to fertilization. J Reprod Fert 44: 167–184, 1975.
52. Bedford JM: Changes in fine structure of the rabbit sperm head during passage through the epididymis. J Anat 99: 891–906, 1965.
53. Fawcett DW, Phillips DM: Observations on the release of spermatozoa and on changes in the head during passage through the epididymis. J Reprod Fert, Suppl 6: 405–418, 1969.
54. Jones RC: Studies on the structure of the head of boar spermatozoa from the epididymis. J Reprod Fert, Suppl 13: 51–64, 1971.
55. Bedford JM, Rifkin JM: An evolutionary view of the male reproductive tract and sperm maturation in a monotreme mammal – the echidna, tachyglossus aculeatus. Am J Anat 156: 207–230, 1979.
56. Bedford JM: Changes in the electrophoretic properties of rabbit spermatozoa during passage through the epididymis. Nature 200: 1178–1180, 1963.
57. Moore HDM: The net negative surface charge of mammalian spermatozoa as determined by isoelectric focusing. Changes following sperm maturation, ejaculation, incubation in the female tract, and after enzyme treatment. Int J Androl 2: 244–462, 1979.
58. Hammerstedt RH, Keith AD, Hay S, Deluca N, Amann RP: Changes in ram sperm membrane during epididymal maturation. Arch Biochem Biophys 196: 7–12, 1979.
59. Cooper GW, Bedford JM: Acquisition of surface charge by the plasma membrane of mammalian spermatozoa during epididymal maturation. Anat Rec 169: 300–301, 1971.
60. Yanagimachi R, Noda YD, Fujimoto M, Nicolson GL: The distribution of negative surface charges on mammalian spermatozoa. Am J Anat 135: 497–520, 1972.
61. Fléchon JE: Ultrastructural and cytochemical modifications of rabbit spermatozoa during epididymal transport. In: The Biology of Spermatozoa. Transport, survival, and fertilizing ability. Hafez ESE, Thibault CG (eds), Karger, Basel, 1975, pp 36–45.
62. Courtens JL, Fournier-Delphech S: Modifications in the plasma membrane of epididymal ram spermatozoa during maturation and in-

cubation in utero. J Ultrastruct Res 68: 136–148, 1979.
63. Cooper GW, Bedford JM: Asymmetry of spermiation and sperm surface charge patterns over the giant acrosome in the musk shrew Suncus Murinus. J Cell Biol 69: 415–428, 1976.
64. Edelman GM, Millette CF: Molecular probes of spermatozoan structure. Proc Nat Acad Sci 68: 2436–2440, 1971.
65. Nicolson G, Lacorbiere M, Yanagimachi R: Quantitative determination of plant agglutinin membrane sites on mammalian spermatozoa. Proc Soc Exp Biol Med 141: 661–663, 1972.
66. Olson GE, Danzo BJ: Surface changes in rat spermatozoa during epididymal maturation. Biol Reprod 24: 431–443, 1981.
67. Nicholson GL, Usui N, Yanagimachi R, Yanagimachi H, Smith JR: Lectin-binding sites on the plasma membranes of rabbit spermatozoa. Changes in surface receptors during epididymal maturation and after ejaculation. J Cell Biol 74: 950–962, 1977.
68. Gordon M, Dandekar PV, Bartoszewicz W: The surface coat of epididymal, ejaculated and capacitated sperm. J Ultrastruct Res 50: 199–207, 1975.
69. Bedford JM, Millar RP: The character of sperm maturation in the epididymis of the ascrotal hyrax, procavia capensis and armadillo, dasypus novemcinctus. Biol Reprod 19: 396–406, 1978.
70. Olson GE, Hamilton DW: Characterization of the surface gylcoproteins of rat spermatozoa. Biol Reprod 19: 26–35, 1978.
71. Jones R, Pholpramool C, Setchell BP, Brown CR: Labelling of membrane glycoproteins on rat spermatozoa collected from different regions of the epididymis. Biochem J 200: 457–460, 1981.
72. Lea OA, Petrosz P, French FS: Purification and localization of acidic epididymal glycoprotein (AEG): A sperm coating protein secreted by the rat epididymis. Int J Androl, Suppl 2: 592–607, 1978.
73. Fournier S: Electrophoréses des proteins du tractus génital du rat I. Presence clans le sperme epididymaire d'une glycoprotein migrant vers l'anode à pH = 8.45. CR Soc Biol 162: 568–571, 1968.
74. Garberi JC, Kohane AC, Cameo MS, and Blaquier JA: Isolation and characterization of specific rat epididymal proteins. Mol Cell Endocrinol 13: 73–82, 1979.
75. Fournier-Delpech S, Bayard F, Boulard C: Isolement, extraction et characterisation d'une sialoproteine du sperme epididymaire du rat par électrophorèse sur polyacrylamide. CRS Soc Biol 167: 543–547, 1973.
76. Faye JC, Duquet L, Mazzuca M, Bayard F: Purification, radioimmunoassay and immunohistochemical localization of a glycoprotein produced by the rat epididymis. Biol Reprod 23: 423–432, 1980.
77. Kohane AC, Cameo MS, Pineiro L, Garberi JC, Blaquier JA: Distribution and site of production of specific proteins in the rat epididymis. Biol Reprod 23: 181–187, 1980.
78. Lea OA, French FS: Characterization of an acidic epididymal glycoprotein secreted by principal cells of the rat epididymis. Biochim Biophys Acta 668: 370–376, 1981.
79. Wong PYD, Tsang AVF, Lee WM: Origin of the luminal fluid proteins of the rat epididymis. Intern J Androl 4: 331–341, 1981.
80. Hamilton DW: UDP-Galactose: N-Acetyl-glucosamine galactosyltransferase in fluids from rat rete testis and epididymis. Biol Reprod 23: 377–385, 1980.
81. Hamilton DW: Evidence for α-lactalbumin-like activity in reproductive tract fluids of the male rat. Biol Reprod 25: 385–392, 1981.
82. Hamilton DW, Gould RP: Galactosyltransferase activity associated with rat epididymal spermatozoan maturation. Anat Rec 196: 71a, 1980.
83. Nicolson GL, Brodginski AB, Beattie G, Yanagimachi R: Cell surface changes in the proteins of rabbit spermatozoa during epididymal passage. Gamete Research 2: 153–162, 1979.
84. Vierula M, Rajaniemi H: Changes in surface protein structure of bull spermatozoa during epididymal maturation. Intl J Androl 4: 314–320, 1981.
85. Voglmayr JK, Fairbanks G, Jackowitz MA, Colella J: Post-testicular developmental changes in the ram sperm cell surface and their relationship to luminal fluid proteins of the reproductive tract. Biol Reprod 22: 655–667, 1980.
86. Acott TS and Hoskins DD: Bovine sperm forward motility protein: Binding to epididymal spermatozoa. Biol Reprod 24: 234–240, 1981.
87. Moore HDM: Localization of specific glycoproteins secreted by the rabbit and hamster epididymis. Biol Reprod 22: 705–718, 1980.
88. Moore HDM: Glycoprotein secretions of the epididymis in the rabbit and hamster: Localization on epididymal spermatozoa and the effect of specific antibodies on fertilization *in vivo*. J Exp Zool 215: 77–85, 1981.
89. Feuchter FA, Vernon RB, Eddy EM: Analysis of sperm surface components with monoclonal antibodies: Topographically restricted antigens appearing in the epididymis. Biol Reprod 24: 1099–1110, 1981.

Author's address:
Department of Anatomy
Vanderbilt University
Nashville, TN 37232, USA

CHAPTER 10

The human spermatozoon

BACCIO BACCETTI

1. Introduction

In recent years numerous studies have carried our knowledge of the human spermatozoon to a high level, and constitute a basis for the understanding of many cases of partial or complete infertility, as well as for a useful approach to the general problem of fertility control. Many of the investigations have involved electron microscopy, others have used chemical analysis of isolated cellular components. Useful data, first obtained on sperm in animals, recently has been confirmed in the human. This perhaps is an appropriate moment to gather these results together in a single review.

2. External form

The human spermatozoon has the form of a pin, with a globular apical portion, the head, and the rest being a long flexible cylinder, the tail. In the motile spermatozoon the head is at the front and is pushed forward passively by the tail, which being capable of active worm-like movement, is responsible for the vigorous swimming of the whole cell. The head is the portion that makes first contact with the oocyte, and the recognitive and penetrative capabilities are associated with its front surface, while the tail continues the work of pushing. The head contains the DNA, the transport of which into the oocyte is the primary purpose of the swimming movement, while the tail contains the axoneme which is the actively contractile portion of the spermatozoon.

The human sperm is about 60 μm long. The head is a somewhat flattened, ellipsoid, 4–5 μm long and 2.5–3.5 μm broad. The tail measures about 55 μm overall length, and it has a diameter of more than 1 μm at its base and tapers progressively toward its tip. Even in the light microscope, but especially in the scanning electron microscope (Fig. 1a), it can be seen that the outside of the head and the tail are comprised of several distinct regions. The head appears crossed by a transverse constriction (Fig. 1b) approximately two-thirds along its length, which results in the posterior portion being slightly but abruptly narrower than the anterior. This constriction corresponds to the termination of the acrosomal cap, located beneath the plasma membrane. The tail (Fig. 1a), after a junctional region about 1 μm long, termed the neck, shows a section of a large spiral structure beneath the membrane. The latter consists of the section in which the mitochondria are located and is termed the midpiece. At the end of this region the surface becomes smooth and the diameter decreases abruptly: this portion, about 45 μm long constitutes the principal piece. Following this, after a further abrupt decrease in diameter, is a terminal region about 5 μm long.

3. The fine structure of the head

3.1. *Plasma membrane*

Examination by transmission electron microscopy (TEM) shows the normal trilaminar membranous structure of the plasma membrane. Scattered 9 nm particles are located between the two outer laminae, and can be seen rendily in cryofractured specimens. On the external surface is a fairly thick glycoprotein layer that can be demonstrated by specific binding of concanavalin A conjugated with peroxidase and visualized by diaminobenzidine or fluorescent staining. In rodent and human sperm this layer appears somewhat thicker around the sperm head than around the tail (1), indicating that this region is covered with glycoproteins. Also electroanalytical techniques, have reported the presence of sialic acid and free SH groups in this region. This glycoprotein layer develops in a series of Golgi cisternae that surround the organelle of the future spermatozoon in the anterior of the maturing spermatid.

Van Blerkom, J. and Motta, P.M. (eds.), Ultrastructure of reproduction. ISBN 0-89838-572-5.

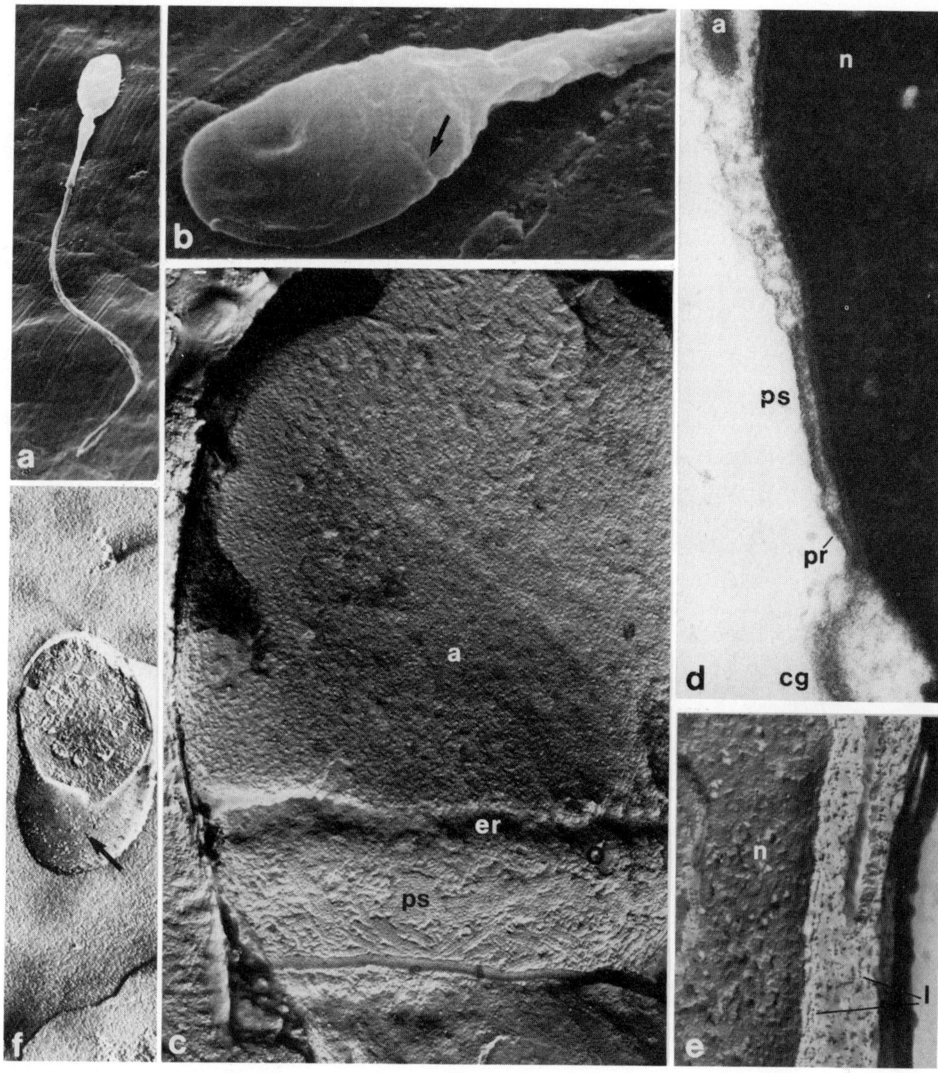

Fig. 1. (a) Human spermatozoon critical point dried and photographed in a field emission scanning electron microscope (× 2,400). (b) Head of a human spermatozoon at the level of the median piece under the same conditions as the preceding. The circular groove is apparent (arrow) (× 12,000). (c) Head region of a human spermatozoon immediately under the plasma membrane, photographed in the transmission electron microscope after freeze-etching. Clearly evident are the structure of the acrosome (a), the equatorial region (er) and the postacrosomal dense sheath (ps) (× 26,400). (d) The same region, in longitudinal section. Posterior to the acrosome (a) and adjacent to the nucleus (n) are seen the postacrosomal dense sheath (ps), the posterior ring (pr) and the circular groove (cg) (× 60,000). (e) Fracture of the nucleus (n) photographed in transmission microscope. The exposed surface shows the disordered arrangement of the laminae (l) (× 36,000). (f) Fracture of the tail photographed in transmission microscope. Disordered particles can be seen in the membrane (arrow) (× 32,000).

The glycoprotein layer is important because it contains the surface antigens, that have a specific localization on the surface of the human sperm head. By using fluorescent markers, differences among the acrosomal region, the equatorial region and the posterior nuclear region have been found. The human sperm membrane interacts with ATP and cAMP, undergoing conformational changes that facilitate the transport of substrates and thus influence drastically the metabolism of the spermatozoon in the male and female genital ducts. cAMP seems, among others, to be one of the most important factors influencing capacitation of the human sperm, and thus its acquisition of fertilizing capability, acting through changes in the sperm membrane during transit of the female genital tract (2). For a general discussion see also (3).

Capacitation has an important influence on the structure of the plasma membrane, for this phenomenon changes the distribution of surface antigens (4) and increases the ATPase activity of the plasma membrane that had diminished during transit through the epididymis (5).

The plasma membrane of the human sperm head contains two differentiated structures (Scheme 1, Figs. 1c, d); an annular lamina, termed the post-acrosomal dense sheath, located metacaudally and crossed by longitudinal striations that connect it to the limiting membrane (6, 7), and a ring, composed of densely packed 20 nm particles, that surrounds the basal area of the nucleus where the plasma membrane fuses with the underlying nuclear membrane (8). The posterior border of the post-acrosomal dense sheath is drawn out and finely striated (6). The posterior ring separates the interstitial spaces of the head from those of the rest of the sperm.

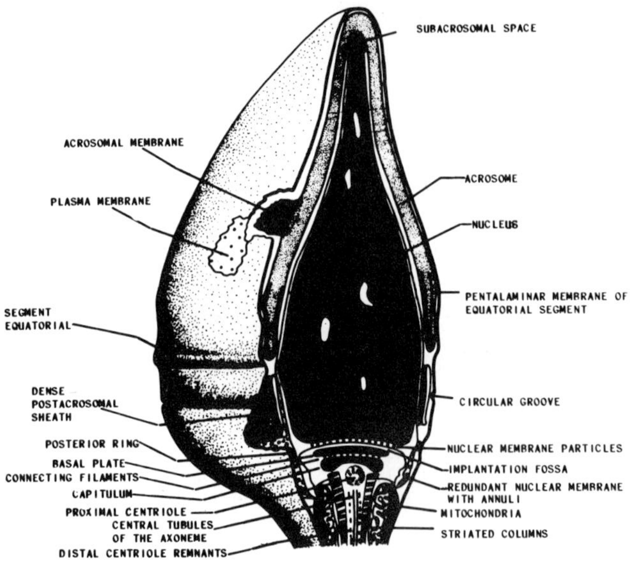

Scheme 1. Schematic diagram of the head of a human spermatozoon.

3.2. *Nucleus*

The nucleus of the mature human spermatozoon is small and very dense compared to that of other cells, being composed of tight packets of chromatin around some vacuoles of various size. Its overall shape is ovular, somewhat flattened, with a conical anterior portion and a convex posterior portion bearing an implantation cavity where the flagellum originates. In the electron microscope (Fig. 1e), it appears almost opaque and fibers 140 to 240 Å in diameter can be distinguished packed in a irregular manner (8).

The essential constituent of the nucleus is the DNA. By definition, the chromosomal complement of the normal sperm must be haploid. In human sperm, Carothers and Beatty (9) have shown, by photometric densitometry of Feulgen-stained material, that 99.37% of the sperm are haploid, 0.56% diploid and 0.07% triploid. Earlier investigations, mostly on invertebrate sperm, have demonstrated that the chromosomes are well separated as in the spermatocyte nucleus and are oriented in a constant manner (10). The same appears true in human sperm. The amount of DNA is nearly constant, although Sarkar, Jones and Shioura (11), measuring the fluorescent intensity of Feulgen-positive material, observed the presence of two classes. The two classes occurred in approximately equal numbers and presumably corresponded to the presence of either the X or Y chromosomes. It was also noted by Summer, Robison, and Evans (12) that sperm carrying the Y chromosome contain 4.23% less DNA than those bearing an X chromosome. In any type of human ejaculate, even oligospermic, the numbers of sperm bearing X or Y chromosomes are equal although the distinction of classes on the basis of DNA content was not confirmed (9). The remaining variability is attributable to chromosomal aberration, the incidence of which amounts to 5% of the gametes capable of fertilization.

Mature spermatozoa do not synthesize RNA. On the other hand, even in man, sperm have been reported to contain an endogenous complex of DNA synthetase associated with high molecular weight RNA and sensitive to ribonuclease. The presence of 18S and 28S RNA's in ejaculated human sperm has been confirmed by many authors, and it has been proposed that the RNA is synthesized prior to meiosis. This RNA may serve for all the post-meiotic differentiation, given that the chromatin is rendered transcriptionally inert by the marked condensation which occurs upon substitution of the somatic histones present in the spermatogonia with the more basic proteins during the spermatid stage. These basic proteins are protamines and histones, rich in arginine (about 50%) and cystine (about 10%), of the type typical of mature sperm of Eutherian mammals including man. Kolk and Samuel (13) have used ion-exchange chromatography and gel filtration to examine this aspect more carefully in human sperm and have isolated two proteins, called protamines 1 and 2, of molecular weights 6280 and 6840 daltons respectively. The former has 22 arginine residues and 5 half-cystine residues, whereas the latter has 24 arginines and 4 half-cystines, as well as about 8 histidine residues. Both these proteins are autoantigens, which opens the interesting possibility of contraception by means of immunization.

The presence of large quantities of cysteine in isolated protamines derives from the presence of numerous intermolecular S-S bridges in association with the deoxyribonucleoprotein of the mature human sperm (14). This characteristic is typical of many types of spermatozoa, particularly of mammals, and yields a stable packing of the nuclear material that favours both transport in the seminal fluid and prevention of damage during the oocyte and zona pellucida penetration (7). In many types of spermatozoa with genetic damage, as in Drosophila (the 'segregation disorder') and man (the round headed'), the normal transformation of substitution of somatic histones appears inhibited. The stabilization of the chromatin by S-S bridges generally is completed during the passage of the epididymis (15), but does not occur uniformly in some human sperm. Bedford, Calvin and Cooper (15), after reducing the S-S linkages, found a variable quantity of vacuoles in the nuclear compartment, and suggested that their technique (treating with sodium dodecylsulfate and dithiothreitol) could be used as a rapid assay of the quality of the sperm nuclei in clinical samples of human sperm. Bearden and Bendet (16) have used polarized light to follow the stabilization process in human sperm as a function of time and temperature.

The stabilized chromatin becomes decondensed after the penetration of the oocyte, as a result of enzymic activity present in the oocyte itself. An intrinsic capability of the sperm for decondensing chromatin has been also suggested, assuming it inhibited during the transit of the genital ducts by the Zn^{2+} present in the prostatic fluid.

The nuclear membrane of spermatozoa is in general unusual in its near absence of pores and perinuclear cisternae (10). In freeze etched samples (Fig. 2f) the nuclear membrane appears completely smooth. However, in human sperm, many pores with annuli remain in the posterior portion of the nucleus (6), and it is not known if these result from migration or the *de novo* formation. In the extreme posterior portion, the nuclear membrane joins the plasma membrane, giving a ring of dense homogeneous material. This ring is the posterior ring. In the region of the flagellar implantation the nuclear membrane of mammalian sperm is traversed by dense particles 6 nm thick with a separation of 6 nm, and its exterior surface is covered by a thick layer, termed the basal plate (Scheme 1).

The formation of the elongated and condensed form of the chromatin occurs gradually over the whole developmental period of spermatids. Some believe that external morphogenetic forces, possibly microtubules or contractile fibrils, mold the nucleus, while others believe that the change in form must be due to the intrinsic properties of the DNA-protein complex (7–10). For human sperm there are as yet no data. The anomalous 'double sperm' (see section 6.5.) in humans shows that often two nuclei contained within a single plasma membrane have the exterior form of a single nucleus, although larger than normal (Fig. 2e), indicating that external morphogenetic forces have molded the whole nuclear complex (17).

3.3. Acrosomal complex

The acrosomal complex is a cap-like structure (Fig. 2a) bound by its own membrane, termed the acrosomal membrane, that covers most of the anterior two-thirds of the nucleus. In the human spermatozoon it is relatively slender. The anterior portion of the acrosome surrounds closely and barely projects beyond the apex of the nucleus. The posterior portion is somewhat constricted and appears, from the exterior, as an annular depression the head, termed the equatorial segment. Its membrane is in contact with the sperm plasma membrane exteriorly, except in the extreme anterior portion where there exists a narrow sub-acrosomal space. A thin layer of subacrosomal material, or 'perforatorium', is also present in human sperm. Thus, any description of the acrosomal complex should include the acrosomal membrane, its contents and the material in the subacrosomal space.

3.3.1. Acrosomal membrane

The acrosome originates from the Golgi apparatus in the spermatid. Numerous Golgi vesicles, filled with material, fuse and form a single cap-shaped body that becomes attached to the nucleus. Thus, the acrosomal membrane is derived from a simple Golgi membrane. Pedersen (6) has described the usual trilaminar structure of the membrane in human sperm, with a uniform thickness of 9 nm, except at the level of the equatorial segment where the acrosomal membrane is five-layered through the interposition of an interior more dense layer. After cryofracture, the acrosomal membrane shows a pattern of tightly packed spheroidal particles, probably protein in nature. In human sperm, these particles measure 20 nm in diameter (8).

During passage of sperm through the female genital tract, the exterior acrosomal membrane invaginates in many places forming vesicles, while the overlying plasma membrane disintegrates, so that the contents of the acrosome are liberated (18). This process accounts for the major part of the capacitation and acrosomal reactions in human sperm,

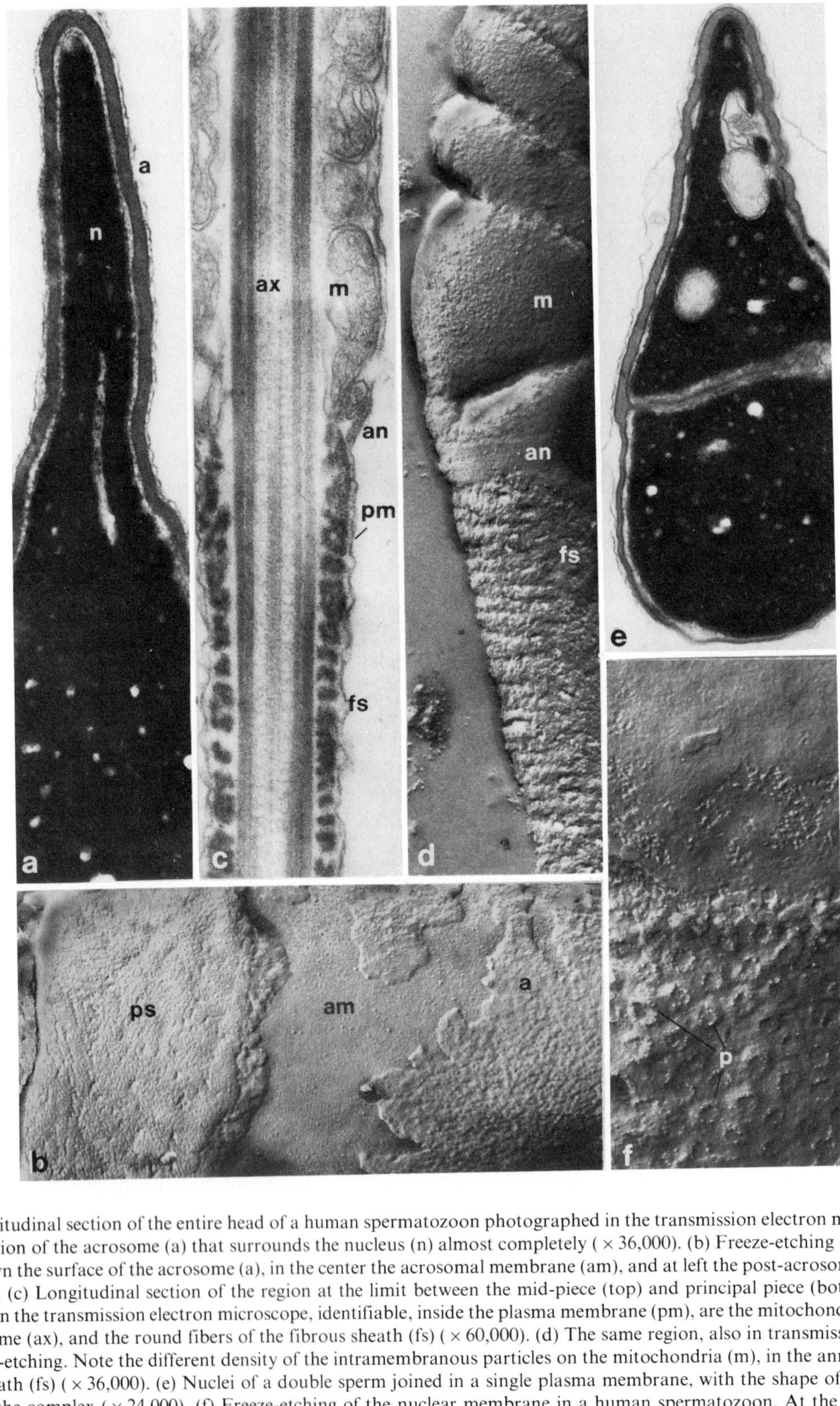

Fig. 2. (a) Longitudinal section of the entire head of a human spermatozoon photographed in the transmission electron microscope. Note the large extension of the acrosome (a) that surrounds the nucleus (n) almost completely (× 36,000). (b) Freeze-etching of the acrosome. At right is shown the surface of the acrosome (a), in the center the acrosomal membrane (am), and at left the post-acrosomal dense sheath (ps) (× 36,000). (c) Longitudinal section of the region at the limit between the mid-piece (top) and principal piece (bottom) of the tail, photographed in the transmission electron microscope, identifiable, inside the plasma membrane (pm), are the mitochondria (m), annulus (an), the axoneme (ax), and the round fibers of the fibrous sheath (fs) (× 60,000). (d) The same region, also in transmission microscopy, but after freeze-etching. Note the different density of the intramembranous particles on the mitochondria (m), in the annulus (an) and in the fibrous sheath (fs) (× 36,000). (e) Nuclei of a double sperm joined in a single plasma membrane, with the shape of a single nucleus maintained in the complex, (× 24,000). (f) Freeze-etching of the nuclear membrane in a human spermatozoon. At the top the uniform region, surrounding the nucleus, at the bottom the redundant nuclear membrane rich in pores (p) (× 45,600).

which is thus somewhat different from the mechanism discribed in other species, where vesiculation is more complicated and extensive. Even in man, however, vesiculation is dependent upon the presence of Ca^{2+}, which is bound to the external acrosomal membrane, as Roomans (19) has been able to demonstrate using an electron microscope with X-ray microanalysis. Immediately after acrosomal reaction the fibrillar intramembranous particles are dispersed.

3.3.2. The acrosome

The acrosomal contents serve to digest the many oocyte envelopes. Varying significantly among different groups of animals, in mammals the acrosome contains highly active proteolytic and glycolytic enzymes. Baccetti (20) has listed the presence of acid phosphatase, β-glucuronidase, arylamidase, arylsulphatase, β-N-acetylglucosaminidase, phospholipase A, non-specific esterase, β-aspartyl-N acetylglucosaminoamido hydrolase, a hyaluronidase, and at least four peptidases characteristic of mammalian sperm: the 'penetrating enzyme' a hydrolase discovered by Zaneveld and Williams, acrosin, collagenase, and neuraminidase, all probably act on the zona pellucida (21). The best described enzymes are hyaluronidase, acrosin, and collagenases. Hyaluronidase, studied in sperm of several animals, has a molecular weight of about 60,000 daltons and consists of 4 polypeptide subunits (21); it is the first enzyme to be liberated and acts on the oocyte cumulus complex. Acrosin can be readily isolated from human sperm by washing sperm in a saline solution containing glucose and freezing in liquid nitrogen, or by extraction at acid pH. This enzyme has a molecular weight of about 30,000 daltons and maximal activity at pH 8. At least 90% of the total acrosin present at the time of ejaculation is in the form of an inactive precursor, the 'proacrosin', and is activated during capacitation when the inhibitor 'acrostatin', earlier bound to the enzyme, is removed (22). An extraction procedure for acrosin, proacrosin and the inhibitor of acrosin from human sperm has been recently developed (23). Acrosin is usually believed to be responsible for the penetration of the zona pellucida. However, it does not facilitate the penetration of the cervical mucus in humans. The collagenase found in the sperm of humans and other mammals by Koren and Milkovic (24) has a molecular weight of about 110,000 daltons and a maximal activity at a pH of 7.5. As the localization of these enzymes in the interior of the acrosome is concerned, most of the research has been carried out on bull, ram or rodents, where several different crystalline structures have been detected in distinct regions of the acrosome. Typically, immunochemical techniques have been used to localize to the hyaluronidase anterior region of the acrosome, and acrosin close to the inner acrosomal membrane (20). Also in human spermatozoa the two enzymes seem to have such localization. These data may explain how hyaluronidase is liberated before acrosin. An important problem (20) is whether the acrosome of mammalian sperm contains 'bindin', the protein responsible for specific sperm-egg attachment that was recently discovered and studied in sperm of marine invertebrates.

An effective inhibitor of acrosin might provide a useful method of contraception (20). A first approach might be immunological, because it has been shown that the acrosins from many mammals possess interesting antigenic similarities, and are consistently different from trypsin. Another approach is pharmacological: Bhattacharyya and Zaneveld (25) have observed that acrosin from primate sperm is particularly inhibited by p-nitrophenyl-guanidino-benzoate, and Reddy, Joyce and Zaneveld (26) have found that hyaluronidase is inhibited by the non-toxic agent, myocrisine.

3.3.3. Perforatorium

In marine sperm and in those of the more primitive terrestrial animals, the perforatorium is a slender longitudinal cylinder beneath the acrosome which at the moment of the acrosomal reaction is pushed forward through the acrosome (20). In sperm of echinoderms, molluscs and arthropods, it contains actin that polymerizes and changes its form of packing at the moment of extension. It is known that it contains also spectrin, profilin, α-actinin and myosin, that play varied roles in the extension of the actin boundle. Little is known about these proteins in vertebrates, including man. The presence of myosin in the acrosomal region of human sperm has been reported by Clarke and Yanagimachi (27) at immunofluorescence level, while an acto-myosin network located between the nucleus and the acrosome has been later demonstrated by immunoelectronmicroscopy.

4. Fine structure of the tail

4.1. Plasma membrane

Viewed in the electron microscope, the plasma membrane of the tail in the human spermatozoon shows the same classical trilaminar appearance as that of the head, with particles of 9 nm (Fig. 1f) inserted between the two phospholipid layers, and a glycoprotein layer exterior. There are, however, clear indications of important differences at the molecular level. First, in both rodents and man, the surface of the tail membrane generally binds lectins with a lower affinity than that of the head (1) but binding is strong in the region surrounding the mitochondria. Also fluorescent probes have been used in order to differentiate human sperm regions: both head and tail membranes bind 8-anilino-1-naphthalenesulphonic acid uniformly, while ethidium bromide (1) stains only the head intensely. These studies indicate that some components of the plasma membrane of the spermatozoon have a non-uniform distribution, and are not able to diffuse freely in the plane of the membrane. This may allow production of antibodies against particular regions of the sperm by using components fractioned by appropriate methods as antigens. In addition, the plasma membrane of the tail interacts with ATP and cyclic AMP (2), and markedly influences the internal metabolism and the regulation of the capacitation of the spermatozoon.

4.2. Organelles in the neck region

In the human spermatozoon, two types of organelles have been described inserted between the flagellar axoneme and the basal lamina that lines the implantation fossa in the nucleus (Scheme 1): a skeleton consisting of the capitulum and nine columns, and the two centrioles.

The capitulum (Fig. 3c, Scheme 1) is a thick, convex plate, composed of compact and strongly osmiophilic material and connected by thin filaments to the basal lamina (7, 29). Extending directly posterior from the capitulum are nine segmented columns (Fig. 3c, Scheme 1), about 0.5 μm in length (30). These columns are paired and subdivided by transverse electron-dense bands about 50 nm in length and separated by short, less dense regions of about 10 nm. These columns are contiguous with the accessory fibers of the axoneme, but have a different origin. Fawcett (7, 29) believes that they may have a chemical resemblance to the satellite centrioles located at the base of cilia. These latter are constituted of a particular protein called 'ankyrin' a dymer of two subunits weighing 230,000 and 250,000 daltons.

There are two centrioles in the spermatid, one called the proximal, oriented perpendicular to the long axis of the nucleus and situated against the capitulum, and a second, more distal one oriented along the head axis. From the former, a short microtubular structure originates, termed the centriolar adjunct; this structure is not conserved in mature human sperm, although the centriole itself persists.

The nine doublets of the axoneme originate from the more distal centriole, after which the latter disappears and is not present in mature sperm, in which only the proximal centriole remains. Both centrioles are constituted of nine triplets of microtubules arranged along the wall of a cylinder about 0.15 μm in diameter and 0.5 μm in length, forming an angle of about 40° with the radius of the cylinder. The three microtubules are called A, B and C, beginning with the most interior one. As in cilia of the tracheal epithelium, centrioles of spermatozoa consist of tubulin (31). Enclosing these centrioles is a radiating system of actin filaments (30). This demonstration impinges on the problem of the initiation of axonemal movement. Even in man, it is thought that the centriole may be the initiator of this movement (7).

4.3. Mitochondria

The section of the tail about 5 μm long and next to the neck, called the middle piece, contains mitochondria (Figs. 2c, 3a) arranged in a spiral sheath that surrounds the axoneme. The mitochondria of human sperm, seen by TEM, have a characteristic structure because each mitochondrion is half-moon shaped and rich in rather tortuous internal cristae. The mitochondria are assembled together into a rather tight helix that completes 12 turns around the axoneme. An elevated cytochrome oxidase activity has been shown by means of the oxidation of diaminobenzidine (32); this is probably due to the presence of cytochrome c in the cristae and cytochrome b_s in the external mitochondrial membrane. This outer mitochondrial membrane consists of five layers in eutherian mammals including man, and becomes strengthened by the formation of S-S cross bridges during maturation of the sperm in the epididymis (14, 15). This stabilization condenses the mitochondrial sheath, which is formed during development of the spermatid by fusion of the isolated mitochondria typical of most cell types. The individual mitochondria of sperm show an unusual resistance to hypotonic conditions. The protein responsible for the strength of the mitochondrial membrane is layered against the outer mitochondrial surface (33) and consists of 3 polypeptide chains of molecular weights around 31,000, 29,000 and 20,000 daltons, with the last being rich in cysteine. These proteins are absent from the mitochondria of other types of cells. The 20,000 dalton protein shows a particular affinity for selenium, which is very abundant in sperm mitochon-

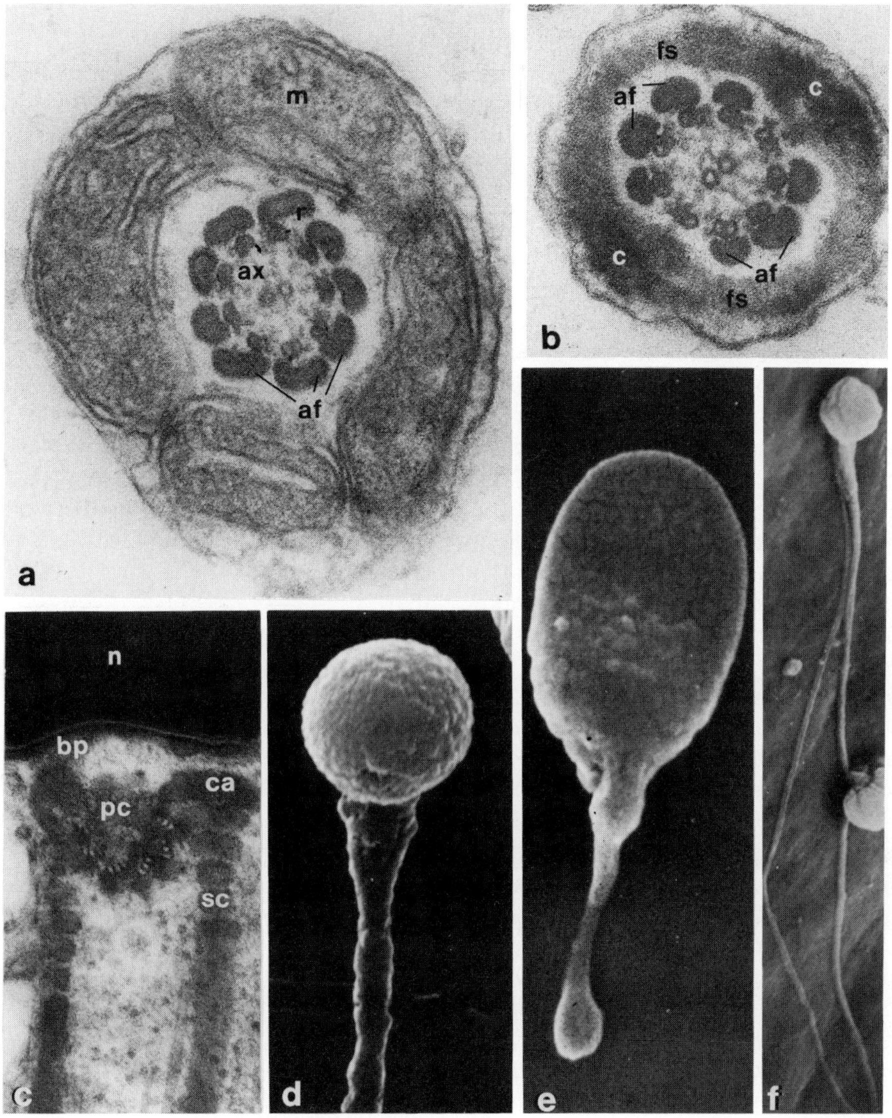

Fig. 3. (a) Transverse section of the median piece. The nine accessory fibers (af) can be seen between the mitochondria (m) and the '9 + 2' axoneme (ax) (× 80,000). (b) Transverse section of the principal piece. The axoneme is surrounded by only 7 accessory fibers (af) which are surrounded by the fibrous sheath (fs) thickened in 2 columns (c) in connection with the missing accessory fibers (× 80,000). (c) Organelles in the neck region, photographed in the transmission electron microscope in the human spermatozoon. Besides the nucleus (n) can be seen the basal plate (bp), the proximal centriole (pc), in which the triplet structure is apparent, a portion of the capitulum (ca) and the striated columns (sc) (× 60,000) (d) 'Round-headed' spermatozoon, scanning micrograph (× 10,400). (e) 'Short tailed' spermatozoon, scanning micrograph (× 11,200). (f) 'Double spermatozoon', scanning micrograph (× 2,800).

dria, and it has been named selenomitin.

Histochemically, it has been shown that glycogen is absent from the midpiece of human sperm, suggesting that this spermatozoon has a metabolism similar to that of sperm of bull and other mammals, in which intracellular phospholipids as well as fructose, polyols and lactic acid derived from the seminal plasma are oxidized aerobically (34). Lactic acid dehydrogenase is located in the mitochondrial matrix and shows a strong affinity for lactate. Lactate is transformed into pyruvate, a substance ultimately oxidized *in situ* (10).

4.4. Structure of the principal piece

The long segment of the tail, 45 μm in length (Fig. 1a), that extends from the midpiece is called the principal piece. Here the sheath of mitochondria is no longer present (Figs. 2c, 3b) and the axoneme is surrounded by a complicated cytoskeleton, placed immediately under the plasma membrane, which extends the length of the principal piece. This cytoskeleton is quite characteristic of mammalian sperm (10) and consists of three elements: the annulus, the fibrous sheath and the two columns of the sheath.

The annulus (Fig. 2c, d) surrounds the axoneme immediately after the last helical turn of the mitochondrial sheath. According to Fawcett (29) it prevents mitochondria displacement and stabilizes the midpiece. It shows a substructure of filaments of about 3–4 nm in diameter and develops in close contact with the plasma membrane. In several mammals it has a triangular cross section, whereas in others it is semicircular, and runs within an invagination of the plasma membrane (29). In man, on the contrary, it is flat in cross section (Fig. 2c), forming a laminar ring of thickness about 10 nm and length about 150 nm. In freeze-etching (Fig. 2d), it has a compact appearance, showing traces of a circular filamentous network. The intramembranous particles are particularly sparse in this region of the membrane.

The *fibrous sheath* in humans (Figs. 2c, 2d, 3b), is a double series of semicircular ribs of fibrous material that enclose the tail on both sides and are joined to the two longitudinal columns (29). In human sperm the ribs of the sheath have a thickness of around 50 nm and are separated from each other by spaces of around 10–20 nm. The protein that constitutes the fibrous sheath is stabilized by S-S bridges formed during traverse of the epididymis (15) and consists of a single chain, of molecular weight around 80,000 daltons (35). Nothing is known about the presence of actin or of specific enzymes in the sheath.

The *two columns of the fibrous sheath* (Fig. 3b) are also structures stabilized by S-S bridges and show a filamentous substructure oriented longitudinally. The columns are rather thick cylindrical structures (around 100 nm in diameter) running longitudinally, and causing an inward thickening of the fibrous sheath in the anterior portion of the principal piece, where they appear fused to accessory fibers numbers 3 and 8. Passing posteriorly they diminish gradually in thickness, and merge into the thickness of the fibrous sheath, which assumes a circular appearance. At this level the accessory fibers disappear, and the connections between the fibrous sheath and axoneme are lost.

Nothing is known about the function of the fibrous sheath and the columns. Certainly they limit the bending of the tail, and probably regulate the form of its beating.

4.5. Axoneme

The axoneme, also called the axial filament, is the portion of the flagellar structure which is capable of active movement. The axoneme is composed of microtubules, fibers and connecting elements. The axoneme of the human spermatozoon contains the typical elements of the classical cilium, arranged according to the well-known '9 + 2' model, as well as nine accessory fibers of substantial thickness.

4.5.1. The structure of the axoneme

The most important elements of the axoneme (Scheme 2) of which nine are present in sperm of most species (10) are the peripheral doublet microtubules. These microtubules are spaced in a regular way, with a distance of 18 nm between them to form a cylinder 0.18 μm diameter. In the center of this cylinder are two singlet microtubules, spaced 9 nm apart.

Each doublet consists of two microtubules. One is termed the A tubule, and is a complete cylinder 20 nm in diameter with a wall thickness of 5.5–8.5 nm. The other, termed the B tubule, shares a part of the wall of the A tubule, and is therefore incomplete, and a little bit wider (20–23 nm). Both tubules are composed of a monolayer of parallel protofilaments,

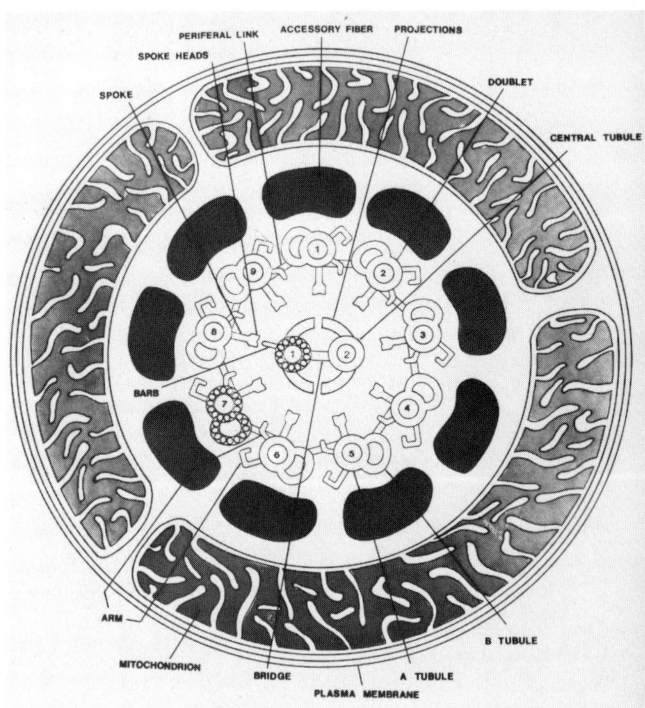

Scheme 2. Schematic diagram of the human sperm axoneme in the region of the mid-piece, in which are present the mitochondrial sheath and accessory fibers.

13 for the A tubule and 10 for the (incomplete) B tubule. Two rows of appendages extend from the A tubule of each doublet toward the B tubule of the adjacent doublet. In cross section these appendages have the form of two arms, an outer one about 35 nm long and an inner one about 20 nm, both turned centripetally. When the axoneme is viewed from the base to the tip, the arms are directed clockwise. This rotation has allowed the doublets to be numbered, with the number 1 being assigned to the doublet that lies in the plane perpendicular to that containing the two central tubules. Successive doublets are numbered progressively according to the direction of the arms.

In addition to the arms, each doublet has a row of radial spokes (Figs. 3a, b, Scheme 2) which extend centripitally from the A-tubule. These spokes are about 35 nm long and terminate in a thickened portion, termed the spoke head, in the region of the central tubules. The spokes on an individual doublet are spaced at an interval of about 27 nm (29). In the axonemal complex the spokes constitute a double helix, with a period of about 90 nm and a pitch of 180 nm according to measurements made on other flagella (36). Olson and Linck (37) revised this measurement, suggesting that the spokes are spaced in groups of three about 32 nm apart, and that within each group spoke 1 is 32 nm from spoke 2, and spoke 2 is 24 nm from 3. The spoke heads also are arranged in a helix around the two central tubules (29) and do not measure more than 20 nm in thickness. In general, in flagella of spermatozoa, two other components have been described emerging from the doublets. The first of these components has been termed the 'Y-links', which originate at the point where the B tubule joins the A tubule, and are attached to the plasma membrane. The second of these components has been termed the 'peripheral links', which join each doublet to the adjacent one passing in the vicinity of the inner arm. Neither of these two types of links has been demonstrated in human sperm, but their presence is probable because they are present in other mammals.

The two central tubules are about 20 nm in diameter, with a wall about 7 nm thick. The wall contains 13 protofilaments lying side by side in a single layer. In cilia of a mollusc, Elliptio, two rows of protrusions are clearly evident along each of the two microtubules. The protrusions are aligned parallel, with the microtubules and are about 18 nm long and 15 nm apart, inclined at about 10° from the tubule, and point toward the basal part of the flagellum on one tubule and distally on the other (36). In human sperm the presence of these protrusions is more difficult to demonstrate but it is probable that the helical structure of 15 nm pitch, described under the term 'central sheath' by earlier authors and to which the spoke heads were reported to be attached (29) is to be interpreted in this sense (10). The two central tubules are, moreover, connected to one another by a system of filamentous bridges spaced longitudinally 16 nm apart (37). Finally, a row of 'barbs', each spaced 32 nm apart protrude from one of the two central microtubules toward doublet 8 (37).

Outside the cylinder of nine doublets, the nine accessory fibers extend the length of the midpiece and the principal piece of the tail (Figs. 3, a, b, Scheme 2). These are large electron-opaque structures, kidney-shaped in transverse section. In the midpiece, where the fibers are thickest, they measure approximately 30 × 100 nm. Seen from the side, they show a banding with a period of about 16 nm with three intraperiod lines (38). According to Baccetti, Pallini and Burrini (39) this period is longer, and two successive periods may be paired. Anteriorly, the fibers are bound to the columns in the neck region; posteriorly they taper gradually until they disappear. The tail has, therefore, an axonemal pattern of 9 + 9 + 2 in the two anterior segments, the mid- and principal pieces, 9 + 0 in the posterior end of the latter, and in a terminal piece at least 5 μm long. Although the doublets originate from the distal centriole during development of the spermatid, the accessory fibers arise as protrusions of the doublets, from which subsequently they detach themselves (40).

4.5.2. *The proteins of the axoneme*

Important information on the nature of the various axonemal components, and hence on the origin of the waves that characterize flagellar movement, has been obtained in recent years. Little of this information derives from human sperm, and much of it results from studies on invertebrate sperm. Nevertheless it can be generalized to human sperm (7).

Tubulin. The most important protein of the flagellum is tubulin (41). Tubulin is an essential constituent of microtubules, both those of the doublets and of the two centrals. It has a globular form of 4 nm that represents the elementary protein subunit, arranged in dimers of 3.5 × 8 nm.

Tubulin is a heterodimer, the properties of which recall those of actin in some respects; for example, both are rich in glutamic and aspartic acids, in hydrophobic amino acids and in free SH groups, and both interact with ATPase proteins, but an important difference is that tubulin contains two molecules of guanylic nucleotide (rather than adenylic as in actin) associated with each dimer. No investigations have been carried out on the tubulin of human spermatozoa, but its properties appear constant, from the flagella of spermatozoa of many different invertebrates to the cilia of the molluscs. Thus, it can be expected that the characteristics of the tubulin from human sperm will not be an exception to the rule.

Dynein. The second protein found in sperm flagella is dynein (42). Gibbons and his associates have shown that, after solubilization of the flagellar ATP-ase, the arms on the doublet tubules of protozoan cilia disappear, and have named the protein that is extracted 'dynein'. A large series of studies which followed have demonstrated that the ATPase activity of dynein is responsible for the flagellar movement of spermatozoa, and that flagella devoid of arms (whether experimentally or pathologically) are immotile. Dynein is a large protein molecule sedimenting at 14 S, with a molecular weight of more than 500,000 daltons and dimensions of 8 × 14 nm. Its enzymatic activity is specific for ATP as deoxy ATP, is activated by Mg^{++}, Ca^{++}, Mn^{++}, Co^{++}, Ni^{++} and K and inhibited by Cd^{++}, Zn^{++}, Hg^{++} and by sulphydryl reagents. There exists, however, different forms of dynein in the axonemes of different cell types. In the axoneme of a single species (sea urchin), it is now possible to distinguish electrophoretically at least 8 polypeptides of high molecular weight (43), forming at least two isoenzymic forms of dynein (dynein 1 and dynein 2). Dynein 1, in turn, can assume two forms. The first, termed the 'latent' form (44) has a low ATPase activity, a total molecular weight of 1.23 × 106 daltons, and can be recombined under appropriate conditions of pH and ionic strength, with the outer doublets to reconstitute perfectly functioning arms. The second, is obtained by various treatment (heat, Triton X-100, p-chloromercuribenzoate) from the latent dynein 1. This form has a higher ATPase activity, lower molecular weight and lacks the capability to recombine with the outer doublets to reform arms. Some data exist regarding the ATPases of mammalian spermatozoa and especially for those of man. Early studies (45) have shown the presence of two ATPases, one sensitive to Na^+ and K^+ and other to divalent cations, but their localization remains to be determined. Subsequent research (46) and unpublished work by Gibbons and Pallini also showed that mammalian ATPases are related to sea urchin ATPases described earlier (43) and that, even in man, there are at least four electrophoretic bands in the dynein heavy chain region (Fig. 4a). The ATPase activity is localized both at the level of the arms on the doublets and at the junction between the radial spoke heads and the protrusions of the central tubules. It is difficult, however, to relate this localization to the 4 electrophoretic bands (46).

The protein of the spokes and central sheath. The central sheath region of the axoneme studied histochemically appears to have ATPase activity and presumably contains at least one dynein. In this region there are, however, other proteins, probably structural in nature. Research involving flagellar mutants of *Chlamydomonas* that lack either entire spokes or just spoke heads, have shown that the spokes contain 6 polypeptides and the spoke heads another 6, not including dynein (47) and, that the major component of the spokes has a molecular weight of 118,000 daltons, while the sheath contains 3 polypeptides with the molecular weight of the largest 200,000 daltons (48).

Nexin. This protein constitutes the peripheral links and was described by Stephens (49) in the sperm of several invertebrates. It represents about 2% of the total of the axonemal protein and has a molecular weight of 150,000–165,000 daltons. No further details are known, nor anything published regarding the characteristics of this protein in human sperm.

Parergins. These proteins termed 'parergin' (39) constitute the accessory fibers of the spermatozoon. Isolated first from bull and rat they have been studied also in man, in other mammals and in molluscs (39). There are two classes of polypeptides. The first class contains those chains of moderately high molecular weight (80,000 and 50,000 daltons) with a high content of leucine and glutamic acid and low in cysteine and proline, with an infrared spectrum indicating an alphahelical configuration formed of two filaments 2 nm thick, visible in the electron microscope. The second class contains those chains of low molecular weight (28,000 and 31,000 daltons), low in leucine and glutamic acid content, high in cysteine and proline, forming the amorphous part of the fiber, capable of binding zinc by reaction with sulphydryl groups (39, 50). In the intact fiber the two types of chains are cross-linked by disulfide bridges. The lateral packing of the filaments agrees with the period observed with the electron microscope. All these characteristics suggest an elastic role for the parergins, and therefore for the accessory fibers. This elasticity may be regulated by disulfide exchanges (39, 50). High levels of triglycerides and small quantities of carbohydrates, present in the mature fibers of mammals, are bound to the parergins.

The presence of ATPase activity in the accessory fibers, although reported histochemically in sperm of various mammals, now appears unlikely. For an extended discussion of this question, see (7) and (10). The high affinity of the fibers for Zn^{++} suggests that this ion may play a role in regulating the number of -S-S- cross bridges. Reports suggest that the substantial quantity of Zn^{2+} present in seminal fluid influences the motility of human sperm.

Calmodulin. Calmodulin is a low molecular weight, heat-stable protein that binds Ca^{2+} and functions as a regulator of many Ca^{2+} dependent enzymes, including dynein. It has been found in the flagella of Chlamydomonas and in the spermatozoa of sea urchin (51, 52). It seems highly probable that it is also present in human spermatozoa, where it may be implicated in the acrosome reaction.

5. Movement of sperm

As in all the mammals, the spermatozoa of man have a three dimensional flagellar movement, consisting of a series of waves that are initiated at the base of the flagellum and are propagated along it in the form of a flattened helix. There is thus a torsional motion that accompanies the wave of curvature: the human sperm rotates by 180° with each beat (53). After cold shock, only a planar motion remains. Progression and speed vary with the pH. Prostate secretion favorably influences motility, yet that of the seminal vesicle depresses it. Velocity is normally 15–20 μm per second. The wavelength differs in sperm of different mammalian species and apparently depends on the thickness of the accessory fibers of the flagellum, which also varies between species (53): in human sperm, which have thin fibers, the wavelength in about 25 μm. This wavelength is significantly less than in hamster sperm, which have relatively thick fibers. In human sperm the motility is reported optimal at 20°, and decreases progressively at temperature up to 37° (54).

After many years of discussion in which attempts were made to develop a model of flagellar mobility based closely upon muscular contraction, the existence of a 'sliding filament' mechanism as proposed by Afzelius in 1959 has been demonstrated in many types of flagella. The most direct evidence has been furnished by demonstrating that ATP is necessary for motility, and that in both echinoderms and mammals the sliding movements are generated by the interaction of the dynein arms with ATP (55). This interaction forms transitory cross bridges between the tubulin subunits of adjacent doublet tubules. The peripheral links, the spokes and the Y-links together constitute a resistive network that, by streaching or by periodic detachment and reattachment during the active movement, governs the precise curved form of the flagellum (36).

A remaining problem is that of the initiation of movement. On this process there is little information available, and no data whatsoever in humans. For a general discussion of this problem see (10).

6. Anomalous spermatozoa

Since the examination of ejaculates by light microscopy became a technically simple matter, the disconcerting conclusion has emerged that man is the species that has the greatest percentage of malformed, anomalous and immotile spermatozoa, even in samples from healthy individuals of proven fertility. A possible explanation for this is that the temperature of the scrotum in which spermatogenesis occurs is raised above its natural physiological level as result of wearing clothing. It has been shown that testicles of rams similarly warmed also begin to yield somewhat aberrant spermatozoa (3, 10). In addition, Appel and Evans (54) have shown that the vitality of human sperm decreases progressively from 20° to 37° C.

The ejaculates of many individuals show a diminished number of spermatozoa. It is usually assumed that normally fertile men have at least 10^7 sperm per ml or 2.5×10^7 per ejaculate. Sperm anomalies can have many causes, many are pathological, but there are also a significant number of known genetic defects that affect the entire sperm population of an individual.

In all cases, a selection process occurs in the female genital tract, particularly as a result of the barrier that the cervical secretions present to the passage of abnormal sperm. However, it appears that there remains a substantial paternal contribution to embryonic mortality, for at least some abnormal spermatozoa are capable of fertilization. Detailed studies of human sperm types may thus lead to the recognition and possible separation of the abnormal from the healthy spermatozoa. Such studies are also fundamental to cases of infertility due to malformed spermatozoa, in as much as they can reveal the extent, type, severity and, in many cases, the origin of the abnormality. Such research has begun only recently, and the relevant literature is still sparse. The known anomalies of human sperm can involve any region of the cell, and can also involve a complete malformation such as incomplete maturation. However, some description of the most characteristic and easily recognizable forms is possible (3).

6.1. Anomalies of the acrosome

6.1.1. Malformed acrosome

This is a very common defect that can affect almost all the sperm in infertile individuals. It usually consists of an abnormal thickening of the organelle, which appears full of vacuoles (56, 57). In addition, the acrosome may be malformed, concave rather than convex, or partially embedded in the nucleus. Such defects are recognizable in developing spermatids and can be detected by biopsy (57). In certain cases, it has been shown that this defect affects just half of the spermatids and so is the result of recessive mutation that becomes expressed phenotypically in the haploid phase. However, all of the spermatids are affected in cases of obstructive azospermia and the anomalies have been interpreted as abnormal differentiation rather than degeneration.

6.1.2. Absence of the acrosome

This anomaly has often been observed in all the spermatozoa of particular individuals in which study of development of spermatids has shown the pro-acrosomal material formed in the Golgi apparatus to become enclosed in large vacuoles that degenerate into multilamellar structures, or become engulfed by ad destroyed within the nucleus itself (Baccetti et al., 1977). The latter is the case for the 'round-headed' spermatozoa. These spermatozoa are characterized also by the abnormal shape of the nucleus, in addition to the total absence of the acrosome and the post-acrosomal lamina (58). During spermatogenesis, it seems that part of the acrosomal material escapes from within the acrosomal vesicle (59).

6.2. Abnormal nucleus

6.2.1. 'Round-headed spermatozoa'

These have been observed by numerous investigations, in both ejaculates and in biopsies (58 and, for a review, 59). They are the most frequent and best described abnormality. Having normal motility but lacking acrosomes, the spermatozoa (Fig. 3d) are characterized by a spherical rather than the normal elongated nucleus, composed of less dense material with numerous vacuoles. This defect can be detected in spermatids by the absence of the microtubular manchette around the nucleus (59) and, during subsequent maturation, the nucleus retains a more or less 'immature' form in almost all the spermatozoa. In fact, it has been demonstrated that the content of phosphorous is diminished, the content of zinc somewhat elevated (indicative of fewer – S-S-bridges), and the amount of lysine in the nuclei remains high, presumably indicating that the substition of the somatic histones by arginine-rich histones has not occurred (59). The genetic basis of this anomaly has been confirmed by Nistal et al. (60) who have described its occurrence in two brothers.

6.2.2. Immature nucleus

This abnormality is probably the most widespread. It occurs even in the sperm of ejaculated of fertile individuals, in which a certain number of immature sperm are always present. However, in some cases almost all the spermatozoa are immature, in which case the individual is infertile. Viewed in the electron microscope, the immature nucleus is recognizable by the persistent granulosity of the chromatin, which never becomes compacted, and by the presence of a greater number of vacuoles than in normal sperm nuclei (56, 61, 62). This defect also has a morphogenetic basis, but it is difficult to recognize in biopsies, where it can be detected only by the absence of condensed chromatin. It seems probable that it corresponds to the same chemical condition as that found in 'round-headed' spermatozoa, but this has not yet been confirmed.

6.3. General cytoplasmic anomalies

6.3.1. Overabundant cytoplasm

Almost all human spermatozoa retain a droplet of cytoplasm attached to the flagellum. In some cases, however, this cytoplasm is present in greater than normal quantities and encloses almost all the axoneme. The axoneme assumes a coiled form within the droplet. In such cases, the overall sperm has a globular exterior form and is immotile (56, 62). Even in 'round-headed' spermatozoa, a certain percentage have the axoneme coiled on itself and included in the cytoplasm (59).

6.3.2. Aberrant mitochondrial helix

An anomaly that frequently accompanies those described above is one which affects the configuration of the head or midpiece. In a high percentage of sperm the helix appears disordered, multilayered (56), or with an abnormal number of turns (63). Holstein (57) has described the absence of mitochondria in many spermatids of subfertile individuals, but the defect has never been reported in ejaculated spermatozoa. It seems probable that this type of cell never achieves maturity.

6.4. Axonemal anomalies

6.4.1. Rotation of the head-tail attachment
In this type of malformation, described in numerous African and European individuals (64), the axoneme is inclined up to 160° with respect to the long axis of the head, and in some cases is even completely separate. Although sperm motility is not affected, the individuals are infertile.

6.4.2. Lacking axoneme
This anomaly (Fig. 4b) has been described recently as affecting all the sperm of one individual (65). The sperm tail, which contains only the accessory fibers and the fibrous sheath, appears flaccid and, as might be expected, is immotile. ATPase activity and the dynein electrophoretic bands are completely absent. In addition, cilia of the respiratory epithelium are also affected, and they appear very short and lacking microtubules.

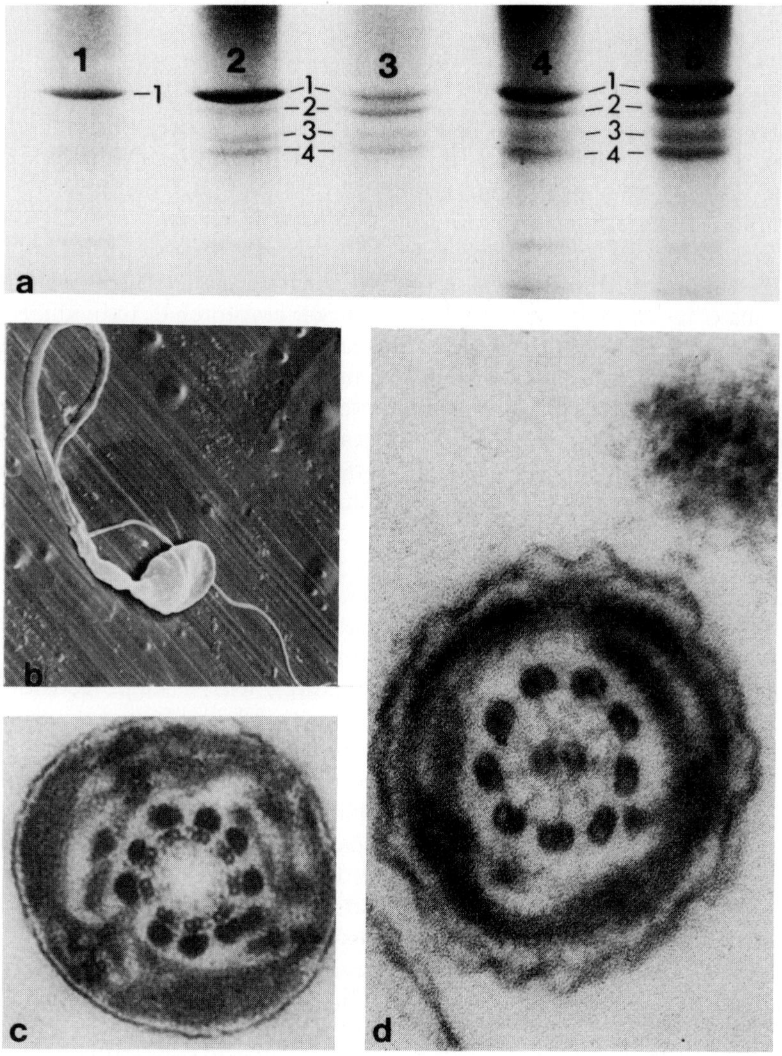

Fig. 4. (a) SDS gel electrophoresis of the high molecular weight polypeptide chains from human spermatozoa. The chains corresponding to the dynein bands have been designated with the numbers 1–4 in the normal spermatozoon (gel 3). Gel 1 contains material obtained from spermatozoa lacking arms in Kartagener's syndrome; gel 2, from '9+0' spermatozoa; gels 4 and 5 show, respectively, spermatozoa lacking arms coelectrophoresed with normal spermatozoa and '9 + 0' spermatozoa coelectrophoresed with normals (from Baccetti, Pallini, Burrini and Renieri, 1981). (b) Spermatozoon lacking the axoneme, photographed with the scanning electron microscope (× 800). (c) Transverse section of the tail of a human '9 + 0' spermatozoon, photographed with the transmission electron microscope (× 60,000). (d) Transverse section of the tail of a human spermatozoon lacking arms, photographed in the transmission microscope (× 120,000).

6.4.3. '9 + 0' and '4 or 5 + 2' axonemes
This anomaly appears to represent an accurate copy of a malformed centriole, and it is present in small numbers even in the ejaculate of normal men (57). Baccetti et al. (66) have recently described an individual in whom all the sperm had a '9 + 0' structure, a lack of the two central tubules, and were immotile (Fig. 4c). These sperm had reduced ATPase activity and lacked electrophoretic band 2 from the dynein heavy chain region (Fig. 4g).

6.4.4. Axoneme with doublets lacking arms
This particular anomaly (Fig. 4d), in which both arms lack from the 9 doublets was discovered simultaneously by Afzelius et al. (68) and by Pedersen and Rebbe (69) in different patients, two of whom were brothers. It has been demonstrated clearly to have a genetic basis and affects all the spermatozoa of the individual. the sperm morphologically appear normal but 'rigid'. The cilia on the respiratory epithelium of affected individuals are also immotile. Baccetti et al. (46) have shown that these sperm lack electrophoretic bands 2, 3 and 4 from the region of the dynein heavy chains (Fig. 4a) and their ATPase activity is greatly reduced. Afzelius and Eliasson (70) have found individuals with sperm lacking only the inner arm or only the outer arm; these sperm were also immotile.

6.4.5. Spermatozoa lacking tails
This anomaly was found to affect 90% of the sperm of one infertile man; it was accompanied by the presence of rounded sperm heads (71).

6.4.6. Spermatozoa with short tails
This anomaly was discovered by Baccetti et al. (72) in two patients and affected all the spermatozoa to some extent. Morphologically these sperm appeared to have short squat tails (Fig. 3e) and were immotile. TEM examination of cross-sections revealed the presence of two imperfectly developed axonemes. The tubules were seldom present as doublets and the dynein arms were always absent, although the accessory fibers were present. The two axonemes were derived from the two centrioles, and thus the anomalous axoneme emerging from the proximal centriole also gained accessory fibers.

6.4.7. Spermatozoa with two or more tails
Sperm with up to 4 tails have been described by Nistal, Paniagua and Herruzo (73) in one infertile patient. The several tails derived from equivalent laminae at the base of a very irregularly shaped nucleus, and were each surrounded by a normal complements of accessory fibers and mitochondrial helix. Motility was almost absent.

6.4.8. Anomalies in accessory axonemal structures
Ross, Christie and Kerr (63) have noted in one infertile man the consistent absence of the fibrous sheath and accessory fibers. They also found, in two oligospermic men, excessive development of the fibrous sheath enclosing the axoneme in several layers. This development was noted in addition to the absence or the presence of an excess number of accessory fibers. In all cases, these defects affected a high percentage of the spermatozoa and rendered them immotile. This anomaly appears to be rare, and, from the literature, seems never to occur in sperm of normal individuals, or in association with other defects. It provides justification for believing that the accessory axonemal structures play an important role in flagellar motility.

6.5. Double sperm

In normal ejaculated, or in biopsied testicular material of cryptorchid individuals, occasional double spermatids are observed in which the two nuclei and much of the tails are enclosed within a single membrane. However, in all these cases, the anomaly did not affect mature spermatozoa nor have a substantial effect on fertility. Uniformly doubled (Fig. 3f) sperm (in rare instances tripled or quadrupled) have been found in ejaculate of an individual with a hypophysial tumor that secretes only prolactin (17). These sperm were mature and motile, and the individual had conceived a son after the manifestation of the tumour. The sperm morphology appeared perfect (Fig. 2e), with a single acrosome covering the two nuclei that had adopted a normal nuclear shape but of double size. The mitochondria, slightly disordered but in a uniform helix, surrounded the two flagella that moved freely within a single membrane. These spermatozoa swam efficiently, as do the numerous biflagellate spermatozoa of invertebrates, a little like an octopus pushed by its own tentacles. When the prolactin level was brought back to within normal levels, the anomalous sperm structure disappeared. This is the single case noted so far of a definite correlation between a hormonal dysfunction and the appearance of a particular morphological defect.

7. Conclusion

From this rapid review it appears that human spermatology may be understood, thus overcoming the usual claim that the typical high cytological variability in normal ejaculates disturbs the recognition of definite malformations associated with single genetic, pathological defects. Most of the results obtained are due to the correlation of electron microscopy with biochemistry, and all the interpretations have been reached only when research carried out on the relatively more simple invertebrate models clarified the general mechanisms of cell recognition, acrosome reaction, and flagellar movement.

8. Summary

In this paper the present status of knowledge concerning the human spermatozoon is reviewed. This cell belongs to the general type present in the higher mammals, but differs in many respects from that of the better known domestic animals, like rodents, bull, ram, boar. The general morphology of the human mature sperm is described and summarized for the different regions (head, neck, tail) in which the cell can be divided. For each component, the chemical structure when known, is described, and a general morpho-functional picture of the human sperm is obtained consequently. A section on human sperm pathology is included, where the best known malformations are described.

References

1. Edelman GM, Millette CF: Chemical dissection and surface mapping of spermatozoa. In: The functional Anatomy of the spermatozoon. Afzelius BA (ed), Oxford, Pergamon Press, 1975, pp 349–357.
2. Mercado E, Hicks JJ, Drago C, Rosado A: A study of the interaction of human spermatozoa membrane with ATP and cyclic-AMP. Biochem Biophys Res Communic 56: 185–192, 1974.
3. Baccetti B: Lo spermatozoo umano. In: I° Congresso Nazionale Società Italia Andrologia, Relazioni. Pacini (ed) 1978, pp 123–167.
4. Koehler JK: The mammalian sperm surface: studies with specific labeling techniques. Int Rev Cytol 54: 73–108, 1978.
5. Gordon M, Dandekar PV: Fine structural localization of phosphatase activity on the plasma membrane of the rabbit sperm head. J Reprod Fert 49: 155–156, 1977.
6. Pedersen H: The postacrosomal region of the spermatozoa of man and Macaca arctoides. J Ultrastr Res 40: 366–377, 1972a.
7. Fawcett DW: The mammalian spermatozoon. Developt Biol 44: 394–436, 1975.
8. Koehler JK: Human sperm head ultrastructure: a freeze-etching study. J Ultrastr Res 39: 520–539, 1972.
9. Carothers AD, Beatty AA: The recognition and incidence of haploid and polyploid spermatozoa in man, rabbit and mouse. J Reprod Fert 44: 487–500, 1975.
10. Baccetti B, Afzelius BA: The biology of the sperm cell. Monographs in developmental biology, 10. Basel, S Karger, 1976.
11. Sarkar S, Jones OW, Shioura N: Constancy in human sperm DNA. Proc Nat Acad Sci USA 71: 3512–3516, 1974.
12. Summer AT, Robinson JA, Evans HJ: Distinguishing between X, Y and YY bearing human spermatozoa by fluorescence and DNA content. Nature New Biol 229: 231–233, 1971.
13. Kolk AHJ, Samuel T: Isolation, chemical and immunological characterization of two strongly basic nuclear proteins from human spermatozoa. Biochim Biophys Acta 393: 307–319, 1975.
14. Bedford J, Calvin HI: The occurrence and possible functional significance of -S-S crosslinks in sperm heads, with particular reference to Eutherian mammals. J Exp Zool 188: 137–156, 1974.
15. Bedford JM, Calvin HI, Cooler GW: The maturation of spermatozoa in the human epididymis. J Reprod Fert Suppl 18: 199–213, 1973.
16. Bearden J, Bendet IJ: Birefringence of spermatozoa. I. Birefringence melting of Squid, Bull, and Human sperm nucleoprotein. J Cell Biol 55: 489–500, 1972.
17. Baccetti B, Fraioli F, Paolucci D, Selmi G, Spera G, Renieri T: Double spermatozoa in a hyperprolactinemic man. J Submicr Cytol 10: 240–260, 1978.
18. Roomans GM, Afzelius BA: Acrosome vesiculation in the human sperm. J Submicrosc Cytol 7: 61–69, 1975.
19. Roomans GM: Calcium binding to the acrosomal membrane of human spermatozoa. Exp Cell Res 96: 23–30, 1975.
20. Baccetti B: The evolution of the acrosomal complex. In: The Spermatozoon. Fawcett DW, Bedford JM (eds) Baltimore-Munich, Urban & Schwarzenberg, 1979, pp 305–329.
21. McRorie RA, Williams WL: Biochemistry of mammalian fertilization. Ann Rev Biochem 43: 777–803, 1974.
22. Bhattacharyya AK, Zaneveld LJD: Kinetic studies on the interaction and specificity of synthetic proteinase inhibitors towards human acrosin. Andrologia 8, suppl 1: 119, 1976.
23. Goodpasture JC, Polakoski KL, Zaneveld LJD: Acrosin, proacrosin and acrosin inhibitor of human spermatozoa: extraction, quantitation, and stability. J Andrology 1: 16–27, 1980.
24. Koren E, Milković S: Collagenase like peptidase in human, rat and bull spermatozoa. J Reprod Fert 32: 319–356, 1973.
25. Bhattacharyya AK, Zaneveld LJD: Release of acrosin inhibitor from human spermatozoa. Fertil Steril 30: 70–78, 1978.
26. Reddy JM, Joyce C, Zaneveld LJD: Role of hyaluronidase in fertilization: the antifertility activity of myocrisin a nontoxic hyaluronidase inhibitor. J Andrology 1: 28–32, 1980.
27. Clarke GN, Yanagimachi R: Actin in mammalian sperm heads. J exp Zool 205: 125–132, 1978.
28. Campanella C, Gabbiani G, Baccetti B, Burrini AG, Pallini V: Actin and myosin in the vertebrate acrosomal region. J Submicr Cytol 11: 53–71, 1979.
29. Fawcett DW: A comparative view of sperm ultrastructure. Biol Reprod 2: 90–127, 1970.
30. Pedersen H: Observations on the axial filament complex of the human spermatozoon. J Ultrastr Res 33: 451–462, 1970.
31. Burrini AG, Baccetti B, Campanella C, Runger-Brändle E, Gabbiani G: Pericentriolar actin in spermatozoa. J Submicr Cytol 12: 161–164, 1980.
32. Novikoff PM, Cohen J, Novikoff AB, Davis C: Cytochemical visualization of the midpiece of ejaculated human spermatozoa. J Microsc Paris 11: 169–174, 1971.
33. Pallini V, Baccetti B, Burrini AG: A peculiar cysteine-rich polypeptide related to some unusual properties of mammalian sperm mitochondria. In: The Spermatozoon. Fawcett DW, Bedford JM (eds) Baltimore-Munich, Urban & Schwarzenberg, 1979, pp 141–152.
34. Mann T: Sperm metabolism. In: Fertilization. Metz C, Monroy A (eds), New York, Academic Press, 1967, pp 99–116.
35. Olson GE, Hamilton DW, Fawcett DW: Isolation and characterization of the fibrous sheath of rat epididymal spermatozoa. Biol of Reprod 14: 517–530, 1976.
36. Warner FD, Satir P: The structural basis of ciliary bend formation. Radial spoke positional changes accompanying microtubule sliding. J Cell Biol 63: 35–63, 1974.
37. Olson GE, Linck RW: Observations of the structural components of flagellar axonemes and central pair microtubules from rat sperm. J Ultrastr Res 61: 21–43, 1977.
38. Pedersen H: Further observations on the fine structure of the human

spermatozoon. Z Zellforsch Mikrosk Anat 123: 305–315, 1972.
39. Baccetti B, Pallini V, Burrini AG: The accessory fibers of the sperm tail. III. High sulfur and low sulfur components in Mammals and Cephalopods. J Ultrastr Res 57: 289–308, 1976.
40. Fawcett DW, Phillips DM: Recent observations on the ultrastructure and development of the mammalian spermatozoon. In: Comparative Spermatology. Baccetti B (ed) New York, Academic Press 1970, pp 13–28.
41. Mohri H: Amino acid composition of 'tubulin' constituting microtubules of sperm flagella. Nature London 217: 1053–1054, 1968.
42. Gibbons IR, Rowe AJ: Dynein: a protein with adenosine triphosphatase activity from cilia. Science NY 149: 424–425, 1965.
43. Bell CW, Fronk E, Gibbons IR: Polypeptide subunits of dynein 1 from sea urchin sperm flagella. J Supramol Struct 11: 311–317, 1979.
44. Gibbons IR, Fronk E: A latent adenosine triphosphatase form of dynein 1 from sea urchin sperm flagella. J Biol Chem 254: 187–196, 1979.
45. Abla A, Mroueh A, Durr IF: The hydrolysis of ATP by the spermatozoa of man. Int J Biochem 5: 787–790, 1974.
46. Baccetti B, Burrini AG, Pallini V, Renieri T: Human Dynein and sperm pathology. J Cell Biol 88: 102–107, 1981.
47. Piperno G, Huang B, Luck DJL: Two dimensional analysis of flagellar proteins from wild-type and paralyzed mutants of Chlamydomonas reinhardtii. Proc nat Acad Sci 74: 1600–1604, 1977.
48. Witman GB, Plummer I, Sander G: Chlamydomonas flagellar mutants lacking radial spokes and central tubules. J Cell Biol 76: 729–747, 1978.
49. Stephens RE: Enzymatic and structural proteins of the axoneme. In: Cilia and Flagella. New York and London, Academic Press 1974, pp 39–76.
50. Calvin HI, Yu CC, Bedford JM: Effects of epididymal maturation, zinc (II) and copper (II) on the reactive sulfhydryl content of structural elements in rat spermatozoa. Exp Cell Res 81: 333–341, 1973.
51. Garbers DL, Hansbrough JR, Radany EW, Hyne RV, Kopf GS: Purification and characterization of calmodulin from sea urchin spermatozoa. J Reprod Fert 59: 377–381, 1980.
52. Jones HP, Bradford MM, McRorie RA, Cormier MJ: High levels of a calcium-dependent modulator protein in spermatozoa and its similarity to brain modulator protein. Biochem Biophys Res Commun 82: 1264–1272, 1978.
53. Phillips DM: Comparative analysis of mammalian sperm motility. J Cell Biol 53: 561–573, 1972b.
54. Appel RA, Evans PR: The effect of temperature on sperm motility and viability. Fertil Steril 28: 1329–1332, 1977.
55. Gibbons IR: Mechanisms of flagellar motility. In: The functional Anatomy of the spermatozoon. Afzelius BA (ed) Oxford, Pergamon Press 1975, pp 127–140.
56. Renieri T: Submicroscopical observations on abnormal human spermatozoa. J Submicr Cytol 6: 421–432, 1974.
57. Holstein AF: Morphologische Studien an abnormen Spermatiden und Spermatozoen des Menschen. Virschows Arch Path Anat Histol 367: 93–112, 1975.
58. Schirren CG, Holstein AF, Schirren C: Ueber die Morphogenese rundköpfiger Spermatozoen des Menschen. Andrologie 3: 117–125, 1971.
59. Baccetti B, Renieri T, Rosati F, Selmi MG, Casanova S: Further observations on the morphogenesis of the round headed human spermatozoa. Andrologia 9: 255–264, 1977.
60. Nistal M, Herruzo A, Sanchez-Corral F: Teratozoospermia absoluta de presentacion familiar. Espermatozoides microcéfalos irregulares sin acrosoma. Andrologia 10: 234–240, 1978.
61. Ross A, Christie S, Edmond P: Ultrastructural tail defects in the spermatozoa from two men attending a subfertility clinic. J Reprod Fert 32: 243–251, 1973.
62. Lacy D, Pettitt AJ, Martin BS: Application of scanning electron microscopy to semen analysis of subfertile man utilizing data obtained by transmission electron microscopy as an aid to interpretation. Micron 5: 135–173, 1974.
63. Ross A, Christie S, Kerr MG: An electron microscope study of a tail abnormality in spermatozoa from a subfertile man. J Reprod Fert 24: 99–103, 1971.
64. Luders G: Ein Defekt der Kopf-Schwanz-Verknüpfung bei menschlichen Spermatozoen. Andrologia 8: 365–368, 1976.
65. Baccetti B, Burrini AG, Pallini V: Spermatozoa and cilia lacking axoneme in an infertile man. Andrologia 12: 525–532, 1980b.
66. Baccetti B, Burrini AG, Maver A, Pallini V, Renieri T: '9 + 0' immotile spermatozoa in an infertile man. Andrologia 11: 437–443, 1979b.
67. Afzelius BA, Eliasson R, Johnsen O, Lindholmer C: Lack of dynein arms in immotile human spermatozoa. J Cell Biol 66: 225–232, 1975.
68. Pedersen H, Rebbe H: Absence of arms on the axoneme of immobile human spermatozoa. Biol of Reprod 12: 541–544, 1975.
69. Afzelius BA, Eliasson R: Flagellar mutants in man: on the heterogeneity of the immotile cilia syndrome. J Ultrastr Res 69: 43–52, 1979.
70. Aughey E, Orr PS: An unusual abnormality of human spermatozoa. J Reprod Fertil 53: 341–342, 1978.
71. Baccetti B, Burrini AG, Pallini V, Renieri T, Rosati F, Menchini Fabris GF: The short tailed human spermatozoa. Ultrastructural alterations and dynein absence. J Submicr Cytol 7: 349–359, 1975.
72. Nistal M, Paniagua R, Herruzo A: Multi-tailed spermatozoa in a case with asthenospermia and teratospermia. Virchows Arch B Cell Path 26: 111–118, 1977.

Author's address:
Institute of Zoology,
University of Siena,
Via Mattioli, 4
53100 Siena, Italy

CHAPTER 11

Structure and evolution of the nucleolus during oogenesis

ANTONIO COIMBRA and CARLOS AZEVEDO

1. Introduction

To a large extent, the morphological organization of nucleoli reflects the degree of ribosomal RNA (rRNA) synthesis taking place in cells. Oocytes are cells in which ribosome formation is extremely active, but only at specific stages of oogenesis. Nucleoli of developing oocytes offer a wealth of morphological information related to the physiological state of the cell. Correlations of nucleolar structure and function during oogenesis can be made on the basis of new formation, in large numbers, intense growth and numerous morphological transformations.

Nucleoli arising in great numbers at the diplotene stage in some amphibia (1), for example, are entirely fibrillar (Fig. 1, E) at the end of the previtellogenic period when rRNA synthesis is lacking, and turn bipartite at the ensuing vitellogenic stage (Fig. 1, F) as soon as ribosome formation starts (1). These bipartite nucleolis have a core fibrillar component which contains the transcribing rDNA molecules (2) and a periphery made up of the granular component in which the immediate precursors of ribosomal RNA accumulate; these nucleoli have also been called spheroidal (3). On the other hand, in the more elaborate, newly formed nucleoli arising at the pachytene stage of meiosis in quail oocytes (4), which also have a bipartite configuration, the fibrillar core contains an electron-lucent region, the fibrillar center (Fig. 11), which corresponds to the nucleolar organizer region (NOR) of the nucleolar chromosomes (5). This aspect of nucleolar organization occurs at a period of incipient transcription but is followed at the diplotene stage (4), during which nucleolar activity is intense (6), by a reticulated organization with the nucleolar mass consisting of the anastomosing strands of the nucleolonema formed by adjacent fibrillar and granular areas. This situation is comparable to that observed in somatic cells with high rates of rRNA synthesis, such as the crypt base cells of rat jejunum which also have reticulated nucleoli (7). Likewise, preceding germinal vesicle breakdown, the largely inactive nucleoli of rodents and pig oocytes become compact, mostly formed by a dense mass of fibrils (8, 9), as in mid-villar cells of the rat jejunum where minimal rRNA synthesis occurs (7). Thus, oocytes are particularly useful for studying correlations between nucleolar form and function. The structural components of nucleoli, however, are identical to those which have been described in

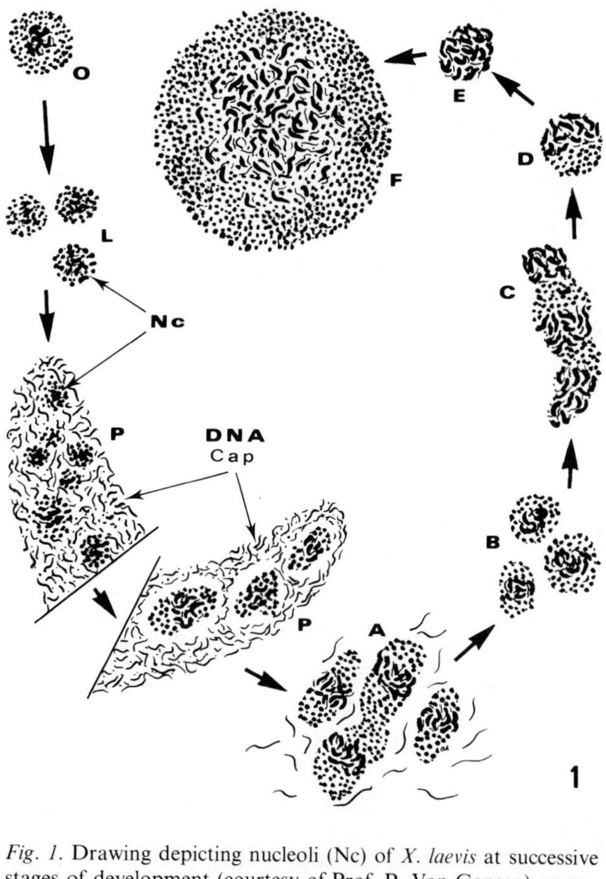

Fig. 1. Drawing depicting nucleoli (Nc) of *X. laevis* at successive stages of development (courtesy of Prof. P. Van Gansen), represented at the same scale in oogonia (O), leptotene (L), pachytene (P, early cap and late cap), diplotene stage, previtellogenic oocytes (A, B, C, D, E) and diplotene stage vitellogenic oocytes (F).

Van Blerkom, J. and Motta, P.M. (eds.), Ultrastructure of reproduction. ISBN 0-89838-572-5.

somatic cells (10, for details on general cytology and functions of the nucleolus). Peculiar to oocytes of certain lower vertebrates and invertebrates are DNA bodies (or caps) which contain extrachromosomal copies of the ribosomal DNA cistrons of the nucleolar organizer (DNA amplification) from which multiple new nucleoli arise. Amplification of DNA will be dealt with extensively in this chapter.

The evolution of the nucleolus will be described in different animal groups, including amphibia, insects and mammals, giving particular attention to those which have been well characterised. The sequence of events in nucleolar evolution throughout the successive stages of meiotic prophase is very similar in the various groups.

2. Premeiotic and early meiotic stages

2.1. Amphibia

Electron microscope (EM) observations show that primordial germ cells and gonia of *Xenopus laevis* (Fig. 1, O) contain 1 or 2 large bipartite, fibrillo-granular nucleoli (1, 11, 12). This finding is in agreement with the general rule in amphibia that one nucleolus is present per haploid chromosomal set (13). Diploid cells therefore should have two nucleoli, or one in the case of occasional fusion. The presence of several extra- or micronucleoli has also been noted in electron micrographs of oogonia (12). Squash preparations are however preferable to determine the exact number of nucleoli. Most often, 4-5 nucleoli per nucleus are counted in *Xenopus* oogonia and, occasionally, sometimes as many as 9 per nucleus have been reported (14), but it is not possible to distinguish between primary and extranucleoli with this method. Such nucleoli, on the other hand, have been shown to be covered by silver grains in autoradiographs after *in situ* hybridization with radioactive rRNA (15). After comparing the grain densities with those over diplotene nuclei in which the number of extra copies of the nucleolar organizer is known (see section 3.1), it was calculated that each oogonial nucleus should contain 20-40 extra copies of rDNA (16). This number implies a modest but non-negligible degree of amplification. The extra rDNA may have remained undetected morphologically up to the pachytene stage requiring as a prerequisite for detection the major amplification phenomena occurring at this time. Conversely, the copied cistrons may have served to form the extranucleoli which have, however, not been studied in detail.

2.2. Invertebrates

In the house cricket *Acheta domesticus*, a large Feulgen-positive intranuclear mass, the DNA-body, is present in oogonia and early prophase cells (Fig. 6). Microphotometric measurements of Feulgen-stained squash preparations reveal a 1.5 C DNA amount in the DNA-body in addition to the expected 4 C value in chromosomes (17). The DNA-body apparently is in contact with one of the 11 bivalents (Figs. 6, 7) and incorporates ^{3}H-thymidine but does not contain RNA (17). Lima-de-Faria et al. (18) later showed that 5 DNA-bodies of different sizes occur at the zygotene stage which, in fact, are large bead-like spiralized regions of certain chromosomes, the chromomeres. These chromomeres were characterized fully in pachytene nuclei (see section 3.2). In the beetle *Dytiscus marginalis* (19) and the fly *Tipula oleracea* (20), the DNA-body also appears for the first time in oogonia, making up the well known Giardina's body in the beetle. The DNA body of *Tipula* is associated with the sex chromosomes in oogonial interphase, and the nucleoli appear inside the DNA body in oocytes at early prophase of meiosis. Formation of nucleoli is also observed inside the DNA body in oogonia of the arthropod *Creophilus maxillosus* (21).

2.3. Higher vertebrates

In mice, examination of ultrathin sections of oogonia shows a single large reticulated nucleolus containing several fibrillar centers each surrounded by a fibrillar zone (22). These zones are sites of RNA synthesis, as shown by their strong labeling with ^{3}H-uridine. The nucleolus becomes smaller and apparently inactive at the leptotene stage since it is devoid of fibrillar centers and no longer incorporates uridine (22). Multiple small nucleoli with mainly a fibrillar composition have also been reported at this stage in mice (23) and rat oocytes (24). However, these may not represent true nucleoli (24). In human ovaries, squash preparations show 1-2 large nucleoli in oogonia and in leptotene and zygotene stage oocytes (25). Oogonia incorporate uridine strongly, leptotene stage oocytes moderately, and zygotene stage oocytes remain unlabeled (25). With the electron microscope, a few large reticulated nucleoli (4-6 in leptotenic oocytes) have been found in such cells, each with several fibrillar centers (7 at the leptotenic stage) which contact the ends of acrocentric chromosomes (26, 27, 28). Wolgemuth et al. (29) hybridized human oogonia obtained from spontaneous abortions with ^{3}H-rRNA and found a grain density in autoradiographs which was twice that obtained in diploid somatic cells. After verifying that a linear correlation existed between the number of silver grains and the DNA ploidy, it was concluded that human oogonia are 4 C and not 2 C as expected in such diploid cells, and leptotenic nuclei, 8 C and not 4 C as expected. A two-fold degree of amplification thus seems to occur in human oogonia and early oocytes (29).

In summary, a moderate amplification of rDNA seems to occur in prepachytene oocytes of most species studied, but it is only manifest morphologically in certain invertebrates. The formation of new nucleoli from amplified DNA is also only apparent in invertebrates. The number of large primary nucleoli of amphibia and mammals is approximately proportional to the number of haploid sets of chromosomes. The nucleoli appear to be derived from chromosomal DNA. It remains to be clarified whether nucleolar bodies (micro- or extranucleoli) are true nucleoli originating in the extrachromosomal DNA of pre-pachytene oocytes, or simply fragments of primary nucleoli.

3. Major amplification of ribosomal DNA at the pachytene stage of oogenesis

3.1. The DNA-cap of certain amphibia and fish

Confirming earlier observations in *Bufo* (30), a large mass of intranuclear material, the cap, has been observed in *Xenopus* oocytes at the pachytene stage in the nuclear area not occupied by chromosomes (1, 30, 31, 32). The cap is Feulgen-positive and stains purple red (Fig. 2) with methyl-green pyronin (Unna-staining). It is a site of rDNA synthesis, as shown by the uptake of ^{3}H-thymidine and the binding of ^{3}H-actinomycin D (31). With methyl green-pyronin, small red bodies are discernible inside the cap which are not stained after RNAse treatment and therefore, contain RNA (1). Electron microscopic observations (Fig. 3) confirm the nucleolar features of these bodies which belong to the spheroidal bipartite type (1). The DNA content of the cap has been estimated to be 10 C, determined as the difference between the microdensidometric readings on Feulgen-stained late pachytene (14 C) and early meiotic nuclei (4 C) (30).

Biochemical demonstration of extrachromosomal DNA has been carried out by Cs Cl density gradient centrifugation of DNA extracted from young ovaries. A dense, heavy DNA fraction anneals almost exclusively with rRNA, in contrast to a major and lighter DNA fraction which does not (33). Brown and Dawid (34) calculated that the rDNA of the denser satellite band is more abundant in oocytes than in somatic cells, while the DNA major lighter band occurs in similar proportions in the two cell types and seems to correspond to the 4 C complement of the chromosomal DNA. The rDNA satellite band is lacking in the anucleolate homozygote *Xenopus* mutant (35), a mutant which is also devoid of secondary constrictions in one pair of chromosomes (36). Secondary constriction is the thin segment of certain chromosomes distinct from the centromere or primary constriction, which may be identified with the NOR because it contains the genes coding for 18 S and 28 S rRNA. The amount of satellite DNA in Xenopus oocyte nuclei has been calculated from the area of the high-density DNA band to be 25 pg (34). Because it anneals at saturation with nearly 18% of its weight of rRNA, the amount of hybridizable rDNA has been estimated to be 4.5 pg, which corresponds to 2500 copies of the nucleolar organizer per oocyte (16). However, based on photometric determinations of extrachromosomal DNA, Perkowska et al. (37) calculated 5200 nucleolar organizers. Hybridization in situ of cytological preparations with radioactive rRNA confirms that in pachytene stage nuclei, silver grains are confined to the cap region (16). On the other hand, the nucleoli of prepachytene stage nuclei seem to undergo fragmentation within the cap (1). The labeling of the intracap bodies with ^{3}H-uridine however suggests that they are foci of transcription on extrachromosomal DNA templates.

Spread preparations of the rDNA satellite peak isolated at this stage reveal the occurrence of linear and circular rDNA molecules (38). Circle lengths fall into size classes which are multiples of a smallest circle whose length corresponds to a molecular mass of 8 million daltons (38). This length is equivalent to the coding unit for the rRNA gene (38, for a discussion of amplification models).

In other amphibians, the cap is either less conspicuous or not apparent as a circumscribed mass, although biochemical studies of *in situ* hybridization show high degrees of amplification (16, for a list of other amphibian species). In *Acipenseridae* fishes a cap of extrachromosomal chromatin occurs at the pachytene stage around the single large nucleolus present from previous stages (39). Amplification of rRNA has been demonstrated in the teleost fishes *Roccus saxatilis* (40), *Salmo irideus* (41) and *Tinca tinca* (42).

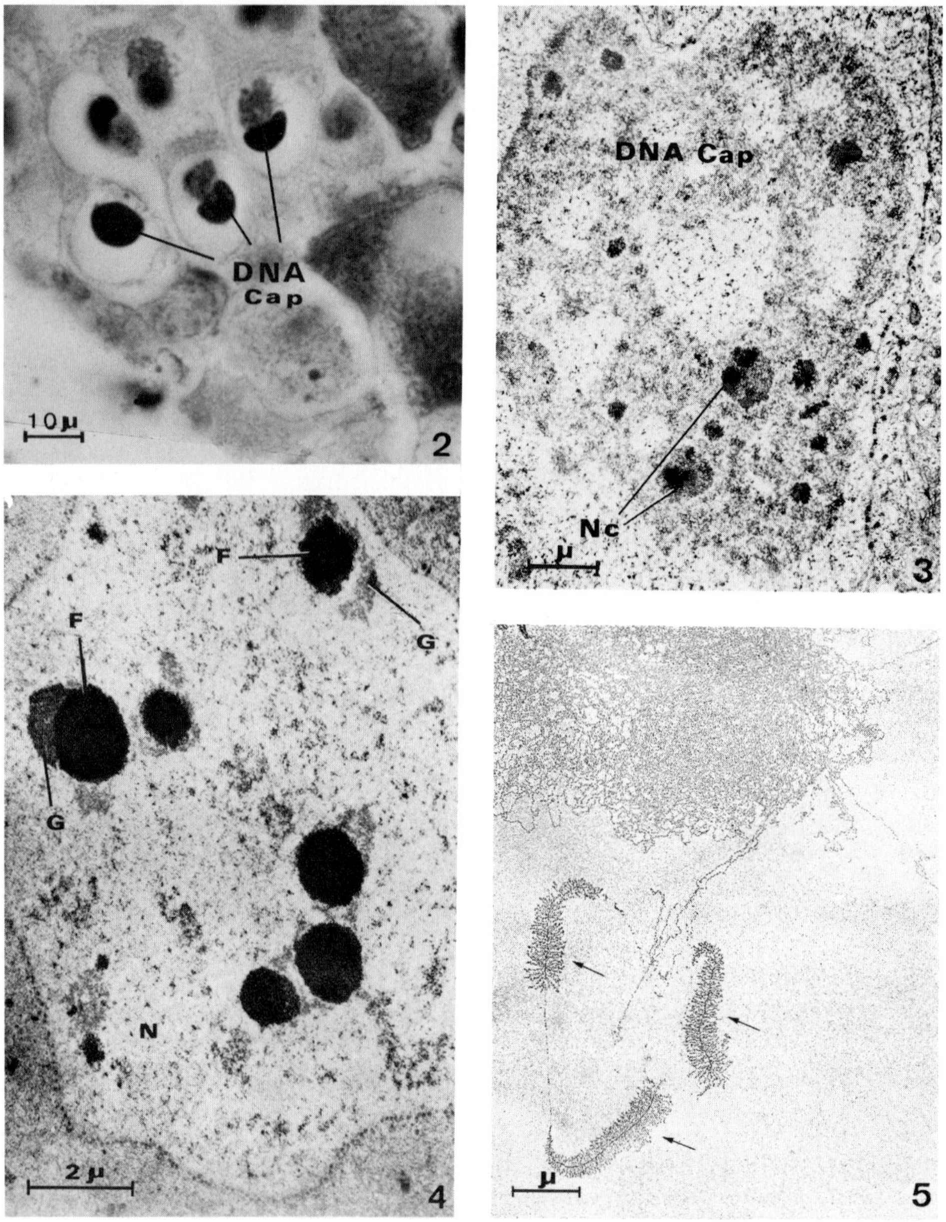

Figs. 2–5. Nucleoli in amphibian oogenesis. (2) Light micrograph of DNA cap (arrows) at the pachytene stage after methyl-green pyronin staining. The cap is purple red and the chromosomes, green. *X. laevis.* From Ficq, 1972 (32). (3) Electron micrograph of the DNA-cap containing spheroidal nucleoli (Nc) at the pachytene stage. *X. laevis.* From Van Gansen and Schram, 1972 (1). (4) Large spheroidal nucleoli in diplotene vitellogenic oocytes. F: Fibrillar component; G: granular component. *X. laevis.* From Van Gansen and Schram, 1972 (1). (5) Spread preparation of nucleoli from postvitellogenic mature oocytes. Most rDNA is not transcribed and there are only three transcription units (arrows). *T. alpestris.* From Scheer et al., 1976 (59).

3.2. The DNA-bodies of invertebrates

According to Lima-de-Faria et al. (18) three of the 5 DNA-bodies or major chromomeres occurring in *Acheta* at the zygotene stage disappear at the early pachytene stage. The remaining two occur on chromosomes 11 and 6 and, as noted in a different experiment by Ullman et al. (43), hybridize *in situ* with rRNA. In squash preparations of mid- pachytene stage nuclei, the chromomere of chromosome 11 has a decreasing core and a puffed fan-shaped expanding periphery which not only maintains its Feulgen staining, but also contains RNA, as shown by Azur B staining and uridine labeling (18). The larger chromomere of chromosome 6 undergoes a similar ‘puffing’ at late pachytene. Electron micro-

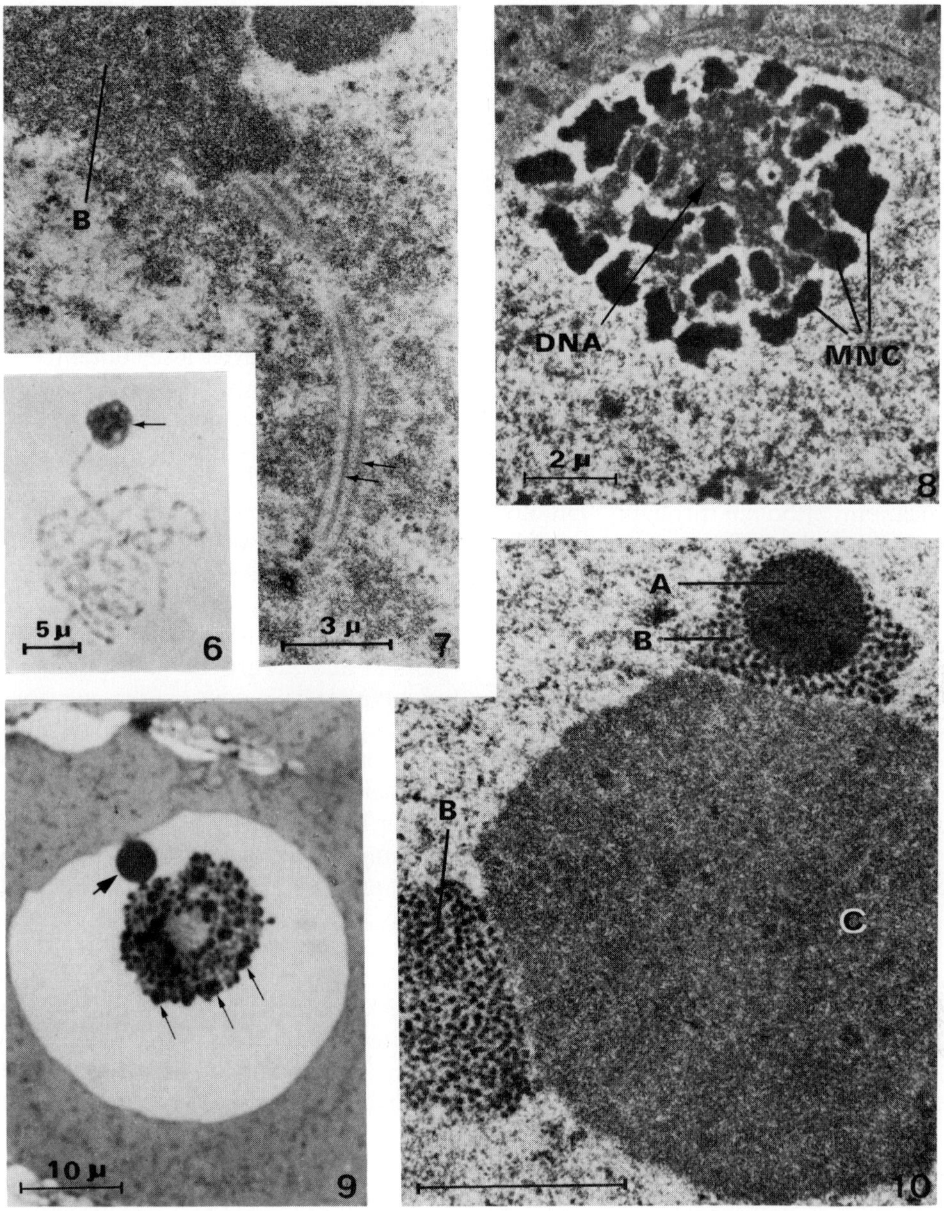

Figs. 6–10. Nucleoli in insect oogenesis (*Acheta*). (6) Light micrograph of Feulgen-stained DNA-body (arrow) in pre-pachytenic oocyte. From Cave and Allen, 1969 (17). (7) Electron micrograph of DNA-body (B) in direct continuity with the synaptonemic complex of one bivalent (arrows). From Cave and Allen, 1969 (17) (8) Electron micrograph of DNA body (major chromomere of chromosome 6) at the late pachytene stage, showing the amplified DNA in the center (DNA), and the budding nucleoli of the main nucleolar component (MNC) at the periphery. From Jaworska and Lima-de-Faria, 1973 (44). (9) Low power electron micrograph of a diplotene stage oocyte showing the mass of newly formed nucleoli made up of MNC nucleoli (long arrows) and one SNC nucleolar body (short arrow). From Jaworska and Lima-de-Faria, 1973 (44). (10) Detail of one SNC nucleolar body with the A, B and C zones described in the text. From Jaworska and Lima-de-Faria, 1973 (66).

scopy (44) shows the 'puff' to be formed by a central fibrillar area containing DNA, surrounded by electron-dense masses which, at the diplotene stage, give rise to new nucleoli (Fig. 8). Equilibrium density centrifugation and hybridization experiments have shown a high degree of amplification of the rRNA genes, 92% of which is spacer DNA (45).

In the dytiscid beetles *Dytiscus marginalis* and *Colymbetes fuscus*, pachytene oocytes are concentrated on the ovariole tips. It is possible to identify their amplified DNA by gradient centrifugation (46) and to make spread preparations for electron microscopy. Like *Xenopus*, the amplified ribosomal DNA occurs in these beetles as circular molecules the sizes of which seem to be multiples of a basic circle which might contain the coding sequence for one precursor rRNA molecule plus the spacer sequence (46). Spread preparations of previtellogenic oocytes in the pachytenic and early diplotenic stages show the DNA to be lacking transcriptional units, as well as beaded due to the presence of nucleosomes (47). Because transcribed chromatin is usually not compacted into nucleosomes (48), this observation indicates the lack of RNA synthesis in previtellogenic oocytes. Amplification of rDNA has been shown in other invertebrates by means of Feulgen staining or DNA-RNA hybridization (16, 49).

In summary, at the pachytene stage of meiosis, rDNA amplification, although probably initiated earlier becomes morphologically apparent in some amphibia under the shape of DNA-caps. The small nucleoli which appear inside the cap are probably derived from transcription on extrachromosomal DNA templates. In the insect *Acheta* the process seems more advanced since new nucleoli already emanate from the puffed chromomeres at the pachytene stage. While the chromosomal sites from which the DNA-bodies of *Acheta* originate are morphologically identified as chromomeres, the DNA-caps of amphibia occur separate from the chromosomal NORs which gave rise to their extrachromosomal rDNA.

3.3. Findings in higher vertebrates

During early and middle pachytene stages, no nucleoli are visible in quail oocytes. At mid-pachytene the microchromosomes, which in this species are the sites of the nucleolar organizers (50), fuse their heterochromatic ends to form 4 to 6 chromocenters (51). One or 2 nucleoli arise at the late pachytene stage, each of them in contact with the end of the euchromatic segment of one microchromosome emerging from the chromocenter (Fig. 11). The fibrillar core is displaced to the periphery of the nucleolus and surrounds a large fibrillar center which is bound to the microchromosome by fibrils (4). The fibrillar center and the fibrillar zone are labeled with ^{3}H-actinomycin D which denotes the presence of rDNA (52). The fibrillar zone alone takes up ^{3}H-uridine (52). These observations suggest that the fibrillar center contains the nucleolar organizer region extending from the nucleolar chromosomes, and that rRNA transcription occurs in the fibrillar zone. In the mouse (22), newly-formed nucleoli also appear at the mid-pachytene stage. Two nucleoli are formed at each side of certain bivalents at the zone of transition between their hetero- and euchromatic segments. This zone corresponds to the secondary constriction. As in quail oocytes, nucleoli have a fibrillar center which is bound to the bivalent by a bundle of fibrils. Transcription, as shown by ^{3}H-uridine labeling, occurs only in the fibrillar zone (22).

In human oocytes it is also at the pachytene stage that intimate contacts are apparent between nucleolar chromosomes and the nucleoli. Nucleoli are only 2-4, fewer than the 4-6 of leptotene stage oocytes. These oocytes also contain a modal number of fibrillar centers per nucleolus, 2.5, which is lower than the 7 per nucleolus occurring in leptotene stage oocytes (26). The fibrillar centers in man are not surrounded by a fibrillar zone, and they are segregated from the remainder of the nucleolar mass (Fig. 12). Each fibrillar center is seen to be penetrated by fibrils emanating from the secondary constriction of one to several bivalents of the D or G group (26). The chromosomes in group D (pairs 13 to 15) and G (pairs 21 and 22) are the chromosomes in the human karyotype which contain the nucleolar organizers and may be termed the nucleolar chromosomes of man. The nucleoli do not take up ^{3}H-uridine until late-pachytene when a fibrillar zone reappears around the fibrillar centers (27). In human oocytes, subjected to *in situ* hybridization (29), DNA ploidy was seen to rise from an 8 C value observed at early pachytene nuclei (a value identical to that estimated for leptotene-zygotene oocytes) to a 16 C value at the late pachytene stage. This finding is suggestive of a further double increase of the DNA amount during pachytene amounting to a four-fold degree of amplification at the end of this stage (a 4 C value was expected in the case of no amplification).

In summary, although rDNA amplification increases in pachytenic oocytes of man and probably other mammals, no DNA-bodies are apparent. However, close relationships are established between the nucleolar chromosomes and nucleoli which in the quail and mouse are formed *de novo* around NORs morphologically identifiable as fibrillar centers. In man, no formation of new nucleoli appears to occur. The decreased number of nucleoli and fibrillar centers in comparison to those observed in the leptotene stage, the fact that up to 3 chromosomes may associated with each fibrillar center, and the absence of RNA transcription in nucleoli in man, suggest a fusion not only of the fibrillar centers that were previously separated at the leptotene stage, but also of the pre-pachytenic nucleoli.

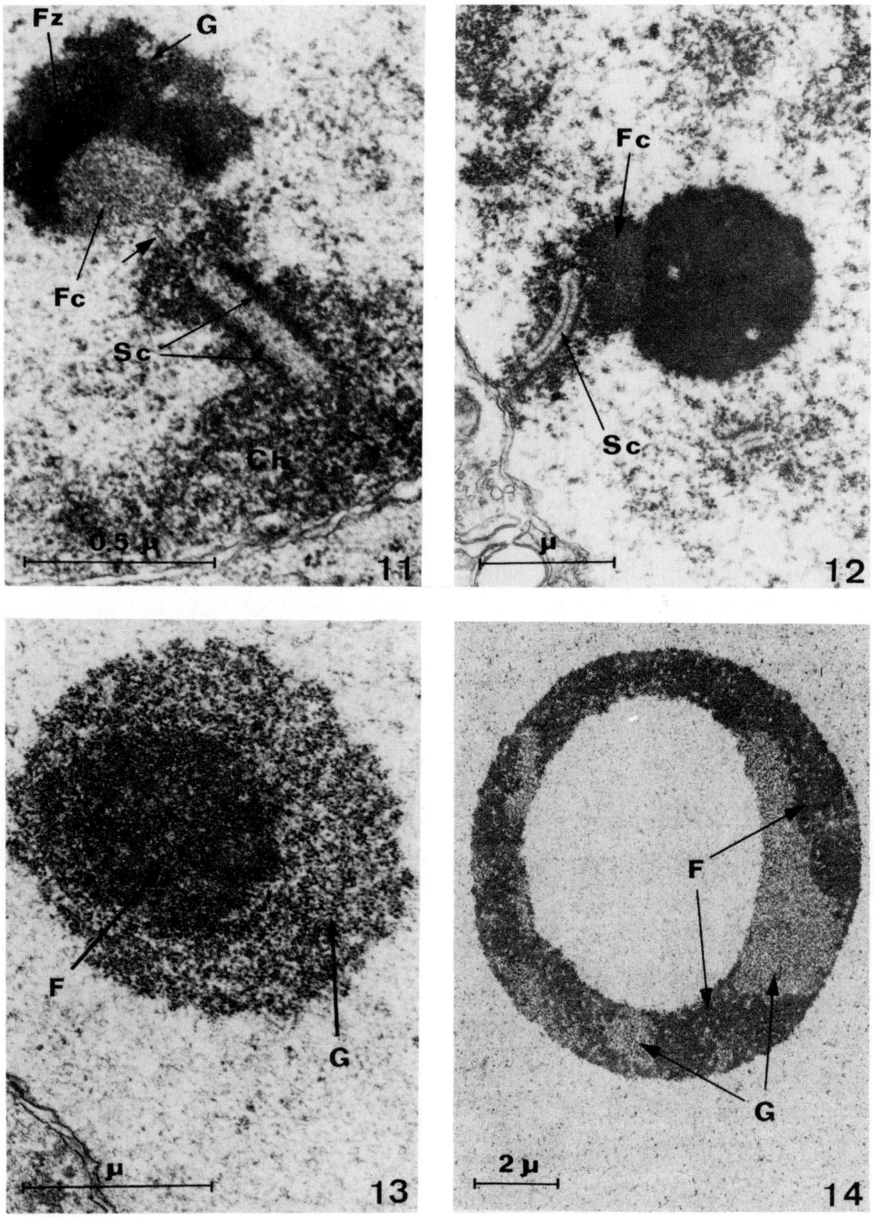

Figs. 11–14. Electron micrographs of oocyte nucleoli from higher vertebrates. (11) Nucleolus of a pachytene stage oocyte in the quail, with one fibrillar center (Fc) surrounded by the denser portion of the fibrillar component or fibrillar zone (Fz) and the granular component (G). Thin fibrils (short arrow) connect the fibrillar center with a microchromosomal bivalent emerging from the chromocenter (Ch) and showing the synaptonemic complex (Sc). From Mirre and Stahl, 1976 (4). (12) Same stage in a human oocyte. The fibrillar center is close to the bivalent. It is not surrounded by, but segregated from, the rest of the nucleolus. From Hartung et al., 1979 (26). (13) Spheroidal single nucleolus in an oogonia. *X. helleri*. F: fibrillar component; G: granular component. From Azevedo and Coimbra, 1980 (78) (14) Ring-shaped nucleolus in *X. helleri* at middictyate stage. From Azevedo and Coimbra, 1980 (78).

4. Nucleoli at diplotene and dictyate stages during oocyte growth

4.1. Nucleolar multiplication in amphibia and fish

Among *Xenopus* oocytes at the diplotene chromosomal stage are some previtellogenic oocytes (the remainder being at the zygotene or pachytene stages), as well as early and mid vitellogenic oocytes (53). Diplotene is a stage of marked oocyte growth (1, 53). In *previtellogenic*, early diplotene *oocytes*, the DNA-cap gradually disintegrates (Fig. 1, A), while numerous spheroidal nucleoli are free in the nuclear sap (Figs. 1, B; 4). Viewed by the light microscope, these nucleoli contain a fluorescent Feulgen-stained granule, which is suggestive of the presence of the cap DNA, and soon become ribbon-like structures formed by 1 to 4 linearly arranged Feulgen-stained granules (54). Viewed by the electron microscope (1), each ribbon consists of alternate fibrillar and granular areas (Fig. 1, C). Because nucleoli with a separate fibrillar and granular component appear in place of the ribbons (Fig. 1, D), it is likely that fragmentation of the latter occurs. However, nucleoli turn entirely fibrillar in late previtellogenic oocytes (Fig. 1, E) and ^{3}H-uridine incorporation which was taking place in previous nucleolar forms (55) is now essentially undetectable (56). Biochemical work shows that previtellogenic oocytes of *Xenopus* and some oviparous teleost fish almost exclusively synthesize 5 S RNA and tRNA which both accumulate in 8 S and 42 S RNA particles (57). The latter are responsible for the basophilia of the cytoplasm in which ribosomes are virtually absent (57). Fibrils dispersed in the cytoplasm may correspond to the 42 S RNP (58). In the newt *Triturus alpestris*, previtellogenic nucleoli also are entirely fibrillar (59) and in spread preparations of them according to Miller's technique, which is explained below, matrix units are rare along the nucleolar rDNA fibrillar axes (59). After labeling these oocytes *in vivo* with ^{3}H-uridine, gel electrophoresis of the ribonucleic acids extracted from them shows a small amount of newly synthesized 18 S + 28 S RNA, and a large quantity of 4-5 S RNA (59).

At the beginning of the *vitellogenic period* the fibrillar nucleolus of *Xenopus* increases enormously in size and gains a granular cortex (Fig. 1, F). When ribosomes first appear in the cytoplasm in mid diplotene oocytes together with yolk of exogenous origin, the bipartite nucleoli are already as large as oogonia in the same animal (1). Using the spreading technique of Miller (2) a separation of the fibrillar dense core and the granular cortex of the nucleoli may be obtained by treating the contents of oocyte nuclei with very dilute saline. The cores are suspended in a formalin solution, placed on a carbon film on an EM grid and spun in a centrifuge tube (2). In vitellogenic oocytes of *Xenopus laevis* and *Triturus viridescens* (2) each spread core appears as a ring-like circular DNA fiber along which there are repeating matrix units in which RNA transcription occurs, separated by spacer segments (Fig. 5). Each nucleolus may contain more than one ring, rings being more numerous in large than in small nucleoli (60). The total number of rings in one vitellogenic oocyte nucleus containing 1400-1600 nucleoli has been seen to be 5200, curiously identical to the number of extra copies of nuclear organizers estimated by Perkowska et al. from photometric measurements of Feulgen dye contents of oocyte nuclei (37). Hence, each extractable ring appears to contain one extra DNA copy of the nucleolar organizer. Again in *Triturus alpestris*, gel electrophoresis of RNA labeled with ^{3}H-uridine from vitellogenic oocytes reveals that 86% of the radioactivity is present in 18S + 28S RNA (59). Transcriptional units are abundant in this material with maximum number of lateral fibrils.

In amphibian and teleost fish, 42 S particles disappear during the vitellogenic period, 5 S RNA incorporates into the ribosomes and the tRNA remains free in the cytosol (61). The 5 S RNA of the oocytes is not amplified and persists in the embryo until gastrulation (57).

Interesting morphological relations between nucleoli and chromosomes have been observed with the phase contrast microscope in some amphibian species in which DNA amplification occurs, but where the excess DNA does not appear as a massive cap. (For example, in *Triturus cristatus* (60), the extra DNA occurs as a series of small Feulgen-stained granules close to the nuclear envelope). In *Amblystoma tigrinum* nucleolar masses have been seen to be attached to the same chromosomal secondary constriction that is associated with the nucleolus in somatic cells (12). In *T. cristatus*, free nucleoli appear at the nuclear periphery only in oocytes above a certain size in which large attached bodies occur on the chromosomes and nucleolar bodies appear to detach from the giant granular loops on chromosomes XII (62). In *Amblystoma mexicanum*, the variable number of nucleolus-like objects attached to the nucleolar organizer region on chromosome III was taken as suggestive of their detachment from that chromosome to become free nucleoli (63). The subsequent evolution of free nucleoli in *A. mexicanum* and *T.viridescens* consists of a series of morphological changes which are rather curious though difficult to explain as to their origin. The spheroidal free nucleoli transform into rings made up of beads of RNA-containing granules, strung out along a string of DNA (3). The string appears to be formed by the same DNA fiber which can be spread from the cores of bipartite nucleoli with Miller's technique. These forms revert to spheroidal nucleoli at late diplotene, which is a stage in which amphibian oocytes complete their growth and yolk accummulation, while chromosomes which had been expanded (lampbrush chromosomes) during all the vitellogenic period retract their loops (53). The nucleoli which have ceased their growth and in some amphibia are vacuolated, move to the center of the nucleus together with the chromosomes. In *T. alpestris*, synthesis of rRNA (Fig. 5) becomes negligible at this period (57).

In *Acipenseridae* fishes the sequence of events leading to the production of extrachromosomal nucleoli from the cap is similar to that in amphibians (39). In several teleost fish there is also multiplication of nucleoli at this stage (64).

In summary, in amphibia and some fish, new nucleoli develop in great numbers from the extrachromosomal DNA during the diplotene stage. The nucleoli become bipartite, and begin to increase in size, when an intense rRNA (28S + 18S) synthesis (and ribosomal formation) starts. Each fibrillar core contains one or more copies of the extrachromosomal DNA. In some newts, there is some evidence for the formation of nucleoli out of the chromosomal NORs. In this case, the extra DNA would be directly associated with the new nucleoli.

4.2. Formation of nucleoli from DNA-bodies in invertebrates

During the diplotene stage, the DNA body arising from the major chromomere of chromosome 6 in *Acheta* (18) gradually loses the Feulgen-positive material (DNA) in its center (17). The Azur B-stained periphery (RNA) increases in size, and by the late diplotene stage a large mass of RNA-containing material occurs at one side of the nucleus (Fig. 9). This mass is a cluster of several hundred small nucleoli (65) and has been designated the 'main nucleolar component' (MNC) (66). At the EM level, the MNC nucleoli are identical to the dense RNA masses observed at the periphery of the 'puff' at late pachytene (Fig. 8) and contain central fibrillar areas interspersed with groups of tightly packed particles. Surrounding the cluster of MNC nucleoli, some larger spherical bodies (Fig. 9) form the secondary nucleolar component (SNC). In the EM (Fig. 10) each SNC body shows a central region containing 5 nm fibrils (zone C) which is RNAse-labile. Zone C is surrounded by several large spheroidal bodies (Fig. 10) each made up of a DNAse-labile center (zone A), and of a granular periphery (zone B) (66). It has been suggested that the DNA of zone A is active in transcription and gives rise to preribosomal RNA which, after association with protein, forms the granules of zone B; these preribosomes would be further processed in zone C (66). Curiously, the two types, main (MNC) and secondary (SNC) nucleoli, are seen to migrate to the cytoplasm after being surrounded by folds of the nuclear envelope (44). In *Dytiscus*, two types of nucleolar granules are also formed, large, 1.5 μm in diameter, with vacuoles, and small, 100 nm in diameter (67). The matrix units of the rDNA circles isolated from diplotene stage nuclei of *Acheta* and *Dytiscus* are relatively homogeneous in length, with the space intercepts showing a pronounced heterogeneity. This heterogeneity differs from one group of rDNA genes to another, suggesting that circles of rDNA could be derived from different regions of the nucleolar organizer (68). Two types of nucleoli have been observed in many other insect species (69). However, in the dragonfly *Cordulia aenea*, only the main or primary nucleoli (PN) appears to form pre-ribosomes, the secondary nucleoli (SN) being probably concerned with the synthesis of 5 S RNA (69).

In summary, the formation of extrachromosomal nucleoli occurs in invertebrates directly at the periphery of the DNA-body, which is itself a chromosomal puff, and there seems to be two types of nucleoli which are often extruded to the cytoplasm.

4.3. Nucleolar events at the diplotene stage in higher vertebrates

In the quail, the nucleoli formed at the pachytene stage become reticulated at the diplotene stage and cease to be attached to the chromocenters (4). Fibrillar centers multiply and move to the center of the nucleolus (4, 6). After hybridization with ^{3}H-rRNA, the number of silver grains over nuclei or nucleoli increased two times over that previously found over pachytene stage nuclei, a finding suggesting that rDNA amplification continues to take place at this stage. RNA synthesis persists in the fibrillar zones (52). In mouse, the diplotene stage of meiosis lasts a few hours at about the time of birth (70). As in quail, large reticulated nucleoli are present and the number of nucleolar fibrillar centers surrounded by fibrillar zones increases (22). After labeling with ^{3}H-uridine, silver grains occur in autoradiographs over both the fibrillar zones and the rest of the fibrillar component, indicating the occurrence of transcription over the entire fibrillar component (71). In the ensuing dictyate stage (postnatal days 1-14) the chromosomes are no longer visible due to the unravelling of the chromatin so that their axes and the lateral projections extend throughout the nucleus (70). Dictyate chromosomes of mammals appear to be homologous of the lampbrush chromosomes of lower vertebrates and exhibit an intense transcriptive activity (70). Fibrillo-granular and fibrillar small bodies made up of fibrils and granules with the same dimensions as in the main nucleoli appear in nuclei (70). During postnatal days 14-28, corresponding to the period of development of the antral follicle during which the lampbrush chromosomes continue expanded, extranucleolar bodies not only increase in size but also exhibit close associations with heterochromatic, Feulgen-positive knobs which stand out from the faintly stained chromosomes (72).

In human oocytes, the 2-4 primary large nucleoli inherited from pachytene oocytes enlarge and turn reticulated at diplotene. The number of fibrillar centers rises to 10, each surrounded by a fibrillar zone (26). RNA synthesis which had resumed in nuclei at the early pachytene stage and in nucleoli at the late pachytene stage, rises markedly in nuclei and nucleoli (25). As in mice, numerous extranucleoli appear. Each shows a silver-stained spot after the silver-NOR method of Goodpasture and Bloom (73). With this technique, acid proteins associated with rRNA at the rDNA sites are silver stained (73), so that the staining indicates the location of active rRNA transcription (26). The extranucleoli also are labeled by hybridization *in situ* with 18S + 28S rRNA (74). However, according to the number of silver grains in the hybridization autoradiographs, the DNA value in diplotene nuclei is about the same (16 C) as that found in late pachytene stage nuclei (29). It has been proposed that the extra DNA copies synthesized at the pachytene stage are located in the silver stained areas of the extranucleoli (26).

In summary, in birds and mammals no primary nucleoli arise at diplotene but nucleoli become reticulated, gain an increased number of fibrillar centers and are sites of maximal RNA synthesis. Extranucleolar bodies which in man contain rDNA, and in mouse seem to detach from the chromosomes, may be foci of transcription on extrachromosomal DNA. It is not known whether the numerous fibrillar centers which appear at this stage in the primary nucleoli, also contain extra DNA.

5. Nucleolar involution in mature oocytes

Upon completion of oocyte growth, the oocytes of amphibians enter the postvitellogenic or maturation period during which nucleoli decrease in size and number, most of them (3) migrate to the center of the nuclei, and rRNA synthesis becomes negligible. These regressive changes are defined here as 'nucleolar involution'. Some large spheroidal nucleoli, sometimes containing vacuoles, remain at the nuclear periphery at the vegetal pole of the nucleus (3). In *Xenopus* oocytes, the fibrillar core of these peripheral nucleoli still contains DNA since it is Feulgen-positive (75) and digestible with DNAse (76), but the DNA amount is small, as shown by the weak labeling after hybridization with rRNA (77). After the breakdown of the nuclear envelope, many Feulgen-positive bodies, also labeled with radioactive RNA and displaying a fibrillar constitution at EM level, appear in areas previously occupied by the intact nucleoli (77). The extrusion of such rDNA material has also been observed in other anurans and urodeles (75). The extruded fibrillar masses appear surrounded by mitochondria in a manner similar to the perinuclear nuage or cement material observed in oogonia and oocytes of lower vertebrates (78, 79).

In mammalian oocytes, maturation is the period of oogenesis from the end of the dictyate stage of the first meiosis to metaphase arrest at the second meiosis (80). During this period nucleolar compaction occurs in mice, rabbit and human oocytes. After compaction the nucleolus shrinks and subsequently disappears when the nuclear envelope dissolves at germinal vesicle breakdown (81). In the mouse, the compact nucleolar mass appears as a feltwork of 6–10 nm fibrils at the time of follicular antrum formation (8). RNA synthesis, which is intense during oocyte growth (82), may continue at a lower rate until nucleolar dissolution (83). In pig oocytes, ^{3}H-uridine labeling is decreased markedly when the nucleolus compacts (9).

6. Oocytes with a single nucleolus

Multinucleolarity has been considered indicative of rDNA amplification, while single nucleoli would imply the lack of amplification (40). However, no numerical proportionality exists between nucleolar numbers and the degree of amplification because extrachromosomal nucleoli differ in size within the same nucleus and contain different amounts of rDNA. On the other hand, the presence of several rDNA cores per nucleolus suggests that amplification may occur in uninucleolated oocytes. Thus in the worm *Urechis caupo* and the clam *Spisula solidissima*, whose oocytes contain a single nucleolus, hybridization experiments showed a five-fold amplification of rDNA (34). In the clam *Mulinia lateralis* it is a two-fold amplification (49). Those cases in which a single nucleolus (Fig. 13) undergoes marked growth, such as in the teleost fish *Xiphophorus helleri* (78), or the mollusc *Helcion* (84), are suggestive of an accumulation of rDNA within the single nucleolus and deserve further study. In *Xiphophorus*, the pachytene stage spheroidal nucleolus is surrounded by a dense chromatin mass which dissipates at the diplotene stage such as is the situation with the cap of amphibians. But the nucleolus, which continues as a single, grows at the diplotene stage and at dictyate undergoes a vacuolization cycle. Although the significance of the vacuolization events leading to the transitory appearance of a large ringshaped nucleolus (Fig. 14) are still not clear, the simultaneous presence of nucleolar-like material on both sides of the nuclear pores suggests that vacuoles may be due to an increased export of pre-ribosomes toward the cytoplasm (78). In some cases of single nucleoli during oogenesis, the occurrence of extranucleoli should be searched, since extranucleoli have been shown to contain rDNA in human oocytes (26, 74).

7. The utilization of oocyte ribosomes in early embryogenesis

With or without proven rDNA-amplification, and whether multiple or single, oocyte nucleoli are agents of an enormous production of ribosomes which are destined to be utilized largely, at least in non-mammalian species, during early embryogenesis and scarcely during oogenesis. In fact, 60% of the ribosomes in *Xenopus* oocytes sediment as monosomes (34), and 90% of the abundant yolk proteins appearing at the same time in the cell (53) are of extraneous origin (85). In *Xenopus* and other amphibians, nucleoli, as well as rRNA synthesis, are generally considered to be lacking until gastrulation (86) even though polymerase I activity which is responsible for the synthesis of rRNA, has recently been shown in some blastulae nuclei (87).

In the mouse, the number of ribosomes is slightly decreased in the 2-cell stage embryo as compared to the 1-cell but it then rises significantly until the early blastocyst stage, an increase due to new synthesis of rRNA (88). Since 40% of the labeled egg RNA is already lost by the 2-cell stage (89), and the qualitative pattern of protein synthesis changes considerably at that time (90), it seems that ribosomes produced by oocyte nucleoli are responsible for protein synthesis only until the completion of first cleavage (88). Nevertheless because the protein synthesis increase from the one- to the eight-cell stage embryo is much smaller than the increase in RNA synthesis (80), others have hypothesized that the ribosomes produced during this period will mainly support protein synthesis in later cleavage stages of the preimplantation embryo (80), while protein synthesis through the first cleavages would be dependent on oocyte RNA (80). The fact that RNA which was synthesized in the nucleoli and chromatin of mouse oocytes persists labeled in the cytoplasm of early blastocyst cells (91) reinforces the possibility that oocyte RNA play some role in protein synthesis beyond the 2-cell stage embryo.

8. Conclusions

Amplification of rDNA appears to occur in oocytes of most species studied throughout most of meiotic prophase, though peaking at the pachytene stage. In some amphibia, fish and invertebrates, extra rDNA is morphologically apparent in the form of DNA bodies which give rise to nucleoli, mainly at the diplotene stage. In those species, as well as in species lacking DNA-bodies but in which DNA amplification occurs, new nucleoli contain extra DNA. In some cases, nucleoli apparently bud from the nucleolar organizer chromosomal regions (some newts, primary nucleoli of quail and mouse, extranucleoli of man an mouse).

In lower vertebrates, new nucleoli are spheroidal with the fibrillar core containing the transcribing DNA. This organization corresponds to a great activity of rRNA synthesis. In insects, two entirely different types of nucleoli are formed, their respective functions being unclear. In some higher vertebrates, the large new nucleoli are firstly spheroidal with a fibrillar center within the core, then reticulated during maximal rRNA synthesis, and finally compact when rRNA synthesis ceases.

Newly-formed extranucleoli are made up of fibrils and granules but lack the elaborate arrangement of these components as seen in large or primary nucleoli. The structure and functions of such simple forms are largely unknown.

The transition from pre-pachytenic to post-pachytenic nucleoli is somewhat obscure. In *Xenopus*, the earlier forms are included in the DNA cap, and in quail and mouse they disappear before pachytene. In man, early nucleoli persist through late meiosis after developing close contacts with the nucleolar chromosomes and undergoing an increase in the number of fibrillar centers. It is possible that these fibrillar centers contain extra DNA.

In some species, oocytes are uninucleolated up to the end of oogenesis and in a few of them, DNA-amplification has been demonstrated. In others, the intense nucleolar growth during pachytene may be an equivalent of nucleolar multiplication, and further studies are needed to find out whether there are extranucleoli and DNA-amplification in oocytes.

The significance of reversible vacuolization in nucleoli of amphibian and fish is not clear, and this phenomenon requires further investigation.

While oocyte ribosomes survive fertilization to

carry out protein synthesis during early embryogenesis, oocyte nucleoli disappear before ovulation, in some cases moving to the cytoplasm before (some insects) or after (amphibia) germinal vesicle breakdown.

Acknowledgements

We thank Mrs. M.T. Laranjeira and Miss A. Guimarães for secretarial work and typing the manuscript. This work was supported by the Center of Experimental Morphology of Oporto University of the I.N.I.C.

References

1. Van Gansen P, Schram A: Evolution of the nucleoli during oogenesis in Xenopus laevis studied by electron microscopy. J Cell Sci 10: 339–367, 1972.
2. Miller OL, Beatty BR, Hamkalo BA: Nuclear structure and function during amphibian oogenesis. In: Oogenesis. Biggers JD, Schuetz AW (eds) Univ Park Press-Baltimore, London, 1972, pp 119–128.
3. Lane NJ: Spheroidal and ring nucleoli in amphibian oocytes. J Cell Biol 35: 421–434, 1967.
4. Mirre C, Stahl A: Ultrastructural study of nucleolar organizers in the quail oocyte during meiotic prophase I. J Ultrastruct Res 56: 186–201, 1976.
5. Goessens G, Lepoint A: The nucleolus–organizing regions (NOR's): recent data and hypotheses. Biol Cell 35: 211–220, 1979.
6. Knibiehler B, Navarro A, Mirre C, Stahl A: Localization of ribosomal cistrons in the quail oocyte during meiotic prophase I. Exptl Cell Res 110: 153–157, 1977.
7. Altmann GG, Leblond CP: Changes in size and structure of the nucleolus of columnar cells during their migration from crypt base to villus top in rat jejunum. J Cell Sci 56: 83–100, 1982.
8. Chouinard LA: A light and electron microscope study of the nucleolus during growth of the oocyte in the prepubertal mouse. J Cell Sci 9: 637–663, 1971.
9. Crozet N, Motlik J, Szöllösi D: Nucleolar fine structure and RNA synthesis in porcine oocytes during the early stages of antrum formation. Biol Cell 41: 35–42, 1981.
10. Fakan S, Puvion E: The ultrastructural visualization of nucleolar and extranucleolar RNA synthesis and distribution. Int Rev Cytol 65: 255–299, 1980.
11. Kalt MR: Ultrastructural observations on the germ line of Xenopus laevis. Z Zellforsch 138: 41–62, 1973.
12. Al-Mukhtar KK, Webb AC: An ultrastructural study of primordial germ cells, oogonia and early oocytes in Xenopus laevis. J Embryol exp Morph 26: 195–217, 1971.
13. Gall JG: Lampbrush chromosomes from oocyte nuclei of the newt. J Morph 94: 283–351, 1954.
14. Coggins LW, Gall JG: The timing of meiosis and DNA synthesis during ovogenesis in the toad, Xenopus laevis. J Cell Biol 52: 569–576, 1972.
15. Pardue ML: Nucleic and hibridization in cytochemical preparations. J Cell Biol 43: 101a, 1969.
16. Gall JG: The genes for ribosomal RNA during oögenesis. Genetics Suppl 61: 121–132, 1969.
17. Cave MD, Allen ER: Synthesis of nucleic acids associated with a DNA-containing body in oocytes of Acheta. Exptl Cell Res 58: 201–212, 1969.
18. Lima-de-Faria A, Jaworska H, Gustafsson T, Daskaloff S: Amplification of ribosomal DNA in Acheta. III – The release of DNA copies from chromomeres. Hereditas 73: 163–184, 1973.
19. Gall JG, Macgregor HC, Kidston ME: Gene amplification in the oocytes of dytiscid water beetles. Chromosoma 26: 169–187, 1969.
20. Lima-de-Faria A, Moses MJ: Ultrastructure and cytochemistry of metabolic DNA in Tipula. J Cell Biol 30: 177–192, 1966.
21. Kloc M: Extrachromosomal DNA and its activity in RNA synthesis in oogonia and oocytes in the pupal ovary of Creophilus maxillosus (Staphylinidae, Coleoptera-Polyphaga). Eur J Cell Biol 21: 328–334, 1980.
22. Mirre C, Stahl A: Ultrastructure and activity of the nucleolar organizer in the mouse oocyte during meiotic prophase. J Cell Sci 31: 79–100, 1978.
23. Palombi F, Viron A: Nuclear cytochemistry of mouse oogenesis. I – Changes in extranucleolar ribonucleoprotein components through meiotic prophase. J Ultrastruct Res 61: 10–20, 1977.
24. Palombi F, Stefanini M: Ultrastructural analysis of nucleolar evolution in the rat primary oocyte. J Ultrastruct Res 47: 61–73, 1974.
25. Hartung M, Stahl A: Autoradiographic study of RNA synthesis during meiotic prophase in the human oocyte. Cytogenet Cell Genet 20: 51–58, 1978.
26. Hartung M, Mirre C, Stahl A: Nucleolar organizers in human oocytes at meiotic prophase I, studied by the silver-NOR method and electron microscopy. Hum Genet 52: 295–308, 1979.
27. Mirre C, Hartung M, Stahl A: Association of ribosomal genes in the fibrillar center of the nucleolus: A factor influencing translocation and nondisjunction in the human meiotic oocyte. Proc Nat Acad Sci, USA 77: 6017–6021, 1980.
28. Stahl A, Mirre C, Hartung M, Knibiehler B: Localisation, structure et activité des gènes ribosomiques dans le nucléole de l'ovocyte en prophase de méiose. Reprod Nutr Dévelop 20: 469–483, 1980.
29. Wolgemuth DJ, Jagiello GM, Henderson AS: Quantitation of ribosomal RNA genes in fetal human oocyte nuclei using rRNA: DNA hybridization in situ. Exptl Cell Res 118: 181–190, 1979.
30. Gall JG: Differential synthesis of the genes for ribosomal RNA during amphibian oögenesis. Proc Nat Acad Sci, USA 60: 553–560, 1968.
31. Ficq A: RNA synthesis in early oogenesis of Xenopus laevis. Exptl Cell Res 63: 453–457, 1970.
32. Ficq A: Meiosis in early amphibian oogenesis: synthesis and location of proteins. Exptl Cell Res 73: 242–248, 1972.
33. Wallace H, Birnstiel ML: Ribosomal cistrons and the nucleolar organizer. Biochim Biophys Acta 114: 296–310, 1966.
34. Brown DD, Dawid IB: Specific gene amplification in oocytes. Science 160: 272–280, 1968.
35. Birnstiel ML, Wallace H, Sirlin JL, Fischberg M: Localization of the ribosomal DNA complements in the nucleolar organizer region of Xenopus laevis. Natl Cancer Inst Monograph 23: 431–448, 1966.
36. Kahn J: The nucleolar organizer in the mitotic chromosome complement of Xenopus laevis. Quart J Microscop Sci 103: 407, 1962.
37. Perkowska E, MacGregor HC, Birnstiel ML: Gene amplification in the oocyte nucleus of mutant and Wild-type Xenopus-laevis. Nature 217: 649–650, 1968.
38. Hourcade D, Dressler D, Wolfson J: The nucleolus and the rolling circle. In: 'Chromosome Structure and Function'. Cold Spring Harbor Symposia on Quantitative Biology 38: 537–550, 1973.
39. Raikova EV: Evolution of the nucleolar apparatus during oogenesis in Acipenseridae. J Embryol exp Morph 35: 667–687, 1976.
40. Vincent WS, Halvorson HO, Chen H-R, Shin D: A comparison of ribosomal gene amplification in uni- and multinucleolate oocytes. Exptl Cell Res 57: 240–250, 1969.
41. Vlad M: Nucleolar DNA in oocytes of Salmo irideus (Gibbons). Cell Tissue Res 167: 407–424, 1976.
42. Denis H, Wegnez M: Biochemical research on oogenesis. Oocytes and liver cells of the teleost fish Tinca tinca contain different kinds of 5S RNA. Dev Biol 59: 228–236, 1977.
43. Ullman JS, Lima-de-Faria A, Jaworska H and Bryngelsson T: Amplification of ribosomal DNA in Acheta. V. Hybridization of RNA complementary to ribosomal DNA with pachytene chromosomes. Hereditas 74: 13–24, 1973.
44. Jaworska H, Lima-de-Faria A: Amplification of ribosomal DNA in Acheta. VII – Transfer of DNA-RNA assemblies from the nucleus to the cytoplasm. Hereditas 74: 187–204, 1973.
45. Pero R, Lima-de-Faria A, Ståhle U, Granström H, Ghatnekar R: Amplification of ribosomal DNA in Acheta. IV – The number of cistrons for 28S and 18S ribosomal RNA. Hereditas 73: 195–210, 1973.
46. Gall JG, Rochaix JD: The amplified ribosomal DNA of dytiscid beetles. Proc Nat Acad Sci, USA 71: 1819–1823, 1974.

47. Scheer U, Zentgraf H: Nucleosomal and supranucleosomal organization of transcriptionally inactive rDNA circles in Dytiscus oocytes. Chromosoma 69: 243–254, 1978.
48. Rindt KP, Nover L: Chromatin structure and function. Biol Zbl 99: 641–673, 1980.
49. Kidder GM: The ribosomal RNA cistrons in clam gametes. Dev Biol 48: 132–142, 1976.
50. Comings EE, Mattoccia E: Studies of microchromosomes and a G-C rich DNA satellite in the quail. Chromosoma 30: 202–214, 1970.
51. Stahl A, Luciani LM, Devictor M, Capodano AM, Hartung M: Heterochromatin and nucleolar organizers during first meiotic prophase in quail oocytes. Exptl Cell Res 91: 365–371, 1974.
52. Mirre C, Stahl A: Peripheral RNA synthesis of fibrillar center in nucleoli of Japanese quail oocytes and somatic cells. J Ultrastruct Res 64: 377–387, 1978.
53. Dumont JN: Oogenesis in Xenopus laevis (Daudin). I – Stages of oocyte development in laboratory maintained animals. J Morph 136: 153–180, 1972.
54. Thiébaud CH: Quantitative determination of amplified rRNA and its distribution during oogenesis in Xenopus laevis. Chromosoma 73: 37–44, 1979.
55. Van Gansen P, Schram A: Incorporation of (^{3}H) uridine and (^{3}H) thymidine during the phase of nucleolar multiplication in Xenopus laevis oögenesis: a high-resolution autoradiographic study. J Cell Sci 14: 85–103, 1974.
56. Van Gansen P, Thomas C, Schram A: Nucleolar activity and RNA metabolism in previtellogenic and vitellogenic oocytes of Xenopus laevis. Exptl Cell Res 98: 111–119, 1976.
57. Denis H: Accumulation du RNA dans les oocytes des vertébrés inférieurs. Biol Cell 28: 87–92, 1977.
58. Thomas C: Évolution des structures ribosomales au cours de l'oogenèse chez Xenopus laevis. Arch Biol (Liège) 78: 347–369, 1967.
59. Scheer U, Trendelenburg MF, Franke WW: Regulation of transcription of genes of ribosomal RNA during amphibian oogenesis. J Cell Biol 69: 465–489, 1976.
60. MacGregor HC: The nucleolus and its genes in amphibian oogenesis. Biol Rev 47: 177–210, 1972.
61. Denis H, Mairy M: Recherches biochimiques sur l'oogenèse. I – Distribution intracellulaire du RNA dans les petits oocytes de Xenopus laevis. Eur J Biochem 25: 524–534, 1972.
62. Callan HG, Lloyd L: Lampbrush chromosomes of crested newts Triturus Cristatus (Laurenti). Phil Trans B 243: 135–219, 1960.
63. Callan HG: Chromosomes and nucleoli of the axolotl. Ambystoma mexicanum. J Cell Sci 1: 85–108, 1966.
64. Buslé S, Buslé J: Les apports de la microscopie électronique à la connaissance des cellules germinales précoces des poissons. Vie Millieu, 28–29: 267–285, 1979.
65. Cave MD: Localization of ribosomal DNA within oocytes of the house cricket, Acheta Domesticus (Orthoptera: Gryllidae). J Cell Biol 55: 310–321, 1972.
66. Jaworska H, Lima-de-Faria A: Amplification of ribosomal DNA in Acheta. VI – Ultrastructure of two types of nucleolar components associated with ribosomal DNA. Hereditas 74: 169–186, 1973.
67. Trendelenburg MF, Franke WW, Scheer U: Frequencies of circular units of nucleolar DNA in oocytes of two insects, Acheta domesticus and Dytiscus marginalis, and changes of nucleolar morphology during oogenesis. Differentiation 7: 133–158, 1977.
68. Trendelenburg MF, Scheer U, Zentgraf H, Franke WW: Heterogeneity of spacer lengths in circles of amplified ribosomal DNA of two insect species, Dytiscus marginalis and Acheta domesticus. J Mol Biol 108: 453–470, 1976.
69. Halkka L: Ultrastructural changes and kinetic relationships of the secondary nucleolus and nuclear bodies in previtellogenic oocytes of the dragonfly Cordulia aenea. Hereditas 95: 259–268, 1981.
70. Chouinard LA: An electron-microscope study of the extranucleolar bodies during growth of the oocyte in the prepubertal mouse. J Cell Sci 12: 55–69, 1973.
71. Mirre C, Stahl A: Ultrastructural organization, sites of transcription and distribution of fibrillar centres in the nucleolus of the mouse oocyte. J Cell Sci 48: 105–126, 1981.
72. Chouinard LA: A light and electron-microscope study of the oocyte nucleus during development of the antral follicle in the prepubertal mouse. J Cell Sci 17: 589–600, 1975.
73. Goodpasture C, Bloom SE: Visualization of nucleolar organizer regions in mammalian chromosomes using silver staining. Chromosoma 53: 37–50, 1975.
74. Wolgemuth-Jarashow DJ, Jagiello GM, Henderson AS: The localization of rDNA in small nucleolus-like structures in human diplotene oocyte nuclei. Hum Genet 36: 63–68, 1977.
75. Brachet J, Hanocq F, Van Gansen P: A cytochemical and ultrastructural analysis of in vitro maturation in amphibian oocytes. Dev Biol 21: 157–195, 1970.
76. Van Gansen P, Schram A: Ultrastructure et cytochimie ultrastructurale de la vésicule germinative et du cytoplasme périnucléaire de l'oocyte mûr de Xenopus laevis. J Embryol exp Morph 20: 375–389, 1968.
77. Steinert G, Thomas C, Brachet J: Localization by in situ hybridization of amplified ribosomal DNA during Xenopus laevis oocyte maturation (a light and electron microscopy study). Proc Nat Acad Sci, USA 73: 833–836, 1976.
78. Azevedo C, Coimbra A: Evolution of nucleoli in the course of oogenesis in a viviparous Teleost (Xiphophorus helleri). Biol Cell 38: 43–48, 1980.
79. Clérot J-C: Les groupements mitochondriaux des cellules germinales des poissons téléostéens cyprinidés. II – Étude autoradiographique à haute résolution de l'incorporation de phénylalanine ^{3}H et d'uridine ^{3}H. Exptl Cell Res 120: 237–244, 1979.
80. Schultz RM, Letourneau GE, Wassarman PM: Program of early development in the mammal: changes in patterns and absolute rates of tubulin and total protein synthesis during oogenesis and early embryogenesis in the mouse. Dev Biol 68: 341–359, 1979.
81. Zamboni L: Comparative studies on the ultrastructure of mammalian oocytes. In: Oogenesis. Biggers JD, Schuetz AW (eds) Univ Park Press-Baltimore, London, 1972, pp 5–45.
82. Moore GPM, Lintern-Moore S, Peters H, Faber M: RNA synthesis in the mouse oocyte. J Cell Biol 60: 416–422, 1974.
83. Wassarman PM, Letourneau GE: RNA synthesis in fully-grown mouse oocytes. Nature 261: 73–74, 1976.
84. Azevedo C, Coimbra A: Nucleolar fine structure in oocytes of Helcion pellucidus (Gastropoda, Prosobranchia). Ciênc Biol (Portugal) 7: 43a, 1982.
85. Wallace RA, Nickol JM, Ho T, Jared DW: Studies on amphibian yolk. X. The relative roles of autosynthetic and heterosynthetic processes during yolk protein assembly by isolated oocytes. Dev Biol 29: 255–272, 1972.
86. Brown DD: The nucleolus and synthesis of ribosomal RNA during oogenesis and embryogenesis of Xenopus laevis. Natl Cancer Inst Monograph 23: 297–309, 1966.
87. Bouloukhère M, Thomas C, Heilporn-Pohl V, Hanocq F, Brachet J: Nucleolar localization of 'template-bound' RNA polymerase I in nuclei of Xenopus laevis blastulae. Exptl Cell Res 130: 291–295, 1980.
88. Pikó L, Clegg KB: Quantitative changes in total RNA, total poly (A), and ribosomes in early mouse embryos. Dev Biol 89: 362–378, 1982.
89. Bachvarova R, De Leon V: Polyadenylated RNA of mouse ova and loss of maternal RNA in early development. Dev Biol 74: 1–8, 1980.
90. Van Blerkom J, Brockway GO: Qualitative patterns of protein in the synthesis in the preimplantation mouse embryo. I. Normal pregnancy. Dev Biol 44: 148–157, 1975.
91. Fourcroy JL: RNA synthesis in immature mouse oocyte development. J Exp Zool 219: 257–266, 1982.

Authors' addresses:
A. Coimbra
Institute of Histology and Embryology
Faculty of Medicine
University of Oporto
4200 Oporto, Portugal

C. Azevedo
Department of Cell Biology
Institute of Biomedical Sciences
University of Oporto
4000 Oporto, Portugal

CHAPTER 12

Ultrastructural changes in ovarian oocytes induced by exposure to ionizing radiations

TERRY G. BAKER and LESLIE L. FRANCHI

1. Introduction

The transmission electron microscope has proved to be a most useful tool to the biologist and over the past twenty five years it has been used to explore the fine structure of a wide variety of cellular types and their organelles. The internal anatomy of the cell varies with function and characteristic changes are detectable by microscopy when its physiology and biochemistry become altered. It is therefore surprising that there have been few studies concerned with the ultrastructure of radiation damage, since this results from changes in cellular biochemistry (1–3). It will be shown in this chapter, however, that electron microscopy has provided little meaningful information about either the nature of the primary lesion or the sequence of events leading to radiation-induced cell death. It can be argued that a major reason for the paucity of such studies is the very complexity of the problem itself. Radiation biology is a complex interdisciplinary subject bridging large areas of physics, chemistry, biology and medicine. As a subject it is thwart with difficulties owing to the diversity of cell types and of processes of differentiation. In the case of the gonads the problems are compounded by the existence of two types of cellular division (mitosis and meiosis), changes related to aging or species differences, and the number of criteria that can be used to determine the effect ('radiosensitivity' in terms of cell death; reproductive capacity; genetic effects; biochemical changes; etc, 4–7).

It is clear, therefore, that many cell biologists may feel a lack of competence to actively pursue research in all but a few narrow aspects of radiobiology when there is a clear need for a multidisciplinary approach. It is with this approach in mind that we have included a synopsis of the basic principles of radiation physics and biology with the hope that we might encourage others to enter this potentially rewarding field of research. The reader requiring further information on aspects of radiobiology and on oogenesis is referred to the specialist books and articles on these subjects (1–10).

1.1. The nature of ionizing radiations

There are many forms of radiation which have continuously bathed the surface of the earth since its formation. Some of these radiations are derived from the sun or from other areas of the solar system, while others arise from within the earth itself by the decay of naturally occurring radioisotopes. Light is an obvious example of the former and is important in many respects, including photosynthesis. However, this chapter is concerned solely with *ionizing radiations;* namely, those that eliminate electrons from stable atoms, thus causing ionizations in the matter through which they pass. Unless the initial lesion is repaired these changes may result in altered cellular biochemistry which may be highly deleterious to the target cells. However, the success and diversity of living organisms may well have resulted from radiation-induced mutations (1).

The so-called ionizing radiations include those resulting from the decay of naturally occurring and man-made radioisotopes (α, β, and γ-rays), those produced under certain electrical conditions (X-rays which are equivalent to γ-rays and electrons which are β-rays), and the fission products produced in nuclear reactors (protons, neutrons, etc.). Expressed in simple terms, β-rays are streams of electrons, α-rays are helium atoms stripped of their two electrons, while protons and neutrons can be displaced from atomic nuclei to form beams of densely penetrating and highly ionizing radiations. These fission products are more efficient at causing damage to cells than the other types of radiation: expressed another way, they have the highest relative biological efficiency (RBE) and the greatest linear energy transfer (LET). The damage caused by X- and γ-rays can also be severe and extend over considerable distances: by contrast,

Van Blerkom, J. and Motta, P.M. (eds.), Ultrastructure of reproduction. ISBN 0-89838-572-5.

α- and β-rays are weakly penetrating and less highly ionizing in tissues. Their effects are largely superficial unless they are incorporated into the cells (eg., during tracer experiments) when they can induce considerable local damage (6, 11). Thus, neutrons are more damaging than X- or γ-rays (especially high energy or 'fast' neutrons), which in turn are more effective than β-rays and α-rays. However, this is an oversimplification, since the energy imparted by an ionizing radiation is proportional to the energy with which it is produced. Furthermore, the electrons that are eliminated from the stable molecules constitute secondary β-rays, the penetration of which is a function of the photon energy of the electromagnetic radiation from which it is produced (1).

All ionizing radiations have the common property of rapidly (within 10^{-13} seconds) imparting their energy to the tissues through which they pass (1). This is the so-called *physical stage* in the production of the radio-lesion. During the subsequent *chemical stage* the free radicals induced by the ionizations exert their effects within cells before again becoming stable. These free radicals are produced by the elimination of electrons from the outer shells of the atoms, have a short half-life, and exert their effects over very short distances. The final *physiological stage* in the production of the radiolesion may be of long duration and can result in cell death, mitotic failure, or more minor biochemical changes in cellular organelles. If these changes occur in chromosomes, profound genetic changes can occur which may seriously affect subsequent generations of cells (eg., cancer induction): in the case of germ cells they may also affect subsequent progeny (1, 6, 7).

It is common practice in radiobiological studies to compare the response of various differentiated cell types (and of different stages in the life cycle of the cell). Thus, for a particular type of radiation (usually X- or γ-rays), cells may be described in terms of varying degrees of 'radiosensitivity' or 'radioresistance'. These terms are meaningless, however, unless the criterion used to assess the effect is clearly stated. Such criteria include cell death (quantitative studies of cellular depopulation with increasing time after irradiation); biochemical changes (in metabolism, secretion, etc); morphological changes (assessed by light and electron microscopy, including histochemistry and autoradiography); reproductive capacity (number and size of litters), and such genetic effects as mutations and chromosomal aberrations. For the criterion of cell death the results often are expressed in terms of LD50/30 doses; that is, the dose of radiation required to kill half the population of cells within 30 days of exposure.

In summary, it is pointless to state that a given cell type is 'radiosensitive' unless one of these criteria is specified and the physical factors of the exposure are also noted to enable the experiment to be repeated or critically evaluated. These factors relate to the irradiation itself (eg., type of radiation, dose-rate and quality, whether a fractionated or acute exposure, filtration of the beam) and also such physical factors as temperature, atmospheric pressure, and oxygen tension.

The overall radiosensitivity of a cell is governed by many aspects of its anatomy, physiology and biochemistry. These aspects include the stage of the cell cycle, type and distribution of cellular organelles, metabolism, secretory activity, and even the type of medium in which it is exposed. Thus cells at the S-phase and at metaphase are often more sensitive in terms of cell death than those at other stages of the cell cycle. Furthermore, cells secreting steroids (eg., luteal cells producing progesterone) are more resistant than their precursors (6).

The dose of irradiation administered is expressed either as that actually delivered (roentgen, R, or 'r' units) or preferably as that actually absorbed by the tissues (measured in rads). One rad is equivalent to 0.01 Joules per kilogram (J/kg) or 0.01 gray (Gy: 3). Furthermore, when reproductive organs are exposed to X- or γ-rays in air at normal temperature and pressure, lR is approximately equivalent to 1 rad. It must be emphasised, however, that this relationship only holds for soft tissues and does not apply to 'hard' tissues (eg., bone) or when fission products are used as the source (especially neutrons). Since the majority of studies to which we will subsequently refer involve exposures of only a fraction of a gray of X- or γ-rays and were carried out before the introduction of SI units we will retain the units R and rad.

2. Changes in germs cells induced by ionizing radiations

The life history of the female germ cell (Fig. 1) is complex and involves many stages of differentiation (Fig. 1). Each of the stages responds differently to irradiation according to chromosomal configuration and distribution of its cellular organelles. Variations in response are also known to exist within stages; for example, between primary oocytes in primordial and growing follicles. The response of a given stage of follicular development can also vary with age (4–7).

In radiobiological studies on the mammalian ovary it is also important to take account of the

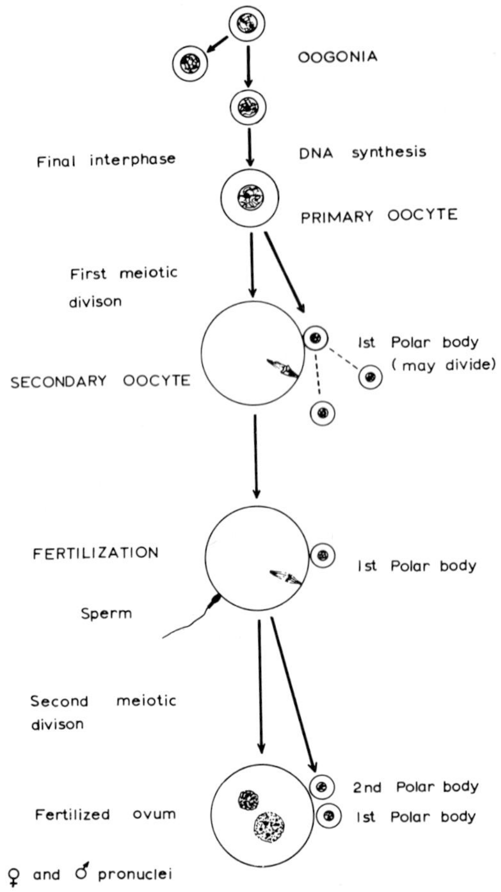

Fig. 1. Life cycle of the female germ cell (from Baker TG: Reproduction in Mammals. Austin CR, Short RV (eds) chapter 2, CUP London 1972).

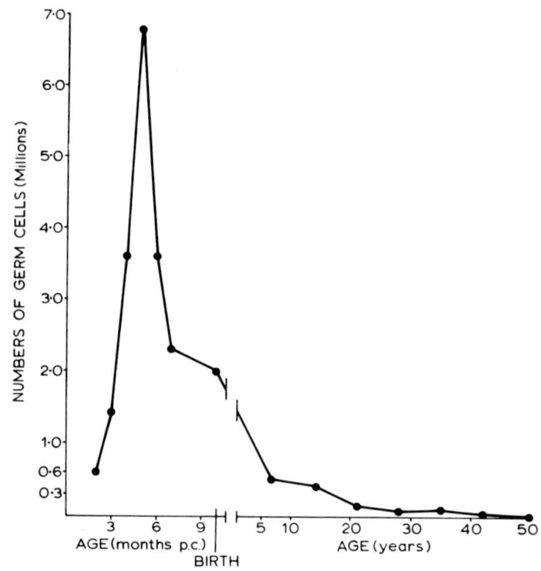

Fig. 2. Changes to the total population of germ cells in the human ovary with increasing age (from Baker, ref. 5).

Table 1. Doses of radiation required to kill all oocytes (Figures in parenthesis are LD50 doses).

Species	Primordial follicles		'Growing' & Graafian follicles
	Juvenile	Adult	
mouse*	15r	50r	2,000r
	(7r)	(10–15r)	–
rat*	100r	315r	4,400r
	–	(100r)	–
guinea pig	–	15,000r	–
	–	(500r)	–
pig	–	–	–
	(500r)	(>500r)	–
cow	–	–	–
	(900r)	(>900r)	–
monkey	72,000r	7,000r	*ca* 5,000r
	–	(5,000r)	–
human+	–	?5,000r	–
	–	(?2,000r)	–

For references and time taken to eliminate germ cells see (4)
* Varies with age, strain and radiation procedure see (4)
+ Depends on authority
– Response not known

characteristic quantitative changes in germ cells. Thus, the relatively small number of primordial germ cells present during migration in the embryo (9, 12–14; see also Chapters 1, 2, 3) typically rises steeply through mitotic proliferation. This process in the female is, however, accompanied by one of cell degeneration (atresia) affecting oogonia and differentiating oocytes. It results in a decline in the germ cell population (Fig. 2) which continues exponentially throughout reproductive life (15–20). In quantitative studies of cell death, therefore, it is essential to compare the total number of oocytes (or follicles per size class) in matched groups of control and irradiated ovaries. The results are expressed in terms of the doses of radiation required to destroy half (LD50) or all ('sterilizing dose') of the germ cells within a given post-irradiation interval. Such studies are beyond the scope of the present review but the results obtained for a variety of post-natal mammals are shown in Table 1 (see 4–7).

A complicating factor for post-natal ovaries concerns the hormonal status of the animal at the time of exposure. It is well established that the rates of spontaneous and radiation-induced atresia vary with the levels of gonadotrophins circulating in the blood. Hypophysectomy markedly reduces the extent of atresia in irradiated and control ovaries of adult rats such that the population of oocytes shows little reduction with increasing age (20, 21). By contrast, injections of pregnant mare's serum gonadotrophin

(PMSG) greatly enhance the rate of depletion of follicular oocytes (20, 21). This effect of PMSG is most marked with respect to primordial follicles. With multilayered and Graafian follicles the hormone can seemingly reverse atresia to the point where the oocytes are 'rescued' and can give rise to viable offspring (22, 23).

The objectives of a qualitative assessment of radiation-induced cell damage or death are to identify patterns of degeneration and/or functional changes. In the case of germ cells one is faced with the added complication that at any given time a proportion of the cells will be undergoing atresia. In the normal course of events this is the major cause of cellular depopulation in the ovary since few of the original stock of oocytes will be ovulated (9, 20). As Ingram (20) has pointed out, the major problem for the radiobiologist is to distinguish between incipiently atretic and healthy oocytes on the one hand, and between spontaneous atresia and irradiation-induced atresia on the other. The morphological signs of atresia in irradiated and control ovaries are superficially similar but the actual processes involved in their production may be different. The problem is compounded by the effect of chemical fixatives on the tissues of the ovary. Those fixatives containing ethanol (eg., Carnoy's, Gendre's) cause marked fixation shrinkage and give the impression to the unwary observer of a higher incidence of atresia than when Bouin's aqueous fluid is employed. Suitably buffered formal-saline and osmium tetroxide cause little swelling or distortion and thus the incidence of atresia appears to be relatively low when these fixatives are used (9, 24, 25).

Clearly, the goal of studies involving the use of light and electron microscopes should be to improve our understanding of the differences between spontaneous and radiation- induced atresia of germ cells.

2.1. Light microscopy

Germ cells are sensitive to ionizing radiations at all stages in development. The purpose of this section is to compare the qualitative responses between stages in the cell cycle (mitosis and meiosis) and between species.

The embryonic ovary contains two types of germinal stem cells, primordial germ cells and oogonia. Exposure to X-irradiation results in a rapid decline in the number of both types of cell, although surviving primordial germ cells are more successful in attempting to repopulate the ovary by mitosis (26, 27). It is not clear whether the observed reduction is due to cell killing alone, to a reduction in the rate of mitosis, losses in migration (primordial germ cells), or to cell death during sex differentiation (oogonia: 26–29). Irradiation of fetal ovaries (ie., after sex differentiation) results in a higher incidence of degenerating germ cells than normal. Examination of ovaries shortly after exposure reveals various abnormalities in appearance which are indistinguishable from those seen in the unirradiated fetus. Thus, primordial germ cells and oogonia show signs of degeneration at interphase ('pyknotic oogonia') or during mitosis ('atretic divisions': 15, 17). The latter are characterised by a defective mitotic apparatus. Oocytes degenerating at various stages during meiotic prophase ('Z cells': 17, 18) are also more numerous. The nuclear chromatin of damaged germ cells becomes highly condensed, nuclear membranes are disrupted or lost, and a marked eosinophilia is developed in the cytoplasm, particularly of the oocytes. The rapid disappearance of these cells, revealed by quantitative studies (17, 29, 30), is probably accomplished as in the normal ovary by the phagocytic activity of neighbouring somatic cells (25, 31, 32).

In the ovaries of mammals at birth or shortly thereafter, the oocytes are at the protracted diplotene stage of meiotic prophase and some 90% are enclosed within primordial (unilaminar) follicles. However, the remainder are contained in growing (multilayered) or Graafian (antral) follicles (8, 9, 33). In all of these follicles the germ cell and its associated somatic cells may be considered as a functional unit. Radiosensitivity will therefore vary between species and with the stage of follicular growth attained by the oocyte.

The formation and structure of primordial follicles and the processes of follicular growth, ovulation and atresia have been the subject of many investigations (19, 20, 33–39) and are beyond the scope of this chapter. By contrast, the literature relating to morphological changes in oocytes and their follicle cells resulting from irradiation is somewhat sparse. Most published reports have been concerned with dose-response relationships. The studies also encompass widely differing doses, dose-rates, post-irradiation intervals and other experimental conditions (eg., whole-body, pelvic or localised ovarian exposures); the influence of these variables is discussed in major reviews (6, 7).

Characteristic signs which represent advanced stages of irradiation damage to primordial oocytes have been reported for a variety of mammalian species. These include the localized condensation of nuclear chromatin leading to the formation of a shrunken pyknotic mass (40–52). At one stage in this process the nucleus appears surrounded by a clear 'halo', probably representing a grossly exaggerated swelling of the perinuclear space (see Section 2.2). The cytoplasm of these cells becomes increasingly eosinophilic and the whole cell shrinks. Such degenerating oocytes are eliminated within 24 h in the mouse and rat (44, 50–52) or over a period of a week or more in the monkey and guinea-pig (6, 7, 49). In some species the surviving primordial follicle cells appear to phagocytose the enclosed oocyte. For example, in mice and guinea pigs (41, 49, 50) irradiation results in the appearance of numbers of small spaces in the ovarian stroma which appear to be 'empty' primordial follicles.

Prior to the onset of pyknosis, on the other hand, there are detectable differences between species in the early responses of primordial follicles to X-irradiation. To what extent these reflect the dose administered and the influence of histological fixatives is not known. Sterilizing doses of X-rays produce the first visible changes in the primordial oocytes of rats and mice in a short time (51, 52). In the mouse, Parsons (52) recorded volume reductions in oocytes within 30 minutes after exposure to 200R (or 6 h after 7R), demonstrating that the time of onset is dose-dependent. Subsequently, the nuclear chromatin underwent partial condensation and eventual pyknosis. However, in a proportion of oocytes the nuclei enlarged and underwent karyolytic changes, accompanied by wrinkling of the nuclear membranes. Parsons claimed that around 50% of the oocytes showing karyolysis induced by 7R reverted to normal morphology within 24 hours of exposure, while the remainder (and all oocytes exposed to 200R) degenerated.

No early changes in oocyte volume were recorded in similar studies in young rats exposed to 100R (44, 51). The fine chromatin network of the dictyate nucleus developed a beaded appearance which gradually changed to a coarse, clumped configuration (Figs. 3–8). This alteration may be equivalent to the karyolytic change seen in mouse oocytes although, as a result of the higher radiation dose used for rats, the majority of these cells soon became pyknotic. In both rats and mice the associated follicle cells appear to remain undamaged by these moderate doses (Figs. 6–8).

In marked contrast, primordial follicles in the rhesus monkey appear unaffected for at least 2 weeks after exposure to 600R X-rays (47) although this dose increased the loss of growing follicles. However, doses of 1000R to 12000R (24, 40) caused changes in small follicles differing in important respects from those already described.

The first signs of radiation damage were detected in follicle cells, where the incidence of cytoplasmic vacuolation and nuclear pyknosis preceded any obvious changes in the enclosed oocytes. In the oocyte, chromosomes appeared thinner and the nucleolus more variable in form than in normal cells (24). With higher doses the nucleolus became enlarged, 'vacuolated' and sometimes stained only weakly; swelling of the whole nucleus and vacuolation of the cytoplasm followed in numerous oocytes after 9000R (Figs. 9–12). The subsequent pattern of degeneration of the oocyte resembled that in other species (40; see above). The incidence and severity of these histological changes were dependent both on dose and post-irradiation intervals. Thus at 4 days after doses of 7000–9000R the few remaining primordial follicles were extensively damaged (Figs. 11–12). Conversely, lower doses (<5000R) may result in a severe reduction in follicle cells well in advance of the associated oocyte.

Primordial follicles in the ovaries of guinea pigs are also relatively resistant to X-irradiation (45, 46, 48), although fairly low doses can elicit an early configurational change in the chromatin of the oocyte (45, 49). There are two sub-populations of primordial follicles ('expanded' and 'contracted'), the proportions of which change with age. The 'expanded' type, in which the chromatin pattern resembles that in primordial follicles of the monkey, is predominant in the ovaries of young animals (Fig. 13). With increasing age the 'expanded' forms become transformed into 'contracted' oocytes in which there is a clear zone between the nuclear envelope and the compact 'knot' of chromosomes (53; Fig. 14). An early response to exposures of 300–500R X-rays is that the contracted oocytes revert to the expanded form. Subsequently many oocytes undergo pyknosis or karyolysis. The result is that 3–32 days after irradiation the adult ovary contains a diminished stock of oocytes with a higher proportion of expanded oocytes than in non-irradiated controls (45).

In a further analysis of this unusual change in nuclear chromatin, Hill (49) found a third form of chromosomal configuration. Following expansion of the chromatin the oocytes either were rapidly eliminated by pyknosis, or entered a phase in which the chromatin resembled the spokes of a wheel (Fig. 15). So-called 'wheel' oocytes are eliminated at a slow rate, during which the nucleus shrinks and eventually the cell collapses as the cytoplasm retracts away from the follicle cells. The latter remained morphologically normal (49).

The radiosensitivity of oocytes in multilayered and Graafian follicles is in general similar in all of the species that have been studied (see Table 1). The

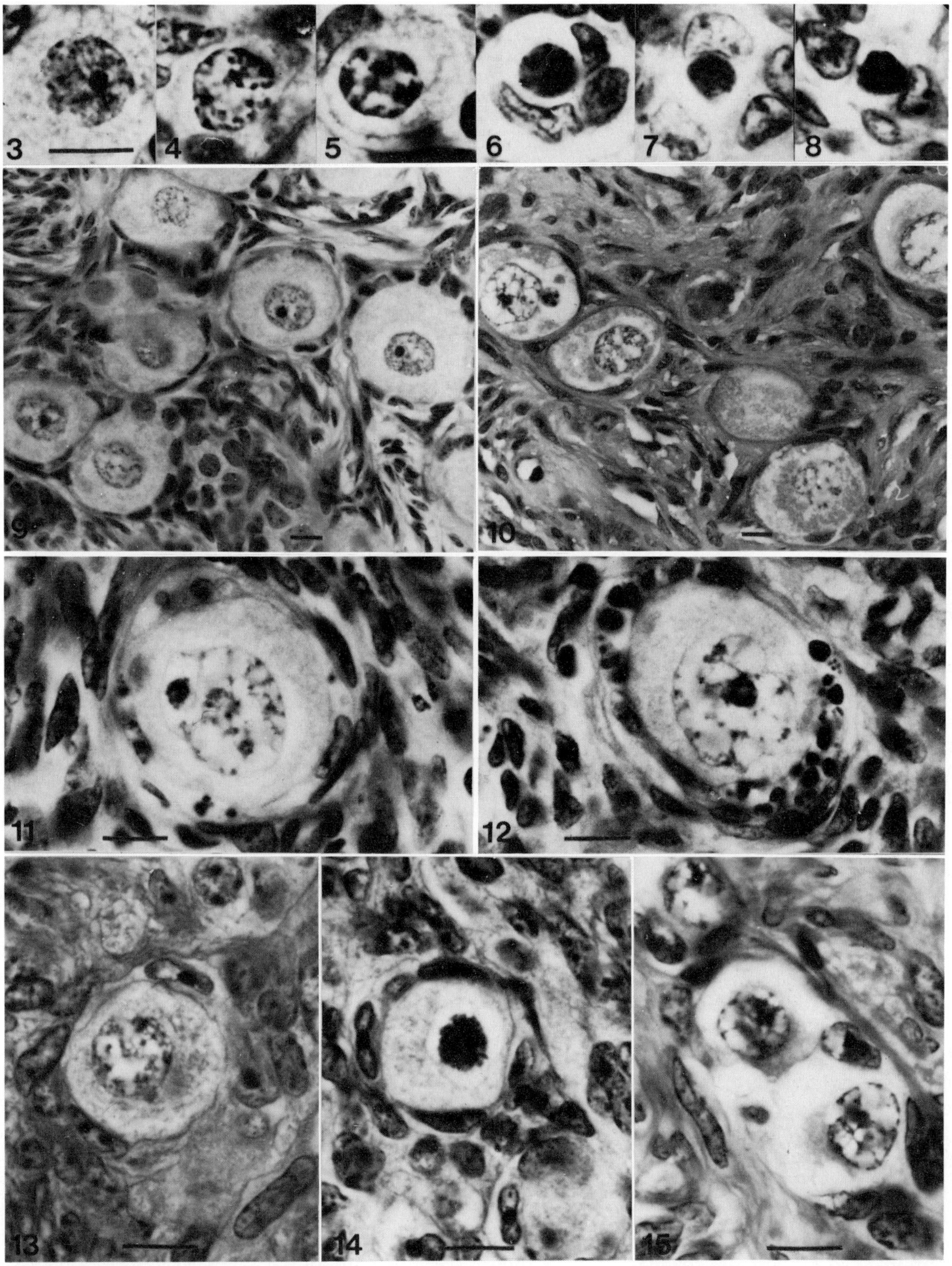

Fig. 3–15. (3–8) Primordial follicles in 7-day old rats: changes in appearance 0–24 hours after exposure to 100R X-rays. The examples illustrate the unirradiated oocyte (3) and increasing severity of changes leading to nuclear condensation, pyknosis and eventual phagocytosis (4–8). Nuclei of the surrounding follicle cells remain seemingly normal. Bar line = 10 μm for all the above figures. (from Parkin, ref. 51). (9–12) Primordial follicles of the rhesus monkey. (9) Unirradiated follicles. (10) Fourteen days after 5,000R X-rays: extreme attenuation of follicular cells, moderate nuclear changes and cytoplasmic clumping in oocytes. (11–12) Two follicles at 4 days after 9,000R showing mild and severe radiation damage respectively in the follicle cells. Characteristic nuclear enlargement and 'vacuolation' of the nucleolus is visible in the oocyte. Bar lines = 10 μm. (13–16) Primordial follicles of guinea pigs (13) Typical oocyte with 'expanded' nuclear chromatin, from a 10-day old animal. (14) Oocyte with 'contracted' chromatin, typical of mature ovary. (15) Mature ovary 16 days after 5,000R X-rays. The surviving oocytes show an expanded 'wheel' configuration and cytoplasmic shrinkage. Bar lines = 10 μm (from material supplied by Hill (49)).

granulosa cells undergo mitotic inhibition followed by widespread pyknosis and some are shed into the antrum where they may become fragmented and/or karyolytic. Eventually the follicle collapses and the debris is eliminated from the ovary, presumably by somatic phagocytes (6, 40, 45, 49). Changes in the oocytes are more variable than in primordial follicles and may not be observed until the follicle cells are at an advanced stage of degeneration. In some cases the oocytes prematurely resume meiosis (pseudomaturation divisions), resulting in the formation of punctate chromatin masses and bizarre fragmentation of the oocyte (pseudocleavage). Phagocytes which penetrate the zona pellucida frequently cause its thickening and folding around the shrunken remains of the egg (20).

Oocytes in growing follicles can survive after the destruction of a proportion of their granulosa cells and some follicles may undergo sufficient repair to persist with a reduced number of layers. These 'giant' oocytes have been observed following relatively large doses of X-rays, (eg., 1,000 R in the rat (44), 5,000R in the monkey (40) and 500R in the guinea pig (45, 49)). Alternatively, the follicle may undergo luteinization with the hyalinized remains of the oocyte at the centre of a 'corpus luteum aberantium' (20, 40).

2.2. Electron microscopy

Molecular damage caused by ionizing radiations theoretically should result in lesions in all cell organelles. However, damage in many sites is capable of repair from a large reserve pool of appropriate molecules. In such instances only very large doses of radiation will significantly impair metabolic functions and lead to cell death. The most likely radiosensitive targets consist of 'key' molecules which are few in number and whose position in organelles is at all times critical (54–56). There is much experimental evidence to show that chromosomal DNA is the most sensitive site, with severe consequences for replication and cell metabolism, although repair through 'unscheduled' DNA synthesis has been reported in both oocytes (57, 58) and spermatocytes (59, 60). In general terms the cytoplasm is much less radiosensitive, with the possible exception of the centriolar apparatus; mitotic inhibition is a frequent result of irradiation (7, 27). Minute changes in lysosomal membranes may cause leakage of potent enzymes resulting in indirect degradation of other cell organelles (56). The aim of electron microscope studies thus should be to detect the onset of visible radiolesions in an attempt to determine the factors which lead to the death or repair of the cell. The few studies that have been carried out on the mammalian ovary using the transmission electron microscope (TEM), and which are described here, deal with oocytes at the arrested diplotene stage of prophase and their surrounding follicle cells.

The accurate diagnosis of early structural radiation damage has been severely hampered by the difficultly of differentiating between oocytes that might have been healthy at the time of irradiation, and those with incipient anomalies associated with spontaneous atresia. Furthermore, slight departures from the accepted appearance of some organelles have been attributed to transient physiological change (61, 62); similar effects can be induced during fixation, dehydration and embedding for TEM (63, 64). The distortion produced in irradiated and unirradiated follicles by polymerization damage is clearly apparent in the illustrations presented in some of these early reports and must be regarded as unacceptable by current standards. By contrast, Baker & Franchi (25) found a lower incidence of cells showing early signs of atresia in normal ovaries prepared for TEM, compared with that in histological material, possibly associated with the differential action of the fixatives.

Mice

The first group of experiments relates mainly to the primordial follicles in young mice, although a few observations relate to other stages of follicular growth. In primordial follicles, the nuclear chromatin of the oocyte is thought to be 'spun out' to form the highly diffuse diplotene or dictyate configuration which is implicated in the high sensitivity of these cells to radiation-induced death (4, 65).

Parsons (52) exposed 4-day old mice to doses of 7–200R of whole-body X-irradiation. The effects on the fine structure of primordial oocytes were analysed semi-quantitatively and their *rate* of appearance was found to be dependent on the dose administered. Two minutes after a dose of 200R there was a transitory reduction in the number of mitochondria, the normal population being restored within the next few minutes. The change was less marked at a dose of 30R and 7R had no significant effect. The restoration of numbers of mitochondria was accompanied by an equally transient drop in their modal diameter; relatively few abnormal types were seen. Parsons concluded that renewal occurred by normal processes of fission from undamaged precursors, the dose being insufficient to cause death of the oocyte by mitochondrial failure. The quantitative changes and the reduction in cell and nuclear volume that he reported must be treated with caution unless they can be confirmed by precise stereometric analysis. Nuclear volumes following exposure to 7R or 200R X-rays were measured from tissues embedded in two different media (Epon and methacrylate). These are known to exert very different shrinkage artifacts, and thus the relative reduction in volume may not be valid.

The first permanent damage in mouse oocytes affected the nucleus – particularly the nucleolus – within about 4 minutes. In primordial follicles, irradiation leads to nucleolar contraction with localized increases in the electron density of the nucleolonema and a reduction in the size of the 'meshes'. A clear narrow zone forms around the nucleolus owing to a redistribution of the associated chromatin fibrils. The resemblance of these changes to those normally seen in the nucleoli of oocytes during normal growth (52, 66, 67) led Parsons to suggest that irradiation might cause precocious ageing in this organelle, leading to degeneration.

More advanced changes in the nucleus result in pyknosis or karyolysis. Chromosomal microfibrils largely disappear from the nuclear matrix during the latter, leaving islands of highly compacted chromatin. In pyknotic nuclei, however, only minor changes in the distribution of the 10 nm fibrils were observed, but there were a number of 'dense aggregates' of granular material which may have resulted from the nucleolar changes. Disruption of cytoplasmic organelles was not evident until 5 to 6 h after 7R (cf. 30–45 min after 200R). Among the changes were vacuolation of the cytoplasm, disruption or loss of mitochondrial cristae, and dilatation of the Golgi saccules, perinuclear space and smooth endoplasmic reticulum. Ribosomes appeared to survive, giving added electron density as the cell shrank in volume. The follicle cells remained unaffected during the period of observation, contrasting with the dense necrotic remains of oocytes at 12–24 h after irradiation. The effects on larger oocytes were not reported but Parsons commented that 'only the young oocytes ... were markedly sensitive to X-radiation'.

Bojadjieva-Mihailova et al. (68, 69) exposed mice to 600–1200R X-rays and noted that, while some oocytes remained essentially normal in appearance, others showed severe structural damage. In contrast to Parsons' (52) observations, mitochondria were found to survive even when the oocytes were grossly abnormal in other respects. Using a simple stereometric method the proportion of the cytoplasm occupied by mitochondria and vesicles of the endoplasmic reticulum was found to have increased by 16% and 78% respectively during the 24 h following the highest dose (1200R). Follicle cells appeared to be unaffected. The survival of oocytes at these relatively high radiation doses contrasts with other observations (41, 48, 52). Presumably a range of follicular sub-stages with markedly different radiosensitivities was included in this study.

The effects of both acute (100–400R X-rays) and chronic irradiation (1.2R/day, γ-rays from a radium source) on mouse oocytes and follicle cells were examined by Jostes and Scherer (70, 71). It is generally accepted that, in terms of cell survival, oocytes are less affected by chronic low doses than by acute exposure (7). However, these authors reported similar symptoms in follicles, the changes taking weeks to develop with chronic irradiation appearing within hours after a single acute whole-body dose. Nevertheless the two types of radiations may differ considerably in their biophysical effect on tissues (3, 55).

In the oocyte, nucleolar enlargement was followed by progressive disturbances in the nuclear chromatin and even by expulsion of the nucleus (70). Vacuolation of the cytoplasm was detected 10 h after exposure to 100R. In addition, the mitochondria and vesicles of the endoplasmic reticulum became vacuolated and swollen. In growing follicles, vacuolated inclusions developed in the processes traversing the zona pellucida and the oocyte microvilli became irregular. Oocytes in tertiary follicles appeared to resist these changes.

In marked contrast to other observations on mouse ovaries, Jostes and Scherer found that the granulosa cells of both primordial and growing follicles were severely damaged only 1 h after 100R X-rays. There was marked distension of all membrane-bound organelles and 'edema' of the cytoplasm. These changes persisted for long periods after irradiation and eventually resulted in a depletion of mitochondria. Ribosomes became relatively abundant. Among the long-term effects, surviving follicles contained a higher proportion of pyknotic nuclei than in spontaneous atresia, the number of layers of follicle cells was reduced, and anovular follicles were found. The tertiary follicles exhibited the same types of atresia found in unirradiated animals, including precocious maturation of the oocyte.

The conflicting results of these studies are not easily explained. Jostes and Scherer (71) also found 'edematous' changes in the thecal and luteal cells, although such cells were generally much more resistant. The possibility that these changes are the result of stimulation by gonadotropic hormones must not be overlooked (72). It is unfortunate that even the unirradiated ovaries examined by these authors, who used methacrylate embedding for electron microscopy, show frequent vacuoles and other electron-lucent spaces in the cellular matrix. These phenomena are reminiscent of polymerization damage (73). On the assumption that this study was marred by technical faults, the finer changes described by Parsons (52) would be masked and some doubts cast upon the validity of the interpretations.

Rats

Matsumoto (72, 74) showed that 4 or 7 months following the exposure of 10-day old rats to 190R X-rays the ovaries were devoid of normal follicles and corpora lutea. Small anovular follicles with essentially normal granulosa cell structure were common, together with some interstitial cells (74). While long-term gonadotrophin-induced changes in the latter were described these studies did not include cellular changes leading to the disappearance of oocytes.

The only observations on short-term damage to rat oocytes appear to be those of Parkin (51). She established that in one week-old Wistar rats, 100R of X-rays (62R/min) resulted in the loss of the majority of primordial follicles within 24 h. The earliest symptoms of abnormality (Phase 1) developed within minutes of exposure and consisted of enlargement of the vesicles of the endoplasmic reticulum adjacent to the mitochondria. The perinuclear cisternae and, occasionally, the saccules of the Golgi complex, also were enlarged (Figs. 16, 17).

Within hours of exposure (Phase 2) qualitative alterations in mitochondria were increasingly obvious: clear areas appeared in the matrix and the cristae became fewer or appeared crowded towards each end of the ovoid profiles. The more grossly affected mitochondria were swollen and lacked cristae. Some enlarged vesicular organelles with variable contents became evident during this phase. The multivesicular bodies, characteristic of rodent oocytes (75–77), appeared to be little affected.

More grossly damaged oocytes (Phase 3) resembled those classed as pyknotic in the mouse (52). Few typical mitochondria persisted and the majority of grossly abnormal organelles were aggregated to one side of the nucleus. The density of the cytoplasmic matrix increased and may be responsible for the cytoplasmic eosinophilia observed in histological preparations. Concurrent with the cytoplasmic changes seen in phases 2 and 3, small condensed masses of chromatin appeared both at the nuclear envelope and elsewhere in the nucleus (Figs. 17, 20). There was otherwise little change in the fibrillar chromatin. The nucleolus, on the other hand, underwent considerable modification, the first signs of which were seen in Phase 2. Portions of the reticulum (nucleolonema) thickened and merged, creating a more compact, electron-dense body. At later stages the nucleolus had condensed to a rounded body in which little fine detail could be discerned. Other regions of the normal nucleolus, notably the granular components (67) appeared to separate from the rest and to merge with the chromatin condensing on the nuclear membrane (Figs. 18, 19). Only rarely was a clear zone seen around the nucleolus (52) in the more advanced stages of contraction.

Oocytes in Parkin's Phase 3 of degeneration were often highly irregular in outline, as though portions were in the process of being 'pinched off' by encroaching follicle cells (Fig. 21). The latter, which were seemingly unaffected by irradiation, appeared to share in the process of elimination of the oocyte by phagocytosis. 'Nests' of these cells containing shrunken necrotic remnants were common in ovaries examined 16–18 h after exposure. This accords with Mandl's estimate (44) for the time taken for the elimination of damaged primordial oocytes in immature and adult rats.

The quantitative changes in mitochondria found by Parsons (52) in mouse oocytes were not detected in the rat owing to the minimum post-irradiation period being 5 minutes. Parkin found no evidence for alterations in chromatin fibrils, nor for the extreme condensation associated with radiation-induced karyolysis in oocytes. On the other hand, the features of spontaneous degeneration in non-irradiated control ovaries were reminiscent of those in the oocytes following the karyolytic route described by Parsons.

In the cytoplasm of unirradiated rat oocytes, early signs of degeneration include clumping and marked vacuolation of mitochondria, followed by the gradual accumulation of aggregates of vesicular and granular components, distortion of nuclear shape and shrinkage of the plasma membrane away from the surrounding follicle cells. At later stages masses of membranous material are observed in the cytoplasm, including myelin figures. Concurrently the nuclear chromatin undergoes extreme condensation and karyolysis. Further cell shrinkage, nuclear collapse, and engulfment by a neighbouring somatic cell represent the later stages in the resorption of the oocyte. Similar features of spontaneous atresia have been described for oocytes in fetal and neonatal ovaries in several mammals (31–33, 78–80). These observations indicate that spontaneous and radiation-induced degeneration of rat primordial oocytes proceed by quite distinct mechanisms.

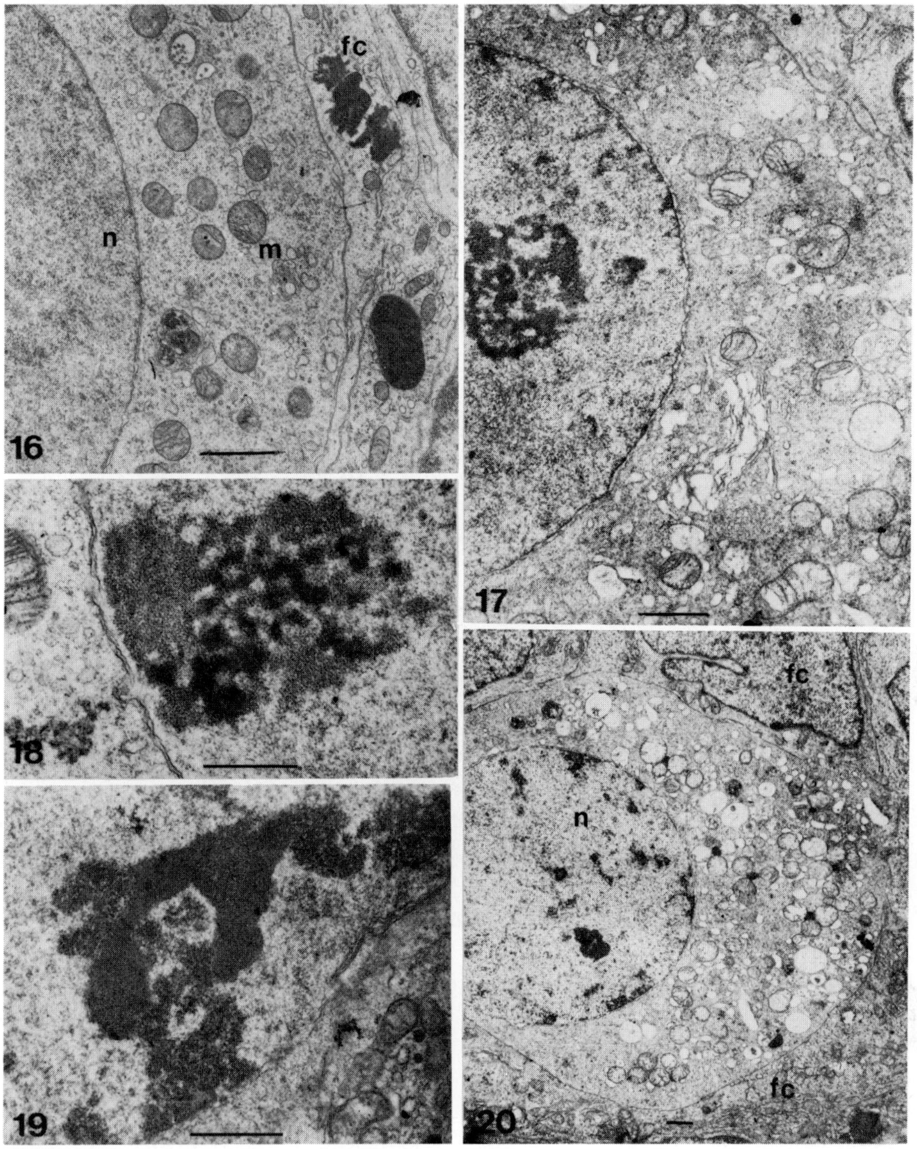

Figs. 16–20. Fine structure of primordial oocytes in week-old rats. (16) Unirradiated oocyte – note appearance of the nuclear envelope, matrix (n) and mitochondria (m). Part of a flattened follicle cell (fc) is at the right. (17) Early changes following a dose of 100R X-rays – condensation of chromatin and distension of cytoplasmic organelles. (18) Nucleolus in an unirradiated oocyte. (19) The condensation and separation of nucleolar components 6 h after irradiation. (20) More advanced stage of irradiation damage; the cytoplasmic organelles are polarized to one side of the nucleus and many are grossly distorted. Chromatin condensation is more widespread. The follicle cells appear normal. Bar lines = 1 μm.

Rhesus monkey

The extreme radiosensitivity of rodent oocytes contrasts with the apparent refractoriness of those in the monkey. Baker (40) found no significant change in the numbers of follicles in the 48 h following exposure to 1000–4000R X-rays. At 4–5 days after exposure, however, many growing follicles were undergoing degeneration, but only with doses in excess of 3000R were there any differences in the histological appearance of atretic follicles between control and irradiated ovaries. Presumably the typical organelles in the cells of one species must resemble those in similar cells in other species; that is, they should thus be equally prone to damage if they represent 'sensitive' targets (see above). Baker & Franchi (24) therefore examined the fine structure of oocytes in adult monkeys at intervals from 7 h to 30 days following localized ovarian exposure to X-rays (1000–12000R). The early response of small follicles was fairly consistent throughout the dose range, large doses eliciting

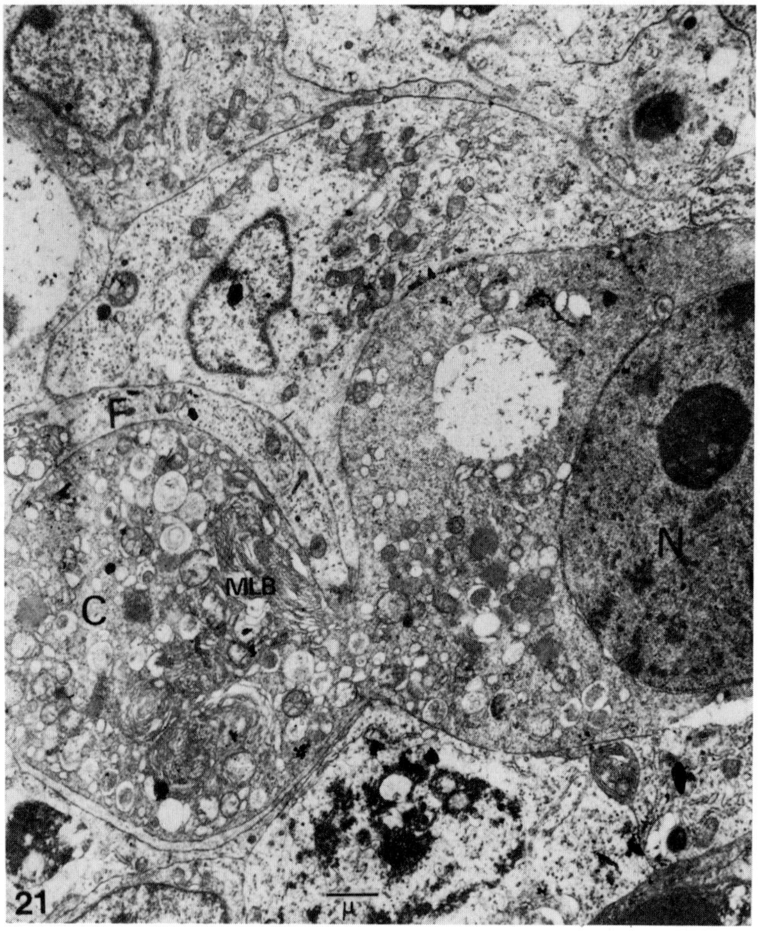

Fig. 21. Electron micrograph illustrating the gross distortion which takes pace during the phagocytosis of a damaged oocyte by adjacent follicle cells (F) in the rat. (C = cytoplasm of oocyte; MLB = multilamellate body: from Parkin, 51). Bar line = 1 μm.

changes in more follicles at shorter time intervals than the smaller doses.

The initial alterations in the appearance of the cytoplasmic organelles in small oocytes were reminiscent of those produced in the rat by 100R (51). Profiles of the smooth endoplasmic reticulum, which are normally flattened against the outer membrane of mitochondria, became dilated vesicles which distorted the shape of these organelles and later fused to form large irregular vacuoles in the cytoplasm (Figs. 22, 23). The small and rather sparse mitochondrial cristae also swelled into pale vesicles in a dense or flocculent matrix and eventually disappeared. In the short term the Golgi elements showed only moderate swelling, but they were rarely identified as aggregated structures after a week. However, the small vesicles scattered in the cytoplasm could have originated from the Golgi apparatus. Ribosomes were difficult to identify after 15 h following exposure to 5000R, although there was a general coagulated appearance to the cytoplasmic matrix which obscured small organelles. Swelling of the perinuclear space and fragmentation of the nuclear envelope were common after high doses (24).

The 'vacuolation' of the nucleolus seen in the histological preparations (Figs. 11, 12) appeared to be due to the loosening of the strands of the nucleolonema and perhaps to some loss of the granular component (Fig. 22). The subsequent fate of this organelle was unclear, although small rounded masses of material of similar appearance to the nucleolonema were scattered in the peripheral regions of the nucleus in grossly damaged oocytes. Detailed examinations of the changes in chromosomal structure proved to be impossible in the electron micrographs. The chromosomal components identified in thin sections of oocytes of young females were hardly discernable in the fibrillar matrix of irradiated nuclei (24). In some of the severely affected cells, peripheral condensation of the

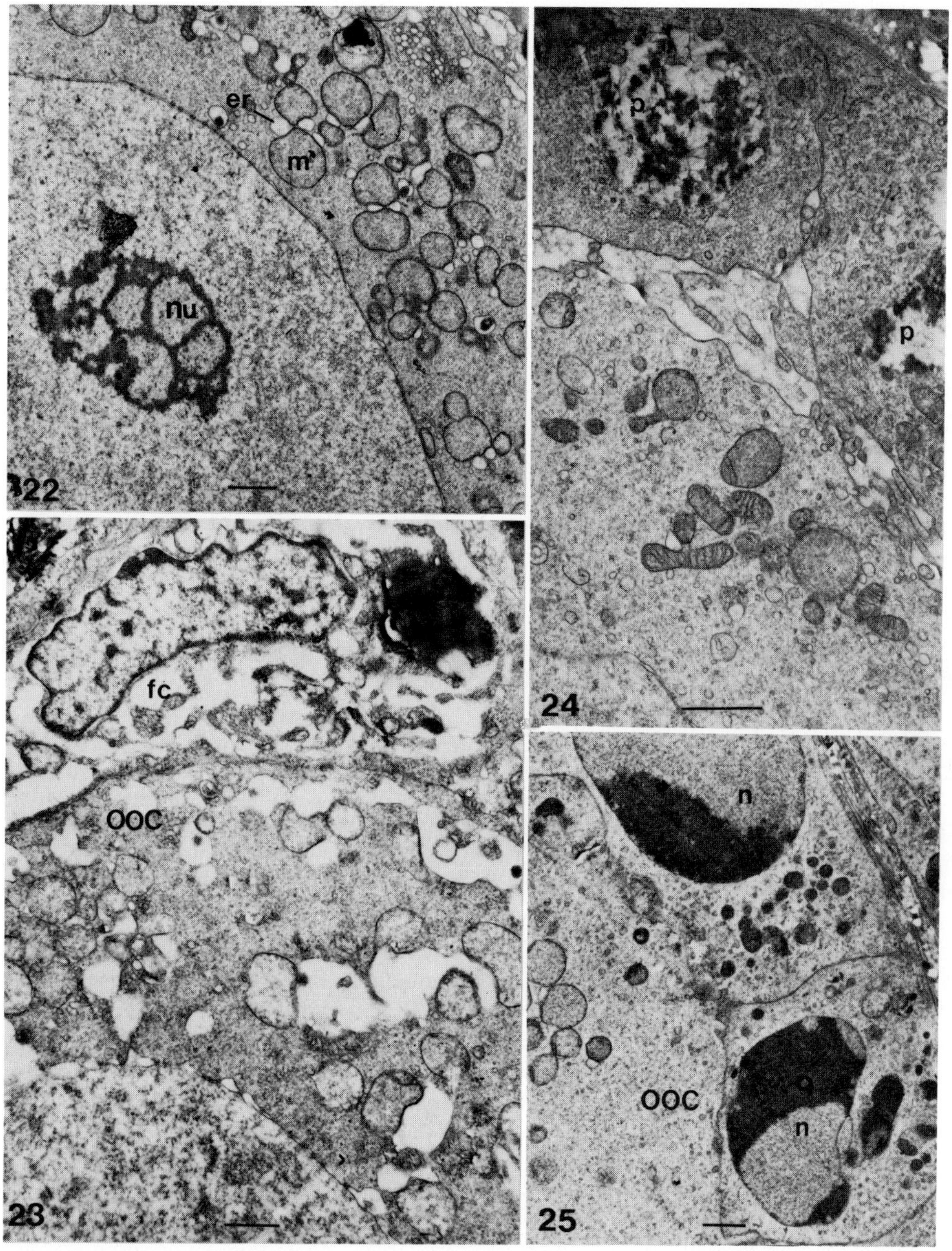

Figs. 22–25. Electron micrographs of irradiated monkey follicles. (22) Oocyte nucleolus (nu) has abnormally large spaces in the nucleolema, mitochondria (m) are devoid of cristae, and there is swelling of associated profiles of the endoplasmic reticulum (er). (23) Oocyte (ooc) and granulosa cells (fc) show severe damage. These two examples were in biopsies recovered 7 days after exposure to 5,000R. (24) Follicle with cuboidal granulosa cells and incipient zona pellucida recovered 7 days after 1,000R X-rays. The cytoplasmic organelles of the oocyte are essentially normal but prominent phagosomes occur in follicle cells (p). (25) Advanced pyknosis in a follicle recovered 20 h after 9,000R. Dense cytoplasmic granules are present but normal organelles are lacking (ooc = oocytes; fc = follicle cell). Bar lines = 1 μm.

chromatin had occurred.

The granulosa cells of primordial and growing follicles were clearly sensitive to doses of 1000R X-rays and showed structural changes in advance of the oocytes (Figs. 24, 25). The dilatation of the endoplasmic reticulum, rupture of membranes, reduction of ribosomes, and changes in mitochondrial shape and loss of cristae, were also identified in atretic unirradiated primordial follicles. However, in the latter the oocyte showed degenerative changes at the same time as the granulosa cells. Large lipid droplets were also formed, especially in the cells of growing follicles where nuclear pyknosis was almost universal following irradiation. Large heterolysosomes were a feature of surviving granulosa cells in the later stages of radiation-induced atresia.

The responses of monkey oocytes to irradiation appear to confirm the claims for mice (52, 70, 71) that high doses of radiation initiate changes in organelles in a shorter time than is evident with lower doses. But none of these observations provides an explanation as to how ultrastructural damage is prevented in monkey oocytes by doses which rapidly destroy all oocytes in rodents (6). Baker and Franchi (24) concluded that an inherent protective mechanism must exist in the nuclei of primate oocytes: a means of rapid repair may also exist for some organelles (see above). With increasing doses the ability of the granulosa cells to provide the necessary supporting role is questionable whether or not they are overtly damaged. By inference, degenerative changes seen in monkey oocytes might not result primarily from radiation damage, but, for example, through the loss of structural and/or physiological links with their follicle cells. The available electron microscope studies provide little supporting evidence for this effect (70, 87). It is known, however, that the cumulus cell extensions are withdrawn through the zona pellucida both during the approach to ovulation (36, 75, 81, 82) and during spontaneous atresia of antral follicles (22, 23, 79, 83). In both instances the separation of the cell contacts is associated with the onset of maturational changes which, in large atretic follicles, result in aberrant nuclear and cytoplasmic divisions in the oocyte.

3. Conclusion

It has long been the hope of radiation biologists that the electron microscope would demonstrate precise intracellular events leading to death or the repair of radiolesions in oocytes and other cell types. In reality the studies made to date have served to extend only slightly the observations of light microscopists and to highlight certain static features in a complex dynamic response. In terms of cytoplasmic effects in oocytes, the use of conventional preparative techniques has demonstrated a radiation-induced transient response in mitochondria (52) which is possibly indicative of repair (cf. 84). The swelling and disruption of membranous structures (early signs which are not uniquely associated with radiation but are recognized also as pointers to spontaneous atresia) are all too easily linked into hypothetical series of changes contributing to the one end, i.e. degeneration. It is tempting to regard them always as an expression of direct damage regardless of dose. Alternatively we err on the side of caution, speculating that they are secondary effects of a nuclear lesion. In fact, there is more evidence for the latter view in experiments on invertebrates, using both very low and very high radiation doses. For example, the normal lethal dose for the *Drosophilia* egg must be increased several hundred-fold in order to kill one in which the nucleus has been shielded from exposure (85). By contrast, a single alpha particle collision from a microbeam aimed at the nucleus of the egg of *Habrobracon* (Hymenoptera) is claimed to be lethal (86).

With respect to nuclear changes, electron microscope studies have so far been unrewarding, having merely confirmed that pyknosis and karyolysis (the fairly advanced stages of chromatin degradation) are typical and more widespread in cells after irradiation. Early structural damage to chromosomes, or clues as to the functional changes, have remained largely undetected except in the region of the oocyte nucleolus. There could be initial mobilisation of the reserves of rRNA precursor molecules, perhaps to participate in repair processes. But the subsequent condensation and loss of integrity of the nucleolus (24, 51, 52) suggests an inhibition of the transcription/translation mechanism for ribonucleoproteins.

These results remain an enigma since variations in the form of the chromosomes within oocytes provide the only reasonable explanation for their differential radiosensitivity. As we have seen earlier, radiosensitivity in terms of cellular lethality and morphological changes varies according to the stage of follicular growth, with changing age (particularly in the perinatal and juvenile periods; 6), between species, and also between strains within one species (especially mouse and rat; 6). The form of chromosomes at the diplotene stage also changes in accordance with these variables, providing some pointers as to the mechanisms underlying this differential radiosensitivity of follicular oocytes, although further studies using a variety of techniques (see below) are required before firm conclusions can be drawn.

Baker et al. (4, 65, 88) have shown that oocytes at all stages of follicular growth in the rhesus monkey, together with those in growing and Graafian follicles in other mammals, have chromosomes which are similar in form to the lampbrush chromosomes of lower vertebrates (Fig. 26). By contrast, primordial oocytes in mice and rats contain highly diffuse chromosomes which, with the electron microscope,

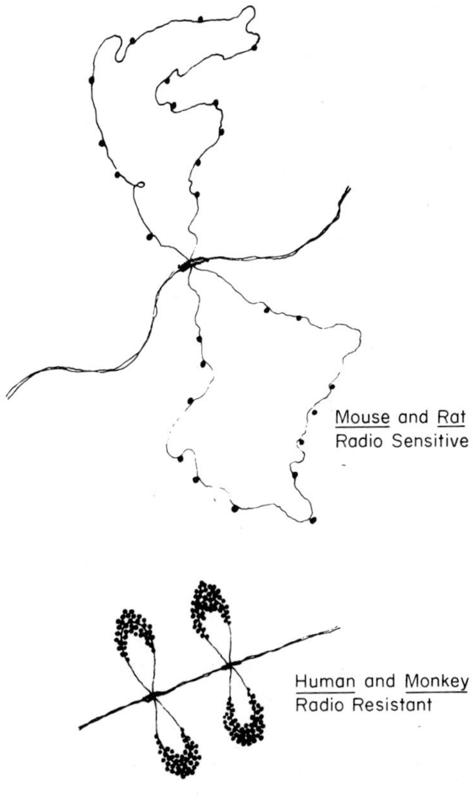

Fig. 26. Diagram to show how chromosomal configuration may influence the differential radiosensitivity of oocytes (from Baker, 4).

appear only as short profiles of thread with little covering of ribonucleoprotein (RNP). Nevertheless, we believe that these oocytes also contain lampbrush chromosomes. It is thus feasible that the radiosensitivity of follicular oocytes is governed by the degree to which the chromosome is 'spun-out', by the size of the lampbrush loops, and by the concentrations of RNP surrounding the loop axis (Fig. 26). The RNP matrix might act as a 'splint' holding the breaks in chromosomes in apposition while repair occurs, or might merely provide the essential components needed in the repair process (4, 89).

When used simply to extend morphological investigations, electron microscopy has only limited value in the radiobiology of the mammalian ovary. Well-designed stereological approaches (e.g. 90; see also 69) would probably yield useful information from thin sections about the chronology and variability of cytoplasmic changes in normal and irradiated oocytes. We have also highlighted on the one hand an inability to distinguish between technical artifact and true structural change, and on the other to determine functional differences between the structurally normal and incipiently degenerating oocyte in the absence of clues about the primary lesions. Several other techniques which have been applied in recent studies of oocyte metabolism could usefully be exploited alongside the TEM to avoid these constraints. Little is known about the synthetic ability or repair phenomena in the nucleus and cytoplasm of oocytes or follicle cells shortly after irradiation. These processes could be investigated by autoradiography (65, 91, 92) and biochemical techniques (93). Cell separation methods and the development of whole-mount chromosome techniques for electron microscopy (94, 95), in combination with specific functional markers, require careful consideration as a more direct means of assessing the relationship between chromosomal structure and radiosensitivity in mammalian oocytes.

References

1. Lawrence CW: Cellular radiobiology. Institute of Biology Series, Studies in Biology 30, London, Arnold, 1971.
2. Puck TT: Radiation and the human cell. Sci Am 202: 142–153, 1960.
3. Alper T: Cellular radiobiology, Cambridge, Cambridge University Press, 1979.
4. Baker TG: Comparative aspects of the effects of radiation during oogenesis. Mutat Res 11: pp 9–22, 1971.
5. Baker TG: Radiosensitivity of mammalian oocytes with particular reference to the human female. Am J Obstet Gynecol 110: 746–761, 1971.
6. Baker TG, Neal P: Action of ionizing radiations on the mammalian ovary. In: The Ovary. Zuckerman S, Weir BJ (eds), New York, Academic Press, 1977 (2nd Edn) Vol. III, pp 1–58.
7. Mandl AM: The radiosensitivity of germ cells. Biol Rev Cambridge Philos Soc 39: 288–371, 1964.
8. Franchi LL, Mandl AM, Zuckerman S: The development of the ovary and the process of oogenesis. In: The Ovary. Zuckerman S, Mandl AM, Eckstein P (eds), New York, Academic Press, 1962, Vol. I, pp 1–88.
9. Baker TG: Oogenesis and ovarian development. In: Reproductive Biology. Balin H, Glasser S (eds), Amsterdam, Excerpta Medica, 1972, pp 398–437.
10. Zuckerman S, Baker TG: The development of the ovary and the process of oogenesis. In: The Ovary. Zuckerman S, Weir BJ (eds), New York, Academic Press (2nd edn), 1977, Vol. I, pp 41–67.
11. Baker TG, McLaren A: The effect of tritiated thymidine on the developing oocytes of mice. J Reprod Fertil 34: 121–130, 1973.
12. Heath JK: Mammalian primordial germ cells. In: Development in Mammals 3: Johnson MH (ed), Amsterdam, Elsevier/North Holland, 1978, pp 267–298.
13. Witschi E: Migration of the germ cells of human embryos from the yolk sac to the primitive gonadal folds. Contr Embryol Carneg Instn 32: 67–80, 1948.
14. Hardisty MW: The numbers of vertebrate primordial germ cells. Biol Rev Cambridge Philos Soc 42: 265–287, 1967.
15. Baker TG: A quantitative and cytological study of germ cells in human ovaries. Proc R Soc London, Ser B 158: 417–433, 1963.
16. Jones EC, Krohn PL: The relationship between age and numbers of oocytes and fertility in virgin and multiparous mice. J Endocr 21: 469–495, 1961.
17. Beaumont HM, Mandl AM: A quantitative and cytological study of oogonia and oocytes in the foetal and neonatal rat. Proc R Soc London, Ser B 155: 557–579, 1965.
18. Baker TG: A quantitative and cytological study of oogenesis in the rhesus monkey. J Anat 100: 761–776, 1966.

19. Baker TG, O W-S: Development of the ovary and oogenesis. In: Clinics in Obstetrics and Gynaecology 3(1). MacNaughton MC, Govan ADT (eds), London, Saunders, 1976, pp 3–26.
20. Ingram DL: Atresia. In: The Ovary. Zuckerman S, Mandl AM, Eckstein P (eds). New York, Academic Press, 1962, Vol I, pp 247–273.
21. Beaumont HM: Effect of hormone environment on the radiosensitivity of oocytes. In: Radiation biology of the fetal and juvenile mammal. Sikov MR, Mahlum DD (eds), CONF-690501, Washington DC, US Atomic Energy Comm, 1969, pp 557–579.
22. Byskov A-G: Atresia. In: Ovarian follicular development and function. Midgley AR, Sadler WA (eds), New York, Raven Press, 1979, pp 41–57.
23. Hay MF, Cran DG, Moor RM: Structural changes occurring during atresia in sheep ovarian follicles. Cell Tiss Res 169: 515–529, 1976.
24. Baker TG, Franchi LL; Electron microscope studies of radiation-induced degeneration in oocytes of the sexually mature rhesus monkey. Z Zellforsch Mikrosk Anat 133: 435–454, 1972.
25. Baker TG, Franchi LL: The fine structure of oogonia and oocytes in human ovaries. J Cell Sci 2: 213–224, 1967.
26. Beaumont HM: The short-term effects of acute X-irradiation on oogonia and oocytes. Proc R Soc London, Ser B 161: 550–570, 1965.
27. Beaumont HM: The effects of acute X-irradiation on primordial germ cells in the female rat. Int J Radiat Biol 10: 17–28, 1966.
28. Mintz B: Continuity of the female germ cell line from embryo to adult. Arch Anat micr Morph exp 48: 155–172, 1959.
29. Beaumont HM: The radiosensitivity of germ cells at various stages of ovarian development. Int J Radiat Biol 4: 581–590, 1962.
30. Baker TG, Beaumont HM: Radiosensitivity of oogonia and oocytes in the foetal and neonatal monkey. Nature (London) 214: 981–982, 1967.
31. Gondos B, Hobel CJ: Germ cell degeneration and phagocytosis in the human foetal ovary. In: The development and maturation of the ovary and its functions. Peters H (ed), Amsterdam, Excerpta Medica, 1973, pp 77–83.
32. Franchi LL, Mandl AM: The ultrastructure of oogonia and oocytes in the foetal and neonatal rat. Proc R Soc London, Ser B 157: 99–114, 1962.
33. Franchi LL, Baker TG: Oogenesis and follicular growth. In: Human reproduction, Hafez ESE (ed), Hagerstown, Harper & Row, 1980 (2nd edn) pp 149–177.
34. O W-S: The interaction between the germinal and somatic cells in gonadal differentiation and development. In: Development in mammals 3. Johnson MH (ed), Amsterdam, Elsevier/North Holland, 1978, pp 299–322.
35. Norrevang A: Electron microscopic morphology of oogenesis. Int Rev Cytol 23: 114–486, 1968.
36. Zamboni L: Modulations of follicle cell-oocyte association in sequential stages of mammalian follicle development and maturation. In: Ovulation in the human. Crosignani PG, Mishell DR (eds), New York, Academic Press, 1976, pp 1–30.
37. Motta P, Hafez ESE (eds): Biology of the ovary; Developments in Obstetrics and Gynecology 2. The Hague, Martinus Nijhoff, 1980.
38. Anderson E: Follicular morphology. In: Ovarian follicular development and function. Midgley AR, Sadler WA (eds), New York, Raven Press, 1979, pp 91–105.
39. Edwards RG: Early human development: from the oocyte to implantation. In: Scientific foundations of obstetrics and gynaecology. Philipp EE, Barnes J, Newton M (eds), London, Wm Heinemann Medical Books, 2nd Edn 1977, pp 175–252.
40. Baker TG: The sensitivity of oocytes in post-natal rhesus monkey to X-irradiation. J Reprod Fertil 12: 183–192, 1966.
41. Murray JM: A study of the histological structure of mouse ovaries following exposure to roentgen irradiation. Amer J Roentgenol 25: 1–45, 1931.
42. Lacassagne A: Etude histologique et physiologique des effects produits sur l'ovaire par less rayons X. These Medecine, Lyon, 1913.
43. Lacassagne A, Duplan JF, Marcovich H, Raynaud A: The action of ionizing radiations on the mammalian ovary. In: The Ovary. Zuckerman S, Mandl AM, Eckstein P (eds), New York, Academic Press, 1962, Vol. II, pp 463–532.
44. Mandl AM: A quantitative study of the sensitivity of oocytes to X-irradiation. Proc R Soc London Ser B 150: 53–71, 1959.
45. Ioannou JM: Radiosensitivity of oocytes in post-natal guinea-pigs. J Reprod Fertil 18: 287–295, 1969.
46. Genther IT: Irradiation of the ovaries of guinea-pigs and its effects on the oestrous cycle. Am J Anat 48: 99–137, 1931.
47. Van Eck GJV: Neo-ovogenesis in the adult monkey. Consequences of atresia of oocytes. Anat Rec 125: 207–224, 1956.
48. Oakberg EF, Clark E: Species comparisons of radiation response of the gonads. In: Effects of ionizing radiation on the reproductive system. Carlson WD, Gassner FX (eds), Oxford, Pergamon, 1964, pp 11–24.
49. Hill RL: Ovarian changes induced by exogenous gonadotrophins and X-irradiation. PhD Thesis, University of Birmingham, Birmingham, UK, 1971.
50. Peters H, Borum K: The development of mouse ovaries after low-dose irradiation at birth. Int J Radiat Biol 3: 1–16, 1961.
51. Parkin PA: The effects of X-irradiation on primordial oocytes in the rat BSc Thesis, University of Birmingham, Birmingham, UK, 1970.
52. Parsons DF: An electron microscope study of radiation damage in the mouse oocyte. J Cell Biol 14: 31–48, 1962.
53. Ioannou JM: Oogenesis in the guinea-pig. J Embryol exp Morph 12: 673–691, 1964.
54. Smith LH, Congdon CC: Biological effects of ionizing radiation. In: Human transplantation. Rapaport FT, Dausset J (eds), New York, Grune & Stratton, 1968, pp 510–525.
55. Casarett GW: Radiation histopathology, Vols I & II, Boca Raton, CRC Press, 1980.
56. Alexander P, Dean CJ, Hamilton LDG, Lett JT, Parkins G: Critical structures other than DNA as sites for primary lesions of cell death induced by ionizing radiations. In: Cellular radiation biology. (18th Symp. fundamental cancer research), Baltimore, Williams & Wilkins, 1965, pp 241–263.
57. Crone M: Radiation stimulated incorporation of (^{3}H) thymidine into diplotene oocytes of the guinea-pig. Nature (London) 228: 460, 1970.
58. Masui Y, Pederson RA: Ultraviolet light-induced unscheduled DNA synthesis in mouse oocytes during meiotic maturation. Nature (London) 257: 705–706, 1975.
59. Kofman-Alfaro S, Chandley AC: Radiation induced DNA synthesis in spermatogenic cells of the mouse. Exp Cell Res 69: 33–34, 1971.
60. Sega GA, Sotomayor RE, Owens JG: A study of unscheduled DNA synthesis induced by X-rays in the germ cells of male mice. Mutat Res 49: 239–257, 1978.
61. Wartenberg H, Stegner H-E: Uber die elektronenmikroskopische Feinstruktur des menschlichen Ovarialeies. Z Zellforsch Mikrosk Anat 52: 450–474, 1960.
62. Wischnitzer S: Intramitochondrial transformations during oocyte maturation in the mouse. J Morph 121: 29–46, 1967.
63. Weakley BS: A comparison of three different electron microscopical grades of glutaraldehydes used to fix ovarian tissue. J Microsc (OXF) 101: 127–141, 1974.
64. Iqbal SJ, Weakley BS: The effects of different preparative procedures on the ultrastructure of the hamster ovary. I. effects of various fixative solutions on ovarian oocytes and their granulosa cells. Histochemistry 38: 95–122, 1974.
65. Baker TG, Beaumont HM, Franchi LL: The uptake of tritiated uridine and phenylalanine by the ovaries of rats and monkeys. J Cell Sci 4: 655–675, 1969.
66. Chouinard LA: A light and electron microscope study of the nucleolus during growth of the oocyte in the prepubertal mouse. J Cell Sci 9: 637–663, 1971.
67. Palombi F, Stefanini M: Ultrastructural analysis of nucleolar evolution in the rat primary oocyte. J Ultrastruct Res 47: 61–73, 1974.
68. Bojadjieva-Mihailova A: Electron microscopical studies of the ovaries of embryos and newborn white mice under the influence of roentgen rays. Izv Inst Morfol Bulg Akad Nauk 9–10, 161–165, 1964. Abstracted in Excerpta Medica Sect 14, Vol 19, Abst No. 2511, 1965.
69. Bojadjieva-Mihailova A, Boneva L, Hadjioloff D: Application of the mathemtical-statistical method for the evaluation of the ultrastructural dimensions of ovocytic mitochondria after X-ray irradiation. In: Proc European Reg Conf Electron Microsc 4th. Bocciarelli DS (ed), Rome, Vol. I, 1968, pp 605–606.
70. Jostes E, Scherer E: Beitrag zur Morphologie röntgen- und radiumbestrahlter Mauseovarien. I. Mitteilung: Die Eizelle. Strahlentherapie 115: 337–365, 1961.
71. Jostes E, Scherer E: Beitrag zur Morphologie röntgen- und radiumbestrahlter Mauseovarien. II. Mitteilung: Beobachtungen an Follikel-Theca- und Luteinzellen. Strahlentherapie 132: 59–78, 1967.
72. Matsumoto A: Behaviour of irradiated ovaries after intrasplenic transplantation in castrated rats. Ann Zool Japon 46: 165–172, 1973.

73. Pease DC: Histological techniques for electron microscopy. New York, Academic Press, 2nd edn, 1964.
74. Matsumoto A: Changes in ultrastructure of rat ovaries after early postnatal X-ray irradiation. Endocrinol Japon 22: 1–15, 1975.
75. Sotelo JR, Porter KR: An electron microscope study of the rat ovum. J Biophys Biochem Cytol 5: 327–342, 1959.
76. Odor DL: Electron microscopic studies on ovarian oocytes and unfertilized tubal ova in the rat. J Biophys Biochem Cytol 7: 567–574, 1960.
77. Odor DL: The ultrastructure of unilaminar follicles of the hamster ovary. Am J Anat 116: 493–522, 1965.
78. Gondos B: Cell degeneration: light and electron microscopic study of ovarian germ cells. Acta Cytol 18: 504–509, 1974.
79. Odor DL, Blandau RJ: Ultrastructural observations on atresia in whole organ cultures of foetal mouse ovaries. In: The development and maturation of the ovary and its functions. Peters H (ed), Amsterdam, Excepta Medica, 1973, pp 63–76.
80. Bonilla-Musoles F, Renau J, Hernandez-Yago J, Torres J: How do oocytes disappear? Arch Gynak 218: 233–241, 1975.
81. Van Blerkom J, Motta P: The cellular basis of mammalian reproduction. Baltimore, Urban & Schwarzenberg, 1979.
82. Tesarik J, Dvorak M: Human cumulus oophorus preovulatory development. J Ultrastruct Res 78: 60–72, 1982.
83. Vasquez-Nin GH, Sotelo JR: Electron microscope study of the atretic oocytes of the rat. Z Zellforsch Mikrosk Anat 80: 518–533, 1967.
84. Baker TG, Franchi LL: The origin of cytoplasmic inclusions from the nuclear envelope of mammalian oocytes. Z Zellforsch Mikrosk Anat 92: 45–55, 1969.
85. Ulrich M: Abtotung von Drosophila-Eiern verschiedenen Alters durch partielle Röntgenbestrahlung. Naturwissenschaften 38: 530–543, 1951.
86. von Borstel RC, Rogers RW: Alpha-particle bombardment of the Habrobacon egg, II. Response of the cytoplasm. Radiation Res 8: 248, 1958.
87. Boyadjieva-Mihailova A, Bakalska-Nesheva M, Kancheva L, Anastosova-Kristeva M: Ultrastructural changes in chick ovaries after X-ray irradiation. Arkh anat Gistol Embriol 73: 42–46, 1977. Abstracted in Biol Absts 65: No 24045, 1978.
88. Baker TG, Franchi LL: The structure of the chromosomes in human primordial oocytes. Chromosoma 22: 358–377, 1967.
89. Miller OL, Carrier RF, von Borstel RC: *In situ* and *in vivo* breakage of lampbrush chromosomes by X-irradiation. Nature (London) 206: 905, 1965.
90. Weibel ER: Stereological methods, Vol I: Practical methods for biological morphometry. New York, Academic Press, 1979.
91. Moore GPM, Lintern-Moore S, Peters H, Faber M: RNA synthesis in the mouse oocyte, J Cell Biol 60: 416–422, 1974.
92. Wolgemuth DJ, Jagiello GM, Henderson AS: Quantitation of ribosomal RNA genes in fetal human oocyte nuclei using rRNA hybridization *in situ*: Evidence for increased mutiplicity. Expl Cell Res 118: 181–190, 1979.
93. Mangia F, Canipari R: Biochemistry of growth and maturation in mammalian oocytes. In: Development in mammals 2. Johnson MH (ed), Amsterdam, Elsevier/North Holland, 1977, pp 1–29.
94. Miller OL Jr, Bakken A: Morphological studies of transcription. Acta Endocrinol 168 suppl: 155–177, 1973.
95. Kierszenbaum AL, Tres L: Transcription sites in spread meiotic prophase chromosomes from mouse spermatocytes. J Cell Biol 63: 923–935, 1974.

Authors' addresses:
TG Baker
School of Medical Sciences
University of Bradford
Bradford BD7 1DP, UK

LL Franchi
Department of Anatomy
University of Birmingham
Birmingham B15 2TJ, UK

CHAPTER 13

Association of oocytes with follicle cells, and oocyte chromatin during maturation

ROBERT W. McGAUGHEY

1. Introduction

Immature mammalian oocytes are derived from primordial germ cells (PGCs) which have migrated to the developing embryonic ovary. These cells proliferated mitotically as oogonia and entered meiotic prophase. In this chapter, the term immature oocyte refers to those stages of female gametogenesis in which the nucleus is at the diplotene stage of meiosis. The diplotene stage is reached in mammalian oocytes at or about the time of parturition. The nuclear stage, germinal vesicle (GV), present in these immature primary oocytes persists to the adult stage. Meiosis is arrested at the GV stage. Resumption of arrested meiosis occurs in the sexually mature female.

From the time of birth, through adolescence and into the period of sexual maturity, some GV-stage oocytes enter a growth phase, characterized by an increased volume until their full size has been attained (1). Accompanying oocyte growth, the number of somatic follicle cells which surround the oocyte increases progressively. The presence of these follicle cells is required for sustained growth of the oocyte (2). During prepubertal ontogeny, fully grown immature oocytes and their surrounding follicular cells do not proceed to ovulation. These oocytes degenerate by follicular atresia. Atresia is characterized by abortive meiotic maturation of the degenerating oocyte (3), or by degeneration of the oocyte in the GV stage (4).

At its earliest stage of development, the immature oocyte is surrounded by very few adherent follicle cells. During the growth phase of the oocyte and follicle, the innermost layer of proliferating follicle cells maintains a close association with the oocyte plasma membrane, or oolemma, by means of junctional complexes (5). With increased follicular growth, the internal follicle cells differentiate into granulosa cells, and the peripheral follicular cells differentiate to form theca layers. During follicular growth, the extracellular zona pellucida is secreted between the oolemma and the inner layer of granulosa cells. Although the zona pellucida forms a gross boundary between the oocyte and granulosa cells, cytoplasmic processes from these somatic cells persist within the zona pellucida, and by means of junctional complexes, maintain contact between the granulosa cell bodies and the oolemma (6). During the follicular growth phase, after oocyte growth is completed, the follicular cavity or antrum develops through a process of cavitation among the granulosa cells. This process results in an antrum filled with follicular fluid. Once the follicular antrum has formed, a distinction can be made between the granulosa cells which directly surround the oocyte (cumulus granulosa cells) and those which peripherally line the antral cavity (membrana or mural granulosa cells). For surviving, competent large follicles in the sexually mature female, growth continues, predominantly through increased volume of follicular fluid, up to the time of ovulation.

During the terminal stages of growth in a preovulatory follicle, the enclosed oocyte proceeds through a complex sequence of developmental changes termed meiotic maturation. In most mammals, oocyte maturation is composed of a reduction in the genetic content of the primary oocyte from the diploid condition (i.e., 2n, 4c) to the haploid state (i.e., n, 2c) which characterizes the oocyte at the time of ovulation and sperm penetration. The classically defined nuclear changes which constitute meiotic maturation include: (1) Condension of discrete bivalents (i.e., synapsed pairs of homologous chromosomes), (2) Dissolution of the gv envelope (i.e., germinal vesicle breakdown or GVB), (3) Establishment of the first meiotic spindle, (4) Disjunction of bivalents at first meiotic anaphase, (5) Segregation of homologous chromosomes at telophase of the first meiotic division, (6) Abstriction of the first polar body by limited cytokinesis, and (7) Arrangement of the remaining group of chromosomes on a spindle for the second meiotic division. Once the first polar

Van Blerkom, J. and Motta, P.M. (eds.), Ultrastructure of reproduction. ISBN 0-89838-572-5.

body has formed and the oocyte chromosomes have become organized for second meiotic metaphase, oocyte meiotic maturation is considered to be complete. The reader is referred to the review by Masui and Clarke (7) for a comparison of meiotic maturation in vertebrates and invertebrates. Although oocyte maturation classically is described as a nuclear process, cytoplasmic events including changes in protein synthesis and in distribution of organelles occur also (7).

The process of meiotic maturation in oocytes is under the general control of gonadotropins, follicle stimulating and luteinizing hormones. In the mature mammalian female, the onset of meiotic maturation occurs subsequent to the peak release of luteinizing hormone from the anterior pituitary (8). The following endocrine conditions exist in the female mammal at the onset of oocyte maturation: (1) High plasma LH levels, (2) Increasing plasma estrogen levels, and (3) Basal plasma progesterone levels (9, 10). In contrast to these plasma levels, the steroid hormonal concentrations of the antral fluid of preovulatory follicles exhibit decreasing estrogen levels and increasing progesterone levels coincidentally with the onset of oocyte maturation (10, 11).

The mechanisms by which mammalian oocyte maturation is regulated are incompletely defined. The mammalian oocyte is capable of spontaneous maturation, unlike the oocytes of other animals (e.g., starfish and amphibians) which require a specific stimulus for their maturation (7). Spontaneous maturation occurs in fully grown mammalian oocytes of adult females when oocytes are removed from the ovarian follicles and maintained under relatively minimal conditions of culture (12, 13). The presence of a small follicular fluid polypeptide, termed oocyte maturation inhibitor (OMI) has been reported to be produced by granulosa cells and to arrest porcine oocytes at the GV stage (14). However, the inhibitory activity of the putative OMI on oocyte maturation has not been observed by other investigators (13, 15, 16). A more probable explanation for the maintenance of the GV stage in fully grown mammalian oocytes includes the steroidogenic changes which occur in the granulosa cells in response to varying concentrations of circulating gonadotropins, the changes in intercellular coupling between the oocyte and the cumulus granulosa cells and changes in the intra-oocyte levels of cyclic nucleotides (17, 18). A complete description of the current hypotheses for the regulation of oocyte maturation in mammals is the subject of a separate review (in preparation).

Spontaneous maturation *in vitro* occurs in isolated, fully-grown oocytes during culture in complex (13, 19) or in minimal, chemically-defined media (20, 21). Both intact (adherent cumulus cells surrounding the oocyte) and denuded (adherent cumulus cells removed chemically or mechanically) oocytes will mature *in vitro* (22). The questions of normality and postfertilization developmental ability for oocytes maturing *in vitro* have been raised (23, 24). Although preimplantation embryos and advanced fetuses have been obtained from oocytes matured *in vitro* (25, 26), various abnormalities also have been reported (19, 27, 28). The degree to which isolated mammalian oocytes mature normally *in vitro* probably is dependent upon the methods by which isolated oocytes are selected to avoid atretic or otherwise developmentally incompetent ovarian gametes (4). Oocytes which exhibit a complete covering of surrounding cumulus cells, and which exhibit normal gross morphology undergo apparently normal meiotic maturation at a high incidence *in vitro* (4).

Two specific aspects of mammalian oocyte maturation have been approached extensively by ultrastructural analysis. The first aspect concerns the significance of intercellular coupling to the regulation of oocyte development and maturation. The second concerns ultrastructural changes in the oocyte nuclear material during meiotic maturation. Electron microscopy has been fundamental to these two areas of oocyte development. This chapter has attempted to correlate ultrastructural analyses with physiological and molecular findings in order to ascertain a more thorough understanding of the maturation process.

2. Establishment of associations between somatic cells and female germ cells

Primordial germ cells (PGCs) have been identified in mammalian embryos no earlier than the late presomite stage of development. When first detected, they are located in the yolk sac endoderm. (see chapters 1, 2, 3). The ultrastructure of early PGCs is well known for the mouse (29–32). The collective findings of these studies confirmed associations between PGCs and the somatic cells of the endoderm, dorsal mesentery and developing gonadal ridge. During residence in the endodermal tissues, and during migratory movements to gonadal ridges, PGCs do not exhibit junctional associations with surrounding somatic cells. However, in one study (30), a small number of regions (termed 'focal junctions') were observed between migrating PGCs and adjacent mesodermal cells. Upon reaching the gonadal ridges, however, PGCs were closely apposed to one another, as well as with surrounding somatic cells. Both focal (32) and gap junctions (30) have been reported in these areas of apposition.

The developmental stage at which mammalian oocytes become associated with ovarian somatic cells and begin to form distinct follicles varies with species. For example, follicular growth occurs in the mouse as early as the first week postpartum (33), whereas in the rabbit, formation of distinct follicles begins at the end of the second week postpartum (34). Rat oocytes are surrounded by closely apposed cells in both granulosa and thecal layers on day 8 postpartum (35), and primary follicles have been reported on day-one postpartum in the rat (5). Although a delay in formation of definitive ovarian follicles may exist in some mammals, from the time of their arrival at the developing ovary, germ cells appear to be associated closely with ovarian somatic cells (35). The ontogeny of ovarian cells which initially become associated with germ cells has not been established (33), but their associations with germ cells are heterologous (5). The general conclusions relative to the role of somatic cell-oocyte associations during follicular development is that granulosa cells provide growing oocytes with energy sources and other macromolecular precursors (2, 6, 36, 37, 38).

3. Associations of follicle cells with fully grown oocytes

During follicular development, the zona pellucida forms as a secreted product of granulosa cells (39), the oocyte (40) or both (41). Granulosa cells nearest the oocyte (the cumulus granulosa cells) are in contact with the oolemma before zona formation (see above). After its deposition, cytoplasmic processes from cumulus cells are present within the zona pellucida, and maintain close association with the oolemma. It therefore is nearly certain that these cumulus granulosa cells remain in continuous association with the oocyte throughout follicular growth and development. The structural association between these cell types is interrupted shortly before or at ovulation (6).

Morphological, physiological and molecular events occurring in the mammalian follicle shortly before ovulation have been studied in great detail. One aspect of follicular development, the intercellular coupling of granulosa cells and oocytes, has received much recent attention (6). Intercellular coupling has been described as contact between cells by junctions which are permeable to small informational and nutrient molecules (6). In summary, the collective findings indicate that the oocyte, cumulus and membrana granulosa cells probably are all coupled. However, it has not been proved definitively that they are coupled by means of gap junctions in all species. Three fundamental questions regarding oocyte-cumulus cell associations in the large, preovulatory Graafian follicle must be addressed. First, are these two cell types coupled physiologically? Second, are the cells coupled structurally by gap junctions as is the situation in other systems? Third, if coupled by gap junctions, what are the physiological and molecular correlates of the coupling?

The study of Gilula, Epstein and Beers (42), in particular, provides rather compelling evidence that rat cumulus cells and oocytes are coupled. Both electrophysiological coupling of the entire cumulus-oocyte complex, and transport of a small (323 dalton) fluorescent molecule (sodium fluorescein) between cumulus cells and between the oocyte and cumulus cells was shown. Optimal transport into the oocyte of radiolabeled nucleic acid precursors and of choline is dependent upon the integrity of the association between cumulus cells and the oocyte in sheep (43), mice (38) and pigs (personal observations). The transport of these relatively small molecules to the oocyte from the surrounding cumulus cells has been interpreted as evidence that the cells exhibit metabolic cooperativity (see below).

The original studies of metabolic cooperativity described the transport of intermediary nucleic acid metabolites from the somatic cells in which they are synthesized to mutant cells in co-culture which were unable to synthesize purines or pyrimidines (44). Because of metabolic cooperativity, deficient mutant cells could grow in co-cultures with normal cells in a medium devoid of the otherwise required precursors. The mammalian oocyte-cumulus cell coupling system may be analogous to somatic cell co-culture systems. Therefore, on the basis of demonstrated molecular transport, the terms metabolic cooperativity and metabolic coupling appear to be applicable.

In most coupled cells which exhibit metabolic cooperativity or low-resistence electrical conductivity, electron microscopy has demonstrated the presence of highly specialized intercellular channels. These channels are functional constituents of the gap junctional complex. In somatic cells structural components of these channels consist of substructures of about 6 nm in diameter which are packed in hexagonal arrays with center-to-center spacing of about 9 nm (44, 45). The distance between cell membranes in regions containing arrays of these channels is about 2 nm (46). The arrays of channels constitute gap junctions, and may have a variety of appearances in different tissues (47). Analyses of gap junction ultrastructure have utilized freeze-fracture techniques (48). From such analyses, models for gap junctional channels have been constructed. These models describe the communicating channels as protein hemichannels of two apposed cell membranes. The corresponding hemichannels (particles and pits on the P and E faces, respectively, observed in replicas of the freeze-fractured membranes) align symmetrically in gap junctional regions to allow for intercellular communication through an intercytoplasmic pore or channel (approximately 1.5 nm wide) located within the aligned hemichannels (47, 49).

Because ultrastructural features of gap junctions and their subunit organization are well defined, stringent constraints have been placed on the identification of gap junctions. An accurate interpretation of areas of close intercellular apposition as gap junctions should include evidence for the existence of as many known chracteristics of the gap junction as possible.

The presence of gap junctions has been described at the interface between the oolema and the cytoplasmic processes of surrounding cumulus cells in the rat (4, 8, 42, 50), rabbit (5, 51), mouse (5) calf (51), and monkey (5). Evidence of gap junctional particle arrays, based on freeze-fracture analysis has been obtained for some of these species (5, 36, 42, 50). Because ultrastructural studies which are limited to thin section analysis cannot provide definitive proof of the presence of gap junctions, a question still exists concerning the universality of gap junctional coupling between cumulus cells and oocytes in mammals. This caveat has been rather strongly emphasized by Moor and Cran (6), who, by thin sectioning methods, were unable to identify gap junctions between the cumulus cells and oocytes of ewes. In only one mammalian species to date, the laboratory rat, have the techniques of electrophysiological and metabolic coupling, thin sectioning with lanthanum labeling and freeze fracturing and optical-diffraction analysis been combined to provide results which have offered compelling evidence both morphologically and physiologically for the presence and expected functions of gap junctions (5, 36, 42).

The means by which trans-zonal cumulus cell processes maintain association with the oocyte plasma membrane is generally thought to involve intermediate junctions or facia adherens (such as in sheep) (6) and macula adherens (such as in the rabbit and cow) (51). In the rat, a very similar or identical junctional complex was identified as a desmosome (42). These terms have been applied quite interchangeably among studies of cumulus cell processes and oocytes, and the characteristics of the desmosomal complexes have been described similarly in all mammalian species examined. These so-called anchoring junctions appear in thin section as electron dense material on the cytoplasmic side of the apposed membranes. They also have been reported to display regions of intermembrane electron density (16).

Investigations in our laboratory of maternal regulation of oocyte maturation in the pig (22) have involved ultrastructural analyses of the cumulus-oocyte complex. The results are necessarily preliminary with regard to intercellular coupling via gap junctions. The findings are, however, consistent with data from other mammals; cumulus cells exhibit distinct trans-zona processes which terminate in junctional apposition with the oolemma (Figure 1). For uncultured, immature oocytes collected from medium-sized follicles (i.e., 3 to 5 mm diameter), both cumulus cell processes and junctional associations with the oolemma are numerous and apparently complex in ultrastructure. Intramembranous particles, similar to those described for desmosomes (6, 42), are apparent. Positive identification of gap junctions, however, awaits application of freeze-fracturing techniques. Parallel physiological studies (personal observations), indicate that the cumulus-oocyte complex exhibits the characteristics of metabolic coupling. Coupling is evidenced by a significantly more efficient transport of radiolabeled tracers into the ooplasm in the presence of adherent cumulus granulosa cells than in their absence.

A third question concerns the function of gap junction coupling between oocytes and cumulus cells. Although many functions, such as electrical coupling and transport of metabolites are known for permeable membrane junctions, evidence for specific physiological or developmentally significant communication is indirect and conclusions therefore remain tentative (47). For example, demonstration of electrophysiological coupling among all cells in the oocyte-cumulus complex (42) has proved the ability of these cells to communicate with one another, but has not defined the developmental role of electrical coupling in this system. Likewise, metabolic coupling between cumulus cells and the oocyte, demonstrated by transport of tracer molecules (38, 43), is insufficient evidence to prove oocyte dependency on cumulus cells for metabolites during growth and development. However, in the case of growing mammalian oocytes contained in small follicles, available corroborating evidence suggests that growth of mammalian oocytes requires the presence of adherent cumulus granulosa cells (2, 38, 52).

In fully grown mammalian oocytes contained

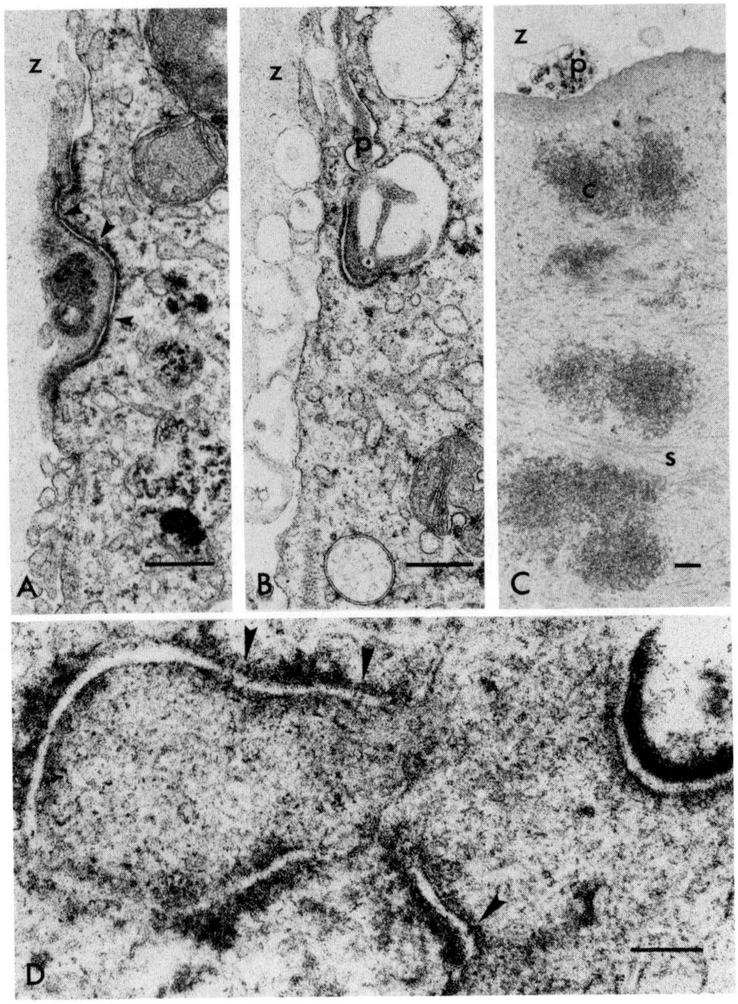

Fig. 1. Electron micrographs of thin-sectioned porcine oocytes. Junctional complexes are shown at the surfaces of the oocytes in A and B; a degenerating terminus of a cumulus cell process (p) is shown in C, and a region of junctional association near the surface of an oocyte is shown at high magnification in D. Noncultured, gv-stage oocytes are exhibited in A, B and D. C depicts an oocyte undergoing maturation after 24 h of culture. Small (A) and large (D) arrowheads indicate regions of close membrane apposition and intermembrane electron density, respectively. The zona pellucida is indicated by Z (A, B, C), and is located just outside the field at the upper left in D. Chromosomes (c) and spindle tubules (s) of the first meiotic division are shown in C. The bars in A, B and C represent 0.5 μm, and 0.1 μm in D, respectively.

within large preovulatory follicles, the significance of cumulus cell-oocyte coupling is even less clear. Demonstrations of electrical and metabolic coupling among the cells of the cumulus-oocyte complex do not provide definitive evidence for developmental and physiological regulation and cooperativity. It has been proposed that cumulus cells coupled with oocytes may regulate cytoplasmic changes which lead to and continue during oocyte maturation. By contrast, coupled cumulus cells may not influence directly nuclear events of oocytes during meiotic maturation (6, 51). Coupling between cumulus cells and the oocyte provides a means by which changes in membrane potentials and permeability can be propagated. These changes could be fundamental to the regulation of oocyte development (23).

One current model for the regulation of mammalian oocyte maturation (18) suggests that elevated levels of cyclic adenosine monophosphate (cAMP) in the oocyte are responsible for maintenance of the GV stage, with intra-oocyte cAMP concentrations controlled through intercellular coupling with cumulus cells. However, indirect evidence now exists for the rat (18) and pig (Rice and McGaughey, unpublished) to indicate that the oocyte is incapable of producing sufficient cAMP to maintain the GV stage. The notion that cumulus cells transport cAMP to the oocyte via junctional coupling, although still an attractive hypothesis, currently lacks direct experimental support. However, cumulus cells are capable of producing cAMP, and of transporting this molecule into other somatic cells to which they are coupled (53).

4. Chromosomal changes during meiotic maturation

Although the mechanisms which regulate meiotic development in oocytes remain obscure (7), many molecular and cellular correlates of oocyte meiotic maturation have been described. Changes in patterns of polypeptide synthesis during oocyte maturation have been described for the pig (54), mouse (55, 56), rabbit (57) and sheep (58). Protein synthesis in maturing oocytes undergoes both qualitative and quantitative change, which are correlated (at least chronologically) with nuclear events of meiotic maturation. Transcriptional activity during maturation in mammalian oocytes has been investigated less vigorously. A general finding is that RNA synthesis continues during the earliest stages of maturation, but is not detected after GVB (59, 60).

Characteristic ultrastructural changes which accompany oocyte maturation include increased microvilli on the oolemma (13, 51, 61, 62), accumulation of cortical granules in a peripheral location (13, 51, 61), redistribution of mitochondria (61, 63) and changes in structure and location of membrane vesicles and fibrous arrays (61, 63). Although identification of the individual stages of maturation are based upon the meiotic changes which occur in the chromatin, surprisingly few investigations have been directed toward the analysis of the ultrastructure of mammalian oocyte chromatin. Technical difficulties related to the very large size of oocytes perhaps have been the most limiting factor for the investigation of chromatin changes. One of the more detailed ultrastructural studies of oocyte chromatin during meiotic maturation was centered on the identification of kinetochores of mouse oocytes during chromosomal condensation and the first meiotic division (64). A more recent investigation described the ultrastructural changes in nucleoli of developing pig oocytes (65).

Comprehensive analyses of chromatin ultrastructure in pig oocytes during transitions from GV configuration through the first meiotic division have been undertaken in our laboratory. These analyses involve examination of ultrastructural characteristics of dispersed GV-stage oocyte chromatin to determine whether the organization of meiotic chromatin undergoes observable change during chromosomal condensation and the first meiotic division. This approach involves dispersed oocyte chromatin, and allows the direct visualization of transcriptional products (i.e., ribonucleoprotein fibrils or RNPs). By such means we have obtained evidence of RNA synthesis in oocytes which has escaped detection by light microscopic autoradiography. Methods of formalin fixation and chromatin spreading have been adapted from techniques developed by Miller and Beatty (66). The protocol consists of selecting immature pig oocytes (4, 13), removing adherent cumulus granulosa cells with mouth operated small-bore glass micropipettes, and disrupting the oocytes in a buffer which disperses oocyte chromatin. GVs of immature oocytes obtained from ovaries, and of cultured oocytes at early stages of maturation are isolated by disrupting the oocyte with micropipettes, and subjected to dispersal conditions. The dispersed samples are layered onto a dense sucrose solution in a lucite container (67) designed to hold a carbon-coated electron microscope grid beneath the sucrose solution. The lucite container is subjected to centrifugation at 2500 xg for 10 min. at 24° C. During centrifugation, the oocyte chromatin sediments through the sucrose solution and adheres to the activated (i.e., hydrophylic) carbon surface on the copper grid. The grid is removed from its container, rinsed in 0.4% (v/v) Photo-flow (Kodak) and air dried. After staining in ethanolic uranyl acetate and lead citrate (67), the sample is dried and examined by electron microscopy.

The types of information obtained are shown in Figure 2. Chromatin of pig oocytes exhibits distinct organizational changes in ultrastructure during meiotic maturation. The most striking change is the pattern of dispersal of early GV chromatin compared to the pattern of chromatin obtained from GVs cultured for several hours and chromatin from oocytes after GVBD. GVs of uncultured oocytes possess an abundance of an uncharacterized ground substance which acts to interfere noticeably with the dispersal of chromatin (Fig. 2A and C). During the first several hours of culture the substance either disappears from the GV or alters its characteristics to allow much more complete chromatin dispersal (Fig. 2E and F). Even under rigorous dispersal conditions [i.e., 10–20 min in 1% (v/v) dimethylsulfoxide, 0.5% (v/v) Triton X100, in glass-distilled water at pH 9.0] the chromatin of GVs from immature oocytes remains relatively undispersed except at its periphery (Fig. 2A). Nucleolar chromatin does not disperse, presumably because nucleoli lie buried in central positions within the GV. By contrast, after six hr of oocyte culture, GV chromatin is extensively dispersed when subjected to the same conditions described

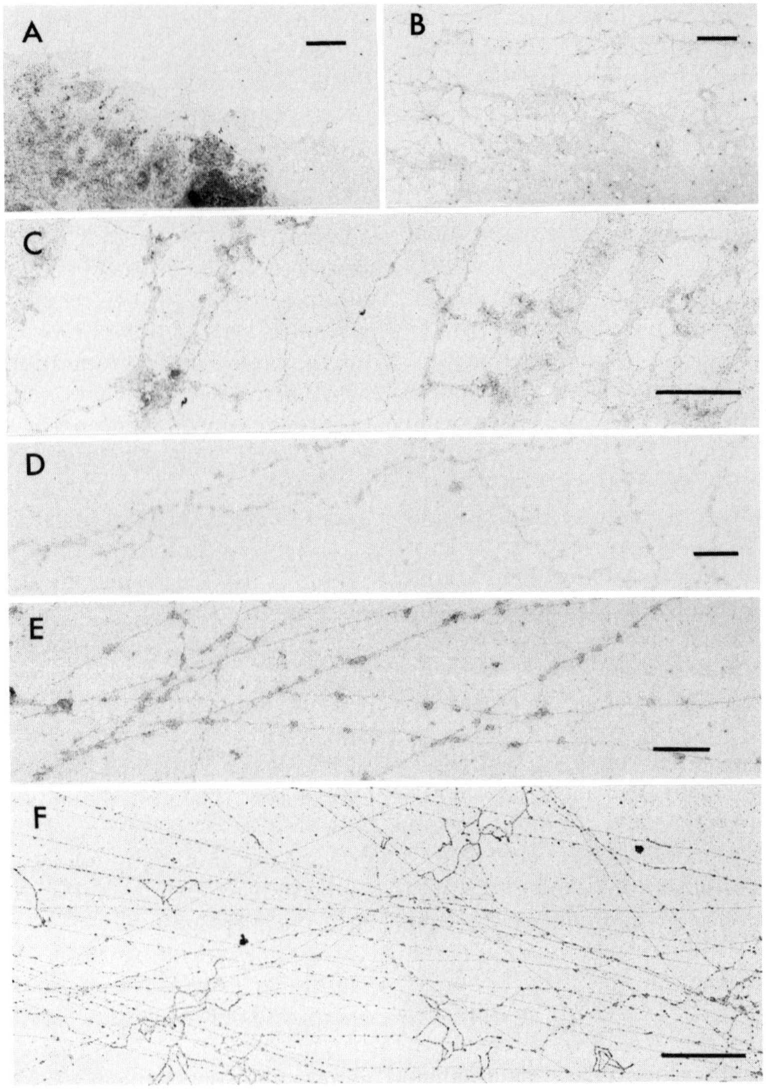

Fig. 2. Electron micrographs of spread, whole-mounted chromatin from porcine germinal vesicles (A and C), from granulosa cell nuclei; (B and D), and from oocytes after culture for 24 h (E and F). A comparison of dispersal characteristics between a gv and a granulosa cell nucleus is shown at low magnification in A and B. Chromatin fibers are less easily or completely dispersed in the gv as compared with the somatic cell nucleus under identical conditions. A region of partially dispersed gv chromatin is shown in C; and exhibits the electron dense 'matrix' ground substance which appears to restrict the dispersal of gv chromatin. Note the beaded, nucleosomal fibers at higher magnification in D and E. The electron-dense, knobby structures in F represent nascent ribonucleoprotein fibrls observed at all stages of oocyte maturation, including those at which condensed meiotic chromosomes were present. The bars in A and B represent 1.0 μm, 0.5 μm in C and F, and 0.1 μm in D and E, respectively.

above; individual chromatin fibers with nucleosomal beads are readily identified (Fig. 2E). With continued culture, the chromatin of maturing oocytes disperses easily. Its fine structure is virtually indistinguishable from the appearance of the control somatic cell type – the cumulus granulosa cells (Fig. 2B and D). The composition of the nuclear ground substance in GVs of immature oocytes is unknown. Both histones (68), and nucleosome assembly proteins (69) are present in great abundance in the GVs of amphibians. A specific polypeptide, GV-associated protein or GVAP, has been described for the mouse (70). GVAP has been reported to be synthesized, phosphorylated and sequestered in the GV (70). It is neither synthesized nor phosphorylated after GVB. Whether or not GVAP is present in the GVs of mammals other than the mouse is unknown. A biochemical analysis of GV-associated proteins of the oocytes from pigs and other mammals clearly is required, and should be correlated with ultrastructural studies to characterize the potentially important ground substance which we have observed. Since the dispersal characteristics of pig oocyte chromatin are directly related to the presence or absence of the ground substance, conceivably it could be of fundamental importance in the structural integrity of oocyte meiotic chromatin, or perhaps even in gene regulation at this crucial stage of development.

We have observed ultrastructural evidence of transcriptional activity throughout spontaneous maturation *in vitro* in pig oocytes. The presence of distinct RNPs, either as nascent fibrils on dispersed chromatin or as detached fibrils, was detected in samples of GVs from uncultured oocytes as well as in samples at each period of culture (Figure 2F). The frequency of appearance of RNPs was highest at the late GV stages (6–8 h culture) and around the time of GVB (10–15 h culture). The lengths of RNPs were greater in samples taken during 8–24 h of maturation than in earlier samples. The RNPs were unstable in the presence of ribonuclease. Their appearance was characteristic of hnRNA transcriptional figures in other systems (71). In the absence of data from experiments with inhibitors of transcription, comments cannot be made on the specific timing of the observed transcription. However, these observations provide rather compelling evidence for the existence of chromatin-associated RNPs at meiotic stages in which the chromatin is condensed into contracted bivalents (metaphase I), and in which the oocyte chromosomes are undergoing the first meiotic division (anaphase I to metaphase II). The presence of RNPs in association with condensed mitotic chromatin has been described both biochemically (72) and ultrastructurally (73). Although the functional significance of stably attached RNPs in condensed meiotic metaphase chromosomes is unknown, it conceivably could be related to the concept of 'stable maternal message', which generally is thought to be present in oocytes and early fertilized embryos (74). Since earlier autoradiographic and biochemical studies of transcription with incorporated radiolabeled nucleotides have not detected transcriptional activity in mammalian oocytes after GVB (59), the level of transcription which we observed ultrastructurally probably is very low. The further characterization of transcriptional activity and of its roles in oocyte maturation, fertilization and cleavage therefore will require experimental approaches with at least the resolving power obtained by ultrastructural analyses of dispersed oocyte chromatin.

5. Summary and conclusions

The mammalian oocyte, its association with the maternal somatic environment and its preparation for fertilization and embryogenesis remain intriguing subjects for future investigations. Knowledge is limited concerning the origin of oocytes before their appearance in the ovary. Investigations into this period of development have yielded results which suggest germ cells associate with ovarian somatic cells for nutritive reasons and that the growth of oocytes is regulated by, and dependent upon, morphologically observable coupling with follicle cells. Generalizations about coupling between granulosa cells and oocytes are premature for universal application to mammals, because too few species in this class have been examined in detail during the ontogenetic stages when oocytes initially associate with ovarian somatic cells. The interactions of mammalian oocytes with ovarian somatic cells relative to genetic changes leading from oogonial mitoses into meiotic prophase and the diplotene GV stage are unknown.

Our state of knowledge is limited concerning the regulatory roles which granulosa cells may play in mediating the development of mammalian oocytes. Hypotheses include: (1) Oocytes progress through their growth, differentiation and maturation according to an endogenous, genetically regulated developmental program with coupled granulosa cells acting to provide permissive conditions such as nutritional requirements; or (2) Oocytes depend upon inductive signals from the coupled granulosa cells for their successful development. The thread of union between these two hypotheses is the underlying assumption that oocytes of mammals are coupled with their surrounding cumulus cells. Coupling, at least electrical and metabolic, has been observed in all species examined. The remaining uncertainty pertains to the absence of conclusive data on the gap junctional complex as the morphological entity which mediates coupling.

An equally challenging problem is the specific nature of the hypothesized 'signals' communicated as developmental regulators between oocytes and the maternal somatic cells. One potential candidate for such a signal, cAMP, has not yet been excluded from a regulatory role in oocyte development. Other candidates might include ions (e.g., calcium), small peptides, or even estrogenic steroids.

The examination of oocyte chromatin in mammals barely has begun. Although chromatin studies of meiotic cells such as oocytes are plagued with difficulties relating to their large amount of chromatin and to the uncharacterized nuclear ground substance, these studies are essential to an understanding of oocyte maturation and of the organization of meiotic chromatin. Further ultrastructural studies, combined with biochemical analyses, make feasible the definitions of chromatin organization in the GV, of changes in chromatin during late meiotic prophase and chromosomal condensation, and of the genomic activity of oocytes at specific stages of pre-ovulatory development. Observation of RNP fibrils in preparations of condensed meiotic chromosomes may provide insight into the existence of preovulatory transcriptional activity in mammalian oocytes and the stabilization of transcriptional products in oocytes as well as other cell types.

Acknowledgements

The author acknowledges the collaborative efforts of Dr. C. Rice-Racowsky in the studies of cumulus cell-oocyte interactions, and of Dr. V. Rider in the studies of oocyte ultrastructure. Some of the work cited in this chapter was supported by grants HDO6532 and HD16788 from the NIH and by a grant from the Whitehall Foundation.

References

1. Peters H, Byskova G, Himelstein-Braw R, Faber M: Follicular growth: The basic event in the mouse and human ovary. J Reprod Fert 45: 559–566.
2. Eppig JJ: A comparison between oocyte growth in coculture with granulosa cells and oocytes with granulosa cell-oocyte junctional contact maintained *in vitro*. J Exp Zoology 209: 345–353, 1979.
3. Himelstein-Braw R, Byskov AG, Peters H, Faber M: Follicular atresia in the infant human ovary. J Reprod Fertil 46: 55–59, 1976.
4. McGaughey RW, Montogomery DH, Richter JD: Germinal vesicle configurations and patterns of polypeptide synthesis of porcine oocytes from antral follicles of different size, as related to their competency for spontaneous maturation. J Exp Zool 209: 239–254, 1979.
5. Anderson E, Albertini D: Gap junctions between the oocyte and companion follicle cells in the mammalian ovary. J Cell Biology 71: 680–688, 1976.
6. Moor RM, Cran DG: Intercellular coupling in mammalian oocytes. In: Development in Mammals, Vol. 4. Johnson MH (ed) Elsevier, North-Holland, 1980, pp 3–37.
7. Masui Y, Clarke HJ: Oocyte maturation. In: Intern Rev Cytol Vol 57. New York, Academic Press, 1979, pp 185–282.
8. Armstrong DT, Goff AK, Dorrington JH: Regulation of follicular estrogen biosynthesis. In: Ovarian follicular development and function. Midgley AR, Sadler WA (eds) New York, Raven Press, 1976, pp 169–181.
9. Cook B, Hunter RHF, Kelly ASL: Steroid-binding proteins in follicular fluid and peripheral plasma from pigs, cows and sheep. J Reprod Fert 51: 65–71, 1977.
10. Eiler H, Nalbandov AV: Sex steroids in follicular fluid and blood plasma during the estrons cycle of pigs. Endocrinol 100: 331–338.
11. Fleming AD, McGaughey RW: A progesterone-binding component in porcine ovarian follicular fluid. J. Endocrinology 94: 69–76, 1982.
12. Donahue RP: The relationships of oocyte maturation to ovulation in mammals. In: Oogenesis. Biggers JD, Schuetz A (eds) Baltimore, University Park Press, 1972, pp 413–438.
13. McGaughey RW: *In vitro* oocyte maturation. In: Methods in mammalian reproduction. Daniel JC, Jr (ed) New York, Academic Press, 1978.
14. Tsafriri A, Channing CP: An inhibitory influence of granulosa cells and follicular fluid upon porcine oocyte meiosis *in vitro*. Endocrinol 96: 922–927, 1975.
15. Leibfried L, First NL: Effect of bovine and porcine follicular fluid and granulosa cells on maturation of oocytes *in vitro*. Biol Reprod 23: 699–704.
16. Rice C, McGaughey RW: Further studies of the effects of follicular fluid and membrana granulosa cells on the spontaneous maturation of pig oocytes J. Reprod Fert 66: 505–510, 1982.
17. Hillier SG: Regulation of follicular oestrogen biosynthesis: A survey of current concepts. J Endocrinol 89: 3P–18P, 1981.
18. Dekel N, Lawrence TS, Gilula NB, Beers WH; Modulation of cell-to-cell communication in the cumulus-oocyte complex and the regulation of oocyte maturation by LH. Develop Biol 86: 356–362, 1981.
19. McGaughey RW, Polge C: Cytogenetic analysis of pig oocytes matured *in vitro*. J Exp Zool 176: 383–396, 1971.
20. Gwatkin RBL, Haidri AA: Requirements for the maturation of hamster oocytes *in vitro*. Exptl Cell Res 76: 1–7, 1973.
21. McGaughey RW: The maturation of porcine oocytes in minimal, defined culture media with varied macromolecular supplements and varied osmolarity. Exp Cell Res 109: 25–30, 1977.
22. Rice C, McGaughey RW: Effect of testosterone and dibutyryl cAMP on the spontaneous maturation of pig oocytes. J Reprod Fert 62: 1–9, 1981.
23. Biggers JD, Powers RD: Comments on the control of meiotic maturation in mammals. In: Ovarian follicular development and function. Midgley AR, Sadler WA (eds) New York, Raven Press, 1979, pp 365–373.
24. Eppig JJ, Koide SL: Effects of progesterone and oestradiol-17β on the spontaneous meiotic maturation of mouse oocytes. J Reprod Fertil 53: 99–101, 1978.
25. Cross CP, Brinster RL: *In vitro* development of mouse oocytes. Biol Reprod 3: 298–307, 1970.
26. Van Blerkom J, McGaughey RW: Molecular differentiation of the rabbit ovum. II. During the preimplantation development of *in vivo* and *in vitro* matured oocyte. Devel Biol 63: 151–164, 1978.
27. Zamboni L: Fine structure of human follicular oocytes maturing *in vitro*. Fourth Annual Meeting Society for the Study of Reproduction, Boston University.
28. Thompson RS, Chakraborty J, Van Pelt L: An ultrastructural study of monkey follicular oocytes maturing *in vitro* and *in vivo*. Fourth Annual Meeting, Society for the study of reproduction. Boston University, 1971.
29. Jeon KW, Kennedy JR: The primordial germ cells in early mouse embryos: Light and electron microscopic studies. Develop Biol 31: 275–284, 1973.
30. Spiegelman M, Bennett D: A light and electron-microscopic study of primordial germ cells in the early mouse embryo. J Embryol Exp Morphol 30: 97–118, 1973.
31. Zamboni L, Merchant H: The fine morphology of mouse primordial germ cells in extra-gonadal locations. Amer J Anat 137: 299–336, 1973.
32. Clark JM, Eddy EM: Fine structural observations in the origin and associations of primordial germ cells of the mouse. Develop Biol 47: 136–155, 1975.
33. Peters H: Some aspects of early follicular development. In: Ovarian Follicular Development and Function. Midgley AR, Sadler WA (eds), New York, Raven Press, 1979, pp 1–13.
34. Gondos B: Granulosa cell-germ cell relationship in the developing rabbit ovary. J Embryol Exp Morph 23: 419–426, 1970.
35. Bjorkman N: A study of the ultrastructure of the granulosa cells of the rat ovary. Acta Anat 51: 125–147, 1962.
36. Amsterdam A, Josephs R, Lieberman M, Linder HR: Organization of intramembrane particles in freeze-cleaved gap junctions of rat Graafian follicles: optical-diffraction analysis. J Cell Science 21: 93–105, 1976.
37. Biggers JD, Whittingham DG, Donahue RP: The pattern of energy metabolism in the mouse oocyte and zygote. Proc Nat Acad Sci USA 58: 560–567, 1967.

38. Heller DT, Cahill DM, Schultz RM: Biochemical studies of mammalian oogenesis: Metabolic cooperativity between granulosa cells and growing mouse oocytes. Develop Biol 84: 455–464, 1981.
39. Chiquione AD: The development of the zona pellucida of the mammalian ovum. Amer J Anat 107: 149–170, 1960.
40. Bleil JD, Wassarman PM: Synthesis of zona pellucida proteins by denuded and follicle-enclosed mouse oocytes during culture *in vitro*. Proc Nat Acad Sci USA 77: 1029–1033, 1980.
41. Martinek J, Kransova H: Development of the zona pellucida in the rat. Folia Morphol 20: 73–75, 1972.
42. Gilula NB, Epstein ML, Beers WH: Cell-to-cell communication and ovulation (A study of the cumulus-oocyte complex). J Cell Biol 78: 58–75, 1978.
43. Moor RM, Osborn JC, Cran DG, Walters DE: Selective effect of gonadotrophins on cell coupling, nuclear maturation and protein synthesis in mammalian oocytes. J Embryol Exp Morph 61: 347–365, 1981.
44. Loewenstein WR: Junctional intercellular communication and the control of growth. Biochim et Biophys Acta 560: 1–65, 1979.
45. Caspar DLD, Goodenough DA, Makowski L, Phillips WC: Gap junction structures. I. Correlated electron microscopy and X-ray diffraction. J Cell Biol 74: 605–628, 1977.
46. Revel JP, Karnovsky JJ: Hexagonal array of subunits in intercellular junctions of the mouse heart and liver. J Cell Biol 33: C7–C12, 1967.
47. Griepp EB, Revel JP: Gap junctions in development. In: Intercellular communication. deMello WC (ed) New York, Plenum Press, 1977, pp 1–33.
48. Kreutziger GO: Freeze-etching of intercellular junctions of mouse liver. 26th Proceedings of the electron microscopy society of America, Baton Rouge, Claitor's Publishing Division, 1968, pp 138–234.
49. Bennett MVL: Electrical transmission: A functional analysis and comparison to chemical transmission. In: Handbook of physiology 1. The Nervous System, Section 1. Cellular Biology of Neurones. Kandel ER (ed) Baltimore, The Williams & Williams Co, 1977, pp 357–416.
50. Fletcher WH: Intercellular junctions in ovarian follicles: A possible functional role in follicle development. In: Ovarian follicular development and function. Midgley AR, Sadler WA (eds) New York, Raven Press, 1979, pp 113–120.
51. Szollosi D, Gerard M, Menezo Y, Thibault C: Permeability of ovarian follicle; corona cell-oocyte relationship in mammals. Ann Biol Anim Bioch Biophys 18: 511–521, 1978.
52. Bachvarova R, Baran MM, Tejblum A: Development of naked growing mouse oocytes *in vitro*. J Exp Zool 211: 159–169, 1980.
53. Lawrence TS, Beers WH, Gilula NB: Transmission of hormonal stimulation by cell-to-cell communication. Nature 272; 501–506, 1978.
54. McGaughey RW, Van Blerkom J: Patterns of polypeptide synthesis of porcine oocytes during maturation *in vitro*. Devel Biol 56: 241–254.
55. Schultz RM, Wassarman PM: Specific changes in the pattern of protein synthesis during meiotic maturation of mammalian oocytes *in vitro*. Proc Nat Acad Sci USA 74: 538–541, 1977.
56. Richter JD, McGaughey RW: Patterns of polypeptide synthesis in mouse oocytes during germinal vesicle breakdown and during maintenance of the germinal vesicle stage by dibutyryl cAMP. Devel Biol 83: 188–192, 1981.
57. Van Blerkom J, McGaughey RW: Molecular differentiation of the rabbit ovum. I During oocyte maturation *in vivo* and *in vitro*. Devel Biol 63: 139–150, 1978.
58. Warnes GM, Moor RM, Johnson MH: Changes in protein synthesis during maturation of sheep oocytes *in vivo* and *in vitro*. J Reprod Fert 49: 331–335, 1977.
59. Wassarman PM, LeTourneau GE: RNA synthesis in fully grown mouse oocytes. Nature 261: 73–74, 1976.
60. Rodman TC, Bachvarova R: RNA synthesis in preovulatory mouse oocytes. J Cell Biol 70: 251–257, 1976.
61. Cran DG, Moor RM, Hay HF: Fine structure of the sheep oocyte during antral follicle development. J Reprod Fert 59; 125–132, 1980.
62. Zamboni L: Fine morphology of the follicle wall and follicle cell-oocyte association. Biol Reprod 10: 125–149, 1974.
63. Zamboni L: Ultrastructure of mammalian oocytes and ova. Biol Reprod 2 (Suppl. 2), 44–63, 1970.
64. Calarco PG: The kinetochore in oocyte maturation. In: Oogenesis, Biggers JD, Schuetz AW (eds) Baltimore, University Park Press, 1972, pp 65–86.
65. Crozet N, Motlik J, Szollosi D: Nucleolar fine structure and RNA synthesis of porcine oocytes during the early stages of antrum formation, Biol Cell 41: 35–42, 1981.
66. Miller OL, Beatty BR: Visualization of nucleolar genes. Science 164: 955–957, 1969.
67. Bakken AH, Hamkalo BA: Techniques for visualizing genetic material. In: Principles and techniques of electron microscopy. Havat MA (ed) New York, Van Nostrand Reinhold Co, 1978, pp 84–106.
68. Woodland HR, Andamson ED: The synthesis and storage of histones during oogenesis of Xenopus laevis. Develop Biol 57: 118–135.
69. Mills AD, Laskey RA, Black P, DeRobertis EM: An acidic protein which assembles nucleosomes *in vitro* is the most abundant protein in Xenopus oocyte nuclei. J Mol Biol 139: 561–568, 1980.
70. Wassarman PM, Schultz RM, LeTourneau GE: Protein synthesis during meiotic maturation of mouse oocytes *in vitro*. Synthesis and phosphorylation of a protein localized in the germinal vesicle. Devel Biol 69: 94–107, 1979.
71. Cotton RW, Manes C, Hamkalo BA: Electron microscopic analysis of RNA transcription in preimplantation rabbit embryos. Chromosoma 79: 169–178, 1980.
72. Zylber EA, Penman S: Synthesis of 5S and 4S RNA in metaphase arrested HeLa cells. Science 172: 947–949, 1971.
73. Hughes ME, Burki K, Fakan S: Visualization of transcription in early mouse embryos. Chromosoma 73: 179–190, 1979.
74. Davidson EH: Gene activity in early development. Second ed. New York, Academic Press, 1976.

Author's address:
Department of Zoology
Arizona State University
Tempe, AZ 85287, USA

CHAPTER 14

Problems in the analysis of mammalian fertilization

DAVID M. PHILLIPS

1. Introduction

This chapter contains a discussion of some of the problems in interpreting literature dealing with mammalian fertilization from the point of view of the experimental systems which have been employed. The discussion is not meant to be a review of the literature in the field. There are several good reviews of the subject, the most recent of which is an excellent review by Yanagimachi (1). A discussion of experimental systems has been chosen because there is a need to consider the difficulties in comparing experiments on many different systems and interpreting the relevance of experiments carried out on *in vitro* systems to fertilization in nature.

Comparisons of results from different studies are difficult. A number of species have been studied. Some workers have examined the fertilization process *in vitro* and others have studied natural fertilization. Some investigations have employed highly polyspermic ova whereas others have looked at monospermic ova. Oocytes from cycling animals as well as oocytes from superovulated immature females or the ovary have been used. In the case of *in vitro* fertilization there are many different types of *in vitro* systems: fertilization systems using zona-intact oocytes or zona-free oocytes, and homologous or heterologous fertilization systems.

In this chapter I will emphasize questions which I feel need to be worked on more extensively. Because workers in the field of fertilization have stressed the similarities among systems, differences among experimental systems will be indicated. We will discuss how the experimental system influences the type of information which is attainable. The aim of this analysis is an understanding of some of the questions that are not well answered and thus require more experimentation.

2. Events in the male

The time when the fertilization process is initiated (e.g., whether it begins when the spermatozoon reaches the cumulus or the oocyte or at some other time) is a subject for semantic discussion. It is important that biologists consider the history of both gametes before their interaction. I will therefore begin with some consideration of sperm maturation. It has been well documented that spermatozoa are incapable, or much less capable, of fertilization when they reach the epididymis than when they have matured during epididymal transmit (2, 3, 4, 5, 6). The maturation process is known to correlate with the development of forward progressive motility (7), with many types of surface changes (8, 9, 10, 11), and in some species with structural changes (12). The meaning of some of the changes that have been reported to occur with epididymal maturation is not clear. For example, surface alterations such as a change in net surface charge or an increase or decrease in lectin binding, could result from glycosylation. Alterations could also result from addition or deletion of a protein(s), or a change in the tertiary structure of a protein which could mask or unmask a portion of an underlying molecule.

The studies of Acott et al. (13, 14, 15) present direct evidence that the addition to spermatozoa of a protein secreted by the epididymis results in the acquisition of forward progressive motility. This type of study suggests that we are beginning to understand the interrelationships between structure and function during the maturation process.

Our understanding of the process of sperm maturation is more complete than our understanding of the phenomena of capacitation and the acrosome reaction. Much of the reason for the progress is that maturation is a relatively easy and satisfying system to explore. One can compare spermatozoa from the caput epididymis with spermatozoa from the cauda epididymis. This is done with *in vivo* spermatozoa so

Van Blerkom, J. and Motta, P.M. (eds.), Ultrastructure of reproduction. ISBN 0-89838-572-5.

there is no problem with possible differences between the natural and the *in vitro* situation. One can examine caput and cauda spermatozoa from the same animal. In studies of maturational changes caput sperm are essentially a control for the cauda gametes. An analogous control is difficult, perhaps impossible, to obtain when one is examining capacitation.

2.1. Capacitation

The phenomenon of capacitation was documented 30 years ago (16, 17). Capacitation may be a universal phenomenon among mammals. A very large number of studies have attempted to demonstrate the morphological, chemical, and motile correlates to capacitation. Recently we have come to understand that a critical part of capacitation may, at least *in vitro*, involve a membrane change which facilitates the influx of calcium (18, 19, 20, 21) which is involved in the acrosomal reaction. However, many studies of capacitation are difficult to interpret partly because of problems in the systems that are available to study the capacitation process.

Capacitation is generally studied *in vitro* where an investigator can vary the environment of the spermatozoa. It should not be assumed *a priori* that the changes that a spermatozoon undergoes during capacitation *in vitro* are the same as those *in vivo*. In fact, Viriyapanich and Bedford (22) recently have demonstrated that rabbit spermatozoa which are capacitated *in vitro* and can penetrate oocytes *in vitro* will not penetrate or penetrate very few oocytes when put into the oviduct and challenged with naturally-ovulated oocytes.

Since the definition of a capacitated spermatozoon is one which will fertilize an oocyte, a proper test for whether or not a population of spermatozoa contains capacitated spermatozoa is to examine their ability to fertilize oocytes. Fertilization can be assessed in several different ways. One method is to allow ova to develop *in vitro* for a day and determine the percentage of two cell embryos. This method may not be accurate because unfertilized ova can cleave and fertilized ova do not always develop into two cells *in vitro*. Some workers examine ova for pronuclei. This is not entirely accurate because two female pronuclei can form in unfertilized oocytes. The best method may be to consider ova as fertilized if one observes two pronuclei in the ovum and sperm tail in the perivitelline space. Blandau (23) has recently written an interesting discussion of the problems of identifying fertilized ova.

Since all methods of assessing fertilization are time-consuming, especially if one has several different samples of possibly capacitated spermatozoa to assess, most studies of capacitation do not employ fertilization as an end point for capacitation. A sample of *in vitro* capacitated hamster spermatozoa, for example, is often judged to be capacitated if the investigator observes some spermatozoa displaying activated motility and observes that many of the motile spermatozoa are missing their acrosomes. There are difficulties with using this type of subjective end point for capacitation. First, it is difficult to determine if the acrosome is lost or present in a motile hamster spermatozoon and, secondly, there is no assurance that spermatozoa in the population would fertilize oocytes and are therefore capacitated.

A proper control for an investigator to employ when correlating some change with capacitation *in vitro* would be to incubate spermatozoa for an equal time in a medium which keeps the spermatoza alive but does not capacitate. Although such a medium has been reported (24, 25), general experience has been that when capacitating factors are left out of the medium, spermatozoa do not survive. Therefore, most studies are controlled only by comparison to fresh epididymal or ejaculated sperm. Reported changes due to 'capacitation' could therefore be caused by incubation time unrelated to capacitation, or could relate to cells being nearer to death.

Another innate difficulty of the *in vitro* capacitation system is that a suspension of *in vitro* capacitated sperm typically is not homogeneous. It contains spermatozoa that do and do not display activated motility. A population of *in vitro* capacitated spermatozoa contain spermatozoa which have as well as those which have not undergone the acrosomal loss (Fig. 1). There are typically motile spermatozoa, immotile spermatozoa and groups of head-to-head associated spermatozoa. The variation in the types of spermatozoa is a considerable problem when observations are made on a few sperm.

Populations of spermatozoa in the female tract at different times after mating also appear morphologically and functionally heterogeneous (26, 27). Recently, the suggestion has been made (Bedford, unpublished) that the male reproductive tract has evolved in such a way to produce spermatozoa which are designed to capacitate at different times after insemination. This presumably would be of selective advantage in most species since insemination and ovulation frequently occur far apart in time. If spermatozoa became capacitated at very different times, male gametes capable of fertilization would be available for a greater period of time. This could also account in part for a selective advantage in having so

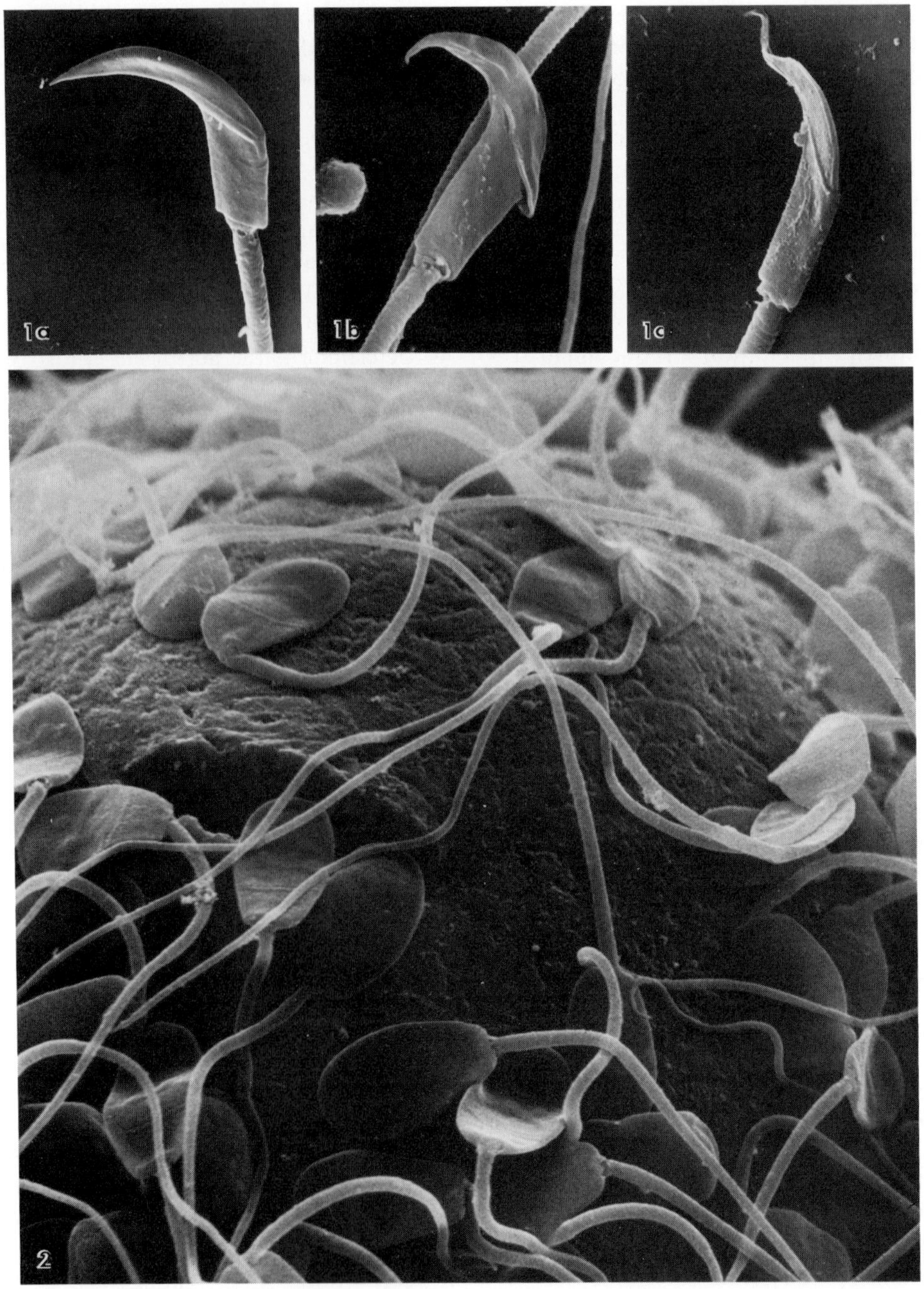

Figs. 1–2. (1) In a population of *in vitro* capacitated spermatozoa one typically observes cells with different morphology and motility. Evidence suggests that spermatozoa from the female tract and even from the cauda epididymis exist as a heterogeneous population in terms of degree or maturation and capacitation. Scanning EMs of hamster spermatozoa from an *in vitro* capacitating medium show (1a) Spermatozoa with acrosomes which appear intact; (1b) Spermatozoa with dissociating acrosomes; and (1c) Spermatozoa with missing acrosomes (×4,800). (2) Uncapacitated sperm typically associate with zonae of heterologous species. This SEM shows fresh epididymal rabbit sperm associated with the pig zona (Micrograph courtesy of Christine Swenson and Bonnie Dunbar). In the analysis of *in vitro* experiments which involve sperm-zona interactions, it is difficult to distinguish between nonspecific zona binding which could be caused by surface charge and a more specific interaction such as may occur *in vivo* (×6,000).

many sperm in an ejaculate as mammals do.

In the usual *in vitro* capacitation system large numbers of spermatozoa must be employed since spermatozoa become quickly immotile when they are diluted. Recently Bavister (25, 28) has experimented with developing more useful *in vitro* capacitation systems. He has been able to make progress in overcoming the problems of sperm concentration by developing a system in which fertilization was achieved with the sperm-oocyte ratio close to unity. Bavister has also been successful in developing a control medium in which spermatozoa obtain motility but do not become capacitated (25, 28). These refined systems will hopefully facilitate more precise and easy to interpret experiments.

2.2. Activation

Since activated motility (also termed whiplash and hyperactivated motility) was described by Yanagimachi (29) for *in vitro* capacitated hamster spermatozoa, activated motility along with the ability to undergo acrosomal loss is employed as a criterion for capacitation. It does, in fact, appear clear that activated motility *in vitro* is correlated closely with the ability of spermatozoa to penetrate oocytes *in vitro*. Our understanding of activated motility, however, is still inadequate. It is not clear whether activated motility is a general phenomenon. In the hamster (30, 31, 32, 33), guinea pig (18), dog (34), and rabbit (27, 35, 36), capacitated sperm have been reported to exhibit activated motility. With respect to mouse sperm capacitated *in vitro*, some workers observe activated motility (37, 38) and others do not (Saling, unpublished). In human sperm capacitated *in vitro*, activated motility is not obvious (Shalgi, unpublished; Katz, unpublished). The relationship between activated motility *in vitro* and the movement of *in vivo* capacitated sperm in the oviduct needs to be examined more closely. I described motility in mouse sperm flushed from the oviduct four hours after mating. The gametes progressed with a snake-like pattern (39). This pattern is similar to whiplash motility in that the flagellar wave displayed a large amplitude *in vivo* but the beat frequency was slower rather than faster than non-activated motility as has been described *in vitro* (37). Overstreet et al., (36) and Cooper and Overstreet (27, 35), have observed motility which appears similar to whiplash movement in rabbit sperm flushed from the oviduct proximal to the ovary at the time when sperm would be expected to be capacitated. Yet Katz and Suarez (unpublished) recently have observed similar motility in rabbit sperm which were removed from the uterus shortly after mating; these sperm should not be capacitated. Katz and Yanagimachi (40) have observed hamster spermatozoa *in vivo* by transilluminating the oviducts of hamsters after natural mating at a time when spermatozoa should have been capacitated. Although it is difficult to obtain clear images when viewing through the oviduct wall, they were able to describe a motility which appeared to be activated motility. In summary, it seems that a motility similar to the activated type of motility observed *in vitro* may occur *in vivo* and that it may be a general phenomenon. However, more work is required before this generalization can be made with certainty. There is no direct evidence for the function of activated motility.

2.3. The acrosomal reaction

The acrosomal reaction was initially defined as the fusion between the plasmalemma and the outer acrosomal membrane (41). The fenestrations produced by the acrosomal reaction allows acrosomal enzymes to be released. Eventually the acrosome is lost entirely. Evidence suggests that the acrosomal reaction and acrosome loss may occur over some period of time and at different locations in the oocyte-cumulus complex (42). In the light microscope one generally can only distinguish between spermatozoa with intact (possibly reacting or reacted) acrosomes and those which have lost their acrosomes. Spermatozoa which have lost acrosomes are generally termed 'acrosome reacted'. I believe this terminology could be misleading as acrosomal reacted spermatozoa could appear to have intact acrosomes when viewed with the light microscope. Therefore, I feel it is important that the term 'acrosome reaction' be applied to the fusion between the sperm plasmalemma and outer acrosomal membrane. After the acrosomal reaction, a fenestrated plasmalemma covers the sperm head over the major portion of the acrosome and an intact plasmalemma is found over the equatorial segment of the acrosome. Acrosomal material may then be lost through fenestrations in the plasmalemma. I feel that 'acrosome loss' should be considered as a subsequent event to the acrosomal reaction in which the remaining acrosomal remnant consisting of the remnants of the fused plasmalemma and outer acrosomal membrane over the major portion of the acrosome and remaining acrosomal contents are lost. After acrosomal loss, the inner acrosome membrane (perhaps with associated material) is exposed to the environment over the major portion of the acrosome. The plasmalemma remains over the equatorial segment of the acrosomes. Evidence (see below) suggests that sperm which have undergone acrosomal reaction can bind to the zona pellucida and subsequently fertilize an ovum, but those which have undergone acrosomal loss in the cumulus cannot bind to the zona and thus will not fertilize an ovum.

Classically, the acrosomal reaction and acrosome loss are believed to occur in or near the cumulus oophorus. Some of the evidence for this conclusion came from observations that sperm flushed from the site of fertilization appear to be missing acrosomes when viewed in the light microscope (27, 35, 43). Katz and Saurez (personal communication) recently have repeated these experiments. They flushed oviducts with oil in lieu of medium and observed spermatozoa in undiluted oviductal fluid. Using high speed video, they observed that activated rabbit spermatozoa had intact acrosomes. They suggest that the introduction of medium could cause spermatozoa in the oviduct to lose their acrosomes. This experiment raises important questions about observations made on the condition of the acrosome of spermatozoa flushed from the oviduct with medium. Flushing of spermatozoa from the oviduct, for example, could cause acrosomal-reacted spermatozoa to undergo acrosome loss.

Hydrolases released during the acrosomal reaction are believed to hydrolyze extracellular mucopolysaccharides known to be responsible for maintenance of the integrity of the cumulus (43, 44, 45, 46). The idea of the acrosomal reaction and loss occurring in the cumulus originates from an assumption that acrosomal enzymes must logically function in hydrolyzing mucus, since acrosomes contain the enzymes for the substrates of mucus. There is also evidence that antibodies against both hyaluronidase and hyaluronidase inhibitors block cumulus dispersion and fertilization (47, 48, 49). Early EM studies reveal that spermatozoa with intact acrosomes, reacted acrosomes and lost acrosomes are found in the cumulus. The interpretation at that time was that the unreacted sperm were dead or abnormal, and the fertilizing sperm were among those which had undergone acrosomal loss (50, 41). Recent observations cast doubts on these conclusions. Working with the mouse, Saling and Story (51) found that sperm bound to the zonae had apparently intact acrosomes. We have observed that acrosome remnants appear to be present in hamster sperm bound to the zona *in vitro* (Figs. 3–6). In contrast, guinea pig spermatozoa which bind to the zona *in vitro* are reported to have undergone acrosomal loss *in vitro*. Therefore, there may be a species difference, at least *in vitro*.

The principal acrosomal protease, a trypsin-like enzyme acrosin, is reported to function in sperm penetration through the zona. The evidence for this is that spermatozoa will not penetrate the zona pellucida *in vitro* in the presence of protease inhibitors (53, 54, 55, 56, 57, 58). This function of acrosin has recently been questioned by Saling (59). By using protease inhibitors she showed that in the mouse, a trypsin-like enzyme appears to be necessary for zona binding *in vitro*, but not zona penetration *in vitro*. Saling proposes the concept that acrosomal enzymes may function in hydrolysis of substrates perhaps to uncover receptors for zona binding. Saling also points out that early experiments which were interpreted to show that sperm proteases function in zona penetration could be interpreted equally well to show that acrosomal enzymes function in zona binding. Bedford and Cross (60) performed a clever experiment which also suggests that proteases do not function in zona penetration. They demonstrated that treatment of zona with wheat germ agglutinin causes the zona to be resistant to trypsin but does not inhibit sperm penetration through the zona.

The interpretation of *in vitro* sperm-zona associations is confused by the fact that sperm clearly form different types of associations with zonae. A few years ago Hartmann and Hutchison published reports which suggested there are two different types of *in vitro* associations which hamster spermatozoa form with zonae; one where spermatozoa form a loose reversible association, termed 'attachment'; and a tighter permanent association which may succeed attachment, termed 'binding' (61, 62, 63, 64). Some workers have tried to distinguish types of associations by treating ova with associated sperm by slow centrifugation, and defining bound sperm as those which remain associated with zona after centrifugation (59). Compounding the problem of attachment vs. binding is the clear evidence that, with the exception of man and apes, uncapacitated epididymal sperm associate with zonae of homologous or heterologous species (Fig. 2) (65). It is likely, therefore, that although sperm association with zonae *in vivo* could be a highly specific reaction, *in vitro* sperm probably are capable of associating to zonae by a number of different mechanisms, including those which could be as non-specific as surface charge. Thus, *in vitro* experiments dealing with sperm associating with zonae are particularly difficult to interpret in terms of the *in vivo* situation.

In summary, the available data suggest that the acrosomal reaction *in vivo* may possibly occur in the cumulus but that loss of the acrosomal remnant may occur on the zona surface. In fact, in the literature of *in vivo* fertilization one can find acrosome-reacted

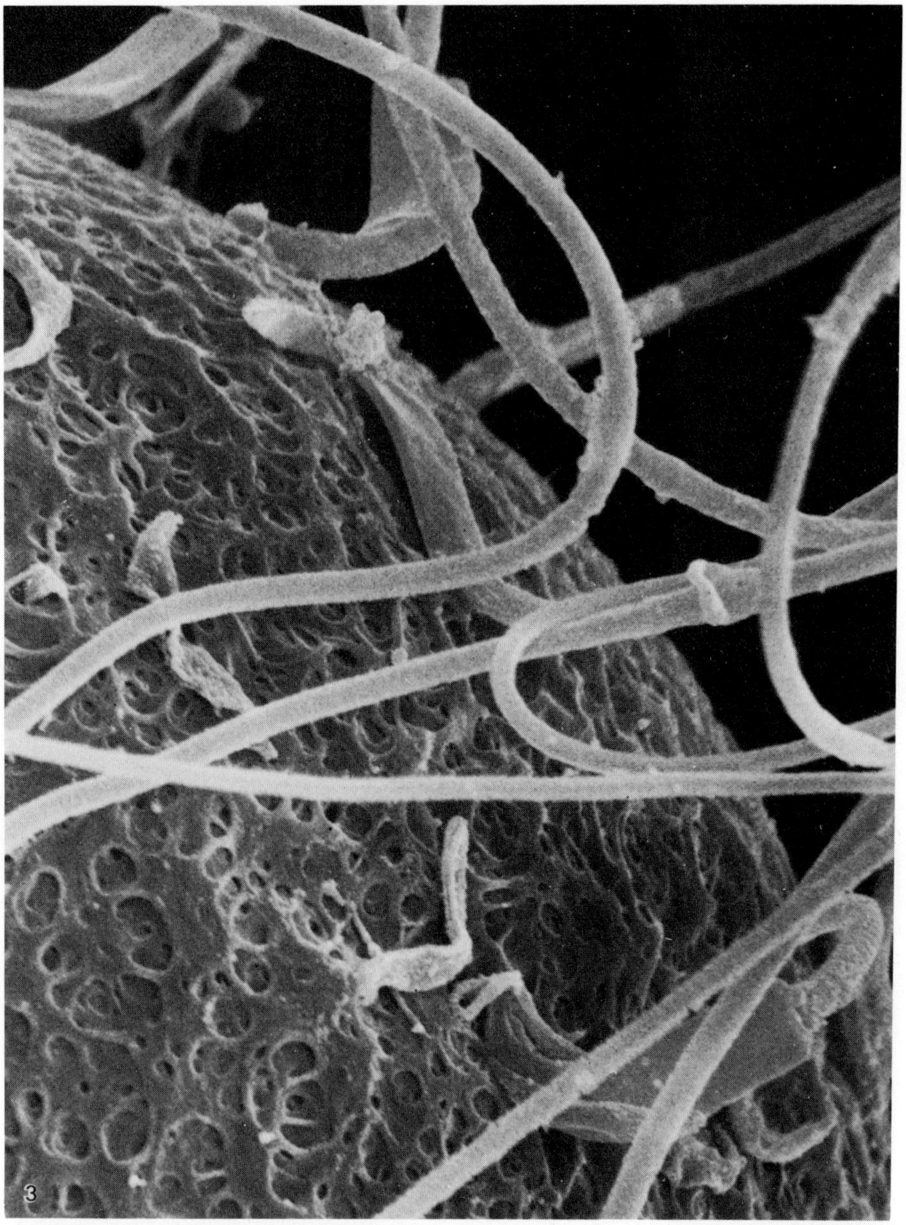

Fig. 3. Zona from an *in vitro* fertilization experiment. Although a population of *in vitro* capacitated hamster spermatozoa typically contains both spermatozoa with acrosome present and those with acrosome missing, we very rarely observe spermatozoa which have undergone acrosome loss associated with the zona. Spermatozoa appear to associate with the zona with acrosomal material which may be the remnants of the reacted acrosome. Often spermatozoa subsequently detach leaving the zona strewn with acrosomal remnants as can be seen in this SEM (× 6,000) (Phillips and Shalgi, unpublished).

sperm undergoing acrosomal loss on the zona surface (50, 66). There is another interesting possibility which could explain why both acrosome-unreacted sperm and sperm which have undergone acrosomal loss are observed in the cumulus oophorus. It may be that sperm which first interact with the zona are not meant to fertilize. These first 'vanguard' sperm lose their acrosomes in the cumulus, thereby dispersing the cumulus mucus and enabling the fertilizing sperm to approach, bind, and loose the acrosomal remnant on the zona surface. An interesting parallel exists in human cervical mucus where 'vanguard' sperm appear to change the properties of the mucus and allow subsequent sperm to traverse the mucus more rapidly (Katz unpublished).

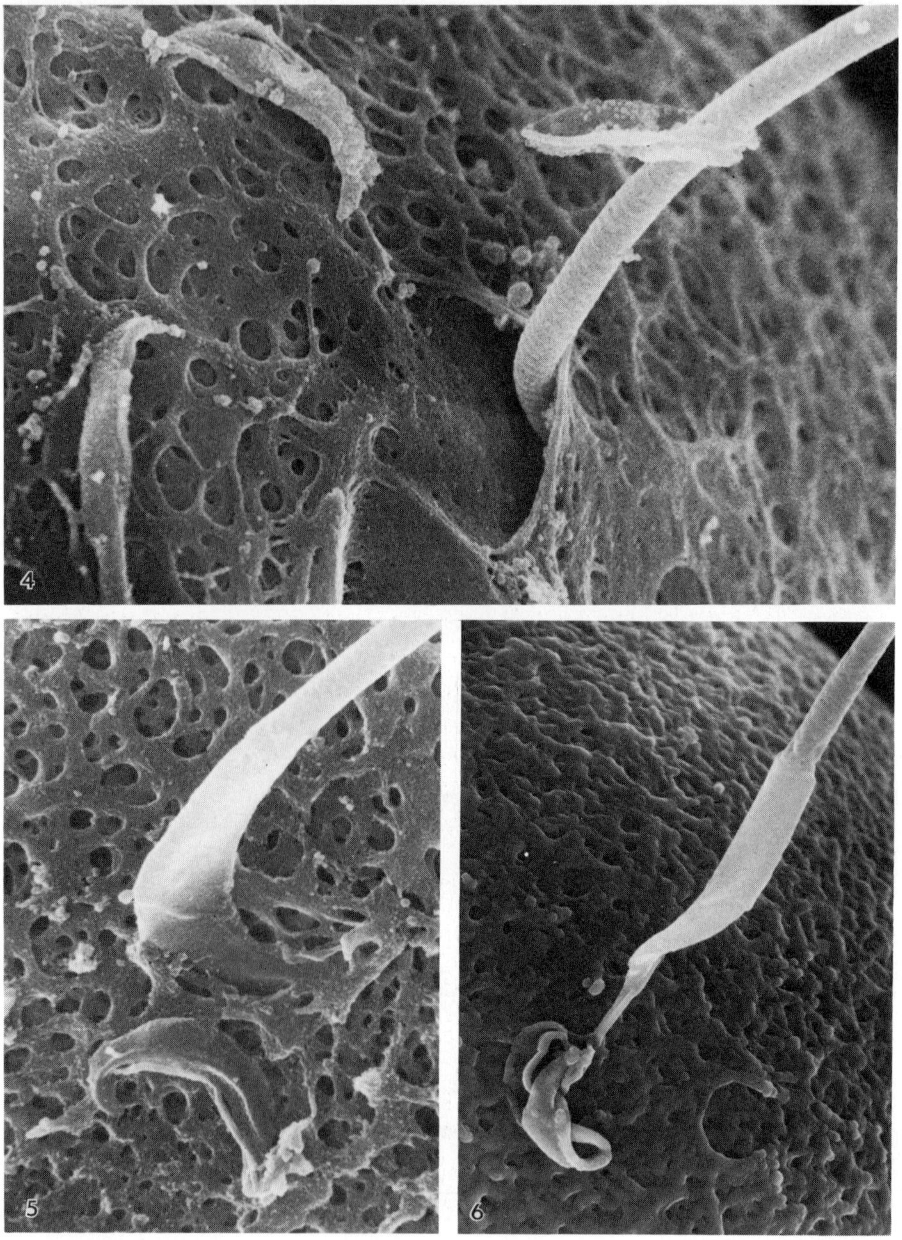

Figs. 4–6. (4) When this spermatozoa penetrated the zona pellucida, the acrosomal remnant apparently became impaled on the sperm tail (× 6,000) (Shalgi and Phillips, unpublished). (5) This sperm appears to be dissociating from its acrosomal remnant which is tightly bound to the zona. With phase microscopy one would probably not see the acrosomal remnant and would judge this cell to be sperm which had associated with the zona after the acrosome loss (× 6,000) (Shalgi and Phillips). (6) Hamster sperm and acrosome remnant bound to the surface of the zona (× 4,800) (Phillips and Shalgi, unpublished).

3. Sperm fusion with the oolemma

Early ultrastructural studies of Bedford (41, 50) and Yanagimachi and Noda (66) of *in vivo* fertilization showed that spermatozoa in the perivitelline space have undergone acrosome loss. The precise region of the sperm plasmalemma which fuses with the oolemma has been difficult to determine, however. Some workers believe that it is the plasmalemma over the equatorial region of the acrosome (1) whereas others believe that it is the plasmalemma over the postacrosomal region which fuses with the oolemma (42). It is also unclear whether fusion is with oocyte microvilli or intervillus regions. The major reason for un-

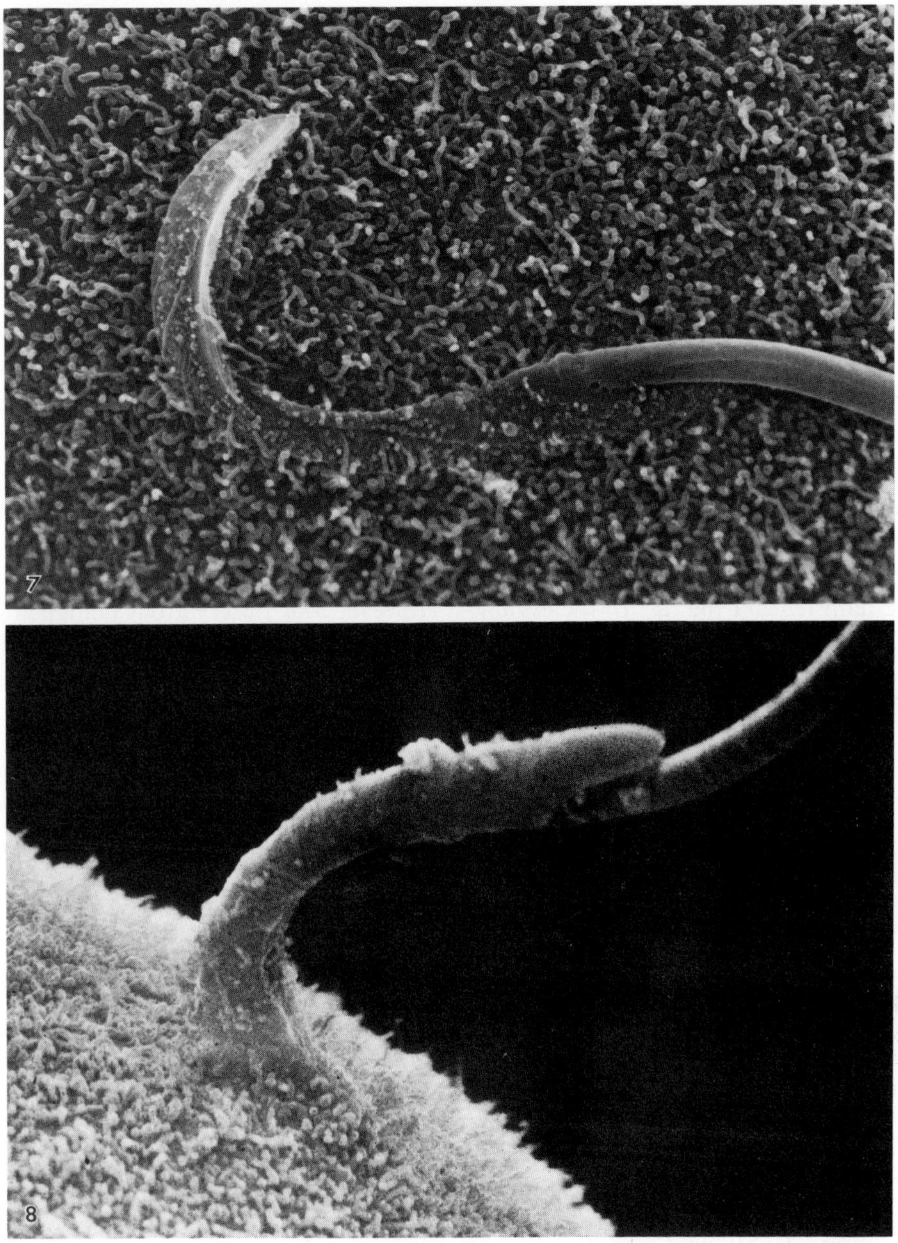

Figs. 7–8. (7) Rat spermatozoon in the early stages of fertilization *in vivo*. The convex surface of the sperm head is closely associated with oocyte microville (× 8,000) (Phillips and Shalgi). (8) Rat spermatozoon in the early stages of fertilization *in vitro*. *In vitro* rat and hamster spermatozoa associate with and penetrate ova tip first (× 8,000) (Shalgi and Phillips).

certainty concerning the site of initial fusion is that it is extremely difficult to 'catch' the single fertilizing spermatozoon at precisely the moment of initial fusion. Micrographs are mostly of later stages when the membrane has fused from the anterior portion of the equatorial segment to the posterior portion of the postacrosomal region. Thus it is not possible to determine which region of the sperm plasmalemma fused first or whether fusion occurred with oocyte microvilli or in the region between microvilli. To alleviate the problems of looking for the single spermatozoon *in vivo*, many workers have examined *in vitro* systems. The denuded hamster ovum has been particularly popular as one can remove the zona which causes the block to polyspermy in this species. A multitude of sperm will then penetrate in a brief period of time. There are differences between this type of system and natural fertilization. Spermatozoa are capacitated *in vitro* rather than *in vivo;* spermatozoa do not pass through the zona pellucida or cumulus; the oocyte membrane interacts with many spermatozoa in a short period of time and the make-up in culture medium is different from ampullar fluid. How much these differences bring about changes in the way sperm and oocytes interact is not clear.

We recently have been examining the relevance of the *in vitro* situation to fertilization in the oviduct (67, 68). We chose to examine the problem with the scanning electron microscope (SEM) where we can examine many ova. We can also observe whether oocytes are associated with more than one spermatozoon. The SEM allows the entire sperm head to be viewed, and thus and a more accurate 3-dimensional perspective of the spatial relationship between the male and female gametes is obtained. Our observations have revealed that there are striking differences between the ultrastructural events which occur during fertilization *in vivo* and those which occur *in vitro* in both the hamster and the rat. In the cycling animal, the SEM revealed that spermatozoa appear to contact oocyte microvilli by the equatorial segment of the acrosome (Fig. 7) (67). Soon after fusion of the oolemma and sperm plasma membrane, microvilli characteristic of the oocyte lie over the broad side of the sperm head. The sperm head undergoes a pronounced flexure as it is incorporated into the ooplasm. On the other hand, when oocytes with intact zonae are fertilized *in vitro* and prepared in exactly the same way for the SEM, the spermatozoon is observed to contact the ovum by the most anterior tip of the sperm head (Fig. 8) (68). Subsequently, the tip of the sperm head is incorporated into the ooplasm first. The importance of these differences between the morphology of sperm incorporation *in vivo* and *in vitro* remains to be established. Since fusion cannot be seen with the SEM, it is possible that membrane fusion occurs over the same region in the sperm plasmalemma even though the mechanics of sperm entry are markedly different. It may be that a zygote has the same chance of developing regardless of the angle at which the sperm penetrates. However, our observations demonstrate that fertilization *in vitro* may not be a precise model for fertilization *in vivo*.

References

1. Yanagimachi R: Mechanism of fertilization in mammals. In: Fertilization and embryonic development. Mastroianni L, Biggers JD (eds) New York Plenum Press, 1981.
2. Young WC: A study of the function of the epididymis. III. Functional changes undergone by spermatozoa during their passage through the epididymis and vas deferens in the guinea pig. J Exp Biol 8: 151–162, 1931.
3. Nishikawa Y, Waide Y: Studies on the maturation of spermatozoa. I. Mechanism and speed of transition of spermatozoa in the epididymis and their functional changes. Bull Jpn Natl Inst Agric Sci Ser G 3: 68–78, 1952.
4. Blandau RJ, Rumery RE: The relationship of swimming movement of epididymal spermatozoa to their fertilizing capacity. Fertil Steril 15: 571–579, 1964.
5. Bedford JM: Development of the fertilizing ability of spermatozoa in the epididymis of the rabbit. J Exp Zool 163: 319–329, 1966.
6. Orgebin-Crist MC: Maturation of spermatozoa in the rabbit epididymis: Fertilizing ability and embryonic mortality in does inseminated with epididymal spermatoza. Ann Biol Anim Biochim Biophys 7: 373–389, 1967.
7. Bedford JM: Maturation, transport, and fate of spermatozoa in the epididymis. In: Handbook of Physiology. Greep RO (ed) Washington, DC, American Physiological Society, pp 303–317, 1975.
8. Gordon M: Cytochemical analysis of the membranes of the mammalian sperm head. In: Male Reproductive System. Yates RD, Gordon M (ed) New York, Masson, pp 15–33, 1977.
9. Cooper GW, Bedford JM: Acquisition of surface charge by the plasma membrane of mammalian spermatozoa during epididymal maturation. Anat Rec 169: 300–301, 1971.
10. Hamilton DW, Gould RP: Galactosyltransferase activity associated with rat epididymal spermatozoon maturation. Anat Rec 196: 71, 1980.
11. Olson GE, Hamilton DW: Characterization of the surface glycoproteins of rat spermatozoa. Biol Reprod 19: 26–35, 1978.
12. Fawcett DW, Phillips DM: Observations on the release of spermatozoa and on changes in the head during passage through the epididymis. J Reprod Fertil Suppl 6: 405–418, 1969.
13. Acott TS, Hoskins DD: Bovine sperm forward motility protein: Partial purification and characterization. J Biol Chem 253: 6744–6750, 1978.
14. Acott TS, Hoskins DD: Bovine sperm forward motility protein: Binding to epididymal spermatozoa. Biol Reprod 234–240, 1981.
15. Acott TS, Johnson DJ, Brandt H, Hoskins DD: Sperm forward motility protein: Tissue distribution and species cross reactivity. Biol Reprod 20: 247–252, 1979.
16. Austin CR: The 'capacitation' of the mammalian sperm. Nature (London) 170: 326, 1952.
17. Chang MC: Fertilizing capacity of spermatozoa deposited into fallopian tubes. Nature (London) 168: 697–698, 1951.
18. Yanagimachi R, Usui N: Calcium dependence of the acrosome reaction and activation of guinea pig spermatozoa. Exp Cell Res 89: 161–174, 1974.
19. Green DPL: Induction of the acrosome reaction in quinea-pig spermatozoa *in vitro* by Ca ionophore A23187. J Physiol 260: 18P–19P, 1976.
20. Green DPL: The induction of the acrosome reaction in guinea pig spermatozoa by the divalent metal cation ionophore A23187. J Cell Sci 32: 137–151, 1978.
21. Talbot P, Summers RG, Hylander BL, Keough EM, Franklin LE: The role of calcium in the acrosome reaction: An analysis using ionophore A23187. J Exp Zool 198: 383–392, 1976.
22. Viriyapanich P, Bedford JM: The fertilization performance *in vivo* of rabbit spermatozoa capacitated *in vitro*. J Exptl Zool 216: 169, 1981.
23. Blandau RJ: *In vitro* fertilization and embryo transfer. Fertil and Steril 33: 1–11, 1980.
24. Barros C, Berrios M: Is the activated spermatozoa really capacitated? J Exp Zool 201: 65–72, 1977.
25. Bavister BD: Recent progress in the study of early events in mammalian fertilization. Develop Growth and Differ 22: 385–402, 1980.
26. Motta P, Van Blerkom J: A scanning electron microscopic study of rabbit spermatozoa in the female reproductive tract following coitus. Cell Tiss Res 163: 29, 1975.
27. Cooper GW, Overstreet JW, Katz DF: The motility of rabbit spermatozoa recovered from the female reproductive tract. Gamete Res 2: 35–42, 1979.
28. Bavister BD: Fertilization of hamster eggs *in vitro* at sperm: egg ratios close to unity. J Exptl Zool 210: 259–264, 1979.
29. Yanagimachi R: The movement of golden hamster spermatozoa before and after capacitation. J Reprod Fertil 23: 193–196, 1970.
30. Yanagimachi R: *In vitro* acrosome reaction and capacitation of golden

hamster spermatozoa by bovine follicular fluid and its fractions. J Exp Zool 170: 269–280, 1969.
31. Yanagimachi R: *In vitro* capacitation of hamster spermatozoa by follicular fluid. J Reprod Fertil 18: 275–286, 1969.
32. Yanagimachi R, Usui N: The appearance and disappearance of factors involved in sperm chromatin decondensation in the hamster egg. J Cell Biol 55: 293a, 1972.
33. Barros C, Fujimoto M, Yanagimachi R: Failure of zona penetration of hamster spermatozoa after prolonged preincubation in a blood serum fraction. J Reprod Fertil 35: 89–95, 1973.
34. Mahi CA, Yanagimachi R: Maturation and sperm penetration of canine ovarian oocytes *in vitro*. J Exp Zool 196: 189–196, 1976.
35. Overstreet JW, Cooper GW: Effect of ovulation and sperm motility on the migration of rabbit spermatozoa to the site of fertilization. J Reprod Fertil 55: 53–59, 1979.
36. Overstreet JW, Katz DF, Johnson LL: Motility of rabbit spermatozoa in the secretions of the oviduct. Biol Reprod 22: 1083–1088, 1980.
37. Fraser LE: Motility pattern in mouse spermatozoa before and after capacitation. J Exp Zool 202: 439–444, 1977.
38. Anouma S, Okabe M, Kawaguchi M, Kishi Y: Studies on sperm capacitation. IX. Movement characteristics of spermatozoa in relation to capacitation. Chem Pharm Bull Jpn 28: 1497–1502, 1980.
39. Phillips DM: Comparative analysis of mammalian sperm motility. J Cell Biol 53: 561–573, 1972.
40. Katz DF, Yanagimachi R: Movement characteristics of hamster spermatozoa within the oviduct. Biol Reprod 22: 759–764, 1980.
41. Bedford JM: Mechanisms involved in penetration of spermatozoa through the vestments of the mammalian egg. In: Physiology and Genetics of Reproduction Part B. Coutinho EM, Fuchs F (ed) New York, Plenum Press, pp 55–68, 1974.
42. Bedford JM, Cooper GW: Membrane fusion events in the fertilization of vertebrate eggs. In: Cell Surface Reviews. Poste G, Nicolson GL (eds), Amsterdam, North-Holland, pp 65–125, 1978.
43. Austin CR: The Mammalian Egg, Springfield, Ill, Thomas, 1961.
44. Austin CR: Capacitation and the release of hyaluronidase from spermatozoa. J Reprod Fertil 1: 310–311, 1960.
45. Piko L: Gamete structure and sperm entry in mammals. In: Fertilization. Metz CB, Monroy A (ed), New York, Academic Press, pp 325–403, 1969.
46. McRorie RA, Williams WL: Biochemistry of mammalian fertilization. Annu Rev Biochem 43: 777–803, 1974.
47. Metz CB, Seiquer AC, Castro AE: Inhibition of the cumulus dispersing and hyaluronidase activities of sperm by heterologous and isologous antisperm antibodies. Proc Soc Exp Biol Med 140: 766–781, 1972.
48. Metz CB: Role of specific sperm antigens in fertilization. Fed Proc 32: 2057–1064, 1973.
49. Dunbar BS, Munoz CG, Cordle CT, Metz CB: Inhibition of fertilization *in vitro* by treatment of rabbit spermatozoa with univalent isoantibodies to rabbit sperm hyaluronidase. J Reprod Fertil 47: 381–384, 1976.
50. Bedford JM: An electron microscopic study of sperm penetration into the rabbit egg after natural mating. Am J Anat 133: 213–254, 1972.
51. Saling PM, Storey BT: Mouse gamete interaction during fertilization *in vitro*. J Cell Biol 83: 544–555, 1979.
52. Huang TTF, Fleming AD, Yanagimachi R: Only acrosome-reacted spermatozoa can bind to and penetrate zona pellucida: A study using the guinea-pig. J Expt Zool 217: 287–290, 1981.
53. Stambaugh R, Brackett BG, Mastroianni L: Inhibition of *in vitro* fertilization of rabbit ova by trypsin inhibitors. Biol Reprod 1: 223–227, 1969.
54. Zaneveld LJD, Robertson RT, Kessle M, Williams WL: Inhibition of fertilization *in vivo* by pancreatic and seminal plasma trypsin inhibitors. J Reprod Fertil 25: 387–392, 1971.
55. Zaneveld LJD, Polakoski KL, Robertson RT, Williams WL: Trypsin inhibitors and fertilization. In: Proteinase and Biological Control. Fritz H, Tschesche H (ed), New York, de Gruyter, pp 236–243, 2975.
56. Zaneveld LJD: Sperm enzyme inhibitors as antifertility agents. In: Human Semen and Fertility Regulation in Men. Hafez ESE (ed), St. Louis, Mosby, pp 570–582, 1976.
57. Bhattacharyya AK, Goodpasteuer JC, Zaneveld LJD: Acrosin of mouse spermatozoa. Am J Physiol 237: E40–E44, 1979.
58. Joyce C, Freund M, Peterson RN: Contraceptive effects of intravaginal application of acrosin and hyaluronidase inhibitors in rabbit. Contraception 19: 95–106, 1979.
59. Saling PM: Involvement of trypsin-like activity in binding of mouse spermatozoa to zonae pellucidae. Proc Natl Acad Sci 78: 6231–6235, 1981.
60. Bedford JM, Cross NL: Normal penetration of rabbit spermatozoa through a trypsin- and acrosin-resistant zona pellucida. J Reprod Fertil 54: 385–392, 1978.
61. Hartmann JF, Hutchison CF: Nature of the pre-penetration contact interactions between hamster gametes *in vitro*. J Reprod Fertil 36: 49–57, 1974.
62. Hartmann JF, Hutchison CF: Mammalian fertilization *in vitro*: Sperm-induced preparation of the zona pellucida of golden hamster ova for final binding. J Reprod Fertil 37: 443–445, 1974.
63. Hartmann JF, Hutchison CF: Contact between hamster spermatozoa and the zona pellucida releases a factor which influences early binding stages. J Reprod Fertil 37: 443–445, 1974.
64. Hartmann JF, Hutchison CF: Nature and fate of the factors released during early contact interactions between hamster sperm and egg prior to fertilization *in vitro*. Dev Biol 78: 390–393, 1980.
65. Bedford JM: Sperm/egg interaction: The specificity of human spermatozoa. Ant Rec 188: 477-488, 1977.
66. Yanagimachi R, Moda YD: Ultrastructural changes in the hamster sperm head during fertilization. J Ultrastruct Res 31: 465–485, 1970.
67. Shalgi R, Phillips DM: Mechanics of sperm entry in cyclic hamster. J Ultrastruct Res 71: 154–161, 1980.
68. Shalgi R, Phillips DM: Mechanics of *in vitro* fertilization in the hamster. Biol Reprod 23: 433–444, 1980.

Author's address:
The Population Council
1230 York Avenue
New York, NY 10021, USA

CHAPTER 15

Ultrastructure of human fertilization

MILAN DVOŘÁK, JAN TESAŘÍK and VÁCLAV KOPEČNÝ

1. Fertilization in man: microanatomy and macromolecular biology

An elucidation of fertilization in mammals emerged only recently. Because of various moral and legal injunctions against the study of fertilization in man, information on this species has been more difficult to acquire than on laboratory and domestic animals. Even if these obstacles are being removed, there will still be many obvious difficulties in the future and data from human investigation will continue to emerge slowly.

Ultrastructural analyses of reproductive tissues and cells in different phases of gamete maturation and at fertilization are of special importance in morphological approach to the investigation of human fertilization. In this review pioneering morphological work will be pointed out (1–6), as well as current knowledge in the field, based on our own observations and results. In discussions of the significance of the morphological data presented, evidence from studies on different mammalian species will be noted wherever the experimental approach has not been feasible in human material. Remaining aware of the danger of overgeneralization, the assumption is made that basic cellular phenomena in fertilization – cellular recognition, cellular adhesion, membrane fusion, exocytosis and endocytosis – as well as fundamental regulatory mechanisms, are probably similar in all mammals (7, 8).

2. Gamete preparation for fertilization

2.1. Spermatozoon

Mammalian spermatozoa, after release from contact with Sertoli cells, undergo major transformations which make them capable of fertilizing the oocyte. Changes taking place during the passage of spermatozoa through the male genital tract are termed maturation while the term capacitation is reserved for changes which the spermatozoa undergo after ejaculation and within the female genital tract (8).

Spermatozoa obtained from the caput epididymis in various mammals are not yet able to fertilize the oocyte. In all mammals studied, maturation of spermatozoa occurs in the mid-portion of the ductus epididymis. Although human spermatozoa very likely undergo a similar transformation during their epididymal passage, the precise topical localization of these changes in the human ductus epididymis remains to be ascertained. Recently the question of the existence of sperm maturation in man has been studied using techniques based on the ability of human spermatozoa to penetrate zona-free hamster oocytes (9). The ability of spermatozoa to bind tightly to oocytes increased in spermatozoa isolated from successive segments of the epididymis. Only spermatozoa recovered from the cauda epididymis were capable of penetrating the oocytes.

Mammalian spermatozoa are able to fertilize the oocyte only after a specific interval of time, which requires residence in the female genital tract. The transformations of spermatozoa occurring within the female genital tract render them capable of fertilization. This process has been shown to be an indispensable step in fertilization in many mammalian species studied, including man (4, 10).

2.1.1. Spermatozoon changes in the male genital tract

Of the changes taking place in mammalian spermatozoa during their epididymal maturation, the most extensive attention has been paid to alterations of the sperm plasma membrane. These alterations affect the ultrastructural appearance, surface charge, distribution and quantity of lectin receptors, membrane antigenicity and composition of surface proteins and glycoproteins (11).

It seems that changes in the composition of surface glycoproteins and glycolipids, the probable carriers of different acidic residues responsible for the acqui-

Van Blerkom, J. and Motta, P.M. (eds.), Ultrastructure of reproduction. ISBN 0-89838-572-5.

sition of negative surface charges during the human spermatozoon epididymal maturation (12), are of paramount importance. These surface modifications are reflected by changes in the pattern of labelling with colloidal iron hydroxide, a marker of negative surface charges on the plasma membrane (13), as well as in lectin-binding properties (14). Since spermatozoa cannot synthesize their own glycoproteins, these macromolecules, constituting the glycocalyx of epididymal spermatozoa, are probably secreted by the epididymis. Such secretion has been demonstrated in mouse, using radioisotopic pre-labelling of secreted glycoproteins (15). Using fine-structural autoradiography, (16) the process of spermatozoon maturation has been shown to clearly involve binding to spermatozoa of radioactive glycoproteins secreted in the epididymis. Epididymal glycoproteins have been localized on the surface of spermatozoa by the fluorescent antibody techniques and by sperm agglutination with antiserum directed against rabbit and hamster (17). In both species, fertilization rates were significantly reduced by specific antibodies directed against specific glycoproteins (17).

A change in the rate and character of flagellar beat of spermatozoa during their transit through the epididymis is associated with ultrastructural transformations of the dense outer fibers, mitochondria and the fibrous sheath of the tail principal piece in human spermatozoa (13). It may also have some relation to the secretion into the epididymal milieu of spermatozoon-binding substances. A protein (Sperm Forward Motility Protein) secreted by bovine epididymal epithelial cells and which binds to spermatozoa has been isolated recently and is presumed to activate their motility (18).

Another important feature of epididymal maturation of the human spermatozoon, shared with other eutherian species, is nuclear condensation, caused by disulfide (S–S) crosslinking between nuclear proteins. This process also takes place in tail structures – dense outer fibers, sheath of the principal piece and outer mitochondrial membrane (13, 19). However, human ejaculated spermatozoa, unlike those of other mammalian species, display a wide variation in the extent of chromatin condensation.

The regulation of maturational changes in the epididymis is currently thought of as a prospective target for contraceptive action in man. It is necessary to consider, however, that human spermatozoa may be different from other mammalian spermatozoa. There is a striking morphological heterogeneity in the human population of spermatozoa which has not been seen in any other mammalian species studied (13). The prominent changes in acrosome morphology during the epididymal passage of spermatozoa, as reported in many mammalian species, are not detectable in human spermatozoa (13). Moreover, human epididymal spermatozoa do not exhibit regional differences in physical, biochemical and immunological properties of their plasma membrane, unlike those of most mammalian species (8). In any case, the extrapolation of knowledge obtained from other mammalian species is necessary, especially since a direct study of human epididymal maturation is obviously difficult.

2.1.2. Spermatozoon changes in the female genital tract

In the female genital tract, spermatozoa are exposed for a relatively long period to the influence(s) of secreted fluids. A direct interaction with different types of cells may represent another possible mechanism of induction of capacitational changes (7).

Physiological studies indicate that the passage of spermatozoa through the female genital tract is accompanied by important functional changes, termed capacitation, which are a necessary prerequisite for the acrosome reaction and spermatozoon-oocyte fusion (20). However, morphological evidence for structural modifications of spermatozoa in the female tract is scarce. Moreover, neither the mechanism nor the molecular nature of capacitation is understood. It is assumed that the capacitational changes involve the spermatozoon plasma membrane and components of its surface (8), as well as the pattern of motility (20) and the state of acrosomal enzymes (21).

There are probably two steps in capacitation (7). As suggested by several studies, the first step is an 'unmasking' of the spermatozoon 'surface' (plasma membrane). The loss of the surface coat of spermatozoa during capacitation detected cytochemically (22) may correspond to such 'unmasking'. As for the nature of the 'coating' material, seminal and epididymal plasma proteins are the most probable candidates.

The second step in capacitation may be represented by alterations in integral proteins of the plasma membrane and perhaps of the outer acrosomal membrane as well (8). The nature of these changes and their relation to capacitation, however, remain to be clarified.

2.2. Oocyte

Quiescent, meiotically arrested oocytes are contained in ovarian follicles for relatively long periods of time before becoming fertilizable. The acquisition of fertilizability occurs no sooner than several hours before ovulation. The process by which an oocyte attains the ability to be fertilized is termed maturation. Maturational changes involve both nucleus and cytoplasm.

2.2.1. *Nuclear maturation*

The prominent feature of nuclear maturation is the resumption of meiosis. Nuclear maturation encompasses transition from the meiotic arrest of the nucleus, in which chromosomes persist in a diffuse diplotene (dictyate) stage, to the final stages of meiosis, which results in the formation of the 1st polar body and chromosomes in a metaphase II configuration. The resumption of meiosis *in vivo* is a consequence of specific actions of FSH and LH on granulosa cells, leading to the loss of their inhibitory influence on oocyte meiosis (23). The disruption of gap junctions between the oocyte and cumulus cells may be a crucial step in the derepression of maturation in the human oocyte (24). The meiotic arrest of the human oocyte is, however, maintained for a certain time period even after the oocyte-cumulus uncoupling, apparently by additional, probably humoral mechanisms present in the follicular fluid (25). The latter may involve both substances released to the follicular fluid by granulosa cells and those passing into the intrafollicular compartment from the circulation. A meiosis-inhibiting substance, called Oocyte Maturation Inhibitor, probably of polypeptide nature, has really been shown to be synthesized and released by pig granulosa cells (26). Contrary to conditions *in vivo*, human oocytes, like those of some other mammalian species, resume meiosis spontaneously in culture (27).

Morphological features of nuclear maturation in the oocyte form a clearcut sequence. After the surge of LH *in vivo*, or after isolation *in vitro* of the oocyte from the antral follicle and removal of associated cumulus cells, the nuclear envelope of the oocyte disintegrates, giving rise to numerous smooth-surfaced vesicles. The process is called germinal vesicle breakdown (GVBD). In mammalian oocytes, GVBD is preceded by chromosomal condensation near the inner membrane of the nuclear envelope (28). Concomitant with the breakdown of the nuclear envelope, the nucleolus disappears. Partially condensed chromosomal bivalents remain in the area previously occupied by the nucleus. This stage of maturation, known as early diakinesis, is followed by further condensation of the bivalents in late diakinesis and arrangement in the equatorial plane at metaphase I. After splitting of bivalents and separation of one set of chromosomes into the 1st polar body at anaphase I, meiosis is arrested at metaphase II. This stage characterizes the ovulated oocyte.

The period preceding nuclear maturation in the mammalian oocyte is associated with high transcriptional activity. Data available on RNA synthesis are mostly based on autoradiographic techniques. These include either examinations of the rate of ^{3}H-uridine incorporation, or an in situ assay for RNA polymerase activity (29). The rate of RNA synthesis increases at the same time when the follicle growth begins. When the follicle reaches its maximal diameter, decrease in RNA synthesis is observed, but it never ceases completely (29). An intensive incorporation of ^{3}H-uridine has been detected in the oocyte nucleus until a few hours before GVBD in mouse (30) and up to GVBD in pig where ^{3}H-uridine label is associated with condensing chromocenters (31). At the ultrastructural level, a marked decrease in the concentration of perichromatin granules, thought to be of ribonucleoprotein nature, is observed at early antral stages, but the nucleoplasm is repopulated by significant amounts of fine granules at late antral stages (32). Incorporation of ^{3}H-uridine has been detected also in human oocytes from large antral follicles, though to a much lesser extent than in associated corona radiata cells (Fig. 1a). After a 30 min pulse with ^{3}H-uridine, nucleoli of pig oocytes are labelled to a greater extent than the nucleoplasm (Fig. 1b), possibly owing to transport of extranucleolar RNA into the nucleolus (33).

It is assumed that the RNA synthesized by the oocyte during its major growth phase is mostly stored for later use in embryogenesis (34). The role of the RNA synthesized just before GVBD is not known. Its role in meiotic maturation itself may be less important, since maturation progresses in the presence of inhibitors of RNA synthesis (35). As is the situation for pig oocytes isolated from prepuberal gilts (31), human oocytes obtained from large antral follicles in the preovulatory period (Fig. 1a) show high diversity of RNA labelling intensity. A defect of RNA synthesis has been found in aged oocytes of the rat (36). We believe that the impairment of RNA synthesis during stages closely preceding GVBD may be important for further viability of the oocyte, and as such, may be an early sign of oocyte predestination to atresia.

2.2.2. *Cytoplasmic maturation*

Important changes in the cytoplasm take place during the relatively brief period of mammalian oocyte maturation. These changes are, however, largely not detectable using a morphological approach. No apparent changes in the appearance and

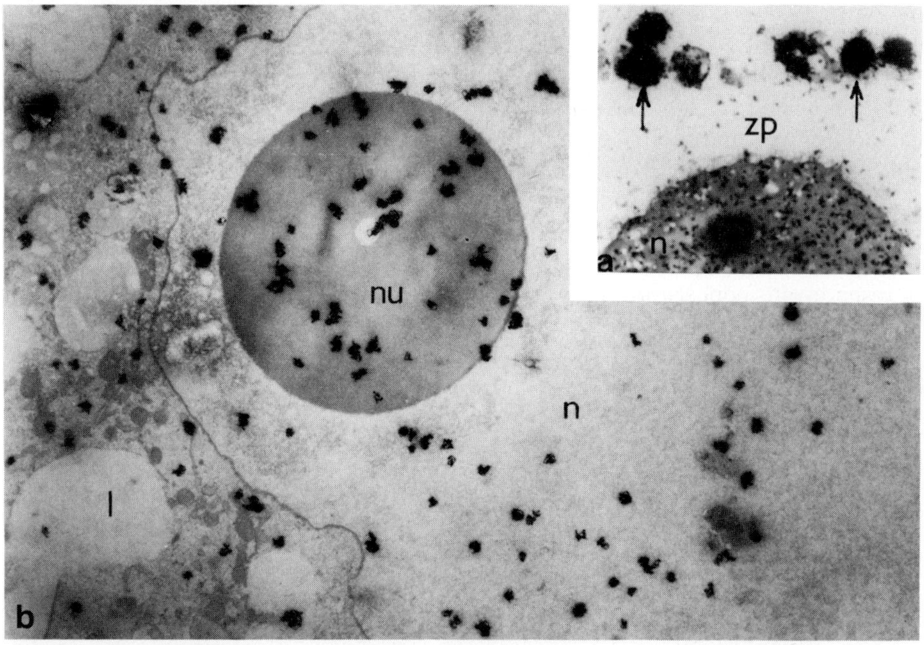

Fig. 1. (a) Autoradiograph of a part of a human oocyte obtained from a large antral follicle and incubated for 30 min with ^{3}H-uridine. Nucleus (n), zona pellucida (zp), corona radiata cells (arrows) (× 440). (b) High resolution autoradiograph of a portion of a pig oocyte from a large antral follicle, labeled for 30 min with ^{3}H-uridine. Nucleus (n), nucleolus (nu), lipid droplets (l) (× 7,040).

distribution of cytoplasmic structures, with the exception of cortical granules, have been observed in the human oocyte at the time of nuclear maturation (37). Cortical granules, previously occupying a rather extensive area, including the subcortical cytoplasm, move towards the cell periphery and form a single layer in proximity to the plasma membrane.

In all mammalian species studied, active protein synthesis and significant qualitative and quantitative alterations in the pattern of newly synthesized proteins have been reported (38). An unusually high proportion of newly synthesized proteins accumulates in the oocyte nucleus at the time close to GVBD (31). It is presumed that proteins synthesized in the preovulatory oocyte are not only destined for immediate requirements, but they are also stored for use in the postfertilization development. A direct relationship between the period just preceding GVBD and the early development of the fertilized egg has been demonstrated by the detection of proteins synthesized during GVBD and localized in the nuclei of rabbit embryos up to the eight-cell stage (39). Although comparable evidence does not exist for human oocytes, it may be presumed that macromolecules synthesized during the preovulatory period may exert an influence on post-fertilization development of the zygote. Taken together, the understanding of such processes may elucidate the difference between *in vivo* and *in vitro* maturation of oocytes and may help to bridge the inability of the majority of human oocytes matured *in vitro* to develop normally after fertilization (40).

3. Spermatozoon-oocyte interaction

A newly ovulated human oocyte is surrounded by the zona pellucida and a large mass of cells, the cumulus oophorus. The fertilizing spermatozoon must first pass through these vestments to establish contact with the oocyte plasma membrane. The following step, fusion between the two gametes, enables the fertilizing spermatozoon to enter the oocyte. The development of pronuclei and the activation of the first cleavage division complete the process of fertilization.

3.1. Spermatozoon passage through the cumulus oophorus

The cumulus oophorus which surrounds the mature human oocyte consists of loosely arranged cells embedded in a viscous intercellular matrix (25). This matrix is glycoprotein in nature, which masks fine mucopolysaccharide fibrils unless the proteins are extracted with pronase (25). Hyaluronic acid is thought to be one of the major components of this intercellular material (7) because the matrix can be dissolved easily with hyaluronidase. During incubation *in vitro* with capacitated spermatozoa, the matrix is rapidly removed and spermatozoa penetrate between cumulus cells (Fig. 2a). The dissolution of the cumulus intercellular matrix by capacitated spermatozoa is due to the action of acrosomal hyaluronidase (7). The release of hyaluronidase from the acrosome was thought to be correlated with the morphological picture of the acrosomal reaction (see below). Certain reservations concerning this concept have been expressed recently since hyaluronidase liberation in some mammalian species is not concomitant with the acrosomal reaction (41, 42). Spermatozoa with both reacted and intact acrosomes are present between cumulus cells that surround human oocytes inseminated *in vitro*. Thus, it is evident that human spermatozoa can gain access to the surface of the zona pellucida prior to the acrosome reaction. A question which remains unresolved is whether the required hyaluronidase is released from morpho-

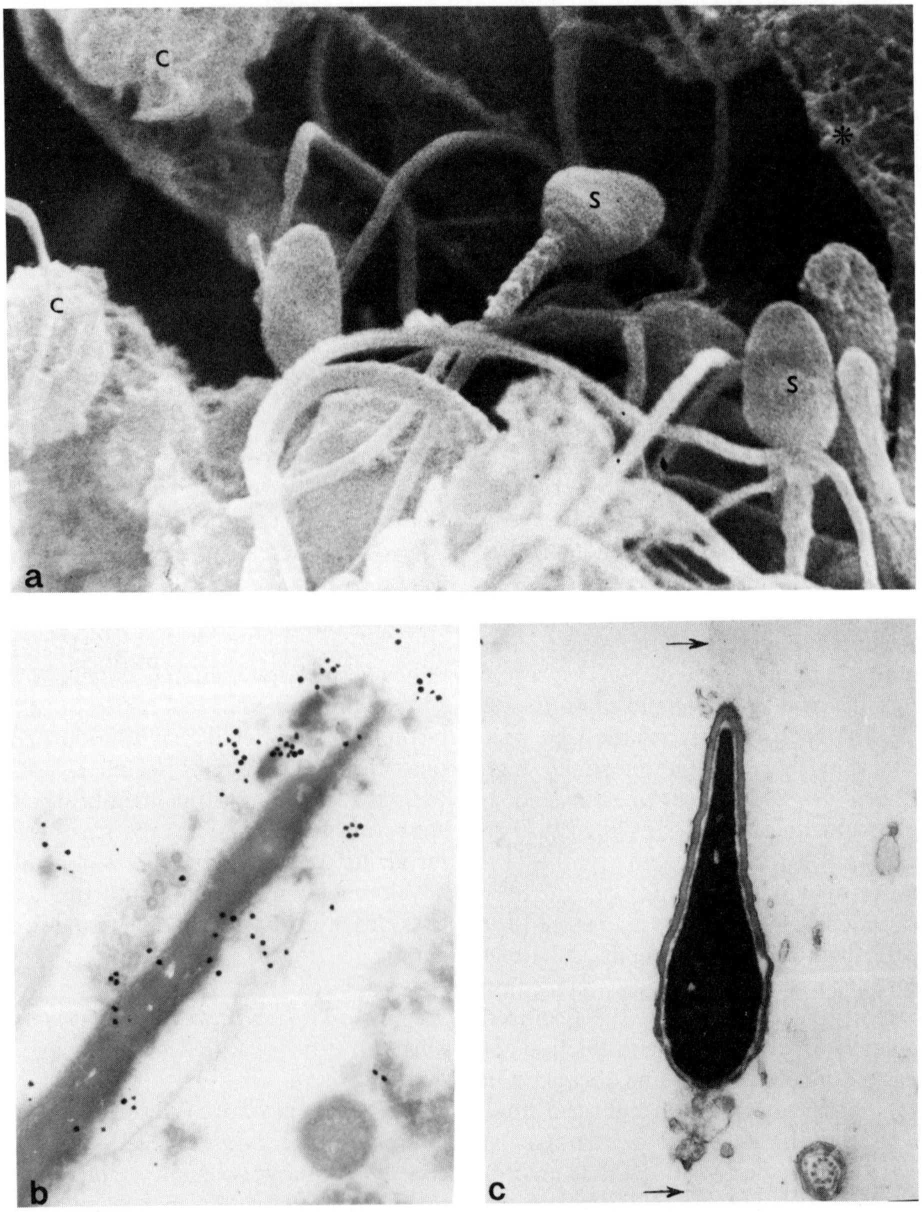

Fig. 2. (a) Scanning electron micrograph of the surface of the human cumulus oophorus with spermatozoa (s) penetrating between cumulus cells (c). Remnants of cumulus intercellular matrix (asterisk) ($\times$ 5,200). (b) High resolution autoradiograph of a head of a rabbit spermatozoon labelled in the acrosome with ^{3}H-fucose, undergoing acrosomal reaction. (from Kopečný, Fléchon Biol Reprod 24: 201, 1981) ($\times$ 14,400). (c) Head of a nonacrosome-reacted human spermatozoon bound to the zona surface (arrowed) ($\times$ 19,200).

logically intact acrosomes or from reacted acrosomes of other spermatozoa located at the site of fertilization.

3.2. *Acrosomal reaction*

In capacitated spermatozoa, under favourable conditions, a substantial change in the acrosome occurs. This change is limited strictly to the rostral region of the spermatozoon. The result of this alteration is denudation of the inner acrosomal membrane, disintegration of the overlying outer acrosomal and plasma membranes, and loss of the acrosomal matrix. This profound morphological change is termed the acrosomal reaction. It is generally accepted that, during the acrosomal reaction, there is a redistribution of acrosomal enzymes which are extraordinarily abundant in this entity which is supposed to represent a modified lysosome.

Heads of human spermatozoa undergoing the acrosomal reaction are surrounded by small vesicles located in proximity to the anterior region (2). These vesicles are thought to originate through multiple point fusions between the plasma membrane and the outer acrosomal membrane. Membrane vesiculation is a characteristic feature of the acrosome reaction in rodent spermatozoa (8) as well as in spermatozoa of other mammalian species studied, e.g. of the boar (41). The vesiculation of membranes is preceded by acrosomal swelling and increased membrane permeability in guinea pig spermatozoa (42). Another mechanism of membrane vesiculation has been described in human spermatozoa and involves a separate loss of plasma and acrosomal membranes (43). However, some doubts have arisen concerning the interpretation of this finding as a normal physiological event in humans (8). In any case, the precise mechanism of the acrosome reaction in human spermatozoa requires further investigation.

The equatorial segment of the acrosome and the postacrosomal surface of the spermatozoon head are not involved in the membrane transformations. The acrosomal reaction in human spermatozoa is easily mimicked by different treatments *in vitro* (20). Under experimental conditions, it is evident that, unlike the outer acrosomal membrane, the inner acrosomal membrane and the equatorial segment are very stable components. Due to the stabilization of its proteins by disulfide bonds (19), the inner acrosomal membrane represents a resilient structure which persists even after decondensation of the spermatozoon nucleus in the egg cytoplasm.

Under physiological conditions of actual *in vivo* fertilization, mammalian spermatozoa undergo the acrosomal reaction in proximity to the oocyte. However, it remains to be clarified, which is the mechanism triggering the acrosomal reaction in capacitated spermatozoa. It is of interest that it does not appear necessary that the oocytes be alive for the acrosomal reaction of human spermatozoa to take place (44).

The described changes in acrosomal membranes suggest a role for lytic acrosomal enzymes in fertilization. This concept, indeed, was almost universally accepted. Extensive research efforts into a possible contraceptive value of the inactivation of acrosomal enzymes were based on this concept. It was believed, based on morphological evidence (45), that acrosin, a proteinase from the acrosome considered to be an essential enzyme for the penetration of the zona pellucida, may remain bound in significant amounts to the inner acrosomal membrane after the acrosomal reaction. This suggestion was followed by biochemical evidence for acrosin binding to acrosome-reacted spermatozoa. Recent research has shown, on the contrary, that there is probably no acrosin left on the inner acrosomal membrane after the fully completed acrosome reaction (46); in fact, no acrosomal glycoproteins can be detected in this location (47). It has been shown that the acrosomal glycoproteins remain associated to a significant degree with acrosomal vesicles (Fig. 2b) and are detectable there until the entry of the spermatozoon into the zona pellucida (47).

In light of current knowledge it seems probable that the primary role of spermatozoon proteolytic enzymes may be the induction of the acrosomal reaction itself (21). The contraceptive action of acrosin inhibitors may therefore be effective at this level. Decreased fertility rates of cryopreserved human spermatozoa, associated with morphologically demonstrable damage to the acrosome, may be also caused by acrosin loss.

3.3. *Spermatozoon binding to and passage through the zona pellucida*

The zona pellucida of the mammalian oocyte is a complex structure of glycoproteins with unique antigenic properties, and serving several roles: (1) protection of the oocyte, (2) recognition of homologous spermatozoa, (3) binding of spermatozoa, (4) protection of the oocyte plasma membrane from premature depolarization and precocious cortical reaction, (5) formation of a barrier against polyspermy, (6) providing a physical containment so that early cleavage stage blastomeres develop appropriate intercellular channels of communication. The zona pellucida constitutes the major obstacle to spermatozoa during their attempts at fertilization.

Observations on the *in vitro* interaction of hamster spermatozoa with the zona pellucida of the same species indicate that early loose- and non-species-specific attachment is followed by species-specific and more firm binding. The attachment phase may be governed by specific factors released during the contact between gametes (7). There is, however, no direct evidence for a similar consequence of events during the spermatozoon-zona interaction in man. Human spermatozoa, in contrast to laboratory rodent species whose spermatozoa adhere readily to the zona pellucida of heterologous oocytes, display an unusual specificity in their interaction with the zona pellucida. Human capacitated spermatozoa adhere only to the zona pellucida of an ape, the gibbon Hylobates lar. They do not even attach to the zona surface of either sub-hominoid primates, or to that of non-primate eutherian species (48).

The location of zona-corresponding binding site(s) in spermatozoon has not yet been established accurately. It remains to be clarified whether the establishment of contact with the zona pellucida is mediated by the intact plasma membrane, by the vesiculated plasma membrane-outer acrosomal membrane complex, or by the inner acrosomal membrane exposed during the acrosomal reaction. Our finding of frequent apposition of nonacrosome-reacted human spermatozoa to the zona surface of human oocytes inseminated *in vitro* (Fig. 2c) is strongly suggestive of the primary role of the plasma membrane covering the acrosome in the binding of human spermatozoa. However, the possibility of some differences of human spermatozoon-zona interaction *in vitro* versus *in vivo* has to be born in mind. Acrosomal reacted spermatozoa found on the surface of the zona pellucida always had begun penetration (Fig. 3a). Vesicles originating from vesiculation of spermatozoon membranes are always situated in proximity to those spermatozoa found on the zona surface, which had undergone the acrosomal reaction. Taken together, there seems to be sufficient evidence to indicate that the acrosomal reaction takes place subsequent to binding to the zona pellucida in those human spermatozoa which retain the capacity for zona penetration.

After the acrosome reaction has taken place, spermatozoa begin their passage through the zona pellucida and leave behind a sharply demarcated penetration slit (Fig. 3b). In certain planes of section, where the penetration slit is cut longitudinally, a 'trace' of vesicles which originate during the acrosome reaction is detectable at margins of the penetration slit (Fig. 3b). Deposition of isotopically labelled acrosomal glycoproteins with assumed enzymatic activity has been observed in morphologically similar vesicles on the zona surface during fertilization in the rabbit (47).

In addition to previously noted doubts concerning the role of acrosin as a 'zona lysin', a new physiological significance in spermatozoon-zona interaction has been attributed to the acrosomal reaction. After primary recognition of complementary receptors on both surfaces, the vesiculated products of the acrosomal reaction possibly form a tenacious contact between the spermatozoon and the zona surface, after which the spermatozoa are able to intrude into the zona substance (8). Based on morphological evidence, it has been further speculated that, due to its configuration, the leading edge of a spermatozoon head could not be active as a carrier of any enzymatic activities during its passage through the zona pellucida but may act as a device for maximizing the force per unit area exerted by the forward motility (49). Regardless of the importance of any enzymatic activity for reducing the physical resistance of zona material to perforation by the spermatozoon, an active mechanism of forcing the sperm cell through the zona must exist. Motility of spermatozoa is then the key factor in zona penetration. In fact, a different form of motility during the final phases of spermatozoon progress towards the oocyte has been detected in a number of mammals (20, 50, 51). The significance of this form of movement may be in an increased activity which forces the sperm head into the zona (51). In this connection, it should be mentioned again that development of the capacity for motility is an important outcome of epididymal maturation of spermatozoa (see above).

Interference with the mechanism(s) of primary interaction between the zona pellucida and the spermatozoon has been postulated as an approach to fertility regulation (7). From this point of view, it would be interesting to ascertain whether acrosomal reacted human spermatozoa still possess any capacity to bind to the zona pellucida and also, whether precocious artificial induction of the acrosomal reaction would prevent zona penetration.

3.4. Spermatozoon-oocyte fusion

Having passed through the zona pellucida, the spermatozoon head quickly traverses the perivitelline space and reaches the surface of the vitellus. Recent concepts of spermatozoon-oocyte fusion presume that the plasma membrane overlying the postacrosomal region of the spermatozoon head (20) or the

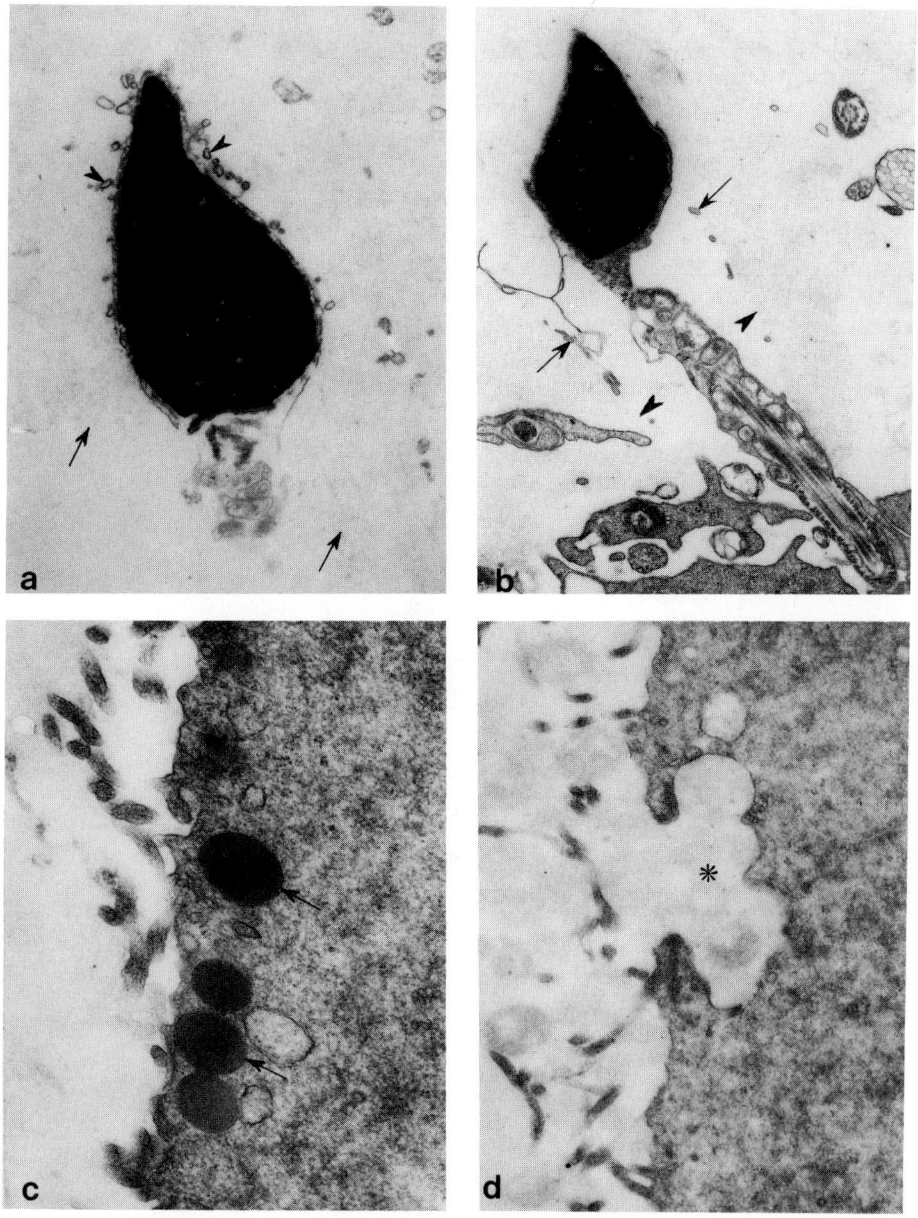

Fig. 3. (a) Head of an acrosomal reacted human spermatozoon starting its penetration into the zona pellucida (surface pointed with arrows). Vesicles originating from the acrosome reaction (arrowheads) (× 16,800). (b) Part of a human spermatozoon within the zona pellucida. Margins of the penetration slit (arrowheads). Vesicles originating through the acrosome reaction (arrows) (× 12,000). (c) Cortical granules (arrows) beneath the plasma membrane of a one-cell human zygote (× 30,400). (d) Cortical cavern (asterisk) opening into the perivitelline space of a human one-cell zygote (× 20,800).

plasma membrane of the equatorial segment of the spermatozoon (8) fuses with the oocyte plasma membrane. Soupart and Strong (3) have described the morphology of spermatozoon-oocyte fusion for zona-free human oocytes inseminated *in vitro*. The continuity of the oocyte plasma membrane with the spermatozoon plasma membrane at the level of the posterior margin of the postacrosomal region has been demonstrated. Fusion also takes place between the oocyte plasma membrane and the plasma membrane covering the spermatozoon tail (3). These authors have reported that spermatozoa both with intact and reacted acrosomes fuse with the oocyte, a situation different from all other mammals studied to date (8). Therefore, it seems that the mode of spermatozoon-oocyte fusion in man will demand a critical reinvestigation because (1) the entry of spermatozoa into the ooplasm may deviate considerably from the physiological events under *in vitro* conditions and (2) unnatural spermatozoon-oocyte interactions *in vitro* could be an early cause of abnormal embryonic development (52). This may be important from a clinical standpoint as present experience with *in vitro* fertilization does warrant concern in this regard.

After the initial membrane fusion, the spermatozoon is incorporated into the ooplasm by means of pseudopodial folds which develop at the oocyte surface (8). Subsequently, the spermatozoon nucleus develops into the male pronucleus (see below) while the other components of the spermatozoon rapidly disintegrate (3). The only identifiable structure of the spermatozoon persisting in the ooplasm for a longer time period are doublets of microtubules from the axial filament complex. These spermatozoon remnants are detected in human eggs up to the two cell stage (53).

3.5. Cortical reaction and block to polyspermy

The block to polyspermy in fertilized human eggs, as in almost all mammalian eggs, is established as a result of exocytosis of cortical granules and discharge of their contents. As a consequence of this process, some physical and biochemical properties of the zona pellucida and/or the egg plasma membrane are modified. The zona pellucida is essential for an effective block to polyspermy in human eggs while the role, if any, of the plasma membrane is of minor significance because polyspermy occurs in human zona-free oocytes (3). Cortical granule discharge, termed the cortical reaction, has been described in man by Tesařík and Dvořák (54). It is triggered by spermatozoon-oocyte fusion. In human oocytes incubated with capacitated spermatozoa, some cortical granules are in contact with each other and with the overlying plasma membrane as well (Fig. 3c). Fusion of cortical granules, both with each other and with the plasma membrane, give rise to typical cortical caverns (Fig. 3d), similar to those described in the rat (55).

In anurans, the fusion of a spermatozoon with the plasma membrane of an oocyte stimulates a rapid and sequential discharge of the cortical granules. Discharge occurs in a wave-like manner, propagated radially over the oocyte surface from the point of sperm contact (8). Although granule discharge is slower in mammals (45, 56), whose oocytes are exposed to comparatively few spermatozoa at fertilization, the wave-like propagation of the cortical reaction may be similar to that in anurans. The recent findings in human oocytes inseminated *in vitro* (54) of both intact cortical granules, and of granules at different phases of discharge, may correspond to observations of different areas of the wave of cortical granule exocytosis.

Although no detailed information about the mechanism of fusion of the cortical granule membrane with the plasma membrane has appeared, it is thought that, preceding the actual fusion, the minimal gap of cytoplasm separating cortical granules from the plasma membrane must be reduced to less than 0.5 nm. This distance is one at which the presumptive net negative charge of the cortical granules would no longer prevent their contact with the negatively charged inner surface of the plasma membrane (8).

Recent experiments have shown that the cortical cytoplasm of mature human oocytes contains at least two types of granules having some morphological similarity but differing cytochemical properties (57). Release of granules rich in glycoprotein has been detected in follicular oocytes *in vivo*. The other population of granules, containing glycoproteins and mucopolysaccharides is considered to represent the 'true' cortical granules. These granules have never been observed to undergo exocytosis prior to fertilization. The former population of granules may be identical with those whose discharge has been reported previously in human follicular oocytes (58). The specific nature of the granules present in the cortical cytoplasm of human oocytes and their physiological role will demand further research.

4. Post-penetration events

Gamete union is followed by changes resulting in the integration of genetic material of both gametes and the activation of early embryonic development. Morphologically this period is characterized by the extrusion of the 2nd polar body, formation of pronuclei and the initiation of the first cleavage division.

Concomitant with the spermatozoon-oocyte fusion, meiosis of the oocyte is completed and the 2nd polar body is extruded at anaphase II (5). Although the 2nd polar body may undergo fragmentation, the fragments can be easily distinguished from fragments derived from the 1st polar body by the absence of cortical granules which have already been discharged at the time of the 2nd polar body formation (2, 53). The maternal chromosomes gradually decondense and become incorporated into the developing female pronucleus (3).

The development of the male pronucleus from the spermatozoon nucleus is a much more complex and delicate process because the resulting structure is morphologically and functionally quite different from its predecessor. The chromatin of spermatozoa differs from the chromatin of most somatic cells. In human spermatozoa, chromatin fibers are present in a highly organized and closely packed arrangement (59). Only as a result of fertilization does the spermatozoon chromatin assume a configuration similar to that seen in somatic cells. This process is accompanied by the replacement of sperm-specific nuclear proteins by histones of oocyte origin. This sequence of events has been well demonstrated in mouse (60).

The transformation of the spermatozoon nucleus into the male pronucleus in man involves breakdown of the nuclear envelope, expansion and decondensation of chromatin, and encapsulation of the diffuse chromatin in a nuclear envelope. The pronuclear envelope forms de novo from vesicles of the endoplasmic reticulum in the egg cytoplasm (3). A similar sequence of events has been described in various mammalian species (7). Numerous mit-

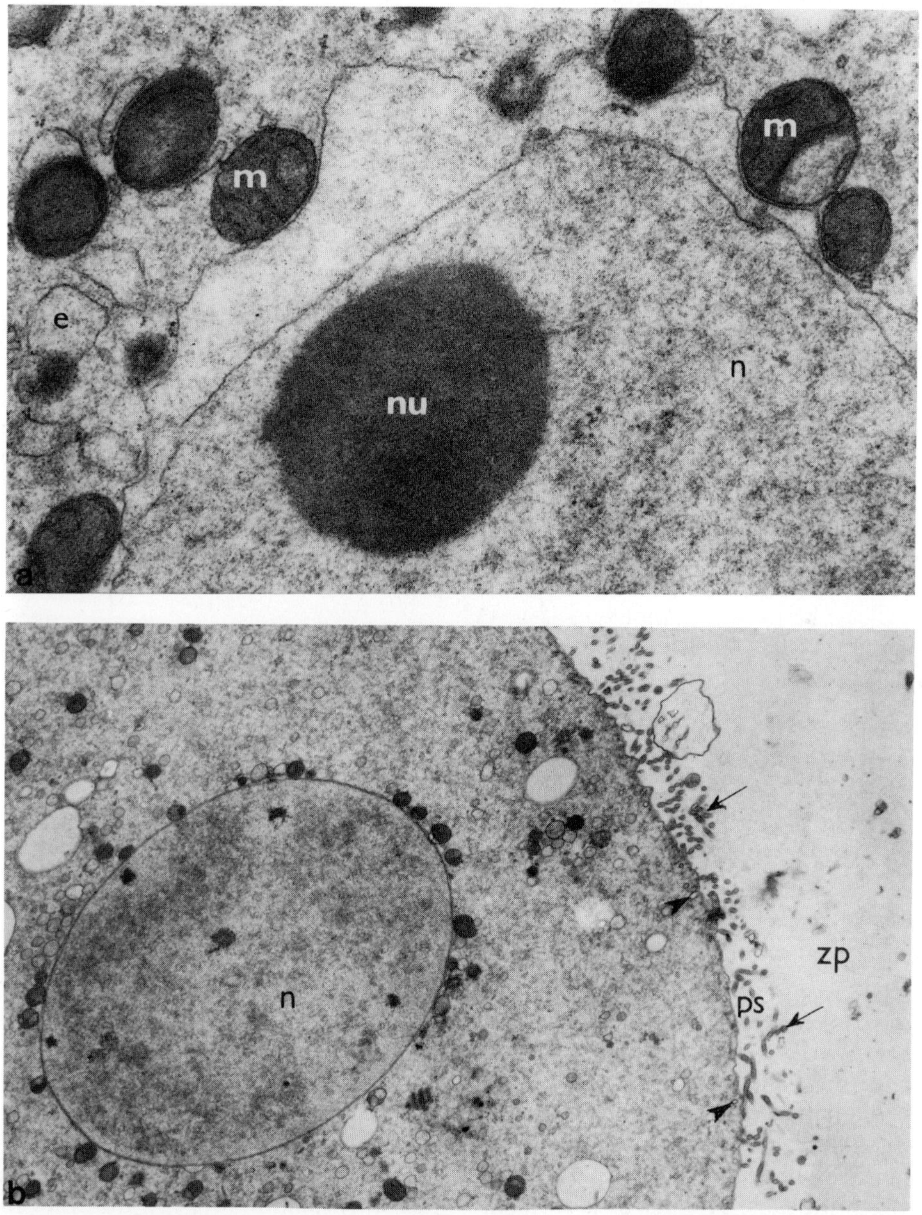

Fig. 4. (a) Part of a human zygote with a developing pronucleus (n) containing a dense nucleolus (nu). Juxtanuclear cytoplasmic region with mitochondria (m) and smooth endoplasmic reticulum (e). (from Tesařik, Dvořák: Folia morph 29: 297, 1981. (× 34,400). (b) Part of a human one-cell zygote with a pronucleus (n), various cytoplasmic structures, microvilli (arrows) and pinocytotic vesicles (arrowheads). Perivitelline space (ps), zona pellucida (zp) (× 6,400).

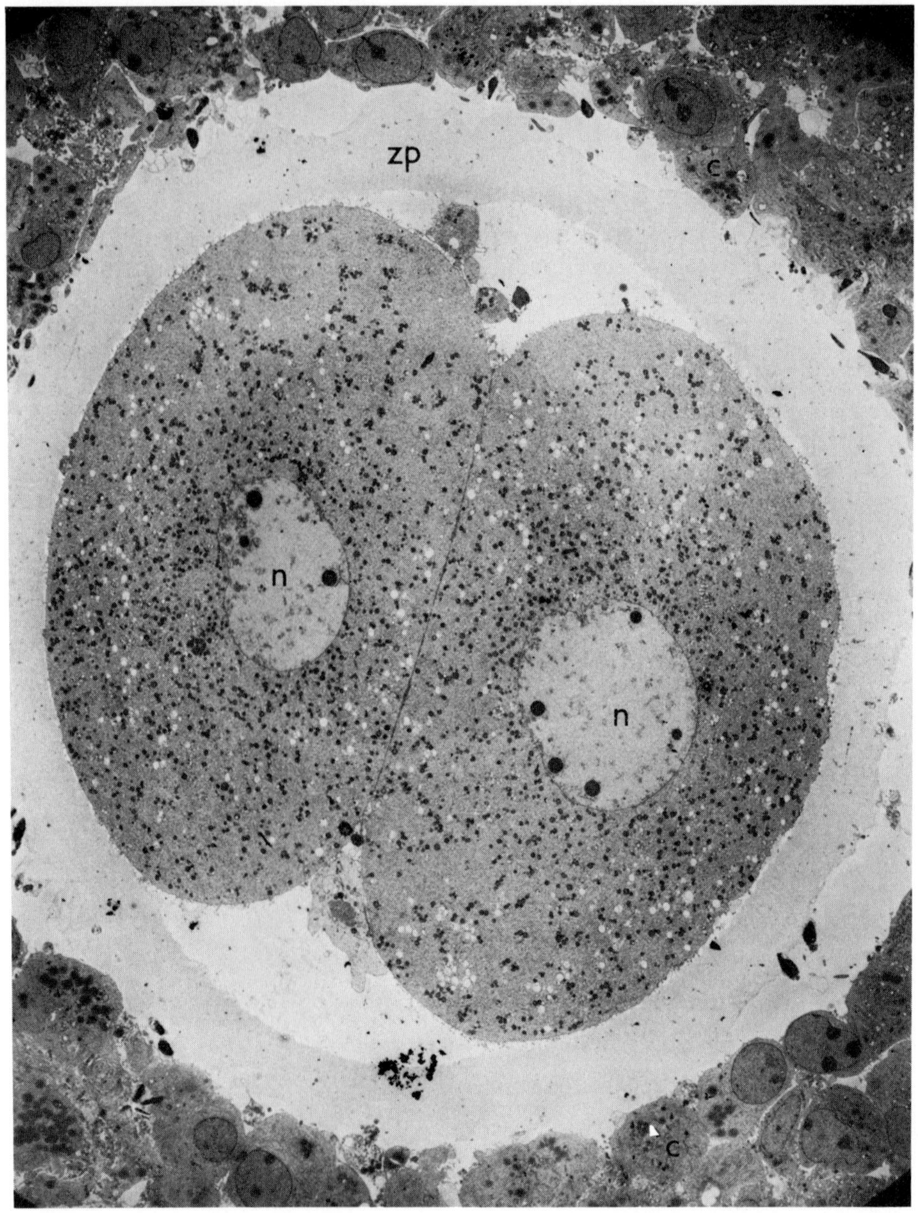

Fig. 5. Survey view of a human two-cell embryo. Blastomere nuclei (n), zona pellucida (zp), cumulus cells (c). (from Dvořák, Tesařík, Pilka, Trávník: Fertil Steril 37: 661, 1982. (× 1,440).

ochondria are in close contact with the developing pronuclear envelope (Fig. 4a) and this relation, typical for both the male and the female pronucleus, is retained even after the process of the pronuclear formation is completed (Fig. 4b). At the same time, dense spherical nucleoli appear (Fig. 4a). As a result of the above events, the male pronucleus is formed. After the completion of its development, the male and female pronuclei are morphologically indistinguishable but for the proximity of the female pronucleus to the 2nd polar body, demonstrable, however, only in certain section planes (1, 6).

Factors required for differentiation of the spermatozoon nucleus into a male pronucleus appear with oocyte maturation and a defect in their constitution, caused for example, by inadequate *in vitro* conditions for oocyte maturation, may lead to abortive pronuclear formation (61, 62). Contrary to observations in different mammalian species (62), in human oocytes matured *in vitro*, the transformation of the spermatozoon head into the male pronucleus occurred normally, as far as morphological aspects are concerned (3, 6). The presence of an enlarged sperm head or male pronucleus, however, has been reported in only 40% of human oocytes matured in a chemically defined medium and exposed to spermatozoa *in vitro* (63). It could not be excluded that failure of cleavage in some human eggs after *in vitro* fertilization may be attributed to an inability of the ooplasm to support a full functional differentiation of the male pronucleus, notwithstanding its normal morphological appearance.

The control of gene activity in pronuclei is clearly exerted by the ooplasm and the principal steps are replacement of sperm protamines by egg histones, DNA synthesis, and reconstitution of chromosomes with subsequent cleavage (62). The explosion of data of molecular, macromolecular and metabolic events associated with post-fertilization development in *mammals*, seen recently, is outside the scope of this review.

The process of fertilization is completed, from the morphological viewpoint, no sooner than after the first cleavage division when the male and female genomes become invested by a common nuclear envelope (Fig. 5), since both pronuclei in almost all mammals studied to date, with a probable exception of the rabbit (64), merely come into close apposition before the disruption of their nuclear envelopes, but their fusion does not really occur.

5. Basic ultrastructural characteristics of the human zygote

The only human zygote fertilized *in vivo* and submitted to a detailed ultrastructural analysis was recovered from the fallopian tube 26 h after the sexual intercourse (1). Following this report, several descriptions of human zygotes produced in culture using various *in vitro* fertilization techniques have been published (2, 5, 6). Minor differences between zygotes developing *in vivo* and those obtained *in vitro* may be due to still not quite appropriate culture conditions. However, typical ultrastructural characteristics of the human zygote may be based on the above studies.

The human zygote is surrounded by the zona pellucida and consists of two pronuclei and cytoplasm whose organelles are mainly mitochondria, components of the endoplasmic reticulum and elements of the Golgi apparatus. Two polar bodies are present in the narrow perivitelline space between the plasma membrane and the inner surface of the zona pellucida.

The pronuclei, after completion of their development, are equal in size, the largest diameter measuring 30 μm, and located eccentrically and close to each other. They appear spherical or ovoid and contain up to eight dense nucleoli, 1–4 μm in size. The nucleoli are composed of highly compacted fibrilar material (Fig. 4a). The perinuclear space may be dilated by evaginations of the outer membrane of the nuclear envelope, in which are contained small vesicles limited by an independent membrane. The functional significance of these vesicles remains unclear. Similar vesicles in pronuclei of the rat have been interpreted as tertiary nucleoli (65), an unequivocal evidence for such a statement, however, is still lacking.

Cytoplasmic structures reveal a tendency to accumulate near the pronuclei. Mitochondria (Fig. 4a, b) are characterized by a marked pleomorphism, however, they predominantly are round or oval with but a few cristae parallel to the outer mitochondrial membrane. The mitochondrial matrix is electron-dense. Mitochondria are often attached to the nuclear envelope, structures of the endoplasmic reticulum and annulate lamellae. The endoplasmic reticulum occurs almost entirely in its smooth form. The granular endoplasmic reticulum is always represented by vesicles whose membranes are only infrequently associated with ribosomes. Annulate lamellae are present in a considerable number. Their occurence, however, is limited to the juxtanuclear region. The Golgi apparatus consists of prominent clusters of closely packed vesicles and parallel tubules containing material of varying electron density. Other cytoplasmic structures such as secondary lysosomes, microtubules, cortical granules and crystalline inclusions are observed infrequently.

The plasma membrane projects through numerous microvilli into the perivitelline space (Fig. 4b). Between the microvilli signs of active pinocytosis are evident. These membrane specializations apparently provide a mechanism for uptake of nutrients into the zygote.

6. Concluding remarks

The advancing knowledge of the events of fertilization at the cellular and subcellular levels in the human is one of the most remarkable contributions to reproductive biology, and offers a solid basis for feasible experimentation with human gametes under *in vitro* conditions. Data concerning the ultrastructural aspects of fertilization events in man remain, however, scanty in comparison with those available for laboratory and farm animals. As a deeper understanding of mechanisms engaged in the control of human fertilization is increasingly dependent on the experimental approach used in non-human species, the concomitant research into fertilization in laboratory and farm animals should stress the most important areas for further microanatomical research in man. These areas should comprise those research goals directed to the processes in which major differences may exist between man and other animals. Therefore, the following topics are suggested: (1) Membrane transformations and morphological changes in the motility apparatus during spermatozoon maturation and capacitation. (2) Nuclear and cytoplasmic maturation of the oocyte and involvement of cumulus oophorus cells in this process. (3) Intraovarian mechanisms of the selection of the ovulatory oocyte. (4) Reinvestigation of the physiological role of the acrosomal reaction in the spermatozoon-oocyte interaction and re-evaluation ot its ultrastructural correlates. (5) Integration of the spermatozoon chromatin into the zygote. (6) Follow-up of the postfertilization development as a criterion of normalcy of *in vitro* fertilization.

It is expected that future research will be facilitated by the accessibility of human eggs fertilized *in vitro* since this technique is being introduced in an increasing number of laboratories, giving a unique opportunity for experimentation and yet adhering to basic moral and ethical principles.

References

1. Zamboni L, Mishell DR, Bell JH, Baca M: Fine structure of the human ovum in the pronuclear stage. J Cell Biol 30: 579–600, 1966.
2. Soupart P, Strong PA: Ultrastructural observations on human oocytes fertilized *in vitro*. Fertil Steril 25: 11–44, 1974.
3. Soupart P, Strong PA: Ultrastructural observations on polyspermic penetration of zona pellucida-free human oocytes inseminated *in vitro*. Fertil Steril 26: 523–537, 1975.
4. McMaster R, Yanagimachi R, Lopata A: Penetration of human eggs by human spermatozoa *in vitro*. Biol Reprod 19: 212–216, 1978.
5. Lopata A, Santhananthan AH, McBain JC, Johnston WIH, Speirs AL: The ultrastructure of the preovulatory human egg fertilized *in vitro*. Fertil Steril 33: 12–20, 1980.
6. Tesařík J, Dvořák M: Ultrastructure of the human ovum fertilized *in vitro*. Folia morph (Prague) 29: 297–304, 1981.
7. Gwatkin RBL: Fertilization mechanisms in man and mammals. New York and London, Plenum Press, 1977.
8. Bedford JM, Cooper GW: Membrane fusion events in the fertilization of vertebrate eggs. In: Membrane fusion. Poste G, Nicolson GL (eds), Amsterdam, Elsevier/North-Holland Biomedical Press, 1978, pp 65–125.
9. Hinrichsen MJ, Blaquier JA: Evidence supporting the existence of sperm maturation in the human epididymis. J Reprod Fertil 60: 291–294, 1980.
10. Edwards RG: Early human development: from the oocyte to implantation. In: Scientific foundations of obstetrics and gynecology. Philipp EE, Barnes J, Newton M (eds), London, Heinemann Medical, 1977, pp 175–252.
11. Olson GE, Danzo BJ: Surface changes in rat spermatozoa during epididymal transit. Biol Reprod 24: 431–443, 1981.
12. Fléchon JE: Ultrastructural and cytochemical analysis of the plasma membrane of mammalian sperm during epididymal maturation. In: Progress in reproductive biology. Epididymis and fertility: biology and pathology. Bolack C, Clavert A (eds), Basel, S. Karger, 1981, pp 90–99.
13. Bedford JM: Components of sperm maturation in the human epididymis. In: Advances in the biosciences 10, Schering workshop on contraception: the masculine gender. Braunschweig, Pergamon Press, 1973, pp 145–155.
14. Koehler JK: The mammalian sperm surface: studies with specific labeling techniques. Int Rev Cytol 54: 73–108, 1978.
15. Kopečný V, Pech V: An autoradiographic study of macromolecular syntheses in the epithelium of the ductus epididymidis in the mouse. II. Incorporation of L-fucose-1-^{3}H. Histochemistry 50: 229–238, 1977.
16. Kopečný V, Fléchon JE, Pivko J: Dynamics of epididymal glycoproteins. Proc IVth Int Symp Spermatology, Seillac, 1982.
17. Moore HDM: Glycoprotein secretion of the epididymis in the rabbit and hamster: localization on epididymal spermatozoa and the effect of specific antibodies on fertilization *in vivo*. J Exp Zool 215: 77–85, 1981.
18. Acott TS, Hoskins DD: Bovine sperm forward motility protein: binding to epididymal spermatozoa. Biol Reprod 24: 234–240, 1981.
19. Calvin HI, Bedford JM: Formation of disulphide bonds in the nucleus and accessory structures of mammalian spermatozoa during maturation in the epididymis. J Reprod Fertil, Suppl 13: 65–75, 1971.
20. Yanagimachi R: Specifity of sperm-egg interaction. In: Immunobiology of gametes. Edidin M, Johnson MH (eds), Cambridge, Cambridge University Press, 1977, pp 255–295.
21. Meizel S: The mammalian sperm acrosome reaction, a biochemical approach. In: Development in mammals. Johnson M (ed), Amsterdam, North-Holland Publishing Co, Vol. 3, 1978, pp 1–64.
22. Gordon M, Dandekar PV, Bartoszewicz W: The surface coat of epididymal, ejaculated and capacitated apermatozoa. J Ultrastruct Res 50: 199–207, 1975.
23. Tsafriri A: Mammalian oocyte maturation: model systems and their physiological relevance. In: Proceedings of workshop on ovarian follicular and corpus luteum function, Miami Beach, Florida, 1978. Channing CP, Marsh J (eds), Plenum Publ Corp.
24. Tesařík J, Dvořák M: Development of intercellular contacts between the oocyte and follicle cells during the differentiation of human ovarian follicles. Scripta med 53: 325–327, 1980.
25. Tesařík J, Dvořák M: Human cumulus oophorus preovulatory development. J Ultrastruct Res 78: 60–72, 1982.
26. Hillensjö T, Channing CP, Pomerantz SH, Kripner AS: Intrafollicular control of oocyte maturation in the pig. *In Vitro* 15: 32–39, 1979.
27. Zamboni L, Moore-Smith D, Thompson RS: Migration of follicle cells through the zona pellucida and their sequestration by human oocytes *in vitro*. J Exp Zool 181: 319–340, 1972.
28. Calarco PG, Donahue RP, Szöllösi D: Germinal vesicle breakdown in the mouse oocyte. J Cell Sci 10: 369–385, 1972.
29. Moore GPM, Lintern-Moore S: Transcription of the mouse oocyte genome. Biol Reprod 17: 865–870, 1978.
30. Wassarman PM, Letourneau GE: Meiotic maturation of mouse oocytes *in vitro*: association of newly synthesized proteins with condensing chromosomes. J Cell Sci 20: 549–568, 1976.
31. Motlík J, Kopečný V, Pivko J: The fate and role of macromolecules synthesized during mammalian oocyte meiotic maturation. I. Autoradiographic topography of newly synthesized RNA and protein in the germinal vesicle of the pig and rabbit. Ann Biol anim Biochem Biophys 3: 735–746, 1978.
32. Palombi F, Viron A: Nuclear cytochemistry of mouse oogenesis. I. Changes in extranucleolar ribonucleoprotein components through meiotic prophase. J Ultrastruct Res 61: 10–20, 1977.
33. Motlík J, Kopečný V, Trávník P, Pivko J: Fine structure cytochemical and autoradiographic analysis of RNA dynamism in the nucleus of the pig oocyte in the germinal vesicle stage. Biol Cell (in press).
34. Bachvarova R: Synthesis, turnover, and stability of heterogeneous RNA in growing mouse oocytes. Dev Biol 86: 384–392, 1981.
35. Crozet N, Szöllösi D: Effects of actinomycin D and α-amanitin on the nuclear ultrastructure of mouse oocytes. Biol Cell 38: 163–170, 1980.
36. Peluso JJ, Butcher RL: RNA and protein synthesis in control and follicularly-aged rat oocytes. Proc Soc Exp Biol Med 147: 350–353, 1974.
37. Dvořák M, Tesařík J: Ultrastructure of human ovarian follicles. In:

Biology of the ovary. Motta PM, Hafez ESE (eds), The Hague, Martinus Nijhoff Publishers, 1980, pp 121–137.
38. Van Blerkom J: Cell and molecular biology of meiotic maturation of the mammalian oocyte. In: Biology of the ovary. Motta PM, Hafez ESE (eds), The Hague, Martinus Nijhoff Publishers, 1980, pp 179–190.
39. Motlík J, Kopečný V, Pivko J, Fulka J: Distribution of proteins labelled during meiotic maturation in rabbit and pig eggs at fertilization. J Reprod Fertil 58: 415–419, 1980.
40. Edwards RG, Donahue RP, Baramki TA, Jones HW: Preliminary attempts to fertilize human oocytes matured *in vitro*. Amer J Obstet Gynec 96: 192–200, 1966.
41. Szöllösi D, Hunter RHF: The nature and occurence of the acrosome reaction in spermatozoa of the domestic pig, Sus scrofa. J Anat 127: 33–41, 1978.
42. Talbot P, Franklin LE: The release of hyaluronidase from guinea-pig spermatozoa during the course of the normal acrosome reaction *in vitro*. J Reprod Fertil 39; 429–432, 1974.
43. Roomans GM, Afzelius BA: Acrosome vesiculation in human sperm. J Submicr Cytol 7: 61–69, 1975.
44. Overstreet JW, Hembree WC: Penetration of the zona pellucida of nonliving human oocytes by human spermatozoa *in vitro*. Fertil Steril 27: 815–831, 1976.
45. Bedford JM: An electron microscopic study of sperm penetration into the rabbit egg after natural mating. Amer J Anat 133: 213–254, 1972.
46. Shams-Borhan G, Huneau D, Fléchon JE: Acrosin does not appear to be bound to the inner acrosomal membrane of bull spermatozoa. J Exp Zool 209: 143–149, 1979.
47. Kopečný V, Fléchon JE: Fate of acrosomal glycoproteins during the acrosome reaction and fertilization: a light and electron microscope autoradiographic study. Biol Reprod 24: 201–216, 1981.
48. Bedford JM: Sperm/egg interaction: the specifity of human spermatozoa. Anat Rec 188: 477–488, 1977.
49. Bedford JM, Cross NL: Normal penetration of rabbit spermatozoa through a trypsin- and acrosin-resistant zona pellucida. J Reprod Fertil 54: 385–392, 1978.
50. Mohri H, Yanagimachi R: Characteristics of motor apparatus in testicular, epididymal and ejaculated spermatozoa. Exp Cell Res 127: 191–196, 1980.
51. Katz DF, Yanagimachi R: Movement characteristics of hamster and guinea pig spermatozoa upon attachment to the zona pellucida. Biol Reprod 25: 785–791, 1981.
52. Shalgi R, Phillips DM: Mechanics of *in vitro* fertilization in the hamster. Biol Reprod 23: 433–444, 1980.
53. Dvořák M, Tesařík J, Pilka L, Trávník P: Fine structure of human two-cell ova fertilized and cleaved *in vitro*. Fertil Steril 37: 661–667, 1982.
54. Tesařík J, Dvořák M: Cortical reaction and block to polyspermy in man (in Czech). Scripta med 52: 494–495, 1979.
55. Szöllösi D: Development of cortical granules and the cortical reaction in rat and hamster eggs. Anat Rec 159: 431–446, 1967.
56. Barros C, Yanagimachi R: Induction of zona reaction in golden hamster eggs by cortical granule material. Nature (London) 233: 268–269, 1971.
57. Šťastná J, Dvořák M, Pilka L: Electron microscopic and cytochemical study of the cortical cytoplasm in preovulatory human oocytes. Cell Tissue Res (in press).
58. Rousseau P, Meda P, Lecart C, Haumont S, Ferin J: Cortical granule release in human follicular oocytes. Biol Reprod 16: 104–111, 1977.
59. Wagner TE, Yun JS: Fine structure of human sperm chromatin. Arch Androl 2: 291–294, 1979.
60. Kopečný V, Pavlok A: Autoradiographic study of mouse spermatozoan arginine-rich nuclear protein in fertilization. J Exp Zool 191: 85–96, 1975.
61. Thibault C: *In vitro* maturation and fertilization of rabbit and cattle oocytes. In: The regulation of mammalian reproduction. Segal SJ, Crozier R, Corfman PA, Condliffe PG (eds), Springfield, Charles C. Thomas, 1973, pp 231–240.
62. Longo FJ: Regulation of pronuclear development. In: Bioregulators of reproduction. Jagiello G, Vogel HJ (eds), New York, Academic Press, 1981, pp 529–557.
63. Nishimoto T, Yamada I, Niwa K, Mori T, Nishimura T, Iritani A: Sperm penetration *in vitro* of human oocytes matured in a chemically defined medium. J Reprod Fertil 64: 115–119, 1982.
64. Gondos B, Bhiraleus P, Conner LA: Pronuclear membrane alterations during approximation of pronuclei and initiation of cleavage in the rabbit. J Cell Sci 10: 61–78, 1972.
65. Szollosi D: Extrusion of nucleoli from pronuclei of the rat. J Cell Biol 25: 545–562, 1965.

Authors' address:
Department of Histology and Embryology
Faculty of Medicine
J.E. Purkyně University
tř. Obránců míru 10
662 43 Brno, Czechoslovakia

CHAPTER 16

The cell surface of the mammalian embryo during early development

LYNN M. WILEY

1. Introduction

In experiments wherein he rearranged the blastomeres of cleavage-stage mouse embryos, Tarkowski (1) made the very simple yet profound observation that those blastomeres residing within the embryo tend to become inner cell mass (ICM) while those on the outside of the embryo tend to become trophectoderm (Fig. 1). He concluded that individual blastomeres possessed the ability to become either ICM or trophectoderm, based on relative cell position within the embryo, at least up to the 8-cell stage. Subsequent experiments have shown that this ability is not lost until the morula stage, when the embryo consists of 16–32 closely apposed cells (2). These observations have led to the 'inside/outside' hypothesis, which states that the relative position of a cell within the morula determines the subsequent developmental fate of that cell (1–3). If this is so, how is such positional information translated into the appropriate genetic activity which will lead to the differentiation of the cell to ICM or trophectoderm? In addition, how is this developmental restriction related to the formation of the blastocyst (i.e., cavitation)?

These questions bring attention to the cell surface as a possible transducer of positional information and have, consequently, generated the current interest in the cell surface properties of preimplantation embryos. The following discussion will analyze the available information on these cell surface properties for possible relevance to mechanisms whereby individual blastomeres could regulate their genetic activity according to their relative position within the morula. In addition, an attempt will be made to relate such positional information to morphogenetic aspects of cavitation. Special attention will be given to those properties that have been revealed by morphological techniques. Unless otherwise indicated, the discussion will be restricted to an evaluation of observations in the mouse.

2. Fertilization

It is necessary to begin with fertilization because in other organisms, most notably in amphibians, there is evidence that the surface of the fertilized oocyte is a developmental mosaic that forms a 'fate map' such that the origin of the germ layers and additional structures that develop from the blastocyst can be traced back to specific regions of the oocyte surface (4). Furthermore, the establishment of the fate map of the amphibian oocyte is believed to depend upon the site of sperm entry during fertilization.

Recent studies have revealed that the surface of the mouse oocyte is morphologically mosaic, consisting of two distinct regions – one region having few microvilli, which overlies the meiotic spindle and comprises about 20% of the oocyte surface, and a second region having many microvilli, under which reside the cortical granules and which comprises the remaining 80% of the oocyte surface (5–7). With fertilization, the smooth region becomes excluded as the surface of the second polar body (6), so that, by morphological criteria, the fertilized egg surface has no structural mosaicism and becomes uniformly and densely microvillous. Sperm rarely adhere to the smooth region of the unfertilized oocyte, leading to the notion that one functional 'raison d'etre' for the smooth region is to prevent sperm entry from occurring near the meiotic spindle. Sperm entry in this region might interfere with the division of the meiotic spindle, with the formation of the second polar body or with the insertion of the appropriate chromosomes into the second polar body (5, 6).

The anatomical mosaicism of the mouse oocyte surface is coincident with the apparent distribution of receptors for fluoresceinated concanavalin A (con A), with the smooth region having considerably fewer receptors than the microvillous region (5). However, further studies with transmission electron microscopy using ferritin-conjugated con A suggest that the number of con A binding sites per unit area

Van Blerkom, J. and Motta, P.M. (eds.), Ultrastructure of reproduction. ISBN 0-89838-572-5.

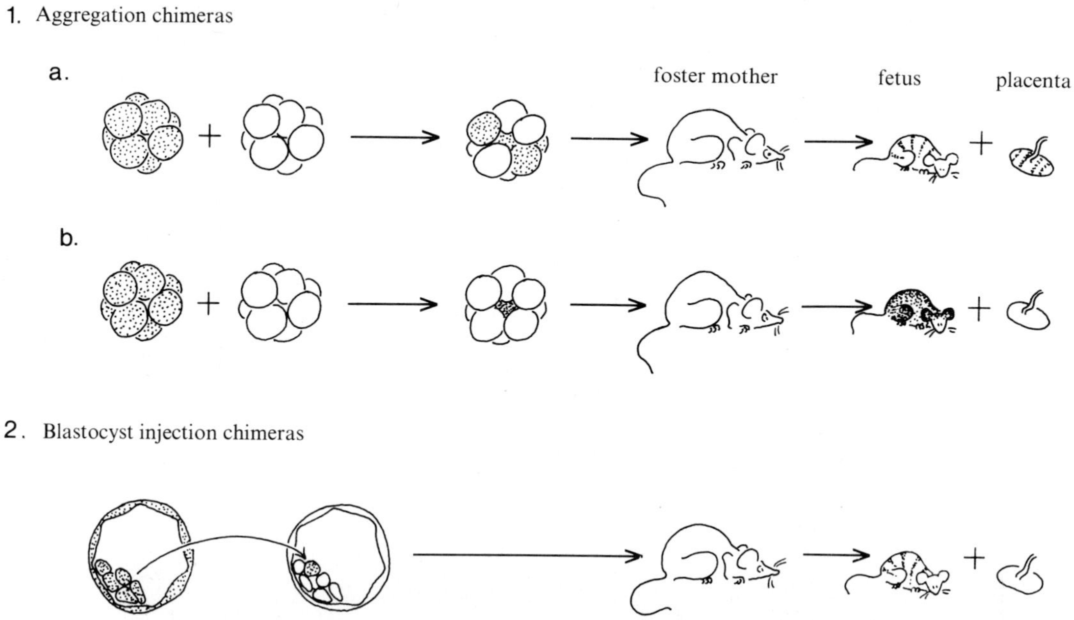

Fig. 1. Inside/outside hypothesis (1.a–1.b) Aggregation chimeras constructed from two different genetically-distinct 8-cell embryo blastomerestype X (stipled blastomeres) and type Y (clear blastomeres). (1.a) If the blastomeres from embryos of genotypes X and Y are randomly reassociated in the aggregate, then both the ICM derivative (the fetus) and the trophectoderm derivative (trophoblastic component of the placenta) will be mosaic (i.e., will consist of X- and Y-derived cells). (1.b) If the blastomeres are arranged so that all of the inside cells are of one genotype (type X) and all of the outside cells are of the other genotype (type Y) then the fetus will consist predominantly of type X-derived cells and the trophoblastic component of the placenta will consist predominantly of type Y-derived cells). (2) Blastocyst injection chimeras: if a cell from the ICM of a type X blastocyst is injected into a host type Y blastocyst, the injected ICM cell will colonize the host type Y ICM so that the resulting fetus will be mosaic – but the trophoblastic component of the placenta will consist solely of the host genotype (type Y).

of cell surface does not vary significantly over the two anatomical regions of the oocyte surface (6). This observation suggests that the apparent reduction of con A binding sites over the smooth region as seen at the light microscopic level is a result of the reduction in the number of microvilli over the smooth region. This conclusion is consistent with the notion that it is the regional difference in the number of microvilli – rather than of specific molecules – that prohibits sperm adhesion-sperm entry in the region overlying the spindle. However, in many instances, the presence of microvilli appears to interfere with cell-cell adhesion, most notably, with the adhesion of microorganisms to the cells lining the intestine (9) and with the cell-cell adhesion that facilitates subsequent cell fusion induced by polyethylene glycol (10). One could argue that the sperm preferentially adhere to the microvillous region of the oocyte because the microvilli act to concentrate a required three-dimensional density array of sperm receptors in the region of sperm-oocyte contact since sperm do seem to 'nestle' amongst the microvilli as a prelude to sperm-oocyte fusion (Fig. 2). At present, there is insufficient information on the identity and the topographical distribution of such putative sperm receptors – if such receptors even exist as unique molecular entities – to assess the validity of this argument.

Microscopic techniques have revealed another interesting feature of the oocyte surface, the molecular properties of the oocyte plasma membrane. Fluorescent microscopy reveals that certain cell surface antigens can be crosslinked into patches by divalent antibody on unfertilized oocytes but not on late 1-cell stage or on 2-cell stage embryos (11). Photobleaching experiments show that the lateral diffusion rates of both lipid probes and of univalent antibody-labelled proteins are 10 to 1000 times more rapid in unfertilized oocytes than in fertilized eggs (12). Taken together, these observations have led to the conclusion that one response to fertilization is an overall reduction in the fluidity of the egg plasma membrane, a reduction which persists into cleavage. This conclusion is further substantiated by the observation of Gabel et al. (13), who found that fluoresceinated

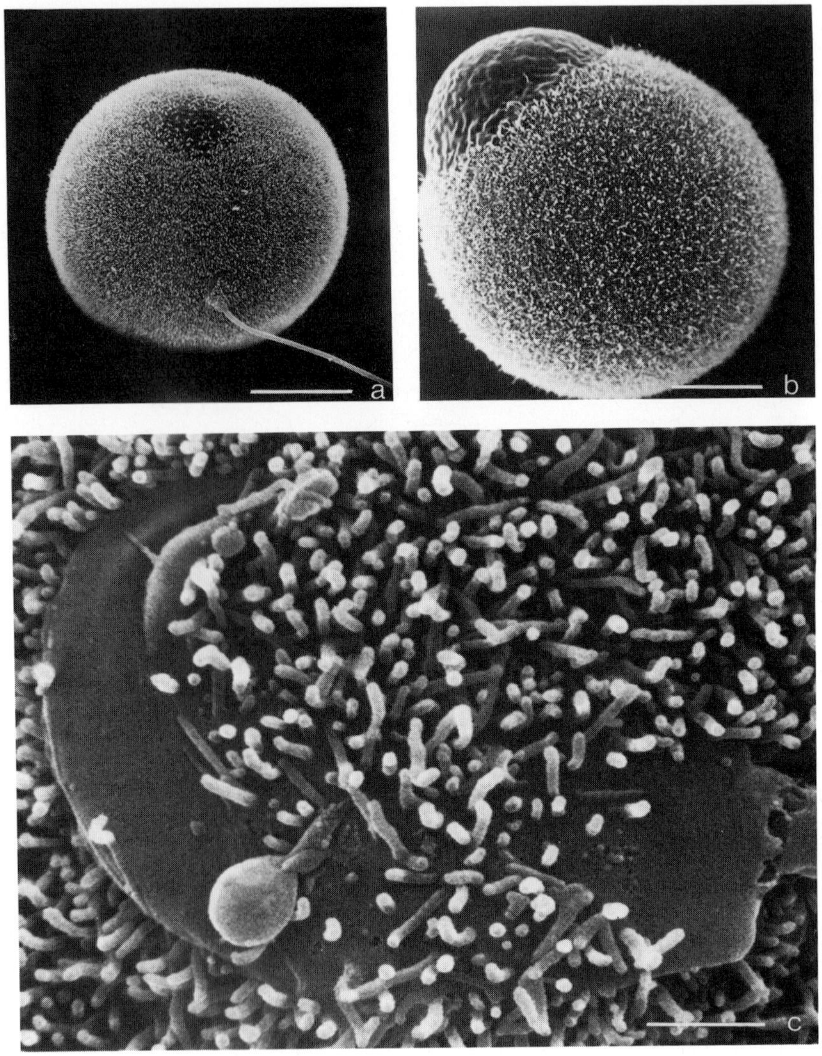

Fig. 2. Scanning electron microscopy of fertilization. (a) hamster oocyte with an attached sperm – notice that sperm attachment is not to the surface region lacking microvilli; (b) unfertilized mouse oocyte with a prominent non-microvillous region, which is somewhat more elevated than what is observed with the hamster oocyte; (c) higher magnification of sperm-oocyte attachment in the hamster, showing that microvilli characterizing the oocyte surface are observed over the equatorial segment of the acrosome and over the anterior region of the postacrosome region-microvilli are not observed over the anterior portion of the sperm head where the acrosome reaction has occurred. Bars in a and in b represent 20 microns; bar in c represents 2 microns. (figures a and c taken from Shalgi and Phillips. J Ultrastruct Res 71: 154–161, 1980; figure b taken from Phillips and Shalgi. J Ultrastruct Res 72: 1–12 1980).

sperm membrane proteins are transferred to the plasma membrane of the oocyte at fertilization where they remain localized as a membrane patch, which persists during early cleavage. This observation, as well as those from the photobleaching experiments, has invoked speculation that the apparent rigidity of the plasma membrane could be involved in the encoding of spatial and temporal information (a fate map?) required for later developmental events (12). Could, in fact, the site of sperm fusion impart a polarity or establish an embryonic axis in the fertilized mouse egg as has been suggested for amphibians, where the site of sperm entry is classically held to govern the subsequent appearance of the grey crescent and the establishment of the dorsal-ventral axis of the embryo (see 14)? Recent observations show that gravitationally induced displacements of the contents of the frog egg – rather than the site of sperm entry – can determine the orientation of the dorsal-ventral axis (15). Furthermore, mammalian embryos exhibit, perhaps, the greatest degree of regulative (versus mosaic) development amongst vertebrate embryos. So if the site of sperm entry – and the persistence of sperm membrane components as a localized membrane patch – do, in fact, impart morphogenetic and spatial/temporal information to the developing mammalian embryo, then such information must be labile and easily reconstructed by individual blastomeres, without a second cueing from another sperm, in order for it to account for the observed regulative capabilities of cleaving mouse embryos.

A most puzzling set of observations regarding fertilization changes in the oocyte surface consists of a series of studies with con A. Several of these studies show that 1) unfertilized oocytes and fertilized eggs bind equal amounts of con A (16), but that 2) con A labelling has a greater tendency to be patchy on fertilized oocytes than on unfertilized oocytes and 3) fertilized eggs are more easily agglutinated with con A than unfertilized oocytes (5, 16). This positive correlation between patchy con A labelling and oocyte agglutination parallels what has been observed for a wide variety of cell types (17, 18). The predominant interpretation of such a positive correlation is that the lateral mobility of con A receptors is required for the ability of con A to agglutinate cells (17, 18). However, recall that fertilization is accompanied by a reduction in apparent membrane fluidity and lateral mobility of membrane components (11–13). Thus, this reduction in lateral mobility with fertilization must either by selective and fortuitously not affect those membrane components that bind con A, or, perhaps, the binding con A *per se* induces an artifactual increase in membrane fluidity and the lateral mobility of oocyte plasma membrane components. Otherwise, it becomes necessary to propose that the characteristics of con A receptors on mouse oocytes differ profoundly from those of con A receptors on other cell types. Of course, there is the possibility that the fertilization-induced reduction in membrane fluidity as assessed by photobleaching (12) is not sufficient to restrict the agglutinability of fertilized eggs as assessed by con A (5, 16).

Finally, there is yet another discrepancy between con A binding on mouse oocytes and that on other cell types. The specific binding affinity of con A receptors and the rate of con A binding both increase with fertilization (16). However, in other cell types, increased binding affinity for and binding rate of con A have a positive correlation with *increased* membrane fluidity (19). Recall that membrane fluidity *decreases* with fertilization.

The above discussion illustrates three points. First, there is insufficient information for any conclusions to be drawn about whether or not the surface of the mammalian oocyte is a developmental as well as a morphological mosaic. Second, if the surface of the mammalian fertilized oocyte is developmentally or morphologically mosaic then such mosaicism has not been resolved by currently available microscopic techniques. Third, the observations regarding the labelling characteristics of con A on unfertilized and fertilized oocytes need to be clarified and assessed for significance to sperm-oocyte interactions and to the subsequent development of the embryo.

3. Cleavage

As cleavage ensues, the surfaces of the blastomeres do not differ from that of the fertilized oocyte insofar as morphological homogeneity and membrane rigidity are concerned. Nevertheless, indirect immunofluorescence shows that the molecular composition of the plasma membrane is changing with respect to the expression of cell surface antigens during cleavage. For example, the fertilized oocyte inherits the major histocompatibility (H-2) antigens of the mother and as the fertilized oocyte nears the first cleavage these H-2 antigens disappear (20). When the embryo attains the 8-cell stage it begins to express several different cell surface antigens, some of which disappear with furter embryonic development to become replaced by yet another wave of different cell surface antigens (Fig. 3) (21, 22). Up to the 8-cell stage, before the embryo undergoes 'compaction' (see the following section), the antigens that are then expressed are distributed homogeneously over the surfaces of individual blastomeres and no consistent differences in antigen quantity have been detected from one blastomere to another. These observations suggest that the lack of any apparent morphological mosaicism of the surfaces of individual blastomeres during the first three cleavages is paralleled by a lack of any apparent molecular mosaicism of the surfaces of individual blastomeres. However, these observations also show that the embryonic surfaces are not qualitatively quiescent insofar as their molecular composition is concerned. In addition, the ability of plant lectins, including con A, to agglutinate or bind to cleavage stage mouse embryos indicates that the nature and/or quantities of cell surface glycoproteins are changing during cleavage (23). Finally, there are distinct changes in the gel profiles of surface glycoproteins that are synthesized by different stages of cleaving mouse embryos (24).

The functional significance of any of these changes is unknown, as is, in most cases, the molecular identity of the changing cell surface molecules whether they be detected by antibodies (histocompatibility antigens being an exception as far as identity is concerned) or by plant lectins.

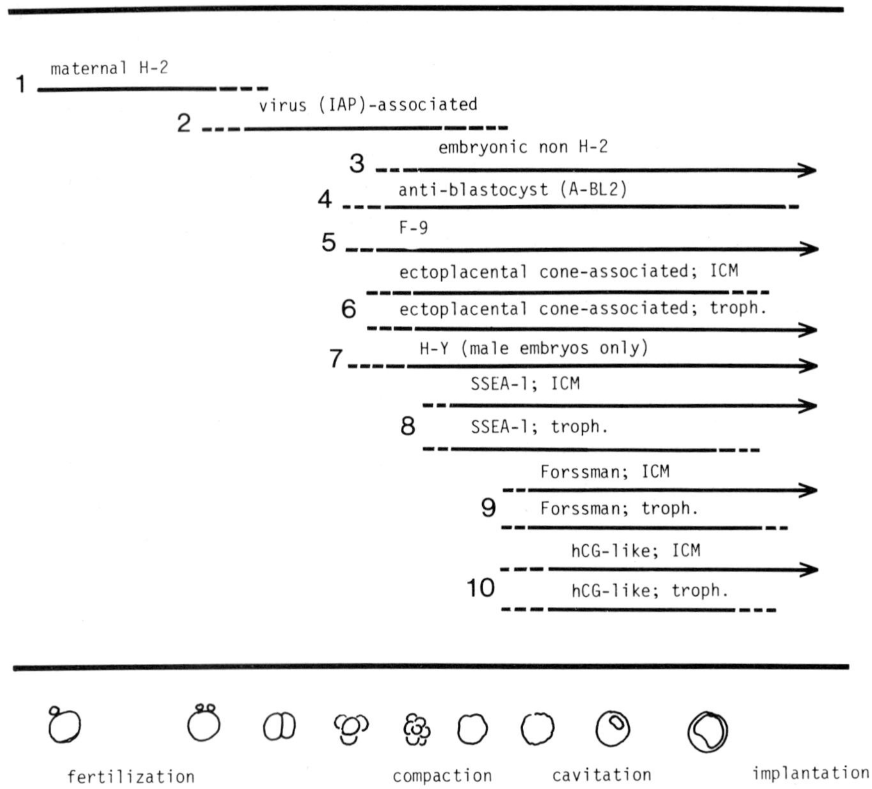

Fig. 3. Cell surface antigens whose expression begins or is transient during mouse preimplantation development. The following points should be noted: 1) 'expression' refers to serological accessibility, which may not correlate with actual physical presence due to 'masking' or to other topographical conditions; 2) the dashed portions of the lines indicate when expression is either weak or is variable from embryo to embryo, whereas the solid portions of the lines indicate when expression is strong and present on the majority of embryos; 3) antigen nos. 6, 8, 9 and 10 become tissue-specific, with antigen no. 6 segregating to the trophectodermal derivatives and antigen nos. 8–10 segregating to the ICM; 4) bibliographical references for the antigens, given as antigen no. (ref. no. or nos.): 1.(20), 2.(84), 3.(85), 4.(11, 86), 5(87, 88), 6.(62), 7.(89, 90), 8.(91, 92), 9(60), 10.(61).

4. Compaction

After the third cleavage, when the embryo consists of 8 blastomeres, it undergoes 'compaction'. The blastomeres flatten against one another so that cell-cell apposition becomes maximized and all intercellular spaces are obliterated (25-28). As this occurs, the 8-cell embryo acquires two new anatomical features. First, two types of cell surfaces are created – free (facing outwards towards the zona pellucida) – and apposed (flattened against an adjacent cell). Second, the compacted 8-cell embryo will, therefore, consist of two anatomically distinct cell types, inner cells having one type of cell surface – apposed, and outer cells having two types of cell surfaces – free and apposed. With further development the inner cells become the ICM while the outer cells differentiate into trophectoderm. It is, therefore, the process of compaction that is preceived (by investigators, perhaps not by the embryo) as being associated with the first overt sign for the existence of (or for the potential to subsequently produce) two developmentally distinct cell types within the mammalian embryo. Consequently, compaction has become a major focus of attention regarding the inside/outside hypothesis in terms of cell surface participation in ICM/trophectoderm differentiation.

It is important to realize that with compaction the surfaces of the outer cells will become polarized and will consist of two regions (domains), one being the free cell surface and the other being the apposed cell surface. The free cell surface domains retain their microvilli while the apposed ones lose their microvilli and acquire specialized cell-cell junctions (Fig. 4). Along with this morphological polarity is a coincident one at the molecular level whose establishment is dependent upon cell contact and which is detectable with anti-mouse species antibodies and with various plant lectins, including con A (see chapter by Johnson et al.; 29, 30). The antibodies and plant

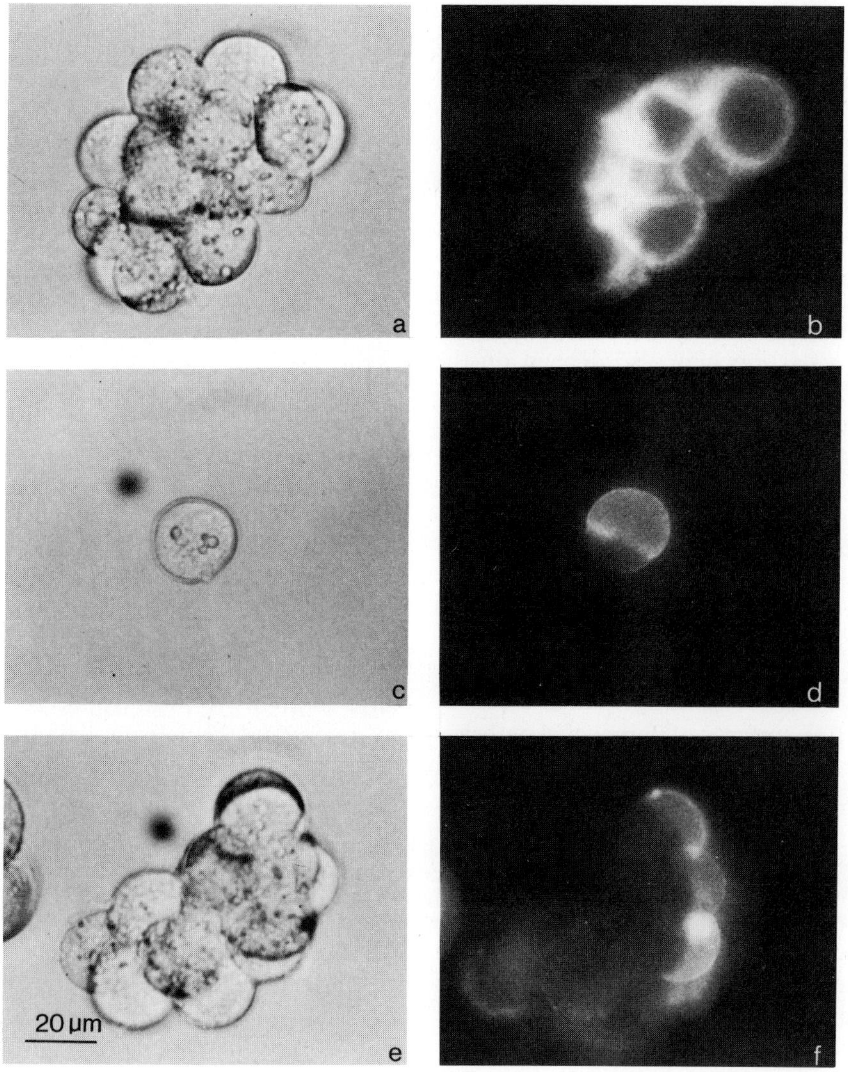

Fig. 4. Indirect immunofluorescence of living dissociated morulae labeled for mouse species cell surface antigens. Panels a, c, and e are phase-contrast images while panels b, d, and f are the same subjects under eipfluorescence illumination. Since the outside (free) surfaces of undissociated morulae are fluorescent, it is assumed that the labeled surfaces of dissociated cells correspond to the free cell surfaces of the intact embryo. A 'raft' of cells (a, b) and a single cell (c, d) from a dissociated calcium-free medium-treated morula – note distinct differences in fluorescence intensity between free (right hand aspect of raft, superior aspect of single cell) and apposed cell surfaces; (e, f) a raft of cells from a control morula (i.e., not dissociated with calciumfree medium) – antigen density differences between free and apposed cell surfaces is similar to that in morulae dissociated with calcium-free medium. (from Wiley and Eglitis, 1981; ref. 45).

lectins bind in greatest quantity to the free cell surface domains as assessed by indirect immunofluorescence. Some of the apparently greater amount of binding of both types of ligands to free cell surface domains may be due to the presence of microvilli on free – but not on apposed – cell surfaces. However, some of the heavier labelling on free cell surfaces that is observed with plant lectins is still observed even when microvilli are reduced in number by certain experimental conditions (31). This last observation suggests that the molecular nature of free cell surface domains is intrinsically different from that of apposed ones.

The major questions that these observations raise are 1) do any of these observed cell surface properties of compacting embryos, particularly the distinctions between free and apposed cell surfaces, have anything to do with the mechanism of compaction, and 2) why does compaction occur insofar as the requirements for ICM/trophectoderm differentiation and cavitation are concerned. For the moment, answers to both questions are speculative. However, these speculations will be described according to available data.

As for the mechanism by which compaction occurs, it seems likely that the cell surfaces become adhesive at the onset of compaction. This conclusion is based on the presence of cell-cell junctions and on the need by compaction for extracellular calcium (32), both of which are known to be important for cell-cell adhesion in other systems (33). There is evidence from studies with an inhibitor of glycosylation, tunicamycin, that is compatible with the idea that N-glycosidically linked glycoproteins are needed for compaction (31, 34) and similar glycoproteins are believed to be essential for cell-cell adhesion in other systems (35). Compaction may also depend on the particular composition of lipids of the plasma membranes of compacting embryos (36), because a similar lipid composition has a positive correlation with cell-cell adhesion in other systems (36). Finally, components of the cytoskeleton – particularly actin-containing structures – may be essential for compaction since the cytochalasins, which inhibit actin-dependent functions, also inhibit compaction (32) and cytochalasin-sensitive actin-containing structures are believed to be essential to cell-cell adhesion in other systems (37).

However, the free surfaces of compacted embryos are also adhesive, which enables zona-free compacted 8-cell embryos to form aggregation chimeras (2, 38). Assuming that compaction involves cell-cell adhesion, then this observation is compatible with at least three situations regarding the relationship of compaction to the molecular distinctions between free and apposed cell surfaces; 1) cell-cell adhesion occurs independently of such distinctions and has no relationship with their establishment or maintenance; 2) cell-cell adhesion occurs independently of such distinctions, but as a component of compaction, may be a cause of their establishment and maintenance, or, 3) cell-cell adhesion and compaction are a result of such distinctions, but these distinctions can be appropriately modulated to accommodate altered cell arrangements. As for the first and second situations, none of the known molecular distinctions between free and apposed cell surfaces have been experimentally correlated with cell-cell adhesion that characterizes compaction; however, the establishment of such distinctions does appear to depend on cell contact, as stated earlier (29), although the maintenance of these distinctions (at least over a period of several hours) appears to be independent of cell contact (29, 45, 49). As for the third situation, because none of the known molecular distinctions between free and apposed cell surfaces have been experimentally correlated with cell-cell adhesion that characterizes compaction, it is not possible to assess the question of modulation.

From the previous paragraph it becomes clear that the establishment of free and apposed cell surface domains during compaction may simply be a temporal correlate of compaction that becomes significant to the formation of the blastocoele (cavitation) and/or to ICM/trophectoderm differentiation. As to whether compaction *per se* is required by cavitation or ICM/trophectoderm differentiation, there is general agreement that neither event occurs if compaction is prevented or reversed (39). There is, however, no agreement on the reason why these events have, so far, proven to be experimentally inseparable from compaction. Actually, it *is* possible to obtain a remarkable degree of trophectoderm differentiation in the absence of compaction (40) – to be precise, it is cavitation and the formation of the pluripotent ICM that appear to have a mandatory need for compaction. One possibility is that compaction promotes cavitation and the creation of an intraembryonic 'microenvironment' (41) that fosters ICM formation by inner cells and supresses their ability to form trophectoderm (2). That inner cells have an ability to form trophectoderm can be shown by removing them from their inside position and culturing them singly or in an outer position of a reconstructed 8-cell embryo, where, in both cases, they will proceed to differentiate into trophectoderm (2, 40). However, an inner position *per se* (i.e., residence within the blastocoele; 73) is not sufficient for the suppression of trophectoderm differentiation, which preliminary evidence suggests may be sensitive to the cell-cell contact that results from an inner position (42). One possibility, then, with respect to ICM/trophectoderm differentiation, could be that compaction initiates an epigenetic cue for ICM/trophectoderm differentiation that arises from a sensitivity on the part of genetic activity within a given cell to the percentage of cell surface that is apposed vs. free (cell apposition hypothesis). Alternatively, compaction could promote the establishment of radial cytoplasmic gradients of certain 'factors' that would become segregated by cell division and would commit cells inheriting such factors to an ICM fate (polarization hypothesis, see chapter by Johnson et al.). This alternative is reminiscent of an earlier 'segregation hypothesis' (43), which suffered from a failure to account for the regulative capacity of the early mouse embryo. Speculation on the necessity of compaction for cavitation will be offered in the following section.

5. Blastocyst formation

Approximately 16–20 hr after compaction the embryo is a 'morula', which consists of 16–32 closely apposed cells. At this stage, the embryo begins to cavitate. As cavitation ensues, randomly distributed cytoplasmic organelles (droplets and mitochondria) become cortically localized to apposed cell surfaces. Shortly thereafter, apposed cell surfaces become separated by fluid-filled intercellular spaces that subsequently coalesce to form the blastocoele (26, 44, 45, 46). Cortical localization of organelles, production of intercellular spaces and formation of the blastocoele are all reversibly inhibited by drugs that disrupt cytoskeletal elements (44, 45, 47) and cytoskeletal elements are believed to function in the vectored movement of organelles to specific regions in other cell types (48). These observations support the hypothesis that cavitation involves some process requiring that certain organelles be specifically loca-

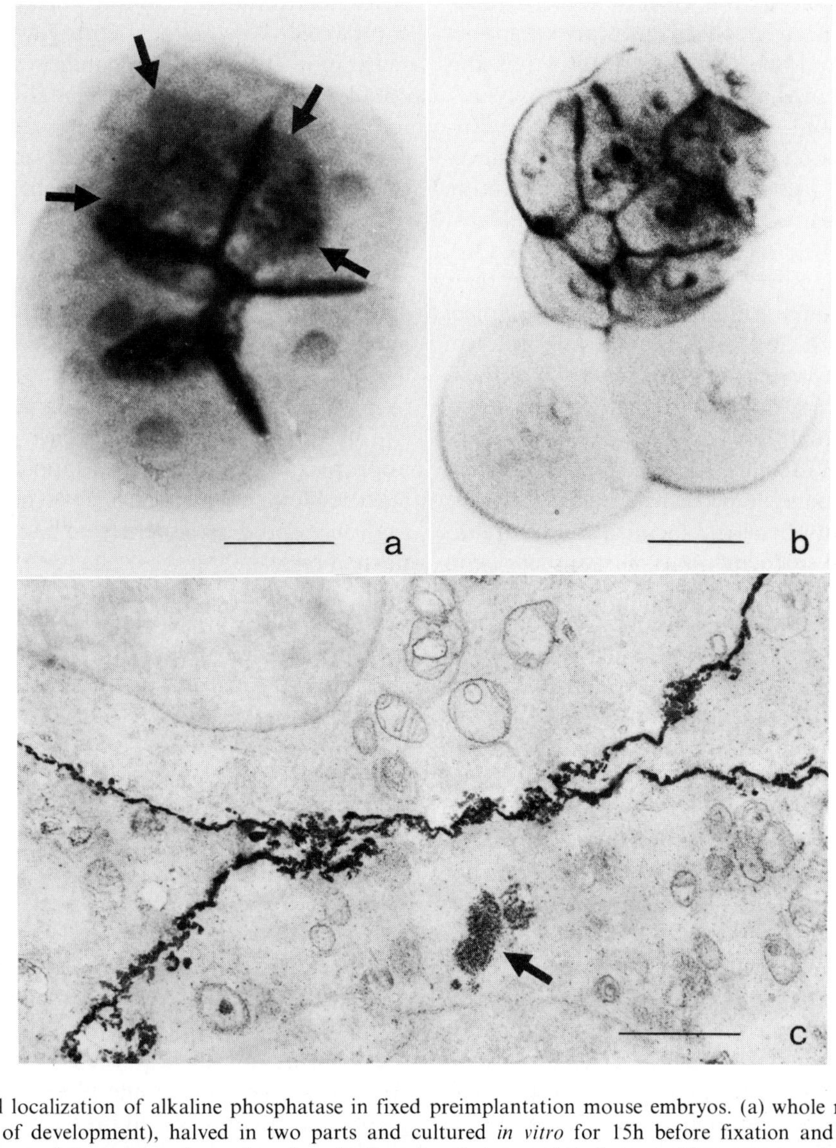

Fig. 5. Cytochemical localization of alkaline phosphatase in fixed preimplantation mouse embryos. (a) whole mount of a morula of about 16 cells (64h of development), halved in two parts and cultured *in vitro* for 15h before fixation and alkaline phosphatase demonstration. The part of the embryo shown here has alkaline phosphatase activity between blastomeres and a patch on the surface exposed by halving (arrows); (b) whole mount of a 2-cell stage (38h of development) fused to a morula of about 16 cells and cultured *in vitro* for 5h before fixation and alkaline phosphatase demonstration. No alkaline phosphatase is observed in the 2-cell stage or in its contact with the morula. The morula shows the typical localization of alkaline phosphatase at this stage; (c) electron microscopy of a section through an advanced morula (80h of development). Alkaline phosphatase is present on the surface of apposed blastomeres and in crystalloid bodies (arrow). Scale bars in a and b represent 15 microns and the scale bar in c represents 1 micron. (photographs kindly provided by Dr. L. Izquierdo and which appeared in Izquierdo et al. J Embryol Exp Morph 59: 89–102, 1980).

lized to the cell cortex of apposed cell surfaces by some cytoskeletal-dependent mechanism (44, 45). How, then is this cortical localization restricted to apposed cell surfaces? At this juncture, it becomes interesting to review the data on the cell surface properties that distinguish free from apposed cell surfaces in cavitating morulae.

The cavitating morula inherits from the compacted 8-cell embryo the same morphological differences that distinguish free from apposed cell surfaces. In addition, the morula exhibits a similar cell surface mosaicism at the molecular level as detected by anti-mouse antibody, with free cell surfaces labelling more heavily than apposed ones (Fig. 4) (45). Cytochemical techniques reveal that at least two ectoenzymes, alkaline phosphatase (49) and the transport Na^+/K^+-ATPase (50) are found only on apposed cell surfaces (Fig. 5). Finally, free cell surfaces of cavitating morulae lose the adhesive properties that enable 8-cell embryos and early morulae to amalgamate (38), while apposed cell surfaces retain some of their adhesiveness, particularly that associated with apical cell junctions, which unite the outside cells into an epithelium (presumptive trophectoderm). All of these observations illustrate the point that there are structural and molecular differences that distinguish free from apposed cell surface domains. Could such differences, in a general sense, be related to the observed restriction of cortical localization of organelle to apposed cell surfaces? At this time there is no difinitive data on this question although there is some compatible data for the participation of calcium-sensitive transmembrane linkages between cytoskeletal elements and apposed membrane domains (45). Another possibility is that the Na^+/K^+-ATPase, which is cytochemically detectable only on apposed cell surfaces (50), could produce a transcellular electrical gradient similar to that described for the fucoid egg (51). Such a gradient could then cause an electrophoresis-like action on certain organelles to result in their accumulation along apposed cell surfaces.

Returning now to speculation on the necessity of compaction for cavitation, three ideas will be proposed here. The first one proposes that compaction results in the formation of zonular tight junctions around the apical ends of outside cells to form a 'permeability seal' that is essential to the ability of the cavitating morula to accumulate nascent blastocoele fluid (32). This idea is based on electron microscopic observations that these apical junctions are impermeable to lanthanum when it is applied during aldehyde fixation (52). However, tight junctional impermeability to lanthanum can be properly assessed only if the tracer is applied before aldehyde fixation because lanthanum rarely enters tight junctions, even those in 'leaky' epithelia, if it is added during or following aldehyde fixation (53). In addition, it is still possible to lyse ICM cells with immune serum and complement that are placed into the extraembryonic milieu (culture medium) 14 h *after* cavitation has begun and the blastocoele has already formed (54). Finally, physiological studies show that the apical functions of rabbit blastocysts are electrically leaky during the early blastocyst stage and do not become tight until the late blastocyst stage (55). Thus, it is unlikely that the apical junctions uniting the outside cells of the compacted 8-cell stage embryo, morula or early blastocyst form a 'permeability seal' as such, but could partially restrict permeability rates of certain ions (and macromolecules) so that gradients were established across them. The second idea is that compaction could simply have the task of maintaining cell-cell apposition to permit the *establishment* of apical cell junctions over a certain required amount of time, for tight junctions do subsequently become essential for blastocoele maintenance (42, 52, 65) and gap junctions and desmosomes become common. The third idea is that a role for compaction could be to provide a means of ensuring that cortical localization of cytoplasmic organelles occurs under the appropriate cell surfaces (apposed) so that blastocoele fluid production results in the efficient formation of the blastocoele. This could be accomplished by invoking the existence of a transmembrane mechanism that is sensitive to the cell-cell apposition produced by compaction (45).

Aside from their possible role in caviation, what role could the cell surface properties of morulae have in the formation of the anatomical distinctions between ICM and trophectoderm? In particular, what mechanism ensures that inside cells are not incorporated into the trophectoderm but remain as a discrete cell cluster? Intuitively, one would suspect that differential cell surface properties are involved, possibly in the form of differences in adhesiveness between ICM and trophectoderm cells as has been described in other embryonic systems (56). That differential cell adhesion could be involved here is supported by two observations. First, zona-free blastocysts will not adhere to each other (38) while ICM's collected from similar-age blastocysts readily amalgamate to form larger ICM's (57). Second, isolated ICM/s – but not zona-free blastocysts – readily agglutinate with con A (58), in spite of the presence of con A receptors on the outer surfaces of the blastocyst (58, 59). These observations are consistent with the notion that selective, ICM-specific cell surface properties enable ICM cells to preferentially adhere to each other and so avoid being incorporated into the trophectoderm. Alternatively, ICM cells could lack a trophoblast-specific cell surface property such that they are passively excluded from the trophectoderm. Perhaps the mechanism of 'ICM exclusion' could be related to (one of?) the numerous cell surface antigens whose expression (serological accessibility) differs on ICM and trophectoderm cells (60, 61, 62).

Finally, what role could the cell surfaces of cavitating morulae have in conveying positional information to cells that might influence their decision about ICM/trophectoderm differentiation? In addition to their possible duty in maintaining the cytoplasmic polarity that is a component of the polarization hypothesis, another postulated role for the cell surfaces in ICM/trophectoderm differentiation arises from asking why does the mammalian embryo cavitate at all? During the discussion on cavitation it was proposed that genetic activity within cells of the compacted embryo could be sensitive to the percentage of cell surface that is apposed (or free). For example, free cell surfaces could cue a cell to differentiate while the lack of free cell surface could suppress differentiation. This postulate would then predict that within a cell aggregate differentiated cells would always occur in the outer cell layer of the aggregate while the inner cells would remain pluripotent 'stem' cells. This is precisely what is observed during cell differentiation in the morula, in isolated ICM's (57) and in embryoid bodies derived from embryonal carcinoma cells (63). In the case of the morula, the outer cell layer becomes trophectoderm while the inner cells remain pluripotent stem cells. In the case with isolated ICM's and with embryoid bodies, the outer cell layer becomes primitive endoderm while the inner cells remain pluripotent stem cells. Thus, in the case of the cavitating embryo, it is proposed that (at least one) purpose of cavitation could be to create new free cell surfaces over one side of the presumptive ICM to provide a cue for endoderm differentiation. Hence, one role for cell surfaces could be to convey positional information in terms of free surface as an epigenetic cue for differentiation commensurate with the position of a cell within a cell aggregate (morula, ICM or embryoid body).

6. The blastocyst: cell surface properties of the trophectoderm

In this section, attention will be limited to cell surface properties of the trophectoderm for two (admittedly arbitrary) reasons. First, by the mature blastocyst stage, ICM/trophectoderm differentiation has already occurred. Second, it is the cell surface properties of the trophectoderm that are related to implantation, which provides a functional endpoint to preimplantation development. As the blastocyst matures its trophectoderm acquires two new features that reflect the maturation of its cell surface properties. The first of these features is active transepithelial transport and the second is a recovery of cell surface adhesiveness.

From fertilization through early cleavage the embryo undergoes a number of changes in electrophysiological and permeability properties (70, 71, 72). During this period the mammalian embryo (mouse) differs from other types of embryos (amphibians, echinoderms and alga) in that internal potassium decreases while internal sodium increases by two-fold over that in the unfertilized oocyte to attain concentrations that are quite different from those in adult cells (70). By the morula stage, however, the internal potassium and sodium concentrations resemble those that are typical of adult mammalian cells and which are compatible with the ion gradient requirements of amino acid transport systems as they occur in adult cells (70). At the morula stage, the embryo activates its plasma membrane transport Na^+/K^+-ATPase, which is ouabain sensitive and is involved with the amino acid transport that also becomes activated at this stage, perhaps because of lowered internal sodium and increased potassium levels (64, 65). It is possible that these transport systems may be inherited in inactive form from earlier embryonic stages and become activated during subsequent development (65, 70). All of these transport properties, together with the morphology of the trophectoderm, are consistent with the description of the mature trophectoderm as a polarized epithelium consisting of cells that are united by semi-permeable apical junctions and whose apical and basolateral cell surface domains are functionally distinct (55, 65). The observation that the ouabain sensitivity of the transport Na^+/K^+-ATPase is best demonstrated when the drug is introduced into the blastocoele (65) further supports this description.

The transport activities of the trophectoderm cause the concentrations in blastocoele fluid of sodium, potassium, calcium and magnesium to be greater than in serum (66), and maintain a transmural electrical resistance of about 2600 ohms cm^2 (66, 67) as the permeability seal formed by apical junctions becomes electrically 'tight' (55). The transport of ions into the blastocoele along with the passive influx of water causes the blastocyst to expand into a cyst, which in the rabbit contains up to 200 μl of blastocoele fluid (68) but in the mouse contains only about 0.4–0.45 nl (69). Of what functional significance is blastocoele fluid accumulation and blastocyst expansion to development of the embryo? It is doubtful, for reasons given earlier, that blastocoele fluid *per se* is essential to ICM/trophectoderm differentiation. Moreover, blastocoele fluid does not appear to be essential to primary germ layer formation by the ICM because all three primary germ layers are formed by ICM's cultured in isolation from trophectoderm and blastocoele fluid (57). However, blastocyst expansion could be important to implantation, perhaps by ensuring proper orientation and/or secure lodging of the blastocyst within a uterine crypt in preparation of trophoblast attachment to the endometrium.

From the morula stage to after the blastocyst has 'hatched' from its zona pellucida, the trophectodermal apical cell surface is nonadhesive. In fact, the trophectoderm does not become adhesive for an additional ten hours after hatching (74, 75, 76).

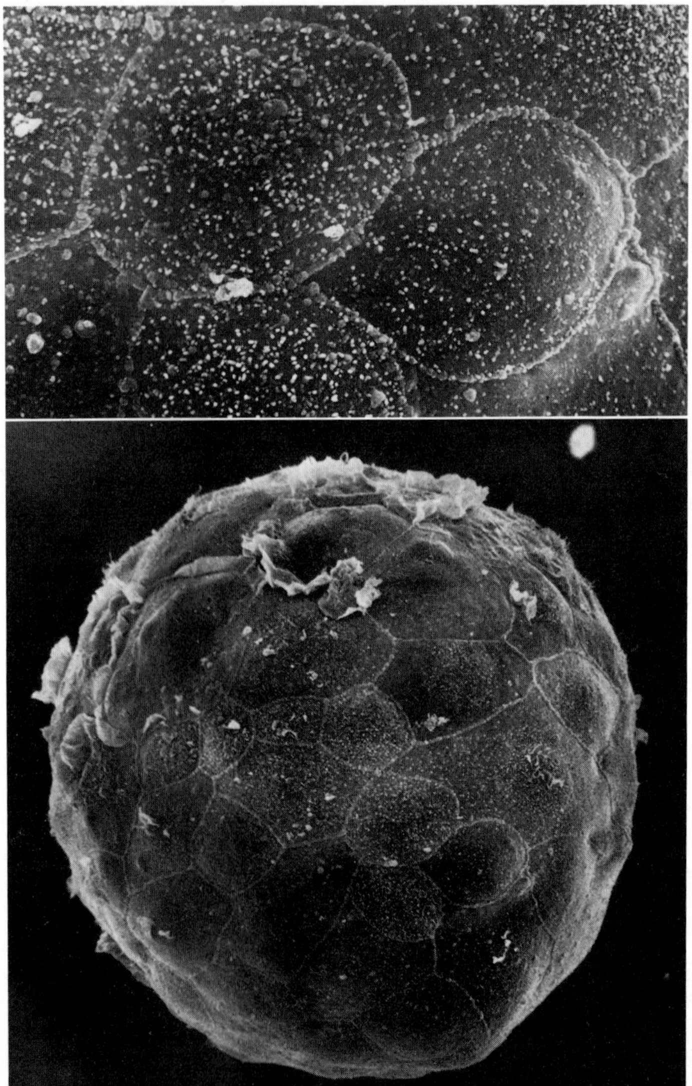

Fig. 6. Scanning electron microscopy of the cell surface of the expanded mouse blastocyst, early in the adhesion stage. This blastocyst was flushed from the uterus on the fourth day of pregnancy and cultured in serum-supplemented medium for 42h. During this period the blastocyst expanded, hatched from its zona pellucida and became adherent to the culture dish. (a) note heterogeneity in cell size and prominent intercellular ridges; (b) higher magnification of intercellular ridges. Scale markers represent 5 microns. (micrographs taken by Dr. Ruth Shalgi and appeared in Sherman et al, Maternal Recognition of Pregnancy. Ciba Foundation Series 64. Excerpta Medica, 1979).

Precocious removal of the zona by enzymatic means (pronase) does not dramatically alter the time at which the blastocyst becomes adherent, which suggests that the acquisition of adhesion reflects discrete changes of the blastocyst surface (77).

Scanning electron microscopy reveals that the surface of the expanded blastocyst is characterized by prominant intercellular ridges, which overlie the apical junctional complexes (Fig. 6) (33, 78). The surfaces of the trophectodermal cells are covered with microvilli, which diminish in number after hatching and during the subsequent ten hours. Microvilli are frequently observed emerging from the intercellular ridges and are more numerous over the mural than over the polar trophectoderm. As the trophectoderm becomes adherent however, there are no obvious changes in trophectodermal surface morphology as revealed by SEM (78).

Aside from the aforementioned regional difference in the distribution of microvilli, there is an additional regional difference between the surfaces of mural and polar trophectoderm. When expanded blastocysts (5 days postcoitum) are coated with con A and then exposed to con A-coated human erythrocytes, the erythrocytes adhere almost exclusively to the mural trophectoderm (79). However, molecular con A binds in similar amounts to both mural and polar trophoblast (79), leading to the conclusion that the failure of polar trophoblast to bind con A-coated erythrocytes could be due to alterations in surface charge and/or membrane rigidity (79). Interestingly, earlier blastocysts (day 4 post-coitum) do not show this regional difference between polar and mural trophectodermal surfaces. It has not yet been possible, however, to correlate the development of this regional difference with any specific advance in embryonic development or in adhesion of the trophectoderm (79).

Changes in cell surface glycoproteins accompany changes in adhesion during phenotypic change (such as oncogenic transformation) in other cell types (80). Consequently, the cell surface glycoproteins of trophectoderm have been examined *in vitro* for changes that correlate with the acquisition of adhesiveness (77). However, treating blastocysts with the enzymes neuraminidase or galactose oxidase or with periodate in an effort to modify cell surface glycoproteins has no effect on blastocyst adhesion to plastic culture dishes (77). In addition, sulfhydral binding agents have no apparent effect on blastocyst adhesion, nor do the enzymes pronase, trypsin or hyaluronidase (77). The only enzyme that so far has been found to affect blastocyst adhesion is collagenase (77). Indirect immunofluorescence suggests that the surface of adhesion-competent blastocysts do have type III collagen (78, 81). These last two observations suggest that collagen may be involved in trophectoderm adhesion, but other factors must also be involved because collagen first becomes detectable on the cell surface at the morula stage, when trophectodermal apical cell surfaces are not yet adhesive (78, 81).

Thus, at this time, we are left without an understanding of the cellular mechanisms whereby the trophectoderm becomes adhesive and the onset of adhesiveness is regulated to coincide with the moment at which the blastocyst is properly ensconced within the uterus. Since the kinetics of the acquisition of adhesiveness are similar *in vivo* and *in vitro*, the onset of adhesion is most likely under intrinsic control by the embryo (77, 78, 82, 83). Of course, successful blastocyst attachment in utero requires an appropriately receptive epithelium, which is certainly under maternal control.

7. Concluding remarks

From the preceding paragraphs it becomes clear that the majority of the observations described therein are phenomenological, and that at this time the onus is on the reader (and authors) to assemble the observations into testable hypotheses. However, the intent here has been to show that enough observations are at hand to construct a number of intriguing working hypotheses on the relationships of cell surface properties to specific morphogenetic events that we choose to recognize as being critical to ICM/trophectoderm differentiation and to blastocyst formation. What now follows is a listing of the major questions and hypotheses regarding these relationships as they were discussed in this chapter.

7.1. Fertilization

There is yet no clear reason why sperm attachment rarely occurs on the cell surface overlying the second meiotic spindle of the oocyte. Although there are morphological differences between the plasma membrane overlying the spindle and that enveloping the rest of the oocyte, there is no direct experimental evidence that establishes a causal relationship between any of the described regional morphological differences and the failure of sperm to attach to the plasma membrane overlying the spindle. Furthermore, there is at present no experimental evidence for the cell surface mosaicism of the oocyte – or for the site of sperm entry – having a role in conferring positional information or in establishing embryonic axes during subsequent development of the mammalian embryo.

7.2. Cleavage

The cell surfaces of cleaving blastomeres lack any apparent morphological or molecular mosaicism until compaction occurs at the 8-cell stage. Furthermore, the cell surface morphology does not change markedly from one cleavage to the next until compaction takes place. However, there are changes at the molecular level in the form of the transient expression (serological accessibility) of a number of cell surface antigens. Additional changes at the molecular level occur in the nature and/or quantity of cell surface glycoproteins as detected by plant lectins. The functional significance of any of these changes is unknown, as is, in most cases, the molecular identity of the changing cell surface molecules.

7.3. Compaction

This event induces (?) the establishment of two morphologically and molecularly distinct cell surface domains over the outside cells of the embryo. These two types of domains consist of the free and apposed cell surfaces that result from compaction. It is likely that close cell-cell apposition, which results from compaction, is produced by cell-cell adhesion. However, none of the known morphological or molecular distinctions between free and apposed cell surfaces have been experimentally correlated with the acquisition or with the maintenance of cell-cell adhesion in the compacted embryo. Close cell-cell apposition has been proposed to be essential to ICM/trophectoderm differentiation and for cavitation, primarily because neither ICM formation nor cavitation has been experimentally separable for compaction. Hypotheses for such a need by ICM formation for compaction include 1) compaction results in (permits?) the development of apical tight junctions that form a 'permeability seal' to create a microenvironment that programs inside cells to form an ICM (microenvironment hypothesis); 2) compaction initiates an epigenetic cue for ICM/trophectoderm differentiation that arises from a sensitivity on the part of genetic activity within a given cell to the percentage of cell surface that is apposed vs. free (cell apposition hypothesis), and 3) compaction could promote the establishment of radial cytoplasmic gradients of certain 'factors' that would become segregated by cell division and would commit cells inheriting such factors to an ICM fate (polarization hypothesis). Hypotheses regarding the role of compaction in cavitation include 1) creation of a 'permeability seal', formed by apical tight junctions, to permit the accumulation of blastocoele fluid during blastocoele formation; 2) maintaining cell-cell apposition to permit the establishment of apical cell junctions over a certain required amount of time, because such junctions subsequently become essential for blastocoele maintainance, and 3) providing a means of ensuring that cortical localization of cytoplasmic organelles occurs under the appropriate cell surfaces (apposed) by invoking the existence of a transmembrane mechanism that is sensitive to cell-cell apposition.

7.4. Blastocyst formation

As cavitation ensues, certain randomly distributed cytoplasmic organelles (droplets and mitochondria) become cortically localized to apposed cell surfaces. For reasons presented in this chapter, it is proposed that such localization is involved in the production of nascent blastocoele fluid. The question that arises, then, is how is this cortical localization restricted to apposed cell surfaces? The cavitating morula inherits from the compacted 8-cell embryo the same morphological differences that distinguish free from apposed cell surfaces. In addition, the morula exhibits a cell surface mosaicism at the molecular level that distinguishes free from apposed cell surfaces. Postulated mechanisms whereby such molecular distinctions could participate in cortical localization or organelles to apposed cell surfaces include 1) Na^+/K^+-ATPase, which is cytochemically detectable only on apposed cell surfaces, could pump sodium out of the cells into the intercellular spaces to establish transcellular electrical gradients to cause an electrophoresis of certain organelles to apposed cell surfaces, and 2) cortical localization occurs by calcium-sensitive transmembrane linkages between cytoskeletal elements and certain components of apposed membrane domains that are absent from free membrane domains. As ICM and trophectoderm become anatomically distinct, their cell surface adhesive and antigenic properties become different. The question is posed as to whether such differences could either promote or maintain anatomically distinct ICM and trophectoderm. Speculation is offered on the relevance of cavitation to subsequent endoderm formation by the ICM in terms of the sensitivity of cell differentiation to the percentage of cell surface that is free vs. apposed.

7.5. The blastocyst: cell surface properties of the trophectoderm

As the trophectoderm matures it acquires two new features that reflect the maturation of its cell surface properties. The first of these features is active trans-epithelial transport and the second is a recovery of cell surface adhesiveness. The active transepithelial transport involves a Na^+/K^+-ATPase, which is present only on the basolateral surfaces of trophectodermal cells and which participates in amino acid transport. The morphological and physiological cell surface properties of the mature trophectoderm are consistent with the features traditionally ascribed to functionally polarized semipermeable transporting epithelia. The transport activities of the trophectoderm result in blastocyst expansion and speculation is offered on the relevance of blastocyst fluid accumulation to ICM development and to implantation. The recovery of cell surface adhesiveness occurs about ten hours after the blastocyst escapes from the zona pellucida. However, none of the reported changes in trophectodermal cell surface properties that are exhibited by the mature blastocyst correlate with the onset of adhesiveness. There is some preliminary data, however, which suggest that cell surface collagen may be a factor in blastocyst adhesiveness.

References

1. Tarkowski AK, Wroblewska J: Development of blastomeres of mouse eggs isolated at the 4- and 8-cell stage. J Embryol Exp Morph 18: 155–180, 1967.
2. Hillman N, Sherman MI, Graham C: The effect of spatial arrangement on cell determination during mouse development. J Emb Exp Morph 28: 263–278, 1972.
3. Gardner RL: Analysis of determination and differentiation in the early mammalian embryo using intra- and interspecific chimaeras. In: The Developmental Biology of Reproduction Markert CL, Papaconstantinou J (eds) Academic Press, New York, 1975, pp 207–236.
4. Vogt W: Gestaltungsanalyse am Amphilbienkeim mit ortlicher Vitalfarbung. II. Gastrulation und mesodermbildung bei urodelen und anuren. Roux Arch 120: 385–706, 1929.
5. Johnson MH, Eager D, Muggleton-Harris A: Mosaicism in organisation of concanavalin A receptors on surface membrane of mouse egg. Nature 257: 321–322, 1975.
6. Eager DD, Johnson MH, Thurley KW: Ultrastructural studies on the surface membrane of the mouse egg, J Cell Sci 22: 345–353, 1976.
7. Nicosia SV, Wolf DP, Inoue M: Cortical granule distribution and cell surface characteristics in mouse eggs. Develop Biol 57: 56–74, 1977.
8. Nicosia SV, Wolf DP, Mastroianni L Jr: Surface topography of mouse eggs before and after insemination. Gamete Res 1: 145–155, 1978.
9. Moon HV, Isaacson RE, Pohlenz J: Mechanisms of association of enteropathogenic Escherichia coli with intestinal epithelium. Am J Clin Nutr 32: 119–127, 1979.
10. Maul GG, Steplewski Z, Weibel J, Koprowski H: Time sequence and morphological evaluations of cells fused by polyethylene glycol. 6000. *In vitro* 12: 787–796, 1976
11. Wiley LM Calarco PG: The effects of anti-embryo sera and their localization on the cell surface during mouse preimplantation development. Develop Biol 47: 407–418, 1975.
12. Johnson MH, Edidin M: Lateral diffusion in plasma membrane of mouse egg is restricted after fertilization. Nature 272: 448–450, 1978.
13. Gabel CA, Eddy EM, Shapiro BM: After fertilization, sperm surface components remain as a patch in sea urchin and mouse embryos. Cell 18: 207–215, 1979.
14. Brachet J: An old enigma: The grey crescent of amphibian eggs. Curr Topics Dev Biol 11: 133–186, 1977.
15. Gerhard J, Ubbels G, Black S, Hara K, Kirschner M: A reinvestigation of the role of the grey crescent in axis formation in xenopus laevis. Nature 292: 511–516, 1981.
16. De Felici M, Siracusa G: Fertilization-induced changes in concanavalin A binding to mouse eggs. Exp Cell Res 132: 41–45, 1981.
17. Inbar M, Huet C, Oseroff AR, Benbassat H, Sachs L: Inhibition of lectin agglutinability by fixation of the cell surface membrane. Biochim Biophys Acta 311: 594–599, 1973.
18. Rutishauser V, Sachs L: Cell-to-cell binding induced by different lectins. J Cell Biol 65: 247–257, 1975.
19. Marshall JD, Heiniger H-J: High affinity concanavalin A binding to sterol-depleted L cells. J Cell Physiol, 100: 539–550, 1979.
20. Heyner S, Hunziker RD: Differential expression of alloantigens of the major histocompatability complex on unfertilized and fertilized mouse eggs. Develop Genet 1: 69–76, 1979.
21. Wiley LM: Early embryonic cell surface antigens as developmental probes. In: Current Topics in Developmental Biology Moscona AA, Friedlander M (eds) Academic Press, New York, Vol 13, 1979, pp 167–197.
22. Heyner S: Antigens of trophoblast and early embryo. In: Immunol Aspects of Infertility & Fertility Regulation. Dhindsa, Schumager (eds) Amsterdam, Elsevier North Holland, Inc, 1980, pp 183–203.
23. Pinsker MC, Mintz B: Changes in cell-surface glycoproteins of mouse embryos before implantation. Proc Nat Acad Sci USA 70: 1645–1648, 1973.
24. Magnuson T, Epstein CJ: Characterization of concanavalin A precipitated proteins from early mouse embryos: A 2-dimensional gel electrophoresis study. Develop Bio 81: 193–199, 1981.
25. Lewis WA, Wright ES: On the early development of the mouse egg. Contrib Embryol Carnegie Inst, No. 148: 115–143, 1935.
26. Calarco PG, Brown EH: An ultrastructural and cytological study of preimplantation development of the mouse. J Exp Zool 171: 253–284, 1969.
27. Calarco PG, Epstein CJ: Cell surface changes during preimplantation development in the mouse. Develop Biol 32: 208–213, 1973.
28. Ducibella T, Ukena T, Karnovsky M, Anderson E: Changes in cell surface and cortical cytoplasmic organization during early embryogenesis in the preimplantation mouse embryo. J Cell Biol 74: 153–167, 1977.
29. Ziomek CA, Johnson MH: Cell surface interaction induces polarization of mouse 8-cell blastomeres at compaction. Cell 21: 935–942, 1980.
30. Reeve WJD, Ziomek CA: Distribution on dissociated blastomeres from mouse embryos: evidence for surface polarization at compaction. J Emb Exp Morph 62: 339–350, 1981.
31. Surani MAH, Kimber SJ, Handyside AH: Synthesis and role of cell surface glycoproteins in preimplantation mouse development. Exp Cell Res,133: 311–340, 1981.
32. Ducibella T, Anderson E: Cell shape and membrane changes in the eight-cell mouse embryo: prerequisites for morphogenesis of the blastocyst. Develop Biol 47: 45–58, 1975.
33. Steinberg MS: Calcium complexing by embryonic cell surfaces: relation to intercellular adhesiveness. In: Biological Interactions in Normal and Neoplastic Growth Brennan MJ, Simpson WL (eds) Little, Brown, Boston, USA 1962, pp 127–140.
34. Atienza-Samols, SB, Pine PR, Sherman MI: Effects of tunicamycin upon glycoprotein synthesis and developmental of early mouse embryos. Develop Biol 79: 19–32, 1980.
35. Rees DA, Lloyd CW, Thom D: Control of grip and stick in cell adhesion through lateral relationships of membrane glycoproteins. Nature, 267: 124–128, 1977.
36. Pratt HPM: Lipids and transitions in embryos. In: Development in Mammals Johnson MH (ed) North-Holland Publishing Co, Amsterdam, Vol 3, 1978, pp 83–129.

37. Helentjaris TG, Lombardi PS, Glasgow LA: Effect of cytochalasin B on the adhesion of mouse peritoneal macrophages. J Cell Biol 64: 407–414, 1976.
38. Burgoyne PS, Ducibella T: Changes in the properties of the developing trophoblast of preimplantation mouse embryos as revealed by aggregation studies. J Embryol Exp Morph 40: 143–157, 1977.
39. Kemler R, Babinet C, Eisen H, Jacob F: Surface antigen in early differentiation. Proc Nat Acad Sci, USA 74: 4449–445, 1977.
40. Sherman MI, Atienza-Samols SB: Differentiation of mouse trophoblast does not require cell-cell interaction. Exp Cell Res 123: 73–78, 1979.
41. Mintz B: Experimental genetic mosaicism in the mouse. In: Preimplantation Stages of Pregnancy Wolstenholme GEW, O'Connor M (eds) Churchill, Ltd, London, 1965, pp 194–207.
42. Eglitis MA, Wiley LM: Tetraploidy and early development: effects on developmental timing and embryonic metabolism J Embryol Exp Morph 66: 91–108, 1981.
43. Dalcq AM: Introduction to general embryology. Oxford Univ Press, Sondon/New York, 1957.
44. Wiley LM, Eglitis MA: Effects of colcemid on cavitation during mouse blastocoele formation. Exp. Cell Res. 127: 89–101, 1980.
45. Wiley LM, Eglitis MA: Cell Surface and cytoskeletal elements: cavitation in the mouse preimplantation embryo. Develop Biol 86: 493–501, 1981.
46. Melissinos K: Die entwicklung des eles der mause. Arch Mikr Anat 70: 577–628, 1907.
47. Granholm NH, Brenner GM: Effects of cytochalasin B (CB) on the morula-to-blastocyst transformation trophoblast outgrowth in the early mouse embryo. Exp Cell Res 101: 143–153, 1976.
48. Malaisse WJ, Malaisse-Lange F, Van Obberghen E, Somers G, Devis G, Ravazzola M, Orci L: Role of microtubules in the phasic pattern of insulin release. Ann N Y Acad Sci 253: 630–652, 1975.
49. Izquierdo L, Lopez T, Marticorena P: Cell membrane regions in preimplantation mouse embryos. J Embryol Exp Morph 59: 89–102, 1980.
50. Vorbrodt A, Konwinski M, Solter D, Koprowski H: Ultrastructural cytochemistry of membrane-bound phosphatases in preimplantation mouse embryos. Develop Biol 55: 117–134, 1977.
51. Nuccitelli R, Jaffe LF: The pulse current pattern generated by developing fucoid eggs. J Cell Biol 64: 636–643, 1975.
52. Ducibella T, Albertini DF, Anderson E, Biggers JD: The preimplantation mammalian embryo: characterization of intercellular junctions and their appearance during development. Develop Biol 45: 231–250, 1975.
53. Machen TE, Erlij D, Wooding FBP: Permeable junctional complexes. The movement of lanthanum across rabbit gallbladder and intestine. J Cell Biol 54: 302–312, 1972.
54. McLaren A, Smith R: Functional test of tight junctions in the mouse blastocyst. Nature 267: 351–353, 1977.
55. Benos DJ: Developmental changes in epithelial transport characteristics of preimplantation rabbit blastocysts. J Physiol 316: 191–202, 1981.
56. Steinberg MS: Does differential adhesion govern self-assembly process in histogenesis? Equilibrium configurations and the emergence of a hierarchy among populations of embryonic cells. J Exp Zool 173: 395–434, 1970.
57. Wiley LM, Spindle AI, Pedersen RA: Morphology of isolated mouse inner cell masses developing in vitro. Develop Biol 63: 1–10, 1978.
58. Rowinski J, Solter D, Koprowski H: Change of concanavalin A induced agglutinability during preimplantation mouse development. Exp Cell Res 100: 404–408, 1976.
59. Sobel JS, Nebel L: Changes in concanavalin A agglutinability during development of the inner cell mass and trophoblast of mouse blastocysts *in vitro*. J Reprod Fert 52: 239–248, 1978.
60. Willison KR, Stern PL: Expression of a forssman antigenic specificity in the preimplantation mouse embryo. Cell 14: 785–793, 1978.
61. Wiley LM: Early mouse embryonic cell surface antigens that are detectable by antisera to human chorionic gonadotropin (hCG). Exp Cell Res., 129: 47–54, 1980.
62. Searle SF, Jenkinson EJ: Localization of trophoblast-defined surface antigens during early mouse embryogenesis. J Emb Exp Morph 43: 147–156, 1978.
63. Martin GR: Teratocarcinomas and mammalian embryogenesis. Science, 209: 768–776, 1980.
64. Borland RM, Tasca RJ: Na^+-dependent amino acid transport in preimplantation mouse embryos. II. Metabolic inhibitors and nature of the cation requirement. Develop Biol 46: 192–201, 1975.
65. Dizio SM, Tasca RJ: Sodium-dependent amino acid transport in preimplantation mouse embryos. III. Na^+-K^+-ATPase-linked mechanism in blastocysts. Develop Biol 59: 198–205, 1977.
66. Borland RM, Biggers JD, Lechene CP: Studies on the composition and formation of mouse blastocoele fluid using electron probe microanalysis. Develop Biol 55: 1–8, 1977.
67. Cross MH, Brinster RL: Transmembrane potential of rabbit blastocyst trophoblast. Exp Cell Res 58: 125–127, 1969.
68. Daniel JC: Early growth of rabbit trophoblast. Amer Naturalist 98: 85–98, 1964.
69. Biggers JD: Mammalian blastocyst and amnion formation. In: The Water Metabolism of the Fetus Barnes AC, Seeds AE (eds) Charles C Thomas, New York, 1972, pp 3–31.
70. Powers RD, Tupper JT: Some electrophysiological and permeability properties of the mouse egg. Develop Biol 38: 320–331, 1974.
71. Powers RD, Biggers JD: Inhibition of mouse oocyte maturation by cell membrane potential hyperpolarization. J Cell Biol 70: 352a 1976.
72. Powers RD, Tupper JT: Developmental changes in membrane transport and permeability in the early mouse embryo. Develop Biol 56: 306–315, 1977.
73. Pederson RA, Spindle AI. Role of the blastocoele microenvironment in early mouse embryo differentiation. Nature, 284: 550–552, 1980.
74. Kirby DRS, Potts DM, Wilson IB: On the orientation of the implanting blastocyst. J Embryol Exp Morphol 17: 527–532, 1967.
75. Potts M, Wilson IB: The preimplantation conceptus of the mouse at 90 hours post coitum. J Anat 102: 1–11, 1967.
76. Potts M: The ultrastructure of implantation in the mouse. J Anat 103: 77–90, 1968.
77. Sherman MI, Atienza-Samols SB: *In vitro* studies on the surface adhesiveness of mouse blastocysts. In: Human Fertilization Ludwig H, Tauber PF (eds) Georg Thieme Pub, Stuttgart, 1978, pp 179–183.
78. Sherman MI, Shalgi R, Rizzino A, Sellens MH, Gay S, Gay R: Changes in the surface of the mouse blastocyst at implantation. In: Maternal Recognition of Pregnancy. Ciba Fdn Series 64, 1979, pp 33–52.
79. Sobel JS, Nebel L: Concanavalin A agglutinability of developing mouse trophoblast. J Reprod Fertil 49: 399–402, 1976.
80. Brady RO, Fishman PH: Membranes of transformed mammalian cells. In: Biochemistry of Cell Walls and Membranes. Fox CF (ed) Butterworths, London, 1975.
81. Sherman MI, Gay R, Gay S, Miller EJ: Association of collagen with preimplantation and peri-implantation mouse embryos. Develop Biol 74: 470–478, 1980.
82. Sherman MI, Matthaei KI: Factors involved in implantation-related events. In: Prog Reprod Biol 7: 43–53, 1980.
83. Sherman MI, Wudl LR: The implanting mouse blastocyst. In: The cell Surface in Animal Embryogenesis and Development Poste G, Nicolson G (eds) North Holland, Amsterdam, 1976, pp 81–125.
84. Huang TTF Jr, Calarco PG: Evidence for the cell surface expression of intracisternal A particle-associated antigens during early mouse development. Develop Biol 82: 388–392, 1981.
85. Heyner S, Hunziker RD, Zink GL: Differential expression of minor histocompatibility antigens on the surface of the mouse oocyte and preimplantation developmental stages. J Reprod Immunol 2: 269–280, 1980.
86. Johnson LV, Calarco PG: Immunological characterization of embryonic cell surface antigens recognized by antiblastocyst serum. Develop Biol 79: 208–223, 1980.
87. Artzt K, Dubois P, Benett D, Condamine H, Babinet C, Jacob F: Surface antigens common to mouse cleavage embryos and primitive teratocarcinoma cells in cultures. Proc Nat Acad Sci USA 70: 2988–2992, 1973.
88. Jacob F: Immunology and differentiation; mouse teratocarcinoma and embryonic antigens. Immunol Rev 33: 3–32, 1977.
89. Goldberg EH, Boyse EA, Bennett D, Scheid M, Carswell EA: Serological demonstration of H-Y (male) antigen on mouse sperm. Nature (London) 232: 478–480, 1971.
90. Krco CJ, Goldberg EH: H-Y (male) antigen: detection on eight-cell mouse embryos. Science, 193: 1134–1135, 1976.
91. Solter D, Knowles BB: Monoclonal antibody defining a stagespecific mouse embryonic antigen (SSEA-1). Proc Nat Acad Sci USA 75: 5565–5569, 1978.
92. Gooi HC, Feizi T, Kapadia A, Knowles BB, Solter D, Evans MJ. Stage-specific embryonic antigen involves 1–3 fucosylated type 2 blood group chains. Nature, 292: 156–158, 1981.

Author's address:
University of California Department of Human Anatomy,
Davis, California, USA

CHAPTER 17

The mosaic organisation of the preimplantation mouse embryo

M.H. JOHNSON, C.A. ZIOMEK, W.J.D. REEVE, H.P.M. PRATT,
H. GOODALL and A.H. HANDYSIDE

1. Introduction

The formation of the inner cell mass (ICM) and trophectoderm of the blastocyst constitutes the first definitive spatial differentiation of cells during embryogenesis. The two tissues show differences in structure, function, biochemistry, prospective fate and developmental potency. An understanding of the origins of these cell subpopulations, and the mechanism by which they are generated during cleavage, has proved to be elusive. It has been agreed generally that whatever mechanism operates, it must accommodate the prodigious regulatory capacity of the cells of the cleaving embryo, the morula and even the early blastocyst itself. Whilst this prolonged period of developmental lability apparently places the mammalian embryo in a unique position, it has also led frequently to the erroneous conclusion that the underlying mechanisms operating within the embryo must be correspondingly unique. Thus, embryos of many lower vertebrates and invertebrates appear to be mosaics, in which spatial differences in the cytoplasmic organisation within the individual egg, zygote or early blastomere are translated by cell divisions into regional differences among the cells of the embryo. In contrast, mammalian embryos, it has been argued, cannot generate cellular differences by the differential inheritance of regionally organised characteristics since their cells are so developmentally labile. The fallacy inherent in this conclusion has been crisply expressed by embryologists of various generations (1, 2, 3), but the misapprehension has persisted nonetheless. The regulatory capacity of the mammalian embryo is not incompatible with a mosaic organisation. It is only incompatible with a mosaic organisation that is rigidly determinate. If cells inherit features that direct one course of development only, they are indeed strict determinate mosaics. There are probably few such types of embryo. If, in contrast, cells inherit features that only guide cell fate but that do not commit them to a single course of development in the face of changing circumstance, they are regulatory mosaics. Probably most classes of embryo are of this type. It is our contention that the mouse embryo also conforms to this pattern (4, 5).

2. Surface and cytoplasmic mosaicism in the mouse embryo

Little evidence exists to support the presence of a developmentally significant mosaicism in the organisation of the ovum, zygote or 2- and 4-cell mouse embryo (6–9). None of the regionally localised features that has been detected so far appears to anticipate subsequent regional differentiation within the embryo. However, this sparsity of positive evidence should not be construed as proof of the absence of developmentally significant mosaicism at these stages, since the studies have been few and often were not focussed directly upon this question.

During the 8-cell stage, in contrast, a massive reorganisation or 'compaction' of the mouse embryo occurs which does appear to have important developmental consequences. Compaction displays three main features, the most obvious, being evident at light microscope level, is the change in shape of the individual blastomeres from discrete spheres, clearly distinguishable from each other, to wedge-shaped cells that flatten against each other to maximise cell contact (compare Figs. 1b & j) (10, 11, 12). The cell-flattening component of compaction is accompanied by the formation of specialised gap, desmosomal and focal tight junctions between adjacent cells (13–18). However, the earliest feature of compaction, and that of more direct relevance to this chapter, is the reorganisation that occurs within each individual cell. The early 8-cell blastomere is spherical and its surface and organelles do not show any obvious orientation (Fig. 1a–h). The late 8-cell blastomere is polarized with clearly defined apical and basolateral features (Fig. 1i–o).

Van Blerkom, J. and Motta, P.M. (eds.), Ultrastructure of reproduction. ISBN 0-89838-572-5.

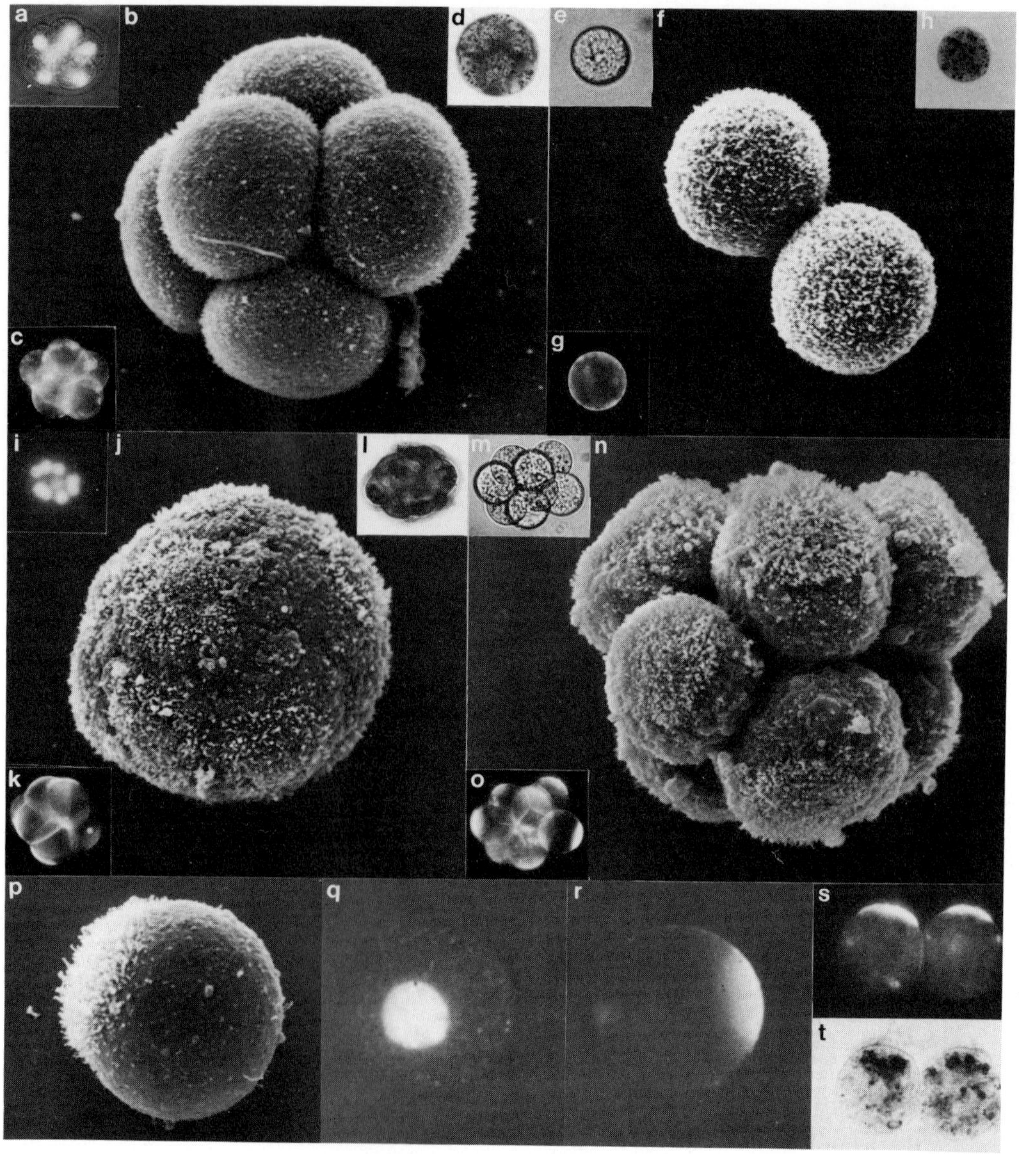

Fig. 1. (a–d) Early precompact 8-cell embryo: (a) Nuclei, stained with Hoechst 33258, are disposed peripherally; (b) Uniform distribution of microvilli viewed by scanning electron microscopy (SEM); (c) homogeneous surface-binding of FITC-concanavalin A (FITC-Con A); (d) embryo fixed after incubation for 3 h in horseradish peroxidase (HRP), note dispersed distribution of ingested enzyme. (e–h) Cells isolated from early precompact 8-cell embryos: (e) under phase; (f) under SEM; (g) after FIIC-Con A binding; (h) after incubation in HRP. (i–l) Late compact 8-cell embryo; (i) nuclei clustered centrally; (j) under SEM showing intercellular flattening and microvillous surface; (k) FITC-Con A binds to a polar region on individual cells; (l) after incubation in HRP, note peripheral localization of ingested enzyme in individual blastomeres. (m–o) Late compact 8-cell embryo decompacted by exposure to medium low in Ca^{2+}: (m) under phase; (n) under SEM, note polar blastomeres; (o) after FITC-Con A binding, note poles of bound ligand. (p–t) Single cells isolated from late decompacted 8-cell embryos: (p) under SEM, note pole of microvilli; (q) nucleus stained with Hoechst 33258 is opposite to (r) pole of FITC-Con A binding; (s) pole of FITC-Con A binding overlies (t) aggregates of ingested HRP.

(Figures 1–6). Live embryos and single cells isolated from them had the following diameters (approx): unfertilised egg, 69 μm; fertilised egg 64 μm, single blastomere from a 2-cell embryo, 47 μm; single blastomere from an 8-cell embryo, 31 μm; intact 8-cell embryo, 83 μm; expanded blastocyst, 93 μm.

The polar nature of the late 8-cell blastomere is observed at several levels of cell organisation. The cell surface is reorganised such that short microvilli are restricted to an apical pole (Fig. 1n, p) (13, 19, 20). Elsewhere longer and thicker microvilli are more sparsely distributed and only persist for as long as the basolateral surfaces of adjacent cells remain in contact (19, 20, 21). The short apical microvilli concentrate a considerable quantity of membrane at one end of the cell, which is reflected in the greater density of binding of a whole range of ligands at this site (Fig. 1r) (20, 22–24).

During the process of polarization, the nucleus migrates towards the basal region of the blastomere (Figs. 1a, i, q, r) (25). The intervening cytoplasm between the basal nucleus and the apical pole of microvilli becomes occupied by various organelles including endocytotic vesicles. Thus whereas the early 8-cell blastomere ingests horseradish peroxidase (HRP) in vesicles that are distributed evenly throughout the cytoplasm (Fig. 1d, h), the polarized blastomere localises these HRP-containing vesicles. Moreover it does so whether the enzyme is ingested by isolated cells or by cells *in situ* in the intact embryo (Fig. 1, l, s, t) (26). The mitochondria become ranged along the basolateral membranes of the polarized cell associated with arrays of microtubules that run parallel to the plasma membrane (21).

During the process of polarization the organisation of each cell is altered profoundly, and although this is most dramatically observed when individual cells are isolated and compared *in vitro*, the process normally occurs within the intact embryo. When polarized cells are observed *in situ*, their axes of polarity radiate from the centre of the embryo. In this way, the microvillous apices of the cells project outwards (Fig. 1m–o) and the basal nuclei cluster centrally (Fig. 1i) (20, 21, 25, 26). The embryo as a whole assumes a regional organisation and becomes a radial mosaic.

3. The regulation of polarization

3.1. Induction

The polarity that develops in individual 8-cell blastomeres could arise autonomously as a result of some internal preprogramming. Alternatively, the polarity could be induced as a result of some externally perceived signal, such as contact with another cell. There is evidence to suggest that the latter mechanism is more likely.

Newly formed 8-cell blastomeres may be isolated and cultured individually in suspension. Under such conditions the single cells are healthy and divide at the appropriate time to generate two 16-cell blastomeres. However, the incidence of surface polarization in such isolated cells is very low (23). If each cell is instead aggregated to a companion, newly formed 8-cell blastomere (Fig. 2a, d) and the couplet placed in culture, then both cells develop a polarized surface phenotype (Fig 2b, c, f.) Moreover, the axis of polarity induced in each cell appears to be determined by the point of intercellular contact (Fig. 2b, c, g, h) (23). When more than two cells are aggregated together, the final axis of polarity that develops in each cell is determined by the total number of cell contacts (Fig. 2i–l), the pole of microvilli developing at the surface most distant from all contact points (27). Perhaps the most spectacular demonstration of the way in which contact not only induces polarity but also determines its axis comes from the following observation (23). A newly formed pair of 8-cell blastomeres derived by division *in vitro* from a single 4-cell blastomere remain connected to each other by their midbody, a remnant of cytoplasm crossing the constriction furrow from the recent division. The blastomeres in this couplet, after culture together, develop microvillous poles opposite to the point of contact at the midbody. In contrast, if such a newly formed pair is first disaggregated, thereby breaking the midbody, and each cell is turned through 180° and the cells reaggregated such that the midbody remnant may be observed opposite to the new point of intercellular contact, then the pole of microvilli develops in the region of the residual midbody. Thus, turning the cells through 180° with respect to their initial point of intercellular contact has also turned the axis of polarity through 180° (Fig. 2e, f).

These results suggest strongly that intercellular contact induces polarity and that the asymmetry of the contact determines the axist of that polarity. The results do not of course preclude the possibility that a change of cell contacts might override a preexisting cell-autonomous programme or modify a preexisting axis.

3.2. The nature of the inducing signal and the induced response

The induction of polarity shows elements of specificity. Thus if a newly formed 8-cell blastomere is aggregated with a companion blastomere from the 16-, 8-, 4- or 2-cell stage (Fig. 3c, d, g, h) then in each case it develops a pole opposite to the point of contact with its companion. However, the frequency

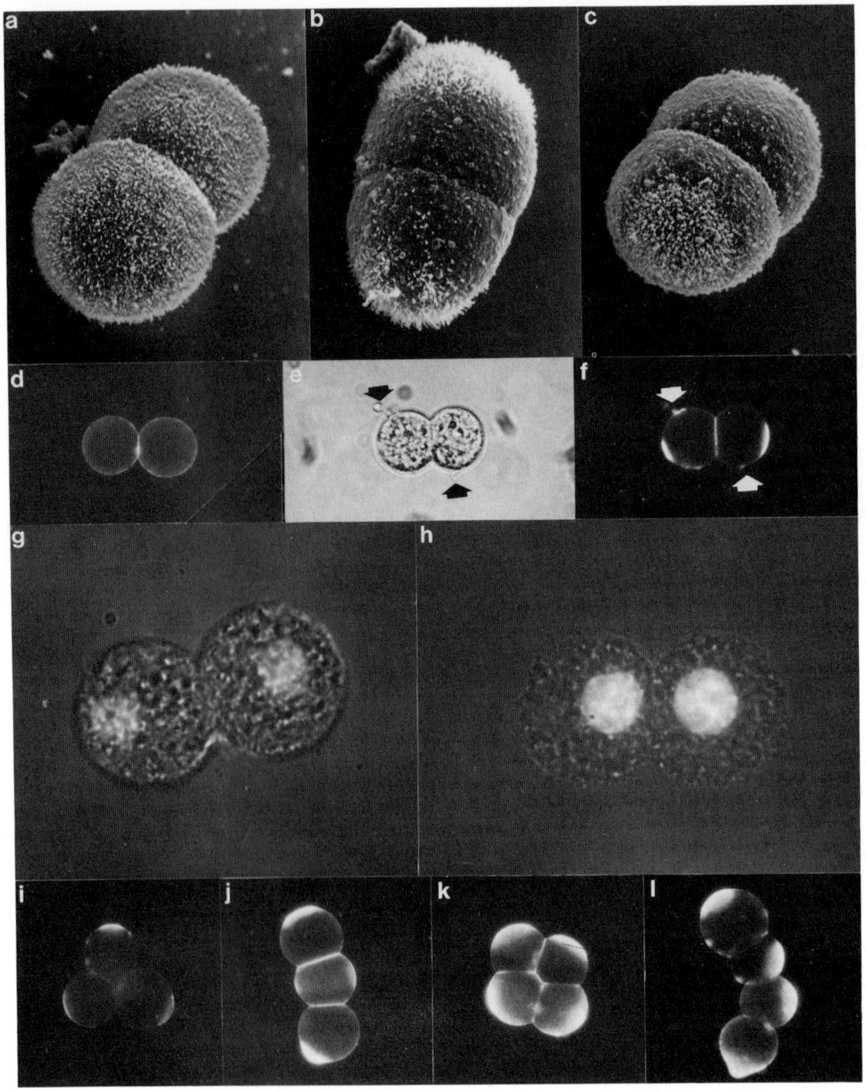

Fig. 2. (a–h) Newly formed 8-cell blastomeres were isolated as single cells and then aggregated in pairs for varying periods of time before analysis: (a) SEM of pair after 1 h culture, showing homogeneous distribution of microvilli; (b) SEM of pair after 8 h culture, showing polar distribution of microvilli and cells flattened on each other; (c) as for (b) but pair of cells exposed to medium low in Ca^{2+} to reduce cell flattening pior to fixation, poles of microvilli persist; (d) FITC-Con A-labelled pair after 1 h in culture, pair cultured for 8 h prior to examination under (e) bright field, and (f) fluorescence; note remains of mid-bodies (arrowed) and polar binding of FITC-Con A; (g) pair cultured for 1 h or (h) for 8 h prior to staining with Hoechst 33258 for nuclear position. (i–l) Newly formed 8-cell blastomeres isolated as single cells and aggregated in groups of 3 or 4; note that poles of FITC-Con A binding develop on surfaces most distant from all contact points.

with which polarity in the 8-cell blastomere is induced declines with 4- and 2-cell blastomeres and is not significant with newly fertilized or unfertilized eggs (Fig. 3a, b, e, f) (27). It seems that the capacity to induce polarity develops at some point during the 2-cell stage. It is during the 2-cell stage that the first major transcription of embryonic genes occurs, and the acquisition of a polarity-inducing capacity may be a direct consequence (29).

Polarity is not induced merely as a consequence of the increased degree of cell contact that occurs during cell flattening at compaction (30). Indeed, induction of polarity precedes cell flattening (20, 23, 27). Moreover, it is possible either to inhibit cell flattening with cytochalasin D (Fig. 3 l, m, n) (31, 32) or medium low in Ca^{2+} (Fig. 3i, j) (14, 30), or to reduce cell flattening with tunicamycin (33, 34, 35), antisera to embryonal carcinomas (36–38), concanavalin A (39) or colcemid (Fig. 3k) (21, 30, 40). In none of these cases is the development of polarity prevented.

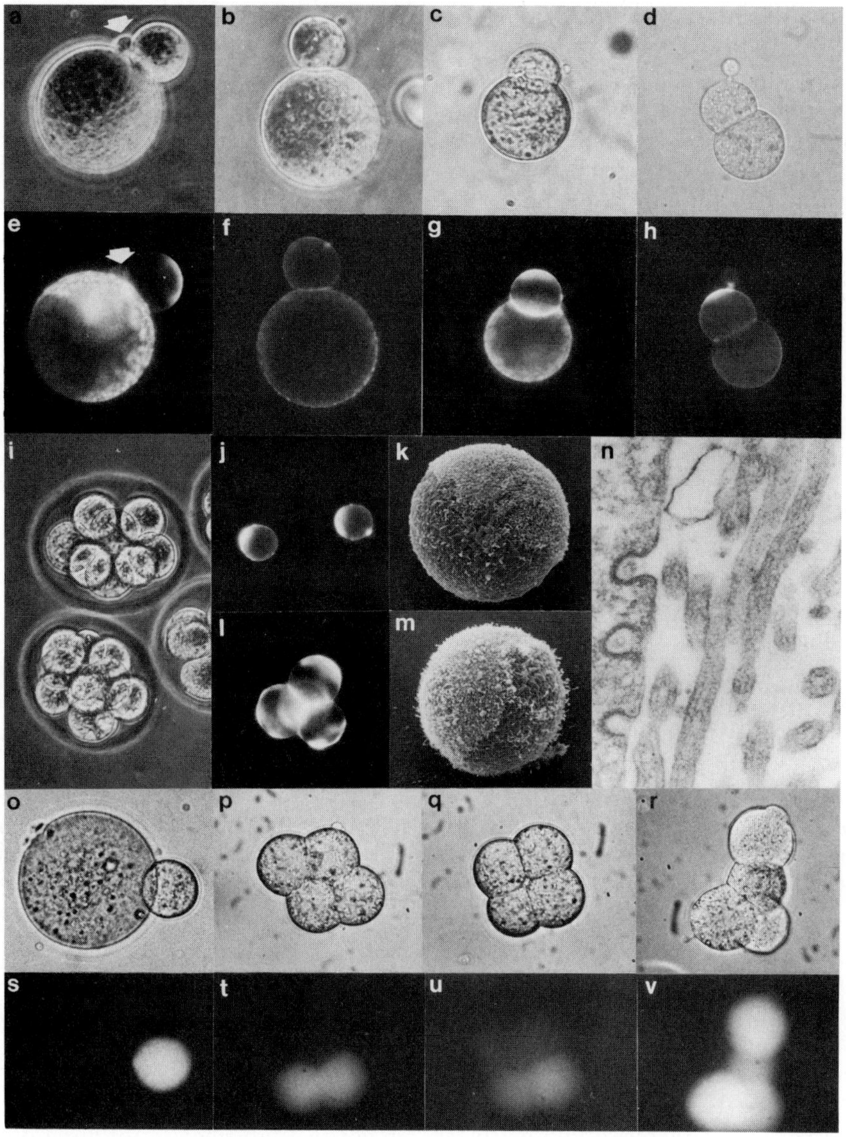

Fig. 3. (a–h) Newly formed 8-cell blastomeres (the smaller cell in each couplet) aggregated for 8 h with (a, e) an unfertilized egg, (b, f) a fertilized egg, (c, g) a 2-cell blastomere, or (d, h) a 4-cell blastomere, and incubated with FITC-Con A (e–h) or viewed under bright field (a–d). Note that poles of FITC-Con A binding develop opposite the point of contact with 2- or 4-cell blastomeres but not with fertilized eggs. Note that a pole of FITC-Con A binding appears on the 8-cell blastomere aggregated with the unfertilized egg but is not opposite the point of contact with the egg. Instead it is opposite the point at which the residual mid-body (arrowed) of the 8-cell contacts the egg. In this example, therefore, the mid-body membrane has induced polarity in the 8-cell blastomere. (i, j) Embryos cultured from the 2-cell stage to the late 8-cell stage in medium low in Ca^{2+} and examined either as whole embryos (i) or as single cells after incubation in FITC-Con A (j). (k, l, m, n) Embryos cultured from the 2-cell or 4-cell stage to the equivalent of the late 8-cell stage in colcemid (k) or in cytochalasin D (l, m, n). Division is suppressed. Note that 2-cell blastomeres (k, m) and 4-cell blastomeres (l) nonetheless develop poles as assessed by SEM or by FITC-Con A binding. (n) Apical surface of CCD-treated embryo (72 h post-hCG) showing localisation of microvilli consisting of CCD-resistant microfilaments and invaginated regions of membrane resembling coated pits ($\times$ 33,600). (o–v) Distribution of carboxy fluorescein between reaggregated mouse blastomeres. Compacted 8-cell mouse embryos were disaggregated in medium low in Ca^{2+} into single cells or pairs. Some were labelled by incubation in medium containing carboxy fluorescein diacetate and were aggregated either with an unfertilized egg or unlabelled, but otherwise identical, blastomeres. After 3 h of incubation, no passage of dye occurred to the egg (o, s), but various stages of junction-mediated transfer can be seen between aggregated 8-cell blastomeres ranging from no passage (p, t) through partial passage (q, u) to complete transfer (r, v). Bright field (o–r) and fluorescence (s–v) photographs are shown.

Specialised intercellular junctions first form at the 8-cell stage (13–18). Present evidence suggests that these junctions do not play an important role in the generation or transmission of a polarizing signal. Formation of tight junctions, for example, is impaired by several treatments that do not prevent polarization (30). Moreover, although functional gap junctions form quite early during the 8-cell stage (18), and are therefore potential participants in the induction process, they are not formed when a 2- or 4-cell blastomere is aggregated to an 8-cell blastomere, a coupling known to induce polarity in the 8-cell blastomere (Fig. 3o–v) (18, 27).

At present, the precise mechanism by which polarity is induced remains obscure. The inducing signal would seem to be specific, developmentally regulated, and not dependent upon extensive intercellular contact or specialised junctions.

The response of the polarizing cell to the inducing signal is also being analysed. Smaller 8-cell blastomeres take less time to polarize than do larger cells (23), and the ability to respond to an inducing signal by polarizing does not develop until the 8-cell stage (27). However, if cytokinesis is inhibited under conditions that permit the temporal programme of the embryo to continue (9, 27, 32) polarization will occur on schedule. Surprisingly, the considerable reorganisation involved in polarization can occur in the presence of colcemid, of cytochalasin D or of a combination of both drugs (30). In blastomeres treated with these drugs an axis of polarity is established but the detailed reorganisation of the cell surface about that axis is atypical (Fig. 3k–n). This result suggests that the integrity of microtubules and CCD-sensitive microfilaments is involved only in the more terminal features of cytoplasmic reorganisation that are responsible for the reorganisation of surface phenotype.

The ability of 8-cell blastomeres to polarize is of interest not only for embryological reasons, as will be discussed below, but also is of intrinsic interest to those who are studying the generation of epithelia. The complete process of polarization in embryos can be studied *in vitro* and after considerable experimental manipulation of the cells involved. It thus provides a suitable model system from which more general conclusions may be drawn concerning assembly of an epithelial phenotype.

4. The stability of polarity

The axis of polarity, once it has been laid down in the early 8-cell blastomere, is stable (27). Blastomeres that have been induced to polarize *in situ* retain their polar phenotype on isolation as single 8-cells (22, 23). Moreover, if two newly formed 8-cell blastomeres are aggregated as a pair, the pair cultured for 3 to 5 hours, separated into its two constituent cells, and these two cells then reaggregated to each other at a new point of intercellular contact, the axis of polarity that was established during the 3 to 5 hour inducing period is not reorientated by the new point of contact (27) (Fig. 4a, b).

The polar axis is also stable throughout cell division. Polarized 8-cell blastomeres, identified and analysed at various points during their division to two 16-cell blastomeres, retain elements of surface (Fig. 4c–n) and cytoplasmic (Fig. 4h, i) polarity throughout the process (41, 42). In seven out of every eight cell divisions, on average, a couplet of 16-cell blastomeres is produced in which one cell incorporates the apical region of the 8-cell and is polar, whilst the other is derived from the basal region and is apolar (Fig. 4o, q). In the remaining division, the cleavage furrow occurs roughly parallel to the axis of polarity, bisecting the pole of microvilli and generating two polar cells the poles of which are contiguous at the connecting midbody (Fig. 4p, r) (4, 41, 43, 44).

The polar 16-cell blastomeres formed from polar 8-cell blastomeres retain their polarity even when isolated and cultured individually for up to eleven hours (45). Moreover, when 87 of these polar 16-cell blastomeres were examined after their division to two 32-cell blastomeres, all without exception generated either two polar cells or one polar and one apolar cell, depending upon the plane of division (46). The polar cells amongst these form the trophectoderm of the early blastocyst (Fig. 4s–v).

Thus, the polar phenotype is impressively stable and once established is not discarded. One important consequence of this stability is that from the 8-cell stage onwards, the inheritance received by each of the mitotic offspring of a polar cell need not be equivalent. It is our hypothesis that this inequality of inheritance marks the foundation of the two distinct cell lineages of the blastocyst (9).

5. Phenotype and position in the 16-cell morula

The division of the polarized 8-cell blastomeres of an embryo generates polar cells (on average 9 per embryo) and apolar cells (on average 7 per embryo) (41, 43). These two cell populations, in addition to having different surface phenotypes, occupy different positions within the embryo (20, 43) and have different properties (45).

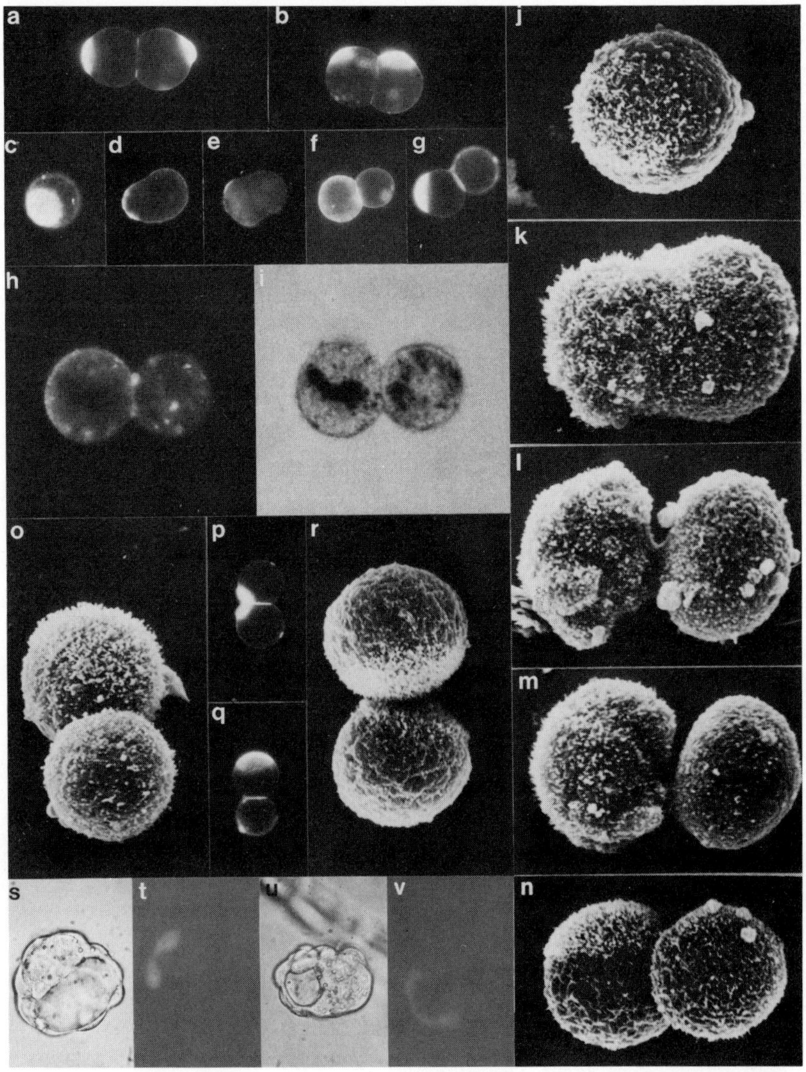

Fig. 4. (a–b) Newly formed 8-cell blastomeres were recovered as single cells and aggregated together in pairs. After 3 h, some were disaggregated into single cells again, reaggregated to each other in a new orientation, and cultured for a further 5 h. Pairs that had *not* been dis- and re-aggregated at 3 h (a) developed poles of FITC-Con A binding opposite to the point of contact. Pairs that *had* been dis- and re-aggregated (b) had poles *not* opposite to the point of contact, suggesting that a stable axis of polarity had been induced during the initial 3 h of contact. (c–g, j–n) Examples of polarized 8-cell blastomeres analysed for either FITC-Con A binding (c–g) or microvillous distribution (j–n) at various stages during division to two 16-cell blastomeres. Note that an axis of polarity is evident throughout. (h–i) A polarized 8-cell blastomere incubated in HRP and then washed and placed in control medium divided to yield two 16-cell blastomeres, a larger one with polar FITC-Con A binding which concentrated the inherited HRP and a smaller one in which the inherited HRP was more diffusely distributed. (o–r) Newly formed pairs of 16-cell blastomeres formed by division of a polarized 8-cell blastomere and analysed under SEM (o, r) or for FITC-Con A binding (p, q). Note that in (o) and (q), a polar and an apolar cell are formed at division whilst in (p) and (r) the two polar cells that formed had poles which were contiguous at the linking residual midbody. These latter pairs appear to have arisen via a division through the pole of the 8-cell blastomere. (s–v) Individual 16-cell polar outer cells were labelled with the short-term fluorescent marker FITC, aggregated with 15 unlabelled polar and apolar 16-cell blastomeres and cultured overnight. Blastocysts formed in which the FITC-labelled polar cells always contributed labelled progeny to the polar (s, t) or mural (u, v) trophectoderm.

The cell membranes exposed on the surface of the intact 16-cell morula are organised into dense, short microvilli (Fig. 5a & b) (19–21). When such embryos are decompacted by brief exposure to medium low in calcium (Fig. 5d–f), these superficial cells pull apart to reveal that their external facing apices bind fluorescent ligand preferentially (compare Figs. 5c & e) and are microvillous (Fig. 5f) (20). The basolateral surfaces of these cells have a few long, thick microvilli which regress when the cells are isolated. These

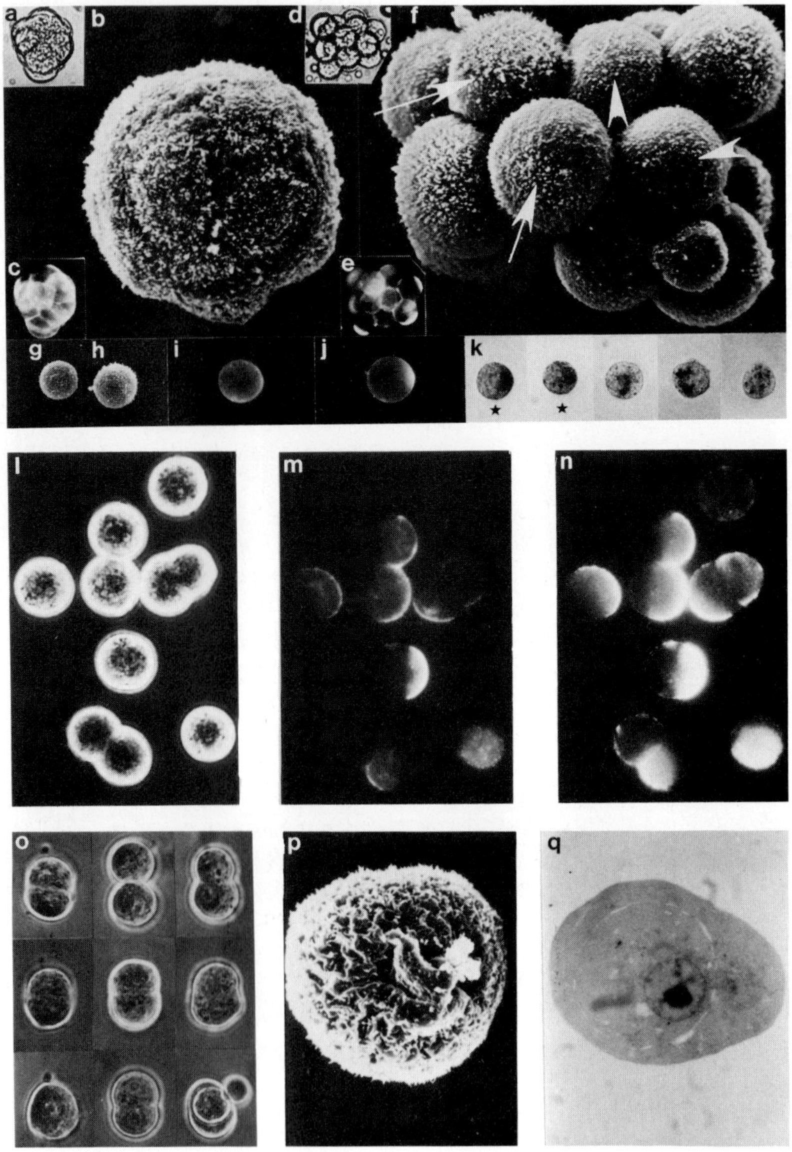

Fig. 5. (a–c) 16-cell embryo: a) under phase; the embryo appeared to have a uniform distribution both of microvilli (b) and of fluorescent-ligand binding sites (c). (d–f) Decompact 16-cell embryo. Individual blastomeres could be observed under phase (d); (e) labelling with fluorescent ligand revealed some polar cells, and under SEM, (f) polar (arrow) and apolar (arrowhead) cells could be identified, the apolar cells appearing from the inside of the morula. Isolated blastomeres showed either (g) apolar or (h) polar microvillous distributions, also revealed after labelling with fluorescent ligand (i, j). (k) patterns of HRP distribution marked * are more common in polar blastomeres; other patterns are more common in apolar cells. (l–n) An intact compact 16-cell embryo was labelled with rhodamine-Con A to mark cells with exposed surfaces, (m) then disaggregated to single cells (phase, 1) that were incubated in fluorescein-labelled Con A to define inhomogeneities in cell surface binding, (n). Note that exposed cells are polar and non-exposed cells are apolar. (o) Phase contrast views of pairs of newly formed 16-cell blastomeres that had been aggregated together immediately after division and cultured for 2 (top row), 4 (middle row) or 9 (bottom row) h before examination. The left-hand column shows morphology of two apolar cells, the middle column two polar cells and the right-hand column a polar + an apolar blastomere in each pair. Note that 2 apolar cells flatten on each other such that the cell outline is lost, unlike 2 polar cells. Note also that a polar cell envelops an apolar cell. The envelopment is shown more clearly by (p) SEM and (o) in section.

superficial cells therefore have a polar phenotype (Fig. 5h, j, k). During decompaction, a new type of cell, not hitherto exposed, is revealed. These inside cells lack the short, dense microvilli characteristic of polar cells (Fig. 5f, g, i, k) and are apolar in surface phenotype (20, 41, 43, 45).

The notion that the two subpopulations of cells with different surface phenotypes occupy different positions within the morula is confirmed by results from an independent type of experiment (43). Intact compact 16-cell morulae may be tagged with rhodamine-labelled concanavalin A, thereby labelling all exposed surfaces 'red'. The embryo may then be decompacted and disaggregated to single cells, and the isolated cells tagged with fluorescein-labelled concanavalin A, thereby labelling the whole of their surface 'green'. When individual cells are then examined, they conform to one of two patterns. Some cells fail to label red but label uniformly green indicating that they are enclosed and apolar in the intact embryo. Other cells have a pole of red label that coincides with a bright pole of green label, while the rest of the cell is labelled weakly green. These cells are therefore polar and have an exposed surface in the intact embryo that coincides with the pole of ligand-binding observed on isolated polar cells (Fig. 5 l–n).

The observation that a polarized 8-cell blastomere, with its apical pole facing outwards, generates a polar daughter cell on the outside and an apolar daughter cell on the inside of the compact embryo is consistent with the notion that the mouse embryo conforms to a mosaic pattern of embryogenesis. However, not only do the two phenotypically distinct cell subpopulations of the 16-cell morula arrive in different positions as a result of the operation of cell division on a mosaic organisation, but the different properties associated with each of the two phenotypes are also important for the maintenance of these initial positional differences.

Outer polar cells tend to be larger than inner apolar cells (Fig. 4g, n, q) (41, 43) and this disparity in size probably arises from the eccentric position of the nucleus in polarized 8-cell blastomeres (Fig. 1q) (25). Additionally, outer polar cells adhere less readily to each other than do apolar cells, and, having adhered, they do not show as extensive a degree of intercellular flattening as do apolar cells (Fig. 5o) (45). However, the most important difference between the two cells is observed when a polar and an apolar cell are aggregated together and then cultured. In most cases the polar cell sends out thin cytoplasmic processes that surround the apolar cell (Fig. 5o, p, q), thereby enveloping it in a small cavity in which more fluid may accummulate subsequently (45). Thus, polar cells will maintain their outside position actively, and in so doing reinforce the inside position of apolar cells. The same process occurs when several apolar and polar cells are aggregated together (Fig. 4s–v). The polar cells always end up on the outside regardless of their initial position in the aggregate (46).

The properties of the polar and apolar cell subpopulations in the 16-cell morula do not merely ensure reinforcement of their relative position. They also represent differentiated features that anticipate the characteristics of trophectoderm and ICM respectively. Trophectodermal cells are polar, enveloping and fluid transporting (12, 47–50), whereas ICM cells are highly adhesive, compact readily on each other and, when aggregated to a morula, move to its centre (47, 51–53).

6. Phenotype and position in the 32-cell morula and early blastocyst

Late morulae and early blastocysts, consisting of about 32 cells, also contain two phenotypically distinct cell subpopulations (Fig. 6a, b, c, f) that are located in different positions (19, 48, 53). The outer trophectoderm cells are, like the outer cells of the 16-cell morula, polar, fluid transporting, and linked to each other by zonular tight junctions of increasing complexity. They enclose a population of inner cells that are characterized by the presence of long, rather thick and sparsely distributed microvilli. This population of inner cells may be isolated immunosurgically as a compact mass (Fig. 6d, e) (53).

There are on average about 20 outside cells and 8 inside cells in late morulae, the ratio changing to around 20:14 in the early blastocyst (unpublished data). This ratio would be expected if the outside and inside cell populations (9:7) at the 16-cell stage bred 'true' (44) and if the outside cells divided slightly ahead of the inside cells. Direct evidence that apolar inside cells at the 16-cell and 32-cell stages have a longer cell cycle than do polar outside cells has been obtained (44, 54, 55, 71).

A direct study of the natural lineage relationships between cells in the same relative positions at the 16- and 32-cell stages requires the vital marking of individual cells at the earlier stage and the analysis of their fate in the later blastocyst. Fluorescein isothiocyanate (FITC) has proved to be a suitable vital, cell-autonomous, short-term marker (56). At slightly alkaline pH, the FITC penetrates blastomeres and labels proteins without having any obvious effect on the developmental capacity of the cells. If an FITC-labelled apolar cell is combined with a group of five

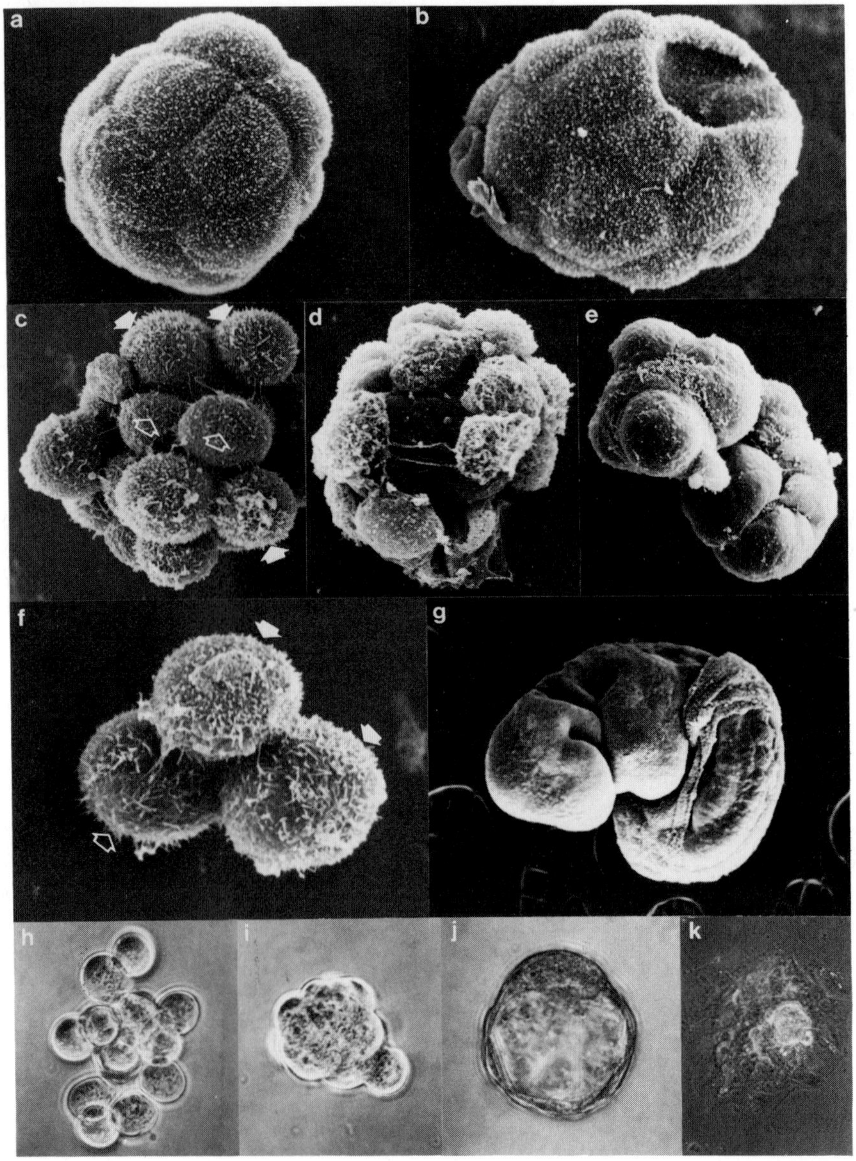

Fig. 6. (a) Intact 32-cell late morula and (b) early blastocyst, (under SEM) in which a partial collapse of the abembryonic trophoblast into the small blastocoel has occurred. (c) A late 32-cell morula decompacted in medium low in Ca^{2+} reveals the presence of an outer subpopulation of polar cells (closed arrowheads) and a newly exposed inside population of apolar cells (open arrowheads). The polar outer cells may be lysed immunosurgically revealing an inside cluster of apolar cells (d) which can be isolated (e). (f) If 32-cell morulae are exposed to medium low in Ca^{2+} and then isolated as single cells or small groups of cells, the polar (closed arrowheads) and apolar (open arrowhead) phenotypes can be seen clearly. (g–k) Photomicrographs to indicate the totipotentiality of 16-cell polar blastomeres. Sixteen such blastomeres were aggregated together (h), and after $9\frac{1}{2}$ h had formed a more compact group of cells (i), which developed by 29 h into a blastocyst with an ICM (j). Such blastocysts attach to the culture dish *in vitro* and outgrow (k) or, on transfer to pseudopregnant recipients, implant to yield a normal conceptus (g).

non-labelled apolar cells, and this group is then surrounded by ten non-labelled polar cells, the whole aggregate will develop after culture overnight into a blastocyst. Two to four cells in this blastocyst are fluorescent and with a few exceptions are confined exclusively to the ICM. In contrast, an aggregate of six non-labelled apolar cells surrounded by ten polar cells, one of which is FITC-labelled, forms a blastocyst in which, in most cases, 2–4 trophectodermal cells are fluorescent (Fig. 4s–v) (46). This result, which is also supported in studies using labelled thymidine or horseradish peroxidase as short-term cell markers (58, 72), suggests that in the 16-cell embryo the prospective fates of apolar, inside cells and of polar, outside cells are indeed ICM and trophectoderm respectively (But see 57). This conclusion supports the proposal that polarization of the 8-cell blastomere has a critical role in the generation of the ICM and trophectodermal cell lineages.

7. The relationship between prospective fate and totipotentiality

The evidence presented in the previous sections is consistent with the polarization hypothesis (9). The cytoplasmic and surface rearrangements that occur at compaction generate a mosaic structure. At division, this mosaicism is conserved, leading to the formation of inside and outside cell populations incorporating the apex or base of each 8-cell blastomere respectively. The different phenotypic and functional features that are inherited by the two subpopulations of the 16-cell embryo reinforce the relative positional differences such that the blastomeres tend to breed true at division to the 32-cell stage and formation of the blastocyst. This progressive reinforcement of relative positional differences also amounts, of course, to the progressive divergent differentiation of ICM and trophectodermal tissues.

However, as was indicated in the Introduction, the cells of the late morula and early blastocyst (or at least some of them) appear to be developmentally totipotent (54, 59–64). It is the difficulty in adequately reconciling this totipotentiality with the mosaic organisation underlying the differentiation of the blastocyst that has caused so many problems for mammalian embryology. It is of the utmost importance to distinguish between what cells will do, if left to themselves, and what they can do if provoked, and to explain the cell dynamics underlying this distinction.

Sixteen polar cells, isolated from 16-cell embryos and aggregated together, will form a blastocyst containing both ICM and trophectoderm (Fig. 6h–k). Sixteen apolar cells will do likewise (54). In each case the blastocyst is able to implant and generate an apparently normal conceptus (Fig. 6g). Since the evidence suggests overwhelmingly that in the normal intact embryos the polar cells form trophectoderm and the apolar cells form ICM, how can these observations be reconciled?

It is clear that depriving polar cells of apolar companions must divert their development from its natural course, and somehow cause the polar cells to generate replacement apolar cells. We have proposed (45, 46) that this alteration in behavior arises from the influence of cell interaction on the plane of cell division. Thus, polar cells will flatten round a core of apolar 16-cell blastomeres so that when the polar cells divide all, or most, cleavage planes will bisect the pole thereby generating two polar cells. The polar cells will tend to breed true. If, however, polar cells are deprived of apolar cells (or if there are too few apolar cells), cell flattening will not be as extensive, and the more rounded polar cells will be less likely to divide through their poles. Any division that is parallel to, rather than at right-angles to, the pole will generate a polar and an apolar cell. The system is regulatory – the fewer apolar cells there are, the greater the chance of generating more (46). Such a system need not operate exclusively in the highly abnormal situation of an aggregate of 16 polar cells but could operate equally well in the intrinsic regulation of the normal embryo to adjust the ICM:trophectoderm ratio if it is distorted by death or loss of inside cells. Data on the orientation of division planes of polar 16-cell blastomeres cultured alone, in combination with a second polar cell or in combination with an apolar cell support the idea that cell interactions do indeed influence cleavage planes as predicted (73).

What of the expression of totipotency in apolar cells? How might a cluster of sixteen apolar cells give rise to a blastocyst? In section 3.1, evidence was presented that polarity in early, non-polar 8-cell blastomeres was induced by asymmetric cell contacts. Moreover, early 8-cell blastomeres surrounded completely by other cells do not polarize (27). If this capacity of non-polar cells to polarize in response to asymmetric cell contacts persisted in inside cells at the 16- and even 32-cell stages, then it would be predicted that any inside cells that became partly exposed should polarize. In other words, placing an apolar cell for a prolonged period on the outside of an aggregate should cause it to polarize. Exactly this result has been obtained both for pairs of apolar 16-cell blastomeres (73) and for apolar cells placed on the outside of an aggregate of apolar cells (45). Again internal regulation of the ICM: trophectoderm ratio is provided; a deficiency of polar cells would lead to an increased chance of exposure of apolar cells and the restorative generation of more polar cells.

Thus, we suggest that in the normal intact morula, inside and outside cell populations will breed true in most cases but minor deviations from this pattern can and will occur in response to atypical cell ratios. The totipotency of these two cell populations may be just as integral to the construction of a blastocyst as is their origin from a mosaic 8-cell embryo.

8. Conclusion

The notion that the prospective fate of cells in the morula might be influenced by their relative positions was proposed over fifteen years ago (65, 66, 67). However, the mechanisms whereby relative position might be recognised have proved to be theoretically and experimentally intractable (67–70). In this chapter we have summarized the evidence that has accumulated over the last three years to suggest that positional recognition can be detected as early as the 8-cell stage.

The data support the critical role proposed for polarization in the generation of cell differences (9) and for cell interactions in their continuing divergence (4, 44), and also allow us to propose precise cellular mechanisms to explain cell totipotency. The cellular description of events leading to the construction of a blastocyst must now be interpreted in molecular terms.

9. Acknowledgements

We wish to thank Tim Crane, Ian Edgar and Tracy Kelly for their assistance in the production of this chapter. The work described was supported by grants to M.H.J. from the Medical Research Council, Cancer Research Campaign and Ford Foundation and to H.P.M.P. from the Medical Research Council. During the course of the work described in this chapter, W.J.D.R. was supported by an MRC Research Training Award and a grant from the Cambridge Philosophical Society, A.H.H. was supported by an MRC Research Training Award and C.A.Z. was supported by a Fellowship from the American Cancer Society.

References

1. Wilson EB: The Cell in Development and Heredity. 3rd edition Macmillan, New York, 1925.
2. Davidson EH: Gene Activity in Early Development. 2nd edition Academic Press, New York, 1976.
3. Graham CF, Wareing PF: The Developmental Biology of Plants and Animals. Blackwell Scientific Publications. Oxford, 1976.
4. Johnson MH: Membrane events associated with the generation of a blastocyst. Int Rev Cytol Suppl 12: 1–37, 1981.
5. Reeve WJD: The generation of cellular differences in the preimplantation mouse embryo. In: Progress in Anatomy Volume 3. The Anatomical Society of Great Britain and Ireland. Harrison RJ, Navaratnam (eds) Cambridge University Press, 1983 (in press).
6. Jones-Seaton A: A study of cytoplasmic basophily in the egg of the rat and some other mammals. Ann Soc Roy Zool (Belgium) 80: 76–86, 1950.
7. Dalcq AM: Introduction to General Embryology. Oxford University Press, London, 1957.
8. Mulnard JG: Studies of regulation of mouse ova *in vitro*. In: Preimplantation Stages of Pregnancy. Ciba Foundation Symposium. Wolstenholme, GEW, O'Connor M (eds) Churchill, London, 1965, pp 123–138.
9. Johnson MH, Pratt HPM, Handyside AH: The generation and recognition of positional information in the preimplantation mouse embryo. In: Cellular and Molecular Aspects of Implantation. Glasser SR, Bullock DW (eds) Plenum Press, New York, 1981, pp 55–74.
10. Lewis WH, Wright ES: On the early development of the mouse egg. Contrib to Embryol Carnegie Inst 148: 115–143, 1935.
11. Lehtonen E: Changes in cell dimensions and intercellular contacts during cleavage-stage cell cycles in mouse embryonic cells. J Embryol exp Morph 58: 231–249, 1980.
12. Ducibella T, Anderson E: Cell shape and membrane changes in the eight-cell mouse embryo: prerequisites for morphogenesis of the blastocyst. Dev Biol 47: 45–58, 1975.
13. Ducibella T, Albertini DF, Anderson E & Biggers SD: The preimplantation mammalian embryo: characterization of intercellular junctions and their appearance during development. Dev Biol 45: 231–250, 1975.
14. Ducibella T, Anderson E: The effects of calcium deficiency on the formation of the zonula occludens and blastocoel in the mouse embryo. Dev Biol 73: 46–58, 1979.
15. Magnuson T, Jacobson JB, Stackpole CW: Relationship between intercellular permeability and junction organization in the preimplantation mouse embryo. Dev Biol 67: 214–224, 1978.
16. Lo CW, Gilula NB: Gap junctional communication in the preimplantation mouse embryo. Cell 18: 399–409, 1979.
17. Lo CW: Gap junctions and development. In: Development in Mammals, vol. 4. Johnson MH (ed) Elsevier/North-Holland Biomedical Press, Amsterdam, 1980, pp 39–80.
18. Goodall H, Johnson MH: The use of carboxy-fluorescein diacetate to study the formation of permeable channels between mouse blastomeres. Nature, 295, 524–526, 1982.
19. Calarco PG, Epstein CJ: Cell surface changes during preimplantation development in the mouse. Dev Biol 32: 208–213, 1973.
20. Reeve WJD, Ziomek CA: Distribution of microvilli on dissociated blastomeres from mouse embryos: evidence for surface polarization at compaction. J Embryol exp Morph 62: 339–350, 1981.
21. Ducibella T, Ukena T, Karnovsky M, Anderson E: Changes in cell surface and cortical cytoplasmic organization during embryogenesis in the preimplantation mouse embryo. J Cell Biol 74: 153–167, 1977.
22. Handyside AH: Distribution of antibody- and lectin-binding sites on dissociated blastomeres from mouse morulae: evidence for polarization at compaction. J Embryol exp Morph 60: 99–116, 1980.
23. Ziomek CA, Johnson MH: Cell surface interation induces polarization of mouse 8-cell blastomeres at compaction. Cell 21: 935–942, 1980.
24. Johnson LV, Calarco PG: Immunological characterization of embryonic cell surface antigens recognized by antiblastocyst serum. Dev Biol 79: 208–223, 1981.
25. Reeve WJD, Kelly FP: Nuclear position in the cells of mouse morulae. J Cell Sci, (in press).
26. Reeve WJD: Cytoplasmic polarity develops at compaction in rat and mouse embryos. J Embryol exp Morph 62: 351–367, 1981.
27. Johnson MH, Ziomek CA: Induction of polarity in mouse 8-cell blastomeres: specificity, geometry and stability. J Cell Biol 91: 303–308, 1981.
28. Johnson, MH: The molecular and cellular basis of preimplantation mouse development. Biol Rev 56: 463–498.
29. Flach G, Johnson MH, Braude PR, Taylor RAS, Bolton V: The transition from maternal to embryonic control of preimplantation mouse development. The EMBO Journal 1: 681–686, 1982.
30. Pratt HPM, Ziomek CA, Reeve WJD, Johnson MH: Compaction of the mouse embryo: an analysis of its components. J Embryol exp Morph 70: 113–132, 1982.
31. Surani MAH, Barton Sc, Burling A: Differentiation of 2-cell and 8-cell mouse embryos arrested by cytoskeletal inhibitors. Exp Cell Res 125: 275–286, 1980.
32. Pratt HPM, Chakraborty J, Surani MAH: Molecular and morphological differentiation of the mouse blastocyst after manipulations of

compaction with cytochalasin D. Cell 26: 279–292, 1981.
33. Surani MAH: Glycoprotein synthesis and inhibition of glycosylation by tunicamycin in preimplantation mouse embryos: compaction and trophoblast adhesion. Cell 18: 217–227, 1979.
34. Surani MAH, Kimber SJ, Handyside AH: Synthesis and role of cell surface glycoproteins in preimplantation mouse development. Exp Cell Res 133: 331–339, 1981.
35. Webb CG, Duksin D: Involvement of glycoproteins in the develop ment of early mouse embryos: effect of tunicamycin and α, α 'dipyridyl *in vitro*. Differentiation 20: 81–86, 1982.
36. Kemler R, Babinet C, Eisen H, Jacob F: Surface antigen in early differentiation. Proc Natn Acad Sci USA 74: 4449–4452, 1977.
37. Johnson MH, Chakraborty J, Handyside AH, Willison K, Stern P: The effect of prolonged decompaction on the development of the pre-implantation mouse embryo. J Embryol exp Morph 54: 241–261, 1979.
38. Ducibella T: Divalent antibodies to mouse embryonal cacinoma cells inhibit compaction in the mouse embryo. Devel Biol 79: 356–366, 1980.
39. Reeve WJD: Effect of concanavalin A on the formation of the mouse blastocyst. J Reprod Immunol 4: 53–64, 1982.
40. Siracusa G, Whittingham DG, De Felici M: The effect of microtubule- and microfilament-disrupting drugs on preimplantation mouse embryos. J Embryol exp Morph 60: 71–82, 1980.
41. Johnson MH, Ziomek CA: The foundation of two distinct cell lineages within the mouse morula. Cell 24: 71–80, 1981.
42. Reeve WJD: The distribution of ingested horseradish peroxidase in the 16-cell mouse embryo. J Embryol exp Morph 66: 191–207, 1981.
43. Handyside AH: Immunofluorescence techniques for determining the numbers of inner and outer blastomeres in mouse morulae. J Reprod Immunol 2: 339–350, 1981.
44. Ziomek CA, Pratt HPM, Johnson MH: The origins of cell diversity in the early mouse embryo. In: Brit Soc Cell Biol Symp No 5. Functional integration of cells in animal tissues. Finbow ME, Pitts JD (eds) Cambridge Univ. Press, 1982, pp 149–165.
45. Ziomek CA, Johnson MH: Properties of polar and apolar cells from the 16-cell mouse morula. W Roux's Arch Dev Biol 190: 287–296, 1981.
46. Ziomek CA, Johnson MH: The roles of phenotype and position in guiding the fate of 16-cell mouse blastomeres. Devel Biol 91: 440–447, 1982.
47. Gardner RL, Johnson MH: An investigation of inner cell mass and trophoblast tissues following their isolation from the mouse blastocyst. J Embryol exp Morph 28: 279–312, 1972.
48. Nadijcka M, Hillman N: Ultrastructural studies of the mouse blastocyst substages. J Embryol exp Morph 32: 675–695, 1974.
49. Borland RM: Transport processes in the mammalian blastocyst. In: Development in Mammals, vol. 1 (MH Johnson, ed.) pp 31–67, North-Holland, 1977.
50. Burgoyne PS, Ducibella T: Changes in the properties of the developing trophoblast of preimplantation mouse embryos as revealed by aggregation studies. J Embryol exp Morph 40, 143–157, 1977.
51. Rossant J: Investigation of inner cell mass determination by aggregation of isolated rat inner cell masses with mouse morulae. J Embryol exp Morph 36: 163–174, 1976.
52. Stewart CL, Kimber SJ: The cell surface and interactions between different cell types of the mouse embryo. J Embryol exp Morph (submitted).
53. Johnson MH, Ziomek CA: Cell subpopulations in the late morula and early blastocyst of the mouse. Devel Biol 91: 431–439, 1982.
54. Ziomek CA, Johnson MH, Handyside AH: The development potential of mouse 16-cell blastomeres. J Exp Zool 221: 345–355, 1982.
55. Kimura S, Kato Y: Cell proliferation and the cell cycle in mouse blastocysts. Proc 51st Annual meeting of the Zoological Society of Japan, 1981.
56. Ziomek CA: The use of fluorescein isothiocyanate (FITC) as a short-term cell lineage marker in the peri-implantation mouse embryo. W. Roux's Arch Dev Biol, 191: 37–41, 1982.
57. Balakier H, Pedersen RA: Allocation of cells to inner cell mass and trophectoderm lineage in preimplantation mouse embryos. Devel Biol 90: 352–362, 1982.
58. Randle, B: Cosegregation of monoclonal reactivity and cell behaviour in the mouse preimplantation embryo. J Embryol exp Morph, 70: 261–278, 1982.
59. Johnson MH, Handyside AH, Braude PR: Control mechanisms in early mammalian development. In: Development in Mammals, vol 2. Johnson MH (ed) North-Holland, Amsterdam, 1977, pp 67–97.
60. Handyside AH: Time of commitment of inside cells isolated from preimplantation mouse embryos. J Embryol exp Morph 45: 37–53, 1978.
61. Hogan B, Tilly R: *In vitro* development of inner cell masses isolated immunosurgically from mouse blastocysts. I. Inner cell masses from 3.5 day p.c. blastocysts incubated for 24 h before immunosurgery. J Embryol exp Morph 45: 93–105, 1978.
62. Spindle AI: Trophoblast regeneration by inner cell masses isolated from cultured mouse embryos. J exp Zool 203: 483–489, 1978.
63. Rossant J, Vijh KM: Ability of outside cells from preimplantation mouse embryos to form inner cell mass derivatives. Devel Biol 76: 475–482, 1980.
64. Rossant J, Lis WJ: Potential of isolated mouse inner cell masses to form trophectoderm derivatives *in vivo*. Devel Biol 70: 255–261, 1979.
65. Mintz B: Experimental genetic mosaicism in the mouse. In: Preimplantation Stages of Pregnancy. Ciba Foundation Symposium. (GEW Wolstenhome, O'Connor M, eds) pp 194–207. Churchill, London, 1965.
66. Tarkowski AK, Wroblewska J: Development of blastomeres of mouse eggs isolated at the 4- and 8-cell stage. J Embryol exp Morph 18, 155–180.
67. Herbert MC, Graham CF, 1974: Cell determination and biochemical differentiation of the early mammalian embryo. Current topics. Dev Biol 8: 151–178, 1974.
68. Ducibella T: Surface changes of the developing trophoblast cell. In: Development in Mammals, vol. 1. Johnson MH (ed) Elsevier/North-Holland Biomedical Press, Amsterdam, 1977, pp 5–30.
69. Johnson MH: Intrinsic and extrinsic factors in preimplantation development. J Reprod Fert 55, 255–265, 1979.
70. Pedersen RA, Spindle AI: Role of the blastocoele microenvironment in early mouse embryo differentiation. Nature 284: 550–552, 1980.
71. MacQueen HA, Johnson MH: The fifth cell cycle of the mouse embryo is longed for smalled cells than for larged cells. J Embryol exp Morph, in press.
72. Geathart J, Shaffed RM, Mussel JM, Oster-Granite ML: Cell lineage analyses of preimplantation mouse embryos after blastomere injections with horse radish peroxidase. Ped Res 16: 111, 1982.
73. Johnson MH, Ziomek CA: Cell interactions influence the fate of mouse blastomeres undergoing the transition from the 16- to the 32-cell stage. Dev Biol 85: 211–218, 1983.

Authors' address:
Department of Anatomy
Downing Street
Cambridge CB2 3DY
United Kingdom

CHAPTER 18

Virus-like particles and related expressions in mammalian oocytes and preimplantation stage embryos

YOSHIO YOTSUYANAGI and DANIEL SZÖLLÖSI

1. Introduction

The presence of virus-like particles in the oocytes and preimplantation stage embryos of several species has perplexed mammalian embryologists since their occurrence was first detected by electron microscopy (1, 2). Interest in this phenomenon was renewed in the early 1970s when three laboratories independently reported the presence of particles resembling RNA tumor viruses in the early cleavage stage embryos of several strains of mice (3, 4, 5). These studies showed that virus-like particle expression was stage-specific, was not the result of exogenous infection and, in spite of relatively large numbers of particles, did not interfere with normal development. Collectively, the observations led to the conclusion that virus-like particle expression was a scheduled event in the normal developmental program of the embryo.

Three recent articles (6, 7, 8) reviewed viral expressions and virus-embryo cell interactions as studied by virological, biochemical, immunological and morphological means in mammalian pre- and postimplantation embryos, as well as in female reproductive tract. The present review is focused on electron-microscopic reexamination of oocytes and preimplantation embryos of the mouse, but data on other mammals also are included. The emphasis on information derived from the mouse rather than from other mammals reflects the fact that, to date, only the mouse has been studied extensively and systematically where viral expression is concerned.

Although the presence of virus-like particles in preimplantation embryos could be documented by electron microscope, a number of fundamental questions remain: 1) are the structures truly viruses and if so; 2) what type, and 3) what is their function(s), if any, in normal development? These questions will be discussed in the light of the recent virological biochemical, and immunogical findings obtained in this and related fields.

2. Morphology of virus-like particles associated with tumors and embryos of the mouse Mus musculus

In neoplastic cells of the mouse *Mus musculus*, electron microscopy reveals various types of virus-like particles (9, 10, 11). Each is spherical, ranging between 70 and 110 nm in diameter, and limited by a unit membrane. Their designation, by ultrastructure and location are as follows:

'Intracisternal A particles' (Figs. 1, 4, 7, 9) consist of two concentric shells with an electron-lucent center and form within the endoplasmic reticulum by budding from the membranes of this latter structure.

'Intracytoplasmic A particles' also are doughnut-like structures with an electron-lucent center, but are located within the matrix of cytoplasm; these particles are not limited by a unit membrane.

'B-type particles' are characterized by the eccentric position of their electron-dense nucleoid; these particles form by budding from the cell surface. During this process, the intracytoplasmic A particle becomes the nucleoid of B-type particle.

'C-type particles' also form by budding from the cell surface. During this process (Fig. 8) or shortly after release from the cell (Fig. 10, arrow) the particles show three-layered internal structure with an electron-lucent center; the term 'immature C-type' (10) designates this form. The released particles subsequently assume another form called 'mature C-type' (10) which possesses a centrally located electron-dense nucleoid (Fig. 10).

Two other types of virus-like structures (12) are observed in addition to C and intracisternal A particles in early mouse embryos.

'Early mouse embryo intracisternal particles' (abbreviated as 'ε particles') display a radial array of spokes projecting from a centrally located nucleoid towards the envelope (Figs. 3, 5, 6). This wheel-like morphology is reminiscent of 'R-type particles' (Fig. 13) which occur within the endoplasmic reticulum of Syrian hamster tumor and embryo cells (13, 14, 15).

Van Blerkom, J. and Motta, P.M. (eds.), Ultrastructure of reproduction. ISBN 0-89838-572-5.

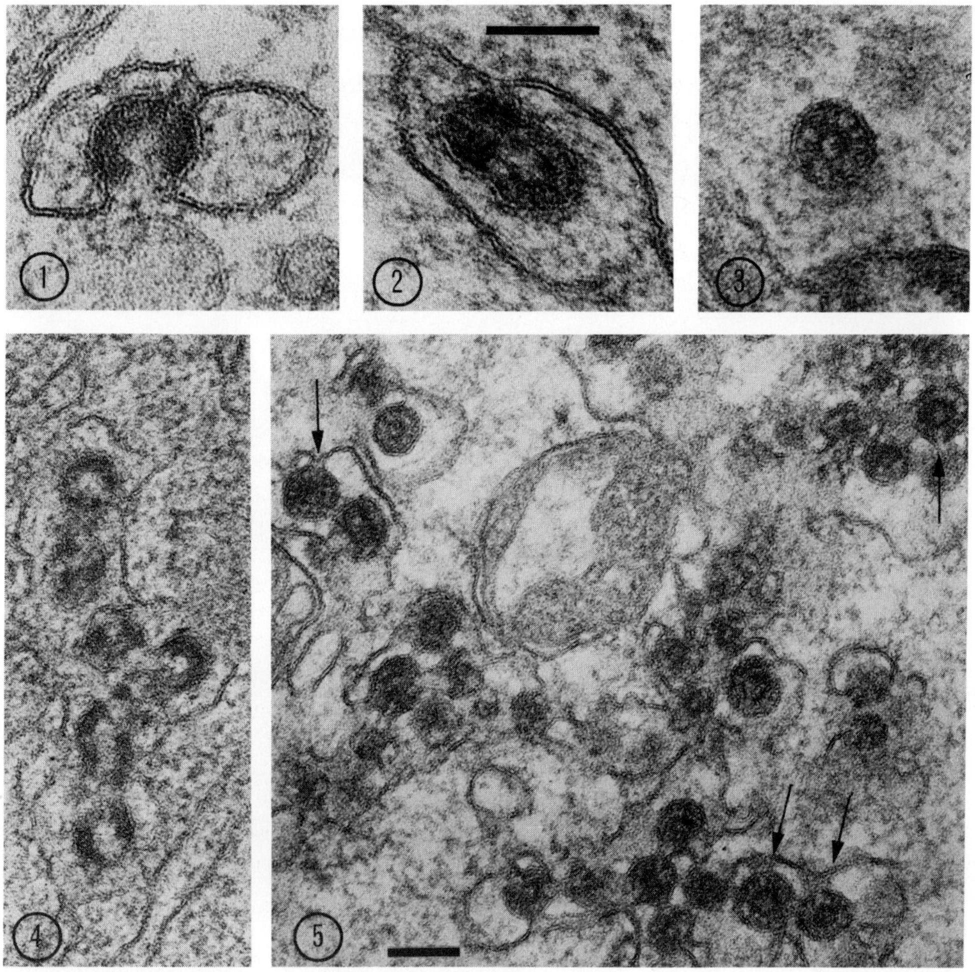

Figs. 1–5. (1) Intracisternal A particle in early embryo of NZB mouse (× 144,000). (2) ε-A mixed form of intracisternal particles in early embryo of DDK mouse. Bar: 100 nm (× 144,000). (3) Intracisternal ε particle in early embryo of AKR mouse. Remark the radial array of spokes between the envelope and the central core (× 144,000). (4) Intracisternal A particles in ovarian oocyte of Balb/c mouse. (× 96,000). (5) Intracisternal ε particles in early embryo of AKR mouse. Arrows indicate particles showing a centrally located nucleoid with a radial projection of spokes. Compare with the A-type morphology with a central lucent area as shown in Figure 4. Bar: 100 nm (× 96,000).

The demonstration of the specific ultrastructural feature of ε particle necessitates a high quality preservation of specimens and high-magnification study. For this reason the presence of two different types of intracisternal particles, ε and A, was not recognized in earlier studies (3, 4, 16–21) except one (5). In this latter report, ε and A particles were called respectively 'small' and 'large' A particles based on their slightly different size. The different terminologies used in the literature are compared in Table 1. The reason for the recommended use of the letter ε rather than R or small A has been discussed in (12).

'Dense-cored tubular structures' (Fig. 11) are sinuous tubular structures approximately 40 nm in diameter and of as yet unknown length. An electron-dense filamentous core occupies their center. These structures were initially described by Chase and Pikó (5) as 'dense-cored vesicles', a term inappropriate due to their three-dimensional configuration.

In somatic cell hybrids harboring two or more different types of intracisternal particles, complex forms comprising heterologous particles within a common envelope are occasionally observed (22, 23). As might be expected from this finding, ε and A particles were seen to form mixed structures within mouse embryos (Fig. 2). This provides the best illustration of the morphological distinction between the two types of particles under conditions that minimize differences due to artifact or sample bias introduced during preparation. The dense-cored tubular structures also form mixed structures either with ε or A particle. This suggests the possibility that these three structures belong to the same category of biological elements.

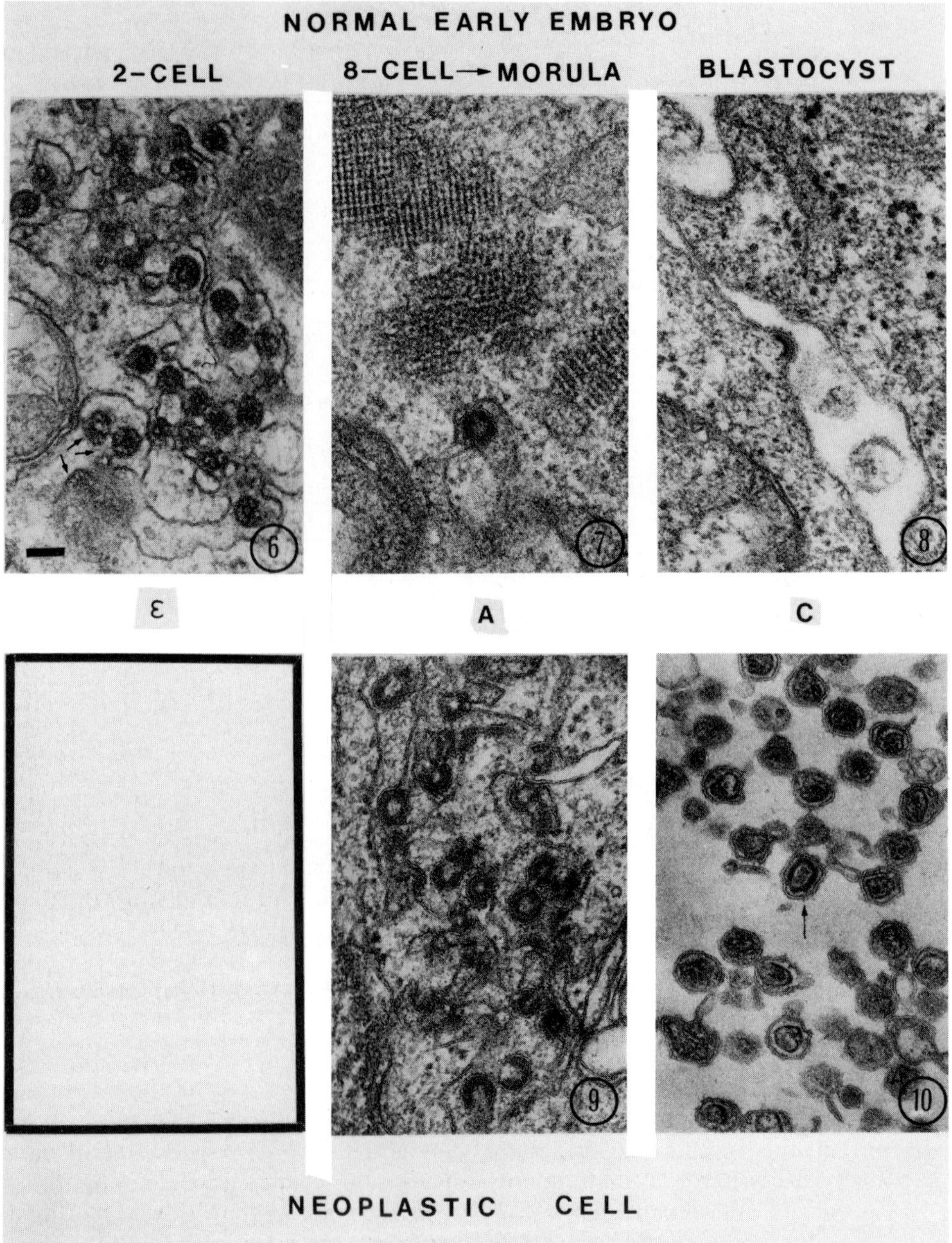

Figs. 6–10. Comparison of the virus-like particles appearing sequentially during the normal embryogenesis (Figures 6, 7 and 8) with those found in neoplastic cells (Figures 9 and 10). Bar in Figure 6: 100 nm (× 48,000). (6) The burst of ε particles. AKR mouse. Arrows indicate particles showing wheel-like morphology. (7) The reappearance of intracisternal A particle. NZB mouse. (8) The first appearance of budding C-type particle. NZB mouse. (9) Intracisternal A particles in hamster melanoma × mouse fibroblast somatic hybrid (ref. 32). (10) C-type particles released from *in vitro* cultured cells of mouse melanoma 440 (cf. ref. 32).

3. Expression of intracisternal virus-like structures and related antigens in oocytes and early embryos of the mouse

3.1. Developmentally regulated formation of intracisternal virus-like structures

Intracisternal A particles have been detected in the oocytes of every strain of mouse so far examined by us, AKR, NZB, Swiss, DDK, Peru, Q (12), and 129 (unpublished observation) and by others, Balb/c, CF-1, Swiss Webster, and NZB (3). ε particles have never been found in oocytes and zygotes. The number of

Table 1. Terminology used for the virus-like particles of early mouse embryo.

Authors	Calarco and Szöllösi (3) Biczysko et al. (4, 16) Calarco (17, 21) Yang et al. (18) Van Blerkom and Runner (19)	Chase and Pikó (5)	Yotsuyanagi and Szöllösi (12)
Mixed population recognized	No	Yes	Yes
Term for the major subpopulation	Intracisternal A particle	Intracisternal small A particle	Early mouse embryo intracisternal particle (ε particle)
Term for the minor subpopulation	Intracisternal A particle	Intracisternal large A particle	Intracisternal A particle
Mouse strains studied	(3) Balb/c, CF-1, Swiss Webster, NZB. (4) Swiss ICR/Ha, AKR, Balb/c, ICR/Ha × C57BL, Balb/cfC3H, GR/Cam, C57BLf/He/Cam, RIII/Cam. (5) Swiss albino. (12) AKR, NZB, Balb/c, C3H/He, Swiss, DDK, Peru, Q, MF1. (16) ICR, AKR. (17) Swiss Webster, Balb/c, CF-1, NZB, ICR. (18) Swiss Webster, ICR, Balb/c, CF-1, NZB, AKR. (19) CURr pc/pc, CURr p+/p+, CURr++/++. (21) Feral mice.		

intracisternal A particles starts decreasing with ovulation and the lowest level is attained at the 2-cell stage (Fig. 16b).

At this stage, many intracytoplasmic foci composed of membranes of endoplasmic reticulum, ribosomes (very often in polysomal form), and patches of electron-dense amorphous material appear around the nucleus. Numerous ε particles are formed by budding from the membranes of the endoplasmic reticulum. The cytological events leading to the appearance of ε particles were described in detail by Calarco (17), though under the name of intracisternal A particles.

ε particles were observed in every strain of mouse examined by us, AKR, NZB, Balb/c, C3H/He, Swiss, DDK, Peru, Q, MF1 (12) and 129 (unpublished observation); their peak expression was invariably around the 2-cell stage regardless of the strain (Fig. 16a). However, the number of ε particles is subject to a wide strain-dependent variation. Three patterns, high (AKR, NZB) intermediate (Balb/c) and low (C3H/He, Swiss, DDK, Peru, Q, MF1) producers of ε particles were noted (Fig. 16a). Three independent groups of investigators agree on the high-producer trait of AKR strain in spite of the difference in the terminologies used (4, 12, 20). Low producers include the 129 strain (unpublished observation).

The number of ε particles decreases progressively during the following stages. In low-producer strains, ε particles disappear generally between the 4-8-cell and 8-16-cell stages, while in high-producer strains, their disappearance is complete only just before implantation, in the 5-day blastocysts.

ε particles occur also in parthenogenetically developing mouse embryos (16, 19). Quantitative study of such embryos is as yet to be done in order to determine whether the particle formation follows exactly the same pattern of stage-dependence as in fertilized embryos.

Intracisternal A particles, once eclipsed around the 2-cell stage, reappear between the 4-cell and early blastocyst stages in several strains of mice, e.g. DDK, Swiss, Balb/c, C3H/He; the reappearance is hardly perceptible in some others, e.g. AKR, NZB (12) and 129 (unpublished observation). In late blastocysts of any strain neither ε nor A particles are observed.

ε and A particles exhibit dramatically contrasted modes of regulation:

a) Intracisternal A particles occur in oocytes (3, 12), pre- (12) and post- (4, 24) implantation embryos, normal (25) and neoplastic (9, 11, 25, 26) tissues of adult mice, whereas ε particles have so far been observed only in early mouse embryos (12).

b) The formation of ε and A particles exhibits inverse patterns of developmental stage dependence in oocytes and during early embryogenesis (12). The critical phase is the 2-cell stage where respectively maximum and minimum numbers of ε and A particles are counted.

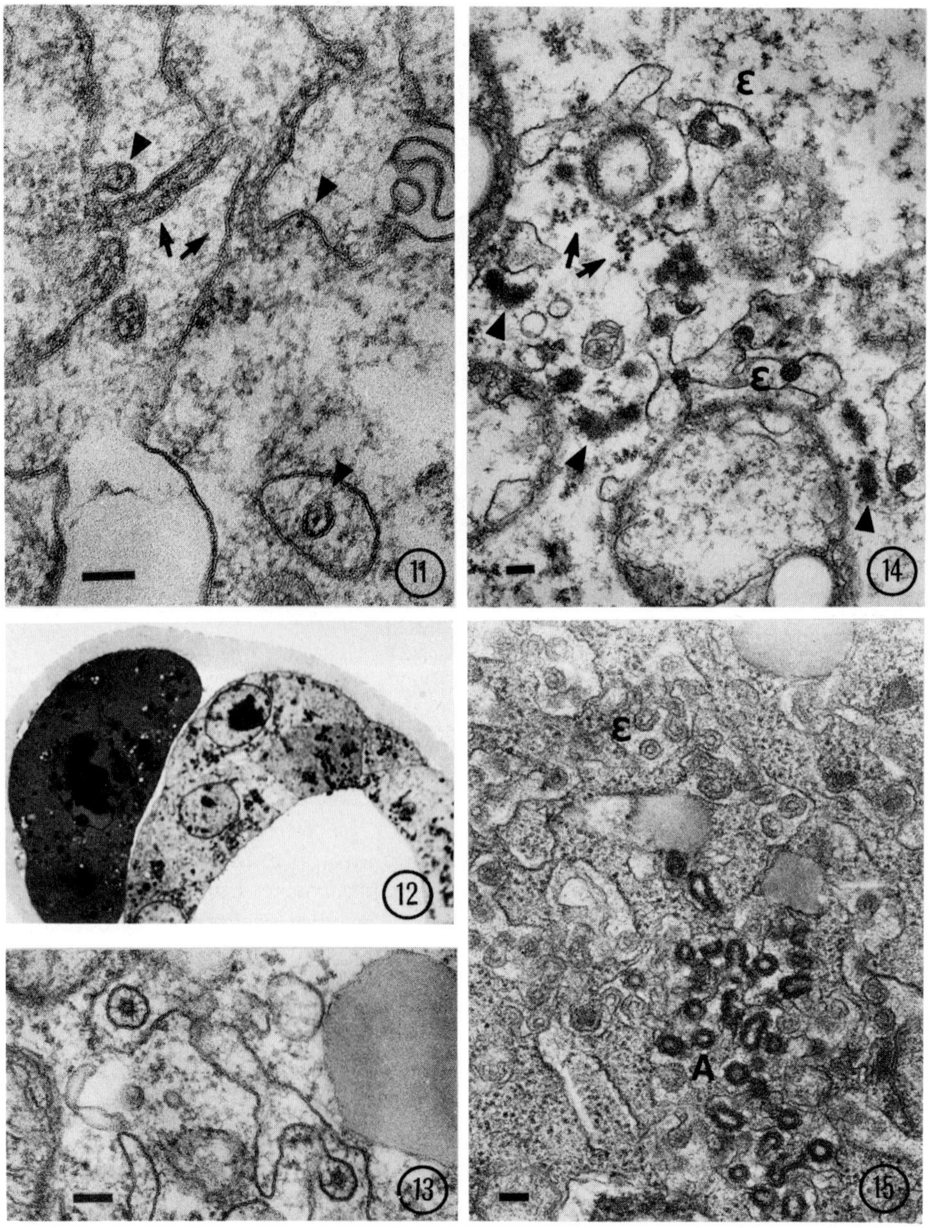

Figs. 11–15. (11) Dense-cored tubular structures in early embryo of NZB mouse. Arrows: longitudinal sections. Arrowheads: cross sections. Bar: 100 nm (× 72,000). (12) Blastocyst (right) with a ε particle-producing giant blastomere (left) excluded from it. AKR mouse. Light microscopic image of 1 μm section stained with Richardson's stain (× 560). (13) R-type particles in cultured melanoma cell of the Syrian hamster (cf. ref. 32). Bar: 100 nm (× 48,000). (14) Initiation stage of the formation of ε particles in NZB mouse embryo cultured in the presence of bromodeoxyuridine. Arrows: polysomes. Arrowheads: electron-dense amorphous material. ε: budding ε particles. Bar: 100 nm (× 32,000). (15) Induction by iododeoxyuridine of ε particles (ε) and intracisternal A particles (A) in a permanent mouse cell line Ki-Balb. Bar: 100 nm (× 32,000).

c) Bursts of large numbers of ε particles can take place in normally developing embryos (Fig. 6) where the production of A particles is constantly restricted to very low levels (Fig. 7; also compare the scales used for the ordinates of Fig. 16a and b). The situation is inverse in neoplastic cells: intracisternal A particles can appear in large numbers (Fig. 9), while ε particles have never been observed (blank under Fig. 6).

Dense-cored tubular structures have been detected in all the 10 strains studied by us, AKR, NZB, Balb/c, C3H/c, Swiss, DDK, Peru, Q, MF1 (12), 129 (unpublished observation), and also in Swiss albino strain studied by others (5). Their peak expression is

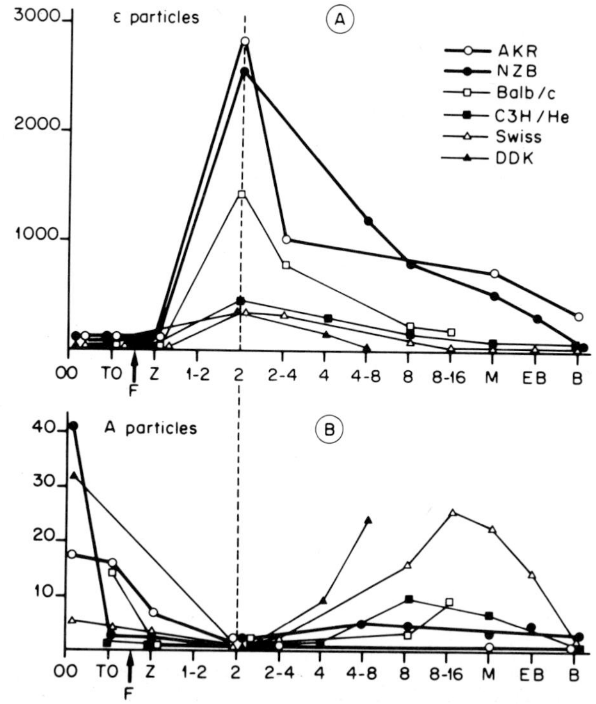

Fig. 16. Variation in number of ε (A) and intracisternal A (B) particles in oocytes and early embryos of the mouse (from Yotsuyanagi and Szöllösi, ref. 12). Number of virus-like particles was counted in 25 randomly chosen sections of oocyte or embryo. Each point in the figure represents mean number of such counts made on 3 to 4 specimens. Developmental stages represented on the abscissa are from left to right, ovarian oocyte (OO), tubal oocyte (TO), fertilization (F), zygote (Z), 1- to 2-cell stage (1–2) through 8- to 16-cell stage (8–16), late morula (M), early blastocyst (EB), and blastocyst (B).

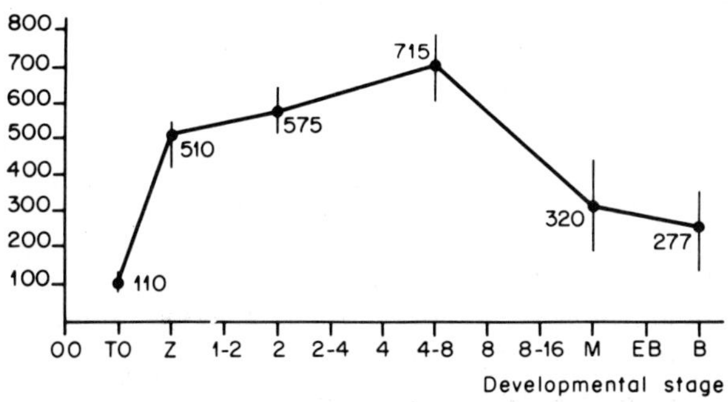

Fig. 17. Expression of intracisternal A particle-associated antigens on the surface of early mouse embryo as revealed by immunoradiolabeling. The amount of ^{125}I-protein A bound to anti-intracisternal A particle IgG-treated embryos of various stages is expressed as cpm/50 embryos. For the developmental stages represented on the abscissa, see the legend of Figure 16 (modified from Huang and Calarco, ref. 36).

at the blastocyst stage in Swiss, AKR (unpublished quantitative result) and Swiss albino (5). NZB strain conversely expresses maximum number of these structures at the 2-cell stage and lower numbers at all other stages including blastocyst stage (unpublished quantitative result).

3.2. Abnormal occurrence of intracisternal particles and embryo development

In morulae or blastocysts in which ε particles generally disappear almost completely, some exceptional blastomeres harboring large numbers of particles are occasionally seen. This unusual phenomenon may or may not be associated with a gross developmental abnormality recognizable at light microscopic level. Questions arise as to the origin and the fate of such cells. Three mechanisms are in principle conceivable: 1) a continued production of ε particles or their accumulation due to a defective turnover within an otherwise normally developing embryo; 2) an arrest of some blastomeres at early cleavage stage while the remainder pursues a normal development; 3) a reinitiation within some blastomeres of ε particles at unscheduled post-cleavage stages. The first mechanism is expected to give rise to (a) clonal sector(s) composed of a certain number of cells recognizable by the presence of numerous particles. Such a situation, of great interest, at present, has not yet been reported. The second mechanism has been illustrated by the following example: in embryos presumed to be homozygous (A^y/A^y) for the yellow allele of the agouti locus, frequent exclusions of blastomeres were noted (27); ultrastructural studies (28, 29) showed large number of ε particles (described under the name of intracisternal A particle) within the excluded blastomeres but not in those constituting the main body of morulae or blastocysts. The excluded blastomeres were generally interpreted as being arrested at an early cleavage stage where ε particles are abundant. A similar event was observed in blastocysts of AKR strain (Fig. 12) following hormonally induced superovulation but not in those obtained through the normal estrus cycle (30). An example suggesting the third mechanism, namely the incidental reinitiation of ε particle production was also provided by AKR strain (30). Some morulae of this strain were found to display within one or two blastomeres the ultrastructural features of the initiation phase of ε particle production similar to those seen at the 2-cell stage. These unusual blastomeres were indistiguishable from other blastomeres either in form or size; the contact was perfectly established between the two classes of blastomeres. It was therefore reasonable to conclude that the embryo had undergone normal development and compaction until the morula stage and that some of their blastomeres subsequently started the production of ε particles anew. Given this observation, the ε particles seen abundantly in the excluded blastomere of AKR blastocysts (Fig. 12) may represent either those newly synthesized at the morula stage or those formed at earlier stages and carried-over within an arrested blastomere.

Yellow agouti and hormone-treated AKR mice make such a study possible because abnormal occurrence of ε particles is detected at sufficiently high frequencies (in the order of 20–30 percent). A similar phenomenon probably takes place in any other strain of mouse but at very low frequency. During the electron-microscopic study of several hundred embryos of other strains, we noted only once the presence of numerous ε particles within a single blastomere of a blastocyst of C3H/He strain (30). This cell looked quite like the other cells in size and form, but had lost contact with these latter cells and displayed pathologic ultrastructural features. Because this cell was not a large excluded blastomere, the reasonable conclusion was an incidental reiniation of ε particle synthesis within a single blastomere of a blastocyst.

A^y/A^y embryos are lethal (27); although the fate of abnormal AKR blastocysts remains as yet to be determined, the excluded ε particle-rich blastomere exhibited cytopathic aspects and the embryo proper lacked the inner cell mass. We can therefore conclude that the unusual occurrence of large numbers of ε particles in post-cleavage embryos is associated with a pathologic feature in development or in cellular ultrastructure, or both. The pathologic state may be either the cause or the consequence of the abnormal expression of ε particles.

Hereditarily transmitted genomes of retroviruses ('endogenous retroviruses', see Section 6) can be activated for expression by halogenated pyrimidines such as 5-bromodeoxyuridine (BUDR) or 5-iododeoxyuridine (IUDR) (31, 32, and also 33). Intracisternal A and ε particles can also be induced in large numbers by such drugs in cultured mouse cells (Fig. 15) initially expressing no ε particles and few A particles (34). When 4-8-cell stage embryos of Swiss or NZB strains are cultured *in vitro* in the presence of BUDR, it is possible to obtain morulae or young blastocysts appearing normal by criteria of cell numbers and general organization, but displaying in certain blastomeres large numbers of ε particles or the ultrastructural features of the initiative phase of their production (Fig. 14) as seen at the 2-cell stage (unpublished observation). Despite many attempts varying the length of the treatment and the concentration of the drug (0.01 to 30 μg/ml), we failed to obtain further embryonic development compatible with induction of large numbers of ε particles: correctly cavitated blastocysts contained few particles; morulae or early blastocysts exhibiting numerous particles were those having the greatest numbers of degenerating and lysed blastomers and appeared destined to die. Thus, neither in mice of particular genotypes (yellow agouti and AKR) nor in BUDR induction experiments, could we expect a normal post-cleavage development in the presence of large numbers of ε particles.

A number of investigators noted the block of cavitation in BUDR-treated early mouse embryos (35). Rather than a specific inhibition of cavitation process, our ultrastructural observation suggested a general pathologic state of the embryo due to the toxic effect of the drug. The parallelism between the abnormal occurrence of ε particles and the inability to achieve blastocyst development probably results from this trivial reason.

3.3. Expression of intracisternal particle-related antigens

Huang and Calarco (36) reported the expression on the surface of early mouse embryos of antigens which react with an antiserum raised against the intracisternal A particle core fraction purified from MOPC 104E plasmacytoma cells. The reaction was positive from zygote to 8-cell stages with a peak at 2- to 8-cell stages, but negative in morula and blastocyst stages (Fig. 17), and also on isolated inner cell masses. The immunoprecipitation study by the same authors (37) revealed 5 proteins of which 3 (67000, 69000, 73000 daltons) were present in 2-cell to 8-cell embryos and absent in morulae and blastocysts, whereas 75000 and 77000 components were detected over all these stages. The 73000 dalton component corresponds to the major structural protein (p73) of intracisternal A particle (26, 38). The significance of the 4 other components is unclear.

The rationale of the experiments and the interpretation given by Huang and Calarco are based on a straightforward logic: since intracisternal A particles are present in large numbers between 2- to 8-cell stages, the expression of intracisternal A antigen on the cell surface can be expected. We disagree with this interpretation, because these stages correspond roughly to the eclipse phase of intracisternal A particles and the peak expression of ε particles (compare Figs. 16a, b and 17). Two alternative interpretations can therefore be conceived: 1) the anti-intracisternal A serum detects effectively intracisternal A specific proteins which are synthesized during 2- to 8-cell stages; these proteins cannot be assembled into electron-microscopically visible particles but migrate to the cell surface where they can be detected immunologically; 2) the antiserum prepared against the intracisternal A particles of tumor origin is cross-reactive with ε particle protein; during the 2- to 8-cell stages, part of this protein enters the composition of the morphologically visible ε particles while another part is expressed as surface antigen.

Huang and Calarco (39) extended their experiments to an immunocytochemical study at electron-microscopic level. The reaction products of peroxidase conjugated anti-intracisternal A serum were localized on the surface of embryos, and over the endoplasmic reticulum cisternae harboring ε particles. Consequently, the available data appear to favor the alternative that ε particles are antigenically related to intracisternal A particles.

4. Expression of C-type virus particles and related antigen in oocytes and early embryos of the mouse

4.1. Occurrence of C-type particles

To date, budding of C-type particles from the surface of oocytes or pre-blastocyst stage embryos has not been reported in any strain of *Mus musculus*, but has been detected in blastocysts of Swiss albino (5), NZB (12) and AKR (4) mice. It is important to note that, even at the blastocyst stage, not all the strains of mice show the budding of C-type particles. This process was observed only in 1 out of 7 strains examined by us (12) and also in 1 out of 8 strains examined by others (4). Moreover the observation made on AKR strain by one group (4) could not be confirmed by another (12), probably due to the difference in sublines used or breeding colonies. Three reported cases compared with the total number of mouse strains examined in different laboratories (>20 strains; see Table 1) indicate the incidental nature of the occurrence of C-type particles in early mouse embryos. The only general rule to emerge is that, if a given strain of mouse shows budding C-type particles, it is never before blastocyst formation. This rule confirms the lack of C-type virus expression in early cleavage-stage mouse embryos as pointed out previously (6).

4.2 Germinal vesicle antigen

Using an antiserum raised against the major core protein (p30) of AKR murine leukemia virus, Pikó (40) detected immunocytochemically (unlabeled antibody peroxidase-antiperoxidase method) an antigen on the germinal vesicles of preovulatory mouse oocytes. Intensity of the staining increased with oocyte growth. In early embryos, the nuclear staining was present from 1-cell to morula stages, but disappeared in expanded blastocysts.

This pattern of developmental stage dependence appears to be fairly well conserved evolutionarily, since not only several mouse strains examined (Swiss, AKR, Balb/c, C3H/He, C57L) showed consistent results, but even the rat (RCS) reacted similarly to the same anti-p30 serum. On the contrary, the antigenic determinant responsible for the germinal vesicle staining is not observed among different virus strains. This staining was abolished by absorption of the antiserum by disrupted virions of Gros (A), AKR, Kirsten, and Moloney strains, but not of Friend and Rauscher strains of murine leukemia virus. In addition the absorption by several other C-type viruses has also no effect on the stainings.

Such conserved pattern among animals and non-conserved pattern among viruses evokes the similar situation shown by src genes: the transforming gene src of sarcoma viruses is considered essentially of cellular and not of viral origin (41; reviewed in 42). By analogy, Pikó proposed that the germinal vesicle antigen reacting with the anti-p30 serum is the product of a cellular gene having a normal function in early embryonic development, and that sequences related to this gene are incorporated into the genome only of AKR-type murine leukemia viruses but not of other murine C-type viruses. The important point is that the positive immunocytochemical reaction observed in cleavage stage embryos is not considered by this author as evidence of a C-type virus gene expression but of a mouse gene expression, although the probe used was prepared from a C-type virus component. According to this interpretation, Pikó's results are not at all contradictory with the lack of C-type expression in cleavage stage embryos.

5. Virus-like particles in mammals other than Mus musculus

The occurrence of virus-like particles in germ and early embryonic cells has been documented also in several other mammals in addition to the mouse *Mus musculus* (Table 2). Reported electron-microscopic studies are not always of high enough quality to permit unambiguous identification. None of the reports concerning C-type particles appears convincing. The frequent reference to intracisternal A par-

Table 2.

Animal	Developmental stage	Type of particle	Author (ref.)	Alternative interpretation or other remark
Mus cervicolor	8-cell	intracisternal A	Calarco et al. (43)	ε, (our unpublished observation)
	morula	intracytoplasmic A	our unpublished observation	
	morula	M432 retrovirus (44, 45)	our unpublished observation	
Mus pahari	2-cell	intracisternal A	Calarco et al. (43)	ε (our unpublished observation)
guinea pig	gonad	intracisternal particle different from A	Black (46)	
	gonad	intracisternal A	Fong and Hsiung (47)	Black's interpretation preferable
	oogonia	intracisternal particle different from A	Black (46)	
	oogonia & oocyte	intracisternal particle	Andersen and Jeppesen (48)	
	8-cell	virus-like particle	Enders and Schlafke (1)	
	blastocyst	virus-like particle	Enders and Schlafke (1)	
Syrian hamster	blastocyst (3 days)	R	Sobis and Vandeputte (15)	
	peak expression at 7 days	R		
rabbit	blastocyst	cytoplasmic or intercisternal A	Manes (49, Van Blerkom, Manes (50)	term intracytoplasmic A should be used
	blastocyst	C	Manes (49)	section of nonviral structure
cat	4-cell to blastocyst	new type intracisternal particle	Bowen (51)	possible identity with the particles found in cat mammary tumor cells, cf. Figs. 9, 10 in ref. 52
	blastocyst	intracisternal A	Bowen (51)	resemble more guinea pig intracisternal particle than murine intracisternal A
baboon	oocyte	C	Kalter et al. (53)	survey pictures do not permit the distinction between C-type and primate placenta particles
	preimplantation embryo (4–8 days)	C	Kalter et al. (54)	(see discussion in ref. 55)
human	oocyte	C	Larsson et al. (56)	micropapillae (see Pederson and Seidel, ref. 57)

ticles also is subject to caution. It is difficult to decide whether the virus-like particles seen in guinea pig and cat are really identical to the murine intracisternal A particles. Moreover the particles described as intracisternal A-type in *Mus cervicolor* and *Mus pahari* are in fact ε particles (our unpublished result).

Nonetheless, taken together with the observations made in *Mus musculus*, the data listed in Table 2 inform us of a very important fact: the viruslike elements most frequently seen associated with the oocytes and early embryos of mammals are those which develop within the endoplasmic reticulum by budding from the cisternal membranes, that is, the so-called intracisternal particles; in contrast, firmly established occurrence of true extracellular virus particles is rare.

6. Nature of the virus-like particles found in germ and early embryo cells of mammals: endogenous retroviruses?

6.1. Endogenous retroviruses

Most vertebrates, including mammals, birds, and reptiles, carry in their genome genetic information for the production of virus particles or of their components (33, 58, 59). These DNA sequences, 'proviruses' (60), can be viewed as components of the normal genetic complement of animals inasmuch as they are transmitted through the germ line from parent to offspring for many millions of years (vertical transmission), segregate as Mendelian determinants (61), and evolve as do other cellular genes

(58). The virus-related sequences can be transcriptionally silent, can be expressed in the form of virus-specific proteins, but can also form complete virions which may or may not be infectious.

The virus particles released from mammalian cells include C- and B-type particles as defined earlier and also M432 virus (44, 45) particles mentioned in Table 2. All these particles are viruses *senso stricto* in that they are infectious agents and some if not all are able to cause neoplasia: for example, leukemia and sarcoma viruses display virion morphology of C-type, and mammary carcinoma virus that of B-type. The major components of virus particles are single-stranded high-molecular-weight RNA, envelope glycoprotein (gp 70), core protein (p30), and RNA-dependent DNA polymerase (reverse transcriptase). Upon infection of a cell double-stranded DNA copies of the genomic RNA is synthesized with the help of this latter enzyme, and becomes integrated into the host cell chromosome. The term 'retrovirus' designates the family of viruses containing reverse transcriptase and using this unusual mode of information transfer RNA→DNA for the synthesis of their replicative intermediate (for further information on retroviruses, the reader is referred to 62). No endogenous virus containing a DNA genome is known to date; all the presently known endogenous viruses are thereby retroviruses. However, there are also retroviruses which survive only by successive infectious cycles of the cells (horizontal transmission). They are called 'exogenous retroviruses'. Upon infection of an animal by this latter category of viruses, proviral DNA is detected within the chromosomes only of some target organs but not of germ line cells.

Three lines of data illustrate how the integration of retroviral genomes into the germ line occurred and still continues to take place at the present time: 1) the infection of the germ line of a species of animal by a retrovirus originating from another phylogenetically distant animal, e.g. primate→cat (58) or mouse→pig (63); 2) experimental infection of early mouse embryos by a leukemia virus, giving rise to a mouse subline which transmits the provirus genetically (64, 65); 3) the continuous accretion of germ line proviruses during the breeding of AKR strains of mice, due to the infection of germ line cells in viremic animals (66, 67, 68).

When virus-like particles or virus-related antigens are detected spontaneously in normally developing mammalian embryos, we have then reason to suspect primarily the expression of endogenous virus. For its confirmation, however, it is necessary to exclude the possibility of a chronic infection by exogenous virus and/or to demonstrate the identity of the observed viral products with those of a known well-characterized endogenous retrovirus. Electron-microscopic observation of C-type particles alone does not constitute evidence for expression of an endogenous C-type virus.

6.2. Intracisternal particles

Such a caution is unneccessary for the intracisternal particles of which extracellular infectious form is not known. However, it is just for this very reason that calling them 'virus' is problematic ald also creates the complex problem of their evolutionary origin. Let us consider for example the best known, the intracisternal A particle. This biological element possesses the principal common features of endogenous retroviruses, i.e. high-molecular-weight RNA (12, 69) presence in mouse genome of corresponding proviral DNA sequences (69–72), reverse transcriptase (12), and retrovirus-like morphology. However, no infectious activity has ever been demonstrated to be associated with it (25, 26).

The biochemical characterization of ε particle is completely lacking at present. The immunologic relatedness between intracisternal A and ε particles as suggested by the series of experiments by Huang and Calarco (36, 37, 39) raises the question of genetic individuality between them: are these particles expressions of two genetically independent elements sharing a small degree of nucleic acid homology or are they polymorphic expressions of one and the same gene family? Sequences which hybridize to intracisternal A particle RNA constitute within the mouse genome a large gene family (500–1000 copies per haploid genome (69, 71)) comprising at least two subpopulations of gene variants (73). Whether these sequences include genes for ε particles or not is an open question.

The R-type particle of Syrian hamster has so far not been extensively studied biochemically: only some reports on its RNA and reverse transcriptase are available at present (23). In Syrian hamster genomic DNA, a recent study detects sequences which strongly hybridize to murine intracisternal A particle gene probe (74) despite the fact that intracisternal particles of the A-type have never been found in this species, but only those of the R-type (13, 15). The question arises of whether the detected intracisternal A particle-related sequences code for R particle-associated RNA or whether they remain unexpressed in Syrian hamster, this latter RNA being transcribed from other separately present proviral sequences. The first alternative is attractive in that there is already a suggestion for immunologic relatedness between intracisternal A particle and R-type-like ε particle in the mouse as we have indicated above.

Taken together, the latest findings on intracisternal particles converge to a working hypothesis that various types of intracisternal particles A, R and ε of mouse and Syrian hamster origins might in fact be closely related genetically; the extent of the distribution among mammals of such genetically related elements might still be larger inasmuch as intracisternal particle expression and the sequence homology of genomic DNA to murine intracisternal A particle genes are found also in other phylogenetically more distant rodents and even in members of other families of mammals (Table 2) (74).

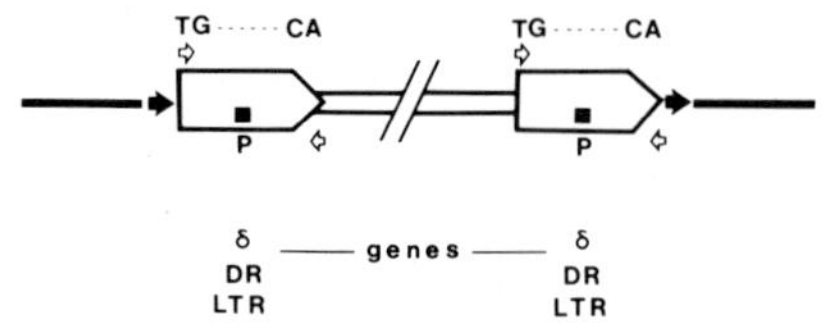

Fig. 18. Generalized structure of integrated retroviruses and eucaryotic transposable elements. A DNA segment of several kilobases (white line) is bounded by a pair of identical sequences (several hundred base pairs long) repeated in the same direction. Various names are in use to designate these repeated sequences: 'long terminal repeat' (LTR) generally used by virologists, 'direct repeat' (DR) more commonly used by others, and 'δ sequence' specifically in the case of yeast Tyl element. These repeats carry small terminal inverted repeats several base pairs long (small white arrows). The terminal dinucleotides of LTR (DR) are 5′TG----CA3′ for copia and Tyl elements respectively of *Drosophila* and yeast and for integrated forms of all the retroviruses thus far studied. This sequence homology may still extend several base pairs further. The complete element is flanked by an identical pair of host DNA sequences (black arrows) repeated in the same direction. This repeat results from the duplication of the host DNA sequence present at the integration site of the movable element. The flanking host DNA is represented by a black line. The presence of promoters of transcription (P), otherwise 'TATA-box', raises the possibility that the element may control the expression of flanking cellular genes.

7. Transposable genetic elements, retroviruses, and intracisternal particles

The genomes of lower organisms such as bacteria, yeast, and *Drosophila* contain sets of discrete DNA segments displaying a striking structural similarity to the integrated forms of vertebrate retroviruses. They are called transposable elements (75) because they are capable of moving to new sites within the genome; the transposition results in various forms of chromosomal rearrangements and in modulations of gene expression. The universal structural schema applicable for both eucaryotic transposable elements and retroviral proviruses is represented in Figure 18 and explained in the figure legend. Not only the similarity in overall sequence organization as shown in this figure, but also a remarkable sequence homology were noted between the vertebrate retroviruses and the transposable elements of phylogenetically distant organisms, such as copia, 412 and B104 elements of *Drosophila* or Tyl element of yeast (76–79). These findings prompted Temin to hypothesize that retroviruses might have evolved from intragenomic transposable elements (76). Several reviews (76, 80, 81) and a symposium (82) provide an account of this recent discovery of the ubiquity of structurally homologous mobile genetic elements in a wide variety of organisms. Here we center our discussion around the problem of intracisternal particles not treated elsewhere.

Whether endogenous retroviruses originated from germ line infections or by evolution from cellular sequences (cf. 33, 59, 60) has been a much debated question during the past decade. The first alternative now appears to be preferred by many of the virologists working on C- or B-type retroviruses, on the basis of a considerable body of recent experimental evidence (59, 60). On the contrary, so far as the intracisternal A particle is concerned, the first hypothesis has no particular advantage over the second. Because the present-day intracisternal A particles are strictly intracellular entites, the germ line infection hypothesis has to assume the existence in evolutionary past of a transmissible progenitor, whereas such a supplementary speculation is unnecessary for the C- and B-type endogenous viruses. The M432 retrovirus (44, 45, 83) which might have been suspected to be candidate for such a progenitor, because of the partial nucleic acid homology between intracisternal A particle and M432 viral genomes, appears now more likely to have originated by a recombination between intracisternal A particle genome and some other viral or cellular sequences (83). The germ line infection hypothesis moreover obliges us to admit that the putative progenitor virus had infected the germ line of mouse ancestor at least 500–1000 times with the subsequent conservation of as many proviral copies within the genome. Such a high number of proviral copies is not known for any of the C- or B-type endogenous viruses (<50 copies; cf. ref. 69) and thereby requires a specific explanation.

The assumption of evolution from cellular sequences such as transposable elements is consistent with many features of intracisternal A particle: 1) the basic organization of intracisternal A particle-related DNA sequence with its long terminal repeats (LTR,

Fig. 18) is indeed transposable element-like (72, 84); 2) these sequences constitute a dispersed multigene family reminiscent of the dispersed middle-repetitive DNA family, for example the copia family (cf. 81, 82, 85) of *Drosophila*, and one might speculate that the same mechanism is responsible for both cases of gene amplification; 3) besides the existence of at least two subpopulations of different genes as mentioned above (73), the major class (ca. 75%) of the intracisternal A particle gene population is composed of relatively well conserved members (72, 73), as are conserved the members of copia gene family (cf. 85), and here again, one might imagine the operation of the same kind of mechanism to maintain the homogeneity of family members; 4) the dramatic difference in copy numbers of intracisternal A particle-related sequences between closely related mouse species *Mus musculus* (500–1000 copies) and *M. cervicolor* or *M. caroli* (20–25 copies) (83) evokes the analogous situation between *Drosophila melanogaster* (30 copies) and the sibling species *D. simulans* (2–4 copies) with respect to the copy numbers of copia (cf. 85); 5) intracisternal A particle-related sequences appear to be conserved within rodent genomes for at least 20 million years (74) and this fact is consistent with the idea of long-term evolution of these sequences, unlike many endogenous C- and B-type viruses studied which seem to have been introduced into the germ line much more recently, subsequent to speciation (59, 60).

The problem of the evolutionary origin of intracisternal particles is far from being settled. At the present status of our knowledge, these elements can be viewed either as retroviruses which have become defective subsequent to their integration into the germ line or as evolutionary intermediates between the transposable elements and the true retroviruses having an extracellular infectious phase. The mouse genome harbors another retrovirus-like dispersed multigene family VL 30 (>100 copies) which raises exactly the same alternative possibilities for its evolutionary origin (86).

The similarity between the retroviruses and the transposable elements is not limited to their sequence organization: data are rapidly accumulating to indicate that retroviruses are also capable to function in similar ways as the transposable elements. The capacity of behaving as cis-acting control element of gene expression was demonstrated with avian leukosis virus (87). Retroviruses also can act as insertion mutagens and appear to promote the deletion of flanking cellular DNA upon reversion (88, 89). The discovery of a mouse α-globin pseudogene, $\alpha\psi 3$, characterized by a clean splicing of intervening sequences and by its location (chromosome 15) far from the cluster of homologous functional genes (chromosome 11) led to a speculation on an eventual involvement of a retrovirus in both processes. Two copies of intracisternal A particle-related sequences were found at both sides of $\alpha\psi 3$ pseudogene (90). Although this may represent a fortuitous spatial relationship, it raises also the possibility that intracisternal A particle was instrumental both in reverse transcription of processed transcript of α-globin gene and reinsertion of the DNA copy at a novel chromosomal site.

8. Does retrovirus play a role in embryogenesis?

The discovery of the hereditary transmission of the RNA tumor virus genomes at the beginning of 70's raised a puzzling problem: why are biological elements which, by definition, are deleterious (since tumor-inducing) for the host, evolutionarily conserved and expressed? A hypothesis was soon proposed postulating a beneficial role of endogenous viruses in embryo development or differentiation, or both (91).

As an approach to this problem Strand, August and Jaenisch (92) compared by radioimmunoassay 3 mouse strains expressing C-type virus markers (p30, gp70). The rationale of their experiment was: if viral expression is necessary for embryogenesis, one should detect developmental stage-specific viral expressions and the pattern of these expressions should be the same in all the 3 strains examined. They failed to demonstrate such a consistent pattern in the materials they examined, i.e. embryos during the second half of gestation (10 to 20 days), newborns and adults. We applied the same methodology to our quantitative electron microscopic study of oocytes and preimplantation embryos (1 to 5 days) of 6 strains of mice (12). Our observation on the expression of C-type particles agreed with those of Strand et al. (92) in that not all the mouse strains studied exhibited C-type particles at the blastocyst stage as expected. However a striking contrast was noted in the behavior of intracisternal A and ε particles: in all the 6 strains analysed quantitatively the pattern of the developmental stage-specific expression of intracisternal particles was basically similar. Moreover, if we include the nonquantitative observations made by other authors, we can conclude that the association of ε particle expression with the initiation stage of embryogenesis has consistently been observed in more than 20 strains of mice (Table 1).

The same methodology was introduced very recently in the study of developmentally regulated transcription (93, 94, 95) of transposable elements of *Drosophila* and, surprisingly, results similar to ours come from this field. Schwartz, Lockett, and Young (95) compared the pattern of stage-specific transcription of copia and 412 elements between 3 strains of *Drosophila*. The patterns shown respectively by copia and 412 were different but were similar between the *Drosophila* strains compared. This work provided further information of basic interest in the developmental regulation of gene expression. The 3 strains used differ in the pattern of 412 integration sites, no chromosomal location common to all strains being detected. Thereby if sequential activation and inactivation of specific chromosomal regions during development were responsible for the expression of mobile elements integrated in the concerned region, one should expect to see the 3 strains display different patterns of stage-dependent 412 expression. This being not the case, the authors suggest that the modulation of expression is controlled by sequences found within 412 DNA segments themselves and capable of responding to regulatory signals.

This interpretation is diametrically opposite to that of Jaenisch et al. (6, 65) who invoked primarily the effect of integration sites on the developmental modulation of C-type virus expression (65, 92). Both interpretations are certainly equally valid in view of the large body of experimental evidence now available on various modalities of the regulation of expression of integrated retroviruses (60, 96).

Three studies based on the same experimental design thus yielded a negative result for C-type virus and positive results for murine intracisternal particles and *Drosophila* transposable elements. Does it mean that the two latter classes of elements play a positive role in development? We further discuss this issue in the next section.

9. Selfish or useful DNA?

Instead of speculating a beneficial role of endogeneous retroviruses, Baltimore (97, 98) regarded them as inert remnants of germ line infection which occurred during the course of evolution, surviving there simply because of the difficulty of their elimination (inasmuch as they are harmless), and serving no function which might confer a selective advantage to the host animals (59, 60, 96). This idea shares some features of what was proposed later as the 'selfish DNA' concept by Orgel and Crick (99) and by Doolittle and Sapienza (100). These authors suggested an inevitable emergence within genomes and subsequent perpetuation of DNA pieces whose only 'function' is their own survival within the genome and which have no or little contribution on phenotype level of an organism. Doolittle and Sapienza (100) in particular developed their theory taking the transposable elements as the most salient example of such intragenomic parasites and further argued that the transposability constitutes their basic strategy against their elimination by deletion or by other mechanisms.

We will now examine whether such a concept provides a satisfactory explanation for the observations we reviewed above, or whether the available data rather favor the alternative view, that is, the evolutionary conservation of endogenous retroviruses by virtue of their beneficial function. Concerning C- and B-type retroviruses, several recent review articles (6, 59, 60) enumerate the experimental evidence against this latter possibility. Also from the study of mammalian oocytes and preimplantation embryos we reviewed here, no significant data emerge in favor of it. The C-type particles reported in these materials are either of doubtful identification or of too incidental occurrence to suggest any developmental significance. As to the only one finding of real interest, namely the conserved expression of 'germinal vesicle antigen' (40), we are obliged to put it aside since the author considers it as an expression of cellular but not of viral gene, in spite of the fact that this discovery became possible by the use of a C-type viral probe.

The data relative to intracisternal particles are obviously more interesting in that these virus-like elements appear regularly in specific stages of early embryogenesis of many mammals. Moreover, the stage-specific expression of ε particles is well conserved among mouse strains (12) so that we know at present of no laboratory strain or feral mouse undergoing early embryogenesis in their absence (Table 1). Certainly attractive are the speculations such as the control of cellular gene expressions by the signal sequences carried in long terminal repeats or the involvement of transient expression of intracisternal particle-related antigen in the determination of the specific surface property of cleavage stage embryos. If intracisternal particles really fulfill such basic functions, one should expect to see the same or very similar types of particles and/or related antigens expressed at a specific developmental stage not only of one mouse species (*Mus musculus*) but also of a wide range of phylogenetically related mammals. The data concerning the evolutionary conservation of the surface expression of intracisternal particle-related antigen is completely lacking at present. As to the electron-microscopically detectable particles, a rather inconsistent pattern emerges from the data listed in Table 2, that is, various types of particles are seen in oocytes and in different stages of preimplantation embryogenesis without suggesting any particular phylogenetical correlation. Even though we agree with the hypothesis of close genetic relatedness of different morphological types of particles, the fact that their expressions peak at variable stages from the 2-cell to blastocyst stages depending on species of animals is not in favor of an important functional contribution of these elements to developmental processes. Moreover some mammalian candidates exist which do not appear to express intracisternal particles during early embryogenesis (101).

The available data may be explained as well by assuming the intracisternal particles as useless and harmless parasites but endowed with a particularly subtle strategy for their survival within the germ line. The cycle of their appearance and disappearance might in fact represent a process of 'intracellular infection' from loci to loci, from chromosome to chromosome, that these virus-like elements evolved as the basic strategy against their elimination by deletion or by other mechanisms. If the cycle of the transcription, reverse transcription, and reintegration of proviral genomes is coordinated with a given cellular process normally programmed in gametogenesis or in very early embryogenesis (this may be achieved, among other possibilities, by acquisition by intracisternal particle genome of a specific cellular regulatory sequence), the new copies of viral genes reinserted into novel cellular chromosomal sites should have much chance of passing through the germ line. Bypassing the process of intercellular transmission, a result ultimately similar to that of the germ line infection may be obtained uniquely by intracellular route. A long survival within the germ line may be assured as long as the rate of this intracellular replication cycle is in equilibrium with the rate of the elimination of useless parasite DNA. At least intracisternal A particles are known to be endowed with the molecular machinery required for such a function, i.e. the reverse transcriptase (see refs. in 12) and the long terminal repeats (LTR) or their genome (72, 84; for the functional role of LTR in integrative process (60)). However, the search for their putative replicative intermediate, free circular double-stranded DNAs comparable to those of retroviruses (60) or copia element (102), has so far been unsuccesful (73). It is furthermore not easy to conceive concretely the nature of selective pressure underlying the dramatic amplification of intracisternal A particle gene merely in one mouse species but not in other closely related species (83). Such a situation rather recalls a 'selfish' amplification of useless DNA which makes rarely phylogenetical sense (99, 100).

10. Intracisternal particles and carcinogenesis

Historically, intracisternal particles were first called to the attention of investigators because of their frequent and abundant occurrence in a wide variety of mammalian tumors. For this simple reason, the intracisternal A particle was included in early morphological classification of RNA tumor viruses (9, 10). Unlike B- and C-type particles, however, the subsequent studies failed to demonstrate the etiological role of intracisternal A particle (26, also cf. 25), rendering these classifications somewhat irrelevant. Nonetheless, intracisternal particles do still present a great interest in the context of cancer research for reasons unexpected before.

The first reason is that these elements constitute a group of very interesting carcino-embryonal markers. In fact, most of the intracisternal particles which occur as regular components of germ or early embryo cells in a given mammal are reexpressed in neoplastic cells of the same animal. This is the case for the murine intracisternal A particle (9, 11, 25, 26), the guinea pig intracisternal A particle (47), the Syrian hamster R particle (13, 14, 15), and feline intracisternal A particle (52). On the contrary, these particles are not or rarely detected in normal tissues of adult animals (15, 25, 48). A better understanding of why these elements are specifically activated only in germ-, early embryo-, and neoplastic cells must contribute to the understanding of carcinogenesis.

The second reason comes from the latest finding of Shen-Ong and Cole (73) who demonstrated in mouse myeloma an exclusive expression; and amplification of a few specific copies (possibly a single among a thousand copies) of intracisternal A particle genes. Their interpretation attributes the observed amplification to reintegration of the genomes of intracisternal A particles abundantly expressed in myelomas studied. This interpretation is consistent with what we proposed in the previous section in terms of 'intracellular infection' cycle in early embryos, and also with the hypothesis on the origin of intronless $\alpha\psi3$ pseudogene (90). If such amplification and reinsertion of DNA segments (several kilobases long) occur extensively in tumor cells, an insertion-mutagenic effect can be expected, provoking progressive chromosomal rearrangements. Several investigators recently pointed to a potential involvement of chromosomal rearrangements in cancer progression (tumors develop by several steps, taking sometimes a very long time), invoking an eventual role of transposable elements in such events (103, 104, 105). An experimental approach to such problems now appears possible using myeloma-intracisternal A particle or other similar mammalian systems. The same mutational event may also occur in early embryos expressing intracisternal particles, but very slowly. In fact, according to the hypothesis of 'intracellular infection' cycle, the efficiency of reinsertion of proviral DNA does not require to be high, because the loss of useless but harmless DNA it should compensate, is a very slow process achieved only in evolutionary scale of time. What is happening very slowly during normal embryogenesis may be accelerated during carcinogenesis so as to become a sizable change at the time scale of an experimental observation period. Let us recall that a 'constitutive' and abundant production of intracisternal A particles is observed in neoplastic cells, whereas only a 'transient' and quantitatively very limited production takes place during normal embryogenesis (compare Figs. 7 and 9).

ε particle represents an exception to the rule: this element constitutes an excellent 'developmental marker' of the 2-cell stage of mouse embryogenesis, but it is not a 'carcino-embryonal marker' in that it has never been observed in mouse tumors. This does not necessarily mean a nonexpression: ε particles may appear to transiently to be detected or somatic cells expressing them accidentally may be immediately destroyed, for instance, by an immuno-surveilance mechanism. It is noteworthy that the high level of ε particle production is associated with the high incidence of spontaneous leukemia in AKR strain of mouse (106, 107).

11. Concluding remarks

That endogenous retroviruses might play a role in mammalian embryogenesis has been a popular idea during the past decade, but has never received firm experimental support. Meanwhile, it has become apparent that this subject constitutes only a small part of a more basic problem in biology. In fact, structurally homologous mobile genetic elements known under the names of transposable elements and endogenous retroviruses inhabit the genomes of a wide variety of organisms from bacteria to mammals, via yeasts, fruitflies, snakes, and birds. Developmental modulation of the expression of these elements is now well documented in transposable elements of *Drosophila* and retroviruses and retrovirus-like elements of the mouse.

The fundamental question is whether such elements are evolutionarily conserved by virtue of their functional role which confers a selective advantage to the host or they are simple parasites adapted to an ecological niche called genome, where they survive, evolve, and where they might have originated. Since harboring such elements appears to be the general property of the genome, one can ask the inverse question of what is the consequence of the presence of the mobile elements in the structural modifications of the genome at the phylogenetic, ontogenetic, and carcinogenic scales of time.

Acknowledgements

We are greatly indebted to Dr. Patricia Calarco who provided us with a series of prepared *M. cervicolor* and *M. pahari* cleavage stage embryos and who shared with us her most recent unpublished immunocytochemical results with Dr. T.T.F. Huang Jr. For the excellent technical assistance of Mrs Cassant we express our thans. This work was supported in part by a grant from ATP 'Génétique et Développement de la Souris' of the Centre National de la Recherche Scientifique.

References

1. Enders AC, Schlafke SJ: The fine structure of the blastocyst: Some comparative studies. In: Preimplantation stages of pregnancy. Wolstenholme GEW, O'Connor M (eds), Ciba Foundation Symp, Londen, J & A Churchill, 1965, pp 29–59.
2. Calarco PG, Brown EH: An ultrastructural and cytological study of preimplantation development of the mouse. J Exp Zool 171: 253–284, 1969.
3. Calarco PG, Szöllösi D: Intracisternal A particles in ova and preimplantation stages of the mouse. Nature New Biol 243: 91–93, 1973.
4. Biczysko W, Pienkowski M, Solter D, Koprowski H: Virus particles in early mouse embryos. J Natl Cancer Inst 51: 1041–1050, 1973.
5. Chase DG, Pikó L: Expression of A- and C-type particles in early mouse embryos. J Natl Cancer Inst 52: 483–489, 1973.
6. Jaenisch R, Berns A: Tumor virus expression during mammalian embryogenesis. In: Concepts in mammalian embryogenesis, Sherman MI (ed), Cambridge, Mass., MIT Press, 1977, pp 267–314.
7. Daniel JC Jr, Chilton BS: Virus-like particles in embryos and female reproductive tract. In: Development in mammals 3. Johnson MH (ed), Amsterdam, Elsevier, 1978, pp 131–187.
8. Kelly F, Condamine H: Tumor viruses and early mouse embryos. Biochim Biophys Acta 651: 105–141, 1982.
9. Bernhard W: The detection and study of tumor viruses with the electron microscope. Cancer Res 20: 712–727, 1960.
10. Provisional Committee for Nomenclature of Viruses: Suggestions for the classification of oncogenic RNA viruses. J Natl Cancer Inst 37: 395–397, 1966.
11. de Harven E: Remarks on the ultrastructure of type A, B, and C virus particles. Adv Virus Res 19: 221–264, 1974.
12. Yotsuyanagi Y., Szöllösi D: Early mouse embryo intracisternal

particle: Fourth type of retrovirus-like particle associated with the mouse. J Natl Cancer Inst 67: 677–685, 1981.
13. Bernhard W, Tournier P: Infection virale inapparente de cellules de hamsters décelée par la microscopie électronique. An Inst Pasteur 107: 447–452, 1964.
14. Shipman C Jr, Van der Weide GC, Ma BI: Prevalence of type R virus-like particles in clones of BHK-21 cells. Virol 38: 707–710, 1969.
15. Sobis H, Vandeputte M: Viruslike particles in hamster embryos, fetuses and tumors. J Natl Cancer Inst 61: 891–895, 1978.
16. Biczysko W, Solter D, Graham C, Koprowski H: Synthesis of endogenous type-A virus particles in parthenogenetically stimulated mouse eggs. J Natl Cancer Inst 52: 483–489, 1974.
17. Calarco PG: Intracisternal A particle formation and inhibition in preimplantation mouse embryos. Biol Reprod 12: 448–454, 1975.
18. Yang SS, Calarco PG, Wivel NA: Biochemical properties and replication of murine intracisternal A particles during early embryogenesis. Eur J Cancer 11: 131–138, 1975.
19. Van Blerkom J, Runner MN: The fine structural development of preimplantation mouse parthenotes. J Exp Zool 196: 113–124, 1976.
20. Nadijcka MD, Hillman N, Gluecksohn-Waelsch S: Ultrastructural studies of lethal c^{25H}/c^{25H} mouse embryos. J Embryol Exp Morph 52: 1–11, 1979.
21. Calarco PG: Intracisternal A particles in preimplantation embryos of feral mice (Mus Musculus). Intervirol 11: 321–325, 1979.
22. Yotsuyanagi Y, Ephrussi B: Behavior of three types of ribovirus-like particles in segregating hamster x mouse somatic hybrids. Proc Natl Acad Sci USA 71: 4575–4578, 1974.
23. Yotsuyanagi Y: Occurrence of mixed forms of intracisternal virus-like particles in somatic hybrid cells. J Ultrastruct Res 60: 71–83, 1977.
24. Vernon ML, Lane WT, Huebner RJ: Prevalence of type-C particles in visceral tissues of embryonic and newborn mice. J Natl Cancer Inst 51: 1171–1175, 1973.
25. Wivel N, Smith G: Distribution of intracisternal A-particles in a variety of normal and neoplastic mouse tissues. Int J Cancer 7: 167–175, 1971.
26. Kuff EL, Lueders KK, Ozer HL, Wivel NA: Some structural and antigenic properties of intracisternal A particles occurring in mouse tumors. Proc Natl Acad Sci USA 69: 218–222, 1972.
27. Pedersen RA: Development of lethal yellow (A^y/A^y) mouse embryos in vitro. J Exp Zool 188: 307–320, 1974.
28. Calarco PG, Pedersen RA: Ultrastructural observations of lethal yellow (A^y/A^y) mouse embryos. J Embryol Exp Morph 35: 73–80, 1976.
29. Cizadlo GR, Granholm NH: Ultrastructural analysis of preimplantation lethal yellow (A^y/A^y) mouse embryos. J Embryol Exp Morph 45: 13–24, 1978.
30. Yotsuyanagi Y, Szöllösi D: Embryo development and intracisternal particles in the mouse. Biol Cell 39: 201–204, 1980.
31. Lowy DR, Rowe WP, Teich N, Hartley JW: Murine leukemia virus: high-frequency activation *in vitro* by 5-iododeoxyuridine and 5-bromo-deoxyuridine. Science 174: 155–156, 1971.
32. Aaronson SA, Todaro GJ, Scolnick EM: Induction of murine C-type viruses from clonal lines of virus-free Balb/3T3 cells. Science 174: 157–159, 1971.
33. Aaronson SA, Stephenson JR: Endogenous type-C RNA viruses of mammalian cells. Biochim Biophys Acta 458: 323–354, 1976.
34. Lasneret J, Canivet M, Bittoun P, Peries J: IdUr induction of a new type of retrovirus-like particle (ε particle) in transformed fibroblastic mouse cells. Ann Virol (Inst Pasteur) 132E: 151–159, 1981.
35. Sherman MI: Developmental biochemistry of preimplantation mammalian embryos. Ann Rev Biochem 48: 443–470, 1979.
36. Huang TTF Jr, Calarco PG: Evidence for the cell surface expression of intracisternal A particle-associated antigens during early mouse development. Dev. Biol 82: 388–392, 1981.
37. Huang TTF Jr, Calarco PG: Immunoprecipitation of intracisternal A particle-associated antigens from preimplantation mouse embryos. J Natl Cancer Inst 67: 1129–1134, 1981.
38. Marciani DJ, Kuff EL: Isolation and partial characterization of the internal structural proteins from murine intracisternal A particles. Biochem 12: 5075–5083, 1973.
39. Huang TTF Jr, Calarco PG: Immunologic relatedness of intracisternal A-particles in mouse embryos and neoplastic cell lines. J Natl Cancer Inst 68: 643–649, 1982.
40. Pikó L: Immunocytochemical detection of a murine leukemia virus-related nuclear antigen in mouse oocytes and early embryos. Cell 12: 697–707, 1977.
41. Stehlin D, Varmus HE, Bishop JM, Vogt PK: DNA related to the transforming gene(s) of avian sarcoma viruses is present in normal avian DNA. Nature 260, 170–173, 1976.
42. Bishop JM: Enemies within: The genesis of retrovirus oncogenes. Cell 23: 5–6, 1981.
43. Calarco PG, Callahan R, Yasamura T, Huang TTF: Preimplantation mouse embryos of Mus cervicolor and Mus pahari express intracisternal A particles. J Cell Biol 87 (Part 2): 140A, 1980.
44. Callahan R, Benveniste RE, Sherr CJ, Schidlovsky G, Todaro GJ: A new class of genetically transmitted retravirus isolated from Mus cervicolor. Proc Natl Acad Sci USA 73: 3579–3583, 1976.
45. Heine UI, Todaro GJ: New type B retrovirus isolates associated with kinetochores and centrioles of the host cell. J Cen Virol 39: 41–52, 1978.
46. Black V: Virus particles in primordial germ cells of fetal guinea pig. J Natl Cancer Inst 52: 545–551, 1974.
47. Fong CKY, Hsiung GD: Oncornavirus of guinea pigs. I – Morphology and distribution in normal and leukemic guinea pig cells. Virol 70: 385–398, 1976.
48. Andersen HK, Jeppesen T: Virus-like particles in guinea pig oogonia and oocytes. J Natl Cancer Inst 49: 1403–1410, 1972.
49. Manes C: Phasing of gene products during development. Cancer Res 34: 2044–2052, 1974.
50. Van Blerkom J, Manes C: The molecular biology of the pre-implantation embryo. In: Concepts in mammalian embryogenesis. Sherman MI (ed), Cambridge, Mass., MIT Press, 1977, pp 37–94.
51. Bowen RA: Expression of virus-like particles in feline preimplantation embryos. J Natl Cancer Inst 65: 1317–1320, 1980.
52. Feldman DG, Gross L: Electron microscopic study of spontaneous mammary carcinomas in cats and dogs: Virus-like particles in cat mammary carcinomas. Cancer Res 31: 1261–1267, 1971.
53. Kalter SS, Heberling RL, Smith GC, Panigel M, Kraemer DC, Helmke RJ, Hellmann A: Vertical transmission of C-type viruses: Their presence in baboon follicular oocytes and tubal ova. J Natl Cancer Inst 54: 1173–1176, 1974.
54. Kalter SS, Panigel M, Kraemer DC, Heberling RL, Helmke RJ, Smith GC, Hellman A: C-type particles in baboon (Papio cynocephalus) preimplantation embryos. J Natl Cancer Inst 52: 1927–1928, 1974.
55. Dalton AJ, Hellman A, Kalter SS, Helmke RJ: Ultrastructural comparison of placental virus with several type-C oncogenic viruses. J Natl Cancer Inst 52: 1379–1381, 1974.
56. Larsson E, Nilsson BO, Sundström P, Widehn S: Morphological and microbiological signs of endogeneous C-virus in human oocytes. In J Cancer, 1982 28: 551–557, 1981.
57. Pedersen H, Seidel G Jr: Micropapillae: A local modification of the cell surface observed in rabbit oocytes and adjacent follicular cells. J. Ultrastruct Res 39: 540–548, 1972.
58. Todaro GJ, Benveniste RE, Callahan R, Lieber MM, Sherr CJ: Endogenous primate and feline type C viruses. Cold Spring Harbor Symp Quant Biol 39: 1159–1168, 1974.
59. Weinberg RA: Origins and roles of endogenous retroviruses. Cell 22: 643–644, 1980.
60. Varmus HE: Form and function of retroviral proviruses. Science 216: 812–820, 1982.
61. Chattopadhyay SK, Rowe WP, Teich NM, Lowy DR: Definitive evidence that the murine C-type virus inducing locus Akv-1 is viral genetic material. Proc Natl Acad Sci USA 72: 906–910, 1975.
62. Weiss RA, Teich NM, Varmus HE, Coffin JM (ed): Molecular biology of tumor viruses (2nd edition); RNA tumor viruses, Cold Spring Harbor, N.Y., Cold Spring Harbor Laboratory Press, 1982.
63. Benveniste RE, Todaro GJ: Evolution of type C viral genes: Preservation of ancestral murine type C viral sequences in pig cellular DNA. Proc Natl Acad Sci USA 72: 4090–4094, 1975.
64. Jaenisch R: Germ line integration and Mendelian transmission of the exogenous Moloney leukemia virus. Proc Natl Acad Sci USA 73: 1260–1264, 1976.
65. Jaenisch R, Jähner D, Nobis P, Simon I, Löhler J, Harbers K, Grotkopp D: Chromosomal position and activation of retroviral genomes inserted into the germ line of mice. Cell 24: 519–529, 1981.
66. Quint W, van der Putten H, Janssen F, Berns A: Mobility of endogenous ecotropic murine leukemia viral genomes within mouse chromosomal DNA and integration of a mink cell focus-forming

virus-type recombinant provirus in the germ line. J Virol 41: 901–908, 1982.
67. Steffen DL, Taylor BA, Weinberg RA: Continuing germ line integration of AKR proviruses during the breeding of AKR mice and derivative recombinant inbred strains. J Virol 42: 165–175, 1982.
68. Herr W, Gilbert W: Germ-line MuLV reintegrations in AKR/J mice. Nature 296: 865–868, 1982.
69. Lueders KK, Kuff EL: Sequences associated with intracisternal A particles are reiterated in the mouse genome. Cell 12: 963–972, 1977.
70. Lueders KK, Kuff EL: Intracisternal A-particle genes: Identification in the genome of Mus musculus and comparison of multiple isolates from a mouse gene library. Proc Natl Acad Sci USA 77: 3571–3575, 1980.
71. Ono M, Cole MD, White AT, Huang RCC: Sequence organization of cloned intracisternal A particle genes. Cell 21: 465–473, 1980.
72. Kuff EL, Smith LA, Lueders KK: Intracisternal A-particle genes in Mus musculus: A conserved family of retrovirus-like elements. Mol Cell Biol 1: 216–227, 1981.
73. Shen-Ong GLC, Cole MD: Differing populations of intracisternal A-particle genes in myeloma tumors and mouse subspecies. J Virol 42: 411–421, 1982.
74. Lueders KK, Kuff EL: Sequences homologous to retrovirus-like genes of the mouse are present in multiple copies in the Syrian hamster genome. Nuc Acids Res 9: 5917–5930, 1981.
75. Calos MP, Miller JH: Transposable elements. Cell 20: 579–595, 1980.
76. Temin HM: Origin of retroviruses from cellular moveable genetic elements. Cell 21: 599–600, 1980.
77. Levis R, Dunsmuir P, Rubin GM: Terminal repeats of the Drosophila transposable element copia: Nucleotide sequence and genomic organization. Cell 21: 581–588, 1980.
78. Will BM, Bayev AA, Finnegan DJ: Nucleotide sequence of terminal repeats of 412 transposable elements of Drosophila melanogaster. J Mol Biol 153: 897–915, 1981.
79. Scherer G, Tschudi C, Perera J, Delius H, Pirrotta V: B104, a new dispersed repeated gene family in Drosophila melanogaster and its analogies with retroviruses. J Mol Biol 157: 435–451, 1982.
80. Finnegan DJ: Transposable elements and proviruses. Nature 292: 800–801, 1981.
81. Shapiro JA, Cordell B: Eukaryotic mobile and repeated genetic elements. Biol Cell 43: 31–54, 1982.
82. Cold Spring Harbor Symp Quant Biol 45: Movable genetic elements, 1980.
83. Callahan R, Kuff EL, Lueders KK, Birkenmeier E: Genetic relationship between the Mus cervicolor M432 retrovirus and the Mus musculus intracisternal type A particle. J Virol 40: 901–911, 1981.
84. Cole MD, Ono M, Huang RCC: Terminally redundant sequences in cellular intracisternal A-particle genes. J Virol 38: 680–687, 1981.
85. Spradling AC, Rubin GM: Drosophila genome organization: conserved and dynamic aspects. Ann Rev Genet 15: 219–264, 1981.
86. Keshet E, Itin A: Patterns of genomic distribution and sequence heterogeneity of a murine 'retrovirus-like' multigene family. J Virol 43: 50–58, 1982.
87. Hayward WS, Neel BG, Astrin SM: Activation of a cellular onc gene by promoter insertion in ALV-induced lymphoid leukosis. Nature 290: 475–480, 1981.
88. Varmus HE, Quintrell N, Ortiz S: Retroviruses as mutagens: Insertion and excision of a nontransforming provirus alter expression of a resident transforming provirus. Cell 25: 23–36, 1981.
89. Jenkins NA, Copeland NG, Taylor BA, Lee BK: Dilute(d) coat colour mutation of DBA/2J mice is associated with the site of integration of an ecotropic MuLV genome. Nature 293: 370–374, 1981.
90. Lueders K, Leder A, Leder P, Kuff E: Association between a transposed α-globin pseudogene and retrovirus-like elements in the Balb/c mouse genome. Nature 295: 426–428, 1982.
91. Huebner RJ, Kelloff GJ, Sarma PS, Lane WT, Turner HC, Gilden RV, Oroszlan S, Meier H, Myers DD, Peters RL: Group-specific antigen expression during embryogenesis of the genome of the C-type RNA tumor virus: Implantations for ontogenesis and oncogenesis. Proc Natl Acad Sci 67: 366–376, 1970.
92. Strand M, August JT, Jaenisch R: Oncornavirus gene expression during embryonal development of the mouse. Virol 76: 886–890, 1977.
93. Flavell AJ, Ruby SW, Toole JJ, Roberts BE, Rubin GM: Translation and developmental regulation of RNA encoded by the eukaryotic transposable element copia. Proc Natl Acad Sci USA 77: 7107–7111, 1980.
94. Scherer G, Telford J, Baldari C, Pirrota V: Isolation of clones genes differentially expressed at early and late stages of Drosophila embryonic development. Dev Biol 86: 438–447, 1981.
95. Schwartz HE, Lockett TJ, Young MW: Analysis of transcripts from two families of nomadic DNA. J Mol Biol 157: 49–68, 1982.
96. Weinberg RA, Steffen DL: Regulation of expression of the integrated retrovirus genome. J Gen Virol 54: 1–8, 1981.
97. Baltimore D: Tumor viruses: 1974. Cold Spring Harbor Symp Quant Biol 39: 1187–1200, 1974.
98. Baltimore D: Viruses, polymerases, and cancer. Science 192: 632–636, 1976.
99. Orgel LE, Crick FHC: Selfish DNA: the ultimate parasite. Nature 284: 604–607, 1980.
100. Doolittle WF, Sapienza C: Selfish genes, the phenotype paradigm and genome evolution. Nature 284: 601–603, 1980.
101. Calarco PG, McLaren A: Ultrastructural observations of preimplantation stages of the sheep. J Embryol Exp Morph 36: 609–622, 1976.
102. Flavell AJ, Ish-Horowicz D: Extrachromosomal circular copies of the eukaryotic transposable element copia in cultured Drosophila cells. Nature 292: 591–595, 1981.
103. Sager R: Transposable elements and chromosomal rearrangements in cancer – a possible link. Nature 282: 447–448, 1981.
104. Cairns J: The origin of human cancers. Nature 289: 353–357, 1981.
105. Klein G: The role of gene dosage and genetic transpositions in carcinogenesis. Nature 294: 313–318, 1981.
106. Furth J, Seibold HR, Rathborne RR: Experimental studies on lymphomatosis of mice. Amer J Cancer 19: 521–604, 1933.
107. Lilly F, Pincus T: Genetic control of murine viral leukemogenesis. Adv Cancer Res 17: 231–277, 1973.

Authors' addresses:
Y. Yotsuyanagi
Centre de Génétique Moléculaire, C.N.R.S.
91190 Gif-sur-Yvette, France

D. Szöllösi
Station Centrale de Physiologie Animale, I.N.R.A.
78350 Jouy-en-Josas, France

Note added in proof

Direct experimental evidence has now been obtained to demonstrate that intracisternal A particle genes indeed are mammalian equivalent of transposon. These genes are capable of changing location via 'intracellular' route and act as insertion mutagen; a consequent alteration in mouse immunoglobuline κ light chain gene expression (Hawley et al, PNAS 79: 7425–7429, 1982; Kuff et al, PNAS 80: 1992–1996, 1983) and an activation of a cellular oncogene c-mos in mouse myeloma (Kuff et al, Nature 302: 547–548, 1983) have been reported.

CHAPTER 19

Crystalline inclusions in embryonic and maternal cells

LOREN H. HOFFMAN and GARY E. OLSON

1. Introduction

A variety of cell types in early mammalian embryos and associated organs of the female genital tract contain inclusion bodies which are crystalline in nature. Although many of these structures were first seen by light microscopy, some have been further analyzed and others seen for the first time in more recent ultrastructural studies. Here, we will review structural aspects of such inclusions in several mammalian species. Evidence is scarce as to the chemical identity of most crystalline inclusions. A comparative look at structural features may allow us to come to tentative conclusions as to their functional importance and/or identity. Hopefully, such a review might stimulate new investigation on certain of these inclusions.

We intend to focus on those inclusions seen in cells of the early embryo or its investments. As will be seen, many of these have been considered to be crystalline 'yolk' material. This may or may not be an accurate descriptive term, implying, as it does, that the material in question was deposited in the oocyte prior to ovulation. Most of the inclusions to be reviewed here have been shown to have highly ordered internal structure based on the ultrastructural demonstration of periodicity. We will, however, also discuss some inclusions which have been referred to as 'crystals' at the light microscopic level even if periodicity has not been reported by electron microscopy. Thus, our definition will include bodies bounded by plane surfaces, symmetrically arranged, giving the external appearance of the presence of highly ordered structures. As such, the oft-used terms 'paracrystalline' and 'crystalloid' will not be employed. Our review is arranged to treat crystalline structures as they have been reported in the following locations; uterus, ovary/oocyte, preimplantation embryos and placenta/fetal membranes.

2. Uterus

Crystalline inclusions have been described in epithelial cells of rabbit and human uteri by Nakao et al. (1). The inclusions were observed in endometrium of ovariectomized rabbits after systemic administration of massive doses of progesterone, put not after estradiol treatment. The authors suggested that the uterine crystals may represent a protein product of progesterone stimulation, by analogy with the progesterone-induced avidin secretion in the chick oviduct (2). The rabbit uterus secretes several specific proteins in response to progesterone stimulation (3) and their appearance in uterine secretions coincides temporally with the first appearance of crystals in glandular epithelium (post-ovulatory days 4–6). Nakao et al. stated that the uterine crystals were actively secreted into the uterine cavity. They reported the occurrence of similar crystals in post-ovulatory human endometrium and in castrated women following progesterone treatment. Wynn (4) makes no mention of such structures in a recent review on the ultrastructure of human endometrium. Further observations were made on the inclusions of rabbit endometrium by Hoffman et al. (5). They were observed to be membrane-bounded and rod-shaped and reached lengths of several micrometers (Fig. 1). They occurred only in a limited population of epithelial cells, the glands of progestational uteri. The inclusions were first seen on day 4 of pregnancy or pseudopregnancy and were present for several days thereafter. This study also demonstrated that near-physiologic levels of progesterone, but not estradiol, induced the appearance of the inclusions. Partial digestion of inclusion contents with pepsin, in epoxy-embedded tissue, suggested their proteinaceous nature. The arrangement of tightly-packed units arranged in the long axis of the inclusions is seen in Figure 1. Other features of these inclusions will be treated in a subsequent section (Preimplantation Embryos) since we view them to be identical with rabbit trophoblast crystalline inclusions.

Van Blerkom, J. and Motta, P.M. (eds.), Ultrastructure of reproduction. ISBN 0-89838-572-5.

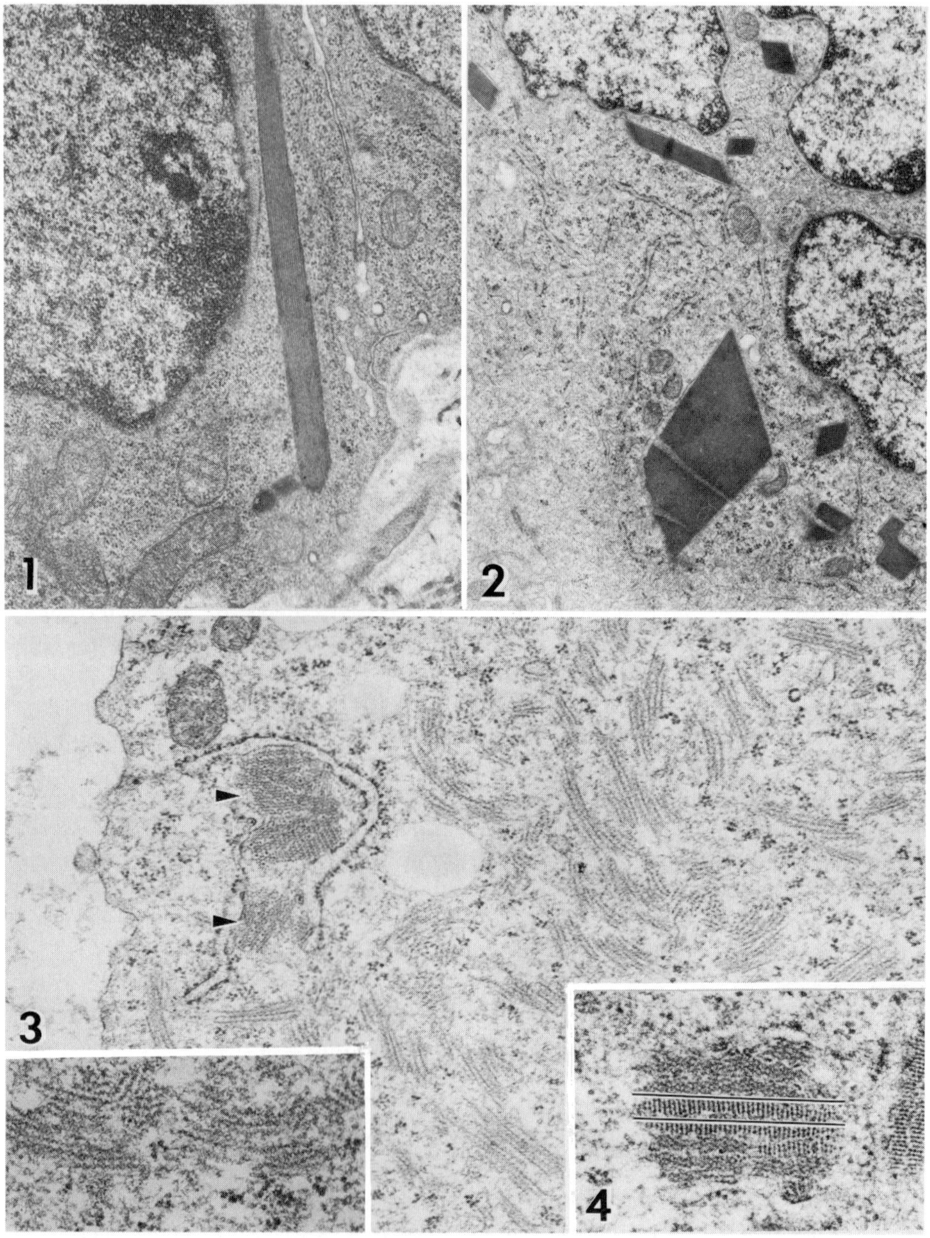

Figs. 1–4. Rod-shaped inclusion in basal cytoplasm of uterine glandular epithelial cell of rabbit during early pseudopregnancy. Note dense lines in longitudinal axis of the inclusion (× 19,200). (2) Membrane-enclosed inclusions of irregular shapes and sizes occurring in symplasmic uterine epithelium of rabbit at mid-gestation (× 12,000). (3) Lattice structures seen as linear arrays of filaments in cytoplasm of mouse blastocyst. Contrast electron density with that of ribosomes. Also seen is large crystalline structure (arrowheads) intimately associated with components of granular ER (× 24,000). Inset: In the plane of section the periodic elements of the lattices appear to have electron lucent centers (× 36,000). (4) Large crystalline structure in cytoplasm of mouse blastocyst cell. Each crystal is composed of side-to-side associations of elongated plates (one plate outlined) which, in turn, may appear as either a series of stacked bilaminar units or, from other angles, as a honeycomb of hexagonal profiles (× 40,000).

We have observed crystalline inclusions, seemingly different from those just described, in rabbit uterine epithelium later in pregnancy. These occur in the luminal epithelium, much of which is symplasmic in nature, and typically assume rectangular or rhombiodal shapes (Fig. 2). Although these bodies are likewise membrane-limited, they exhibit a less prominent series of linear densities with a smaller period than the inclusions of glandular epithelium. Whether or not these could represent a different form of the same material which comprises the rod-shaped bodies has yet to be determined.

Rhomboidal cytoplasmic crystals are also present in uterine epithelial cells of the sheep and goat during early gestation (6). These are presumably identical to the crystalline bodies present in the trophoblast and will be discussed in a subsequent section (Preimplantation embryos).

3. Ovary/Oocytes

Examples of crystalline material in somatic components of the mammalian ovary are rare. One such component appears to be the hilar cell of the human ovary (7). Although it is rare to observe crystals in the normal ovary, they are encountered in about one half of the reported cases of hilar cell tumors of the ovary (8). These tumors are frequently associated with virilism and the cytoplasmic crystals appear identical to the crystals of Reinke in human testicular Leydig cells. Merkow et al. (8) envisioned these crystals to be composed of a tightly packed, hexagonal arrangement of microtubules. Nagano and Ohtsuki (9), however, have arrived at a very different subunit composition of Reinke crystals based on optical diffraction of electron microscopic images.

A number of mammals have crystalline structures in the cytoplasm of their oocytes. In some cases, these structures can be observed in fertilized ova or cleavage stage embryos. The oocytes of several rodents (e.g. rat, mouse, hamster, deer mouse) contain aggregates of filamentous, lamellar or plaque-like material which have a periodic substructure (10, 11). Although there are differences in structure between species, we will refer to all of them as lattice structures. These may occupy a considerable fraction of the cytoplasmic volume and are thought of as yolk material by some investigators (11, 12).

Lattice structures in the rat oocyte or cleaving embryos are seen as closely packed parallel arrays of linear elements. The individual elements are 25 nm in thickness and may be either fibrillar or, in some cases, plate-like (13). A periodicity of 25–30 nm is seen along the linear elements; one author (14) believes the periodic densities to be ribosomes and refers to the lattice elements as fibrillar ribosome aggregates. Schlafke and Enders (13) and Szollosi (11) have cited the lack of fluorescence in ooplasm following acridine orange staining, weak UV absorption and lack of affinity for uranyl salts in thin sections as evidence arguing against the lattice-like structures having RNA as a major component. Dvorak et al. (15), using morphometric techniques, determined that the lattices (lamellar structures) formed the most voluminous component of the cytoplasm of segmenting rat embryos at all stages from 1 cell to blastocyst. They noted a gradual decrease in the volume density of the lattices with successive cleavage divisions and cite this as evidence that the structures represent stored materials being utilized in early development.

The lattice structures of hamster oocytes have been studied extensively by Weakley (16–18). She describes those present in small oocytes as being lamellar structures of 11–21 nm diameter. With growth of the oocyte bilaminar structures appear, apparently from a pairing of individual lamellae. Individual lamellae of the pairs are separated by approximately 40 nm and have cross pieces regularly spaced at 40 nm giving rise to a 'cubic lattice' (18). The lattice structures aggregate to form a series of circular or horseshoe-shaped whorls in the cytoplasm which are associated with cytoplasmic glycogen particles. Hadek (19), who first described such whorls in the hamster, described them as deriving from an assembly of 25 nm particles spaced 4–5 nm apart. Although he noted that the particles were larger and less electron dense than ribosomes, Hadek allowed that the 'denser cores' of such particles could be RNA. Weakley (16) had documented the appearance of 'beaded' structures (prior to pairing to form bilaminar structures), but did not implicate ribosomal particles as a structural component.

As with the rat and hamster, mouse oocyte cytoplasm is heavily laden with lattice inclusions. We have examined the lattice structures of mouse oocytes and blastocysts and our observations are in general agreement with structural features reported by others (12, 17, 20, 21). Individual filaments in the lattice are 8–15 nm in diameter, depending on the plane of section, and adjacent units are separated by a lucent space of 15–20 nm (Fig. 3). Apparent crosslinks are present at 30–38 nm intervals. Calarco and Brown (20) and Nilsson (12) have viewed this lattice as reflecting a helical arrangement of one or more components and Zamboni (21) considered them to represent linear polyribosomes. Burkholder et al. (22) have examined the structure and sensitivity to enzymatic digestion of the lattices in whole mount preparations from mouse oocytes. Such lattices were described as being composed of strands of filaments ~12 nm in diameter connecting a series of 21 nm particles. Adjacent beaded strings were interconnected by crosslinks at ~36 nm intervals. The number of strings crosslinked to form a lattice varied from 2 to 8 or more. Enzyme digestions with RNase destroyed the lattices, while weak trypsin solutions appeared to remove the interconnections between the particles. Burkholder et al. (22) concluded that the mouse lattice structures represent ribosomes held together by proteins and stored in an inactive form until after fertilization. They failed to find the lattices in whole mounts prepared from mouse blastocysts and, from this, suggested a correlated disappearance of lattice structures and an increase in free ribosomes and polysomes during early development. Garcia et al. (23), using quantitative electron microscopic techniques, reported a comparable 'conversion' of free ribosomes to lattice-associated particles during growth of the mouse oocyte. Like Burkholder et al. (22), Garcia et al. (23) performed ribonuclease digestions to substantiate their contention that the lattice structures contained ribosomal material. There is, in fact, biochemical evidence which supports the notion that mouse oocytes contain a store of inactive ribosomes which are mobilized immediately after fertilization for protein synthesis (24). There is also evidence for stable maternal mRNA in mouse oocytes (25) which is activated for translocation shortly after fertilization and Burkholder et al. (22) discussed the possibility that such mRNA may be associated with the lattice structures. Schultz et al. (25), without reference to any particular cytoplasmic structures, likewise considered the possibility that these translationally inactive mRNAs could be packaged in mRNP particles as in sea urchin eggs. An unresolved problem with this view is that the temporal correlation between 'activation' of the maternally derived RNA and the dispersal of cytoplasmic lattice structures after fertilization cited by Burkholder et al. (22) does not hold. Contrary to the belief of these authors, lattice structures are present in the mouse in significant quantities throughout cleavage and on a strictly structural basis, it is difficult to think of the mouse lattice structures as strings of ribosomes. Burkholder et al. (22) in discussing their results noted that free ribosomes were found to be more electron-dense and somewhat larger than 'stored ribosomes' within a chain. They suggest that the association of ribosomes with lattice proteins decreases the electron density of the ribosomes and perhaps alters their configuration. From our limited observations on thin sections of the mouse lattice structures we find it difficult to envision the linear units (filaments?) as strings of ribosomal particles. At moderately high magnifications the thickened regions of the filaments even appear to be somewhat electron-lucent (see Inset, Fig. 3). In a recent study by Piko and Clegg (26), the authors also argue against the notion that the lattice structures are a storage form of ribosomes. They note that the number of recognizable cytoplasmic ribosomes accounts for the bulk of the RNA content in mouse embryos at all stages. They calculated that if the lattice particles trule represented ribosomes, the ribosomal RNA content of the mouse oocyte would be 20 fold greater than the amount actually measured. Furthermore, Piko and Clegg (26) observed that the lattice structures were little affected by postfixation alkali treatment, while the staining of cytoplasmic ribosomes was completely abolished by the treatment.

The lattice structures described above for oocytes of several laboratory rodents appear not to have any regular association with other cytoplasmic organelles. Lattice or lamellar structures found in the pocket gopher and spiny mouse differ in this regard since both are intimately associated with components of the endoplasmic reticulum. In the pocket gopher the lattice takes the form of a series of cylindrical or conical shaped lamellae (27). Although the lattice structures appear to form independent of the endoplasmic reticulum, those in oocytes of larger follicles are wrapped around a single tubular element of the endoplasmic reticulum. Kang and Anderson (28) have described the development of lattice structures ('yolk plates') in the spiny mouse oocyte. In relatively small follicles, stacks of flattened elements of the endoplasmic reticulum are formed into 'lamellar complexes'. Subsequently, filaments exhibiting periodicity appear in the narrow (31 nm) intercisternal space; such filaments are composed of a chain of interconnected particles 15 nm in diameter. These beaded filaments are viewed as precursors of ribosomal fibrils later reported to be observed as stacked structures ('yolk plates') free in the ooplasm, i.e. independent of the ER 'lamellar structures'. The authors discuss this developmental sequence with reference to 'crystallized ribosomes' in hypothermic chick embryo cells (29). That the 'ribosomal fibrils' of the spiny mouse should be considered as a variation of the lattice structure of 'yolk plates' seen in other rodents seems, to us, less then certain. The lattice of particles ('crystallized ribosomes') reportedly set free in the ooplasm as illustrated by Kang and Anderson (28) could also be interpreted as tangential sections through particle-studded ER membranes. If so, this would appear to put these structures in a somewhat different category than the lattice structures described in other rodents.

A crystal of relatively massive proportions is present in the oocytes of the leaf-nosed bat, *Macrotus californicus* (30). Each ovum has a single crystal which increases in size with follicular maturation to reach dimensions of 6.5×20 μm in oocytes contained in Graafian follicles. The crystalline structure is four-sided in the central region, but tapers to a three-sided pyramid on either end. It is not membrane-limited, has a periodic structure at the electron microscopic level and is thought to be proteinaceous based on light microscopic staining characteristics. Of particular interest is Bleier's (30) observation that the crystal is present in cleavage stages of the embryo. It is retained by a single cell following successive mitotic divisions – at least through the morula stage. Thus unequeal segregation of cytoplasmic components occurs from the first cleavage division in the embryo. The crystal disappears from most embryos prior to implantation, however, in the one implanting blastocyst observed which still retained the crystal it was located in the trophoblast.

Crystalline inclusions have been described in the oocytes of certain primates which could conceivably be similar in structure and composition to those reported in rodents. Both Rhesus monkey (31) and human oocytes (32) have lamellae or bundles of filaments with periodic substructure. They occur in much lower frequency than the lattice structures of rodents, however, and meaningful comparison must await more detailed analysis.

4. Preimplantation embryos

Cytoplasmic crystalline inclusions have been described in cleaving embryos, morulae or blastocysts of several mammalian species: leaf-nosed bats, mice, rats, hamsters, rabbits, sheep and the baboon. Of these, only the inclusions of the rabbit and mouse have been studied under experimental conditions which might provide information on their developmental significance.

The large single crystal in oocytes of the leaf-nosed bat is present in a single trophoblast cell when it persists as late as the blastocyst stage (30). Likewise, the lattice structures (lamellae, plaques, etc.) of rodent oocytes are found in cleavage stage embryos and in blastocysts (10, 12), although in no instance do these structures appear to be present after the initiation of ovo-implantation. In at least one of these rodents, the mouse, an additional crystalline structure makes its appearance during the preimplantation period.

Cytoplasmic crystals on the order of 0.5–1.0 μm in diameter appear in mouse embryos at early cleavage stages. The inclusions are prominent from the 8-cell stage on, becoming more numerous in morulae and early blastocysts (20, 33). In thin sections, the crystal presents three different patterns depending on the plane of section, two of which are illustrated in Figure 4. Each crystal contains parallel arrays of elongated plates each of which is ~60 nm wide and up to 1 μm in length. These plates, in turn, may present the appearance of either a series of stacked bilaminar elements or of a honeycomb of hexagonal structures. The bilaminar structure appears to be the major 'unit' of the crystal. In profiles such as that in Figure 4, each unit measures 16–20 nm by ~60 nm, the width of the plates. At higher magnifications, the bilaminar units are seen to be composed of electron dense particles of 6–7 nm diameter attached to one another by thin bridge structures giving the overall appearance of paired beaded filaments. Individual bilaminar units within each stack are separated by electron lucent spaces of 11–12 nm. The three dimensional ordering of these elements required to construct a single crystalline inclusion has not been investigated. To our knowledge, the crystals have not been isolated for either structural or chemical analysis. A study by Izquierdo et al. (34) reported that some, but not all, of the crystals within blastomeres of cleaving mouse embryos had strong alkaline phosphatase activity. The only other site in the embryos with similar activity was the plasma membrane between adjacent blastomeres. These authors postulated a possible relationship between the two alkaline phosphatase-positive structures; the crystals were regarded as being possible 'stores of cortical material indirectly related to the cell membrane through a system of microtubules and microfilaments' (34). All investigators who have noted the crystals have reported a close association with elements of granular endoplasmic reticulum (Fig. 3). Some authors regard them as being derived from the lattice structures (12) although no direct evidence for this has been published.

While the crystals are evident in preimplantation mouse blastocysts they apparently disappear during ovo-implantation. They are absent from mouse embryos studied during experimental delay of implantation (10). The oviductal or uterine environment is not required for development of crystals since they are abundant in blastocysts cultured from the two-cell stage *in vitro* (35). Aggregations of crystalline structures up to 10 μm in diameter appear in superovulated ova exposed to cytochalasin B (36). Moskalewski et al. (36) regard these to be aggregates of the same crystalline structures appearing during normal development. Puromycin did not inhibit the cytochalasin-induced development of the aggregates and they had no relation to components of the endoplasmic reticulum. Crystals are scarce or abnormal at the blastocyst stage of parthenogenetic mouse embryos (37) or those homozygous for the t^{12} mutation (38). In these cases there is a reduced capability for continued development although the correlation of this impairment with absence of crystals is of unknown significance.

Perhaps the best characterized of the embryonic crystalline inclusions are those occurring in trophoblast cells of the rabbit blastocyst. These were first described at the ultrastructural level by Hadek and Swift (39) and more recently by Daniel and Kennedy (40) and Hoffman and Olson (41). Although Hadek and Swift made reference to crystals in all cells of the rabbit blastocyst, we have found them to be uniformly present in trophoblast cells but only rarely, in atypical aggregates, in embryonic shield or endodermal cells. Van Blerkom et al. (42) likewise reported an absence of crystals in the inner cell mass. Light microscopic observations on blastocyst flat mounts stained with fast green FCF showed that the rod-shaped inclusions were present in all trophoblast cells including those in mitosis (41). The crystalline inclusions varied in size, most having diameters of $<1\ \mu m$ and lengths of 4–7 μm. In flat mounts the inclusions were randomly oriented in most cells and the majority was clustered in perinuclear regions leaving the cell margins relatively free of inclusions (Fig. 5). In occasional cells which were elongated, the inclusions appear to be preferentially oriented parallel to the long axis of cells. The inclusions are absent in rabbit embryos of cleavage or morula stages (43); they first appear between 4 and $4\frac{1}{2}$ days p.c., shortly after the blastocyst has entered the uterus (40). Interestingly, the crystals do not develop in embryos grown *in vitro* (42) nor in embryos experimentally retained in the oviduct through day $4\frac{1}{2}$. They do, however, appear in such tube-retained embryos after transfer to the uterine lumen (40), implicating the uterine environment as an essential factor in the development of blastocyst crystals.

Ultrastructural studies had suggested that both rabbit trophoblastic (40, 43, 44) and uterine (45) crystals contained tubular components which were possibly microtubular in nature. Hoffman and Olson (41) provided further evidence for this. As was the case with uterine glandular inclusions, the membrane-enclosed blastocyst inclusions were composed of longitudinally-arranged linear units with angled cross-striations at intervals of ~ 16 nm. They varied in density, seemingly dependent on the tightness of packing of subunits (Fig. 7). In cross section the inclusions have a hexagonal or tubular appearance (Inset, Fig. 7). When preserved with glutaraldehyde containing tannic acid a series of 11–14 electron-lucent filaments was observed to comprise the tubular wall (Fig. 9). Using the Markham rotation procedure, reinforcement of the image was usually observed at 360°/13, suggesting the presence of 13 protofilaments as found in most microtubules (41). Blastocyst homogenate fractions enriched in the crystalline inclusions were subjected to SDS-PAGE and revealed a prominent band at a mobility corresponding to that of purified tubulin (m.w. ~ 55000). Similarly, Van Blerkom and O'Farrell (cited in 45) have observed several protein bands in this molecular weight range in electrophoretic gels of isolated crystals. These results strongly implicate microtuble protein as being a major constituent of the rabbit inclusions. Negative-stained preparations obtained from lysed blastocysts revealed the tubular subunits to be somewhat different than most other (cytoplasmic) microtubules due to an apparent coating of helically-arranged accessory material (Fig. 8). Additional studies were carried out on 6-day blastocysts utilizing cold shock and colchicine treatment. The putative microtubules appeared not to be cold sensitive nor were they rapidly disaggregated by incubation in colchicine-containing media. The fact that some depolymerization of crystals took place in control media (41), may reflect imperfections in the incubation conditions utilized. Apparently, once blastocysts have developed crystalline inclusions *in vivo*, they can be cultured (under more optimal conditions) *in vitro* for up to 48 h without appreciable loss in numbers of the inclusions (Dr. J.C. Daniel, Jr., personal communication).

Both the origin and function of rabbit inclusions are unclear. Daniel and Kennedy (40) suggested that the crystals arise from granular vesicles in the cytoplasm which are present as early as $2\frac{1}{2}$ days p.c. Apparent polymerication of crystalline material was evidenced in such granular vesicles once exposed to the uterine environment. Whether or not these vesicles are equivalent to the 'flocculent vesicles' of rabbit oocytes and early cleavage stage embryos (42, 45, 46) was not discussed by the authors (40). We have likewise interpreted linear aggregates of crystalline material in large vesicles of blastocysts on days 5 or 6 (Fig. 6) to represent polymerization (41), although such images could also represent depolymerization of previously formed crystalline inclusions. Our study did not provide further information on the origin of the granular material in such vesicles. It seems unlikely to us that the mass of material ultimately developing into the inclusions could be present in the oocyte; thus, consideration of such precursors as 'yolk material' may be unwarranted. More acceptable is the notion that the precursor materials for the crystals are synthesized from substrates available in oviductal and/or uterine secretory material. The presence of comparable inclusions in uterine glandular epithelial cells might support this.

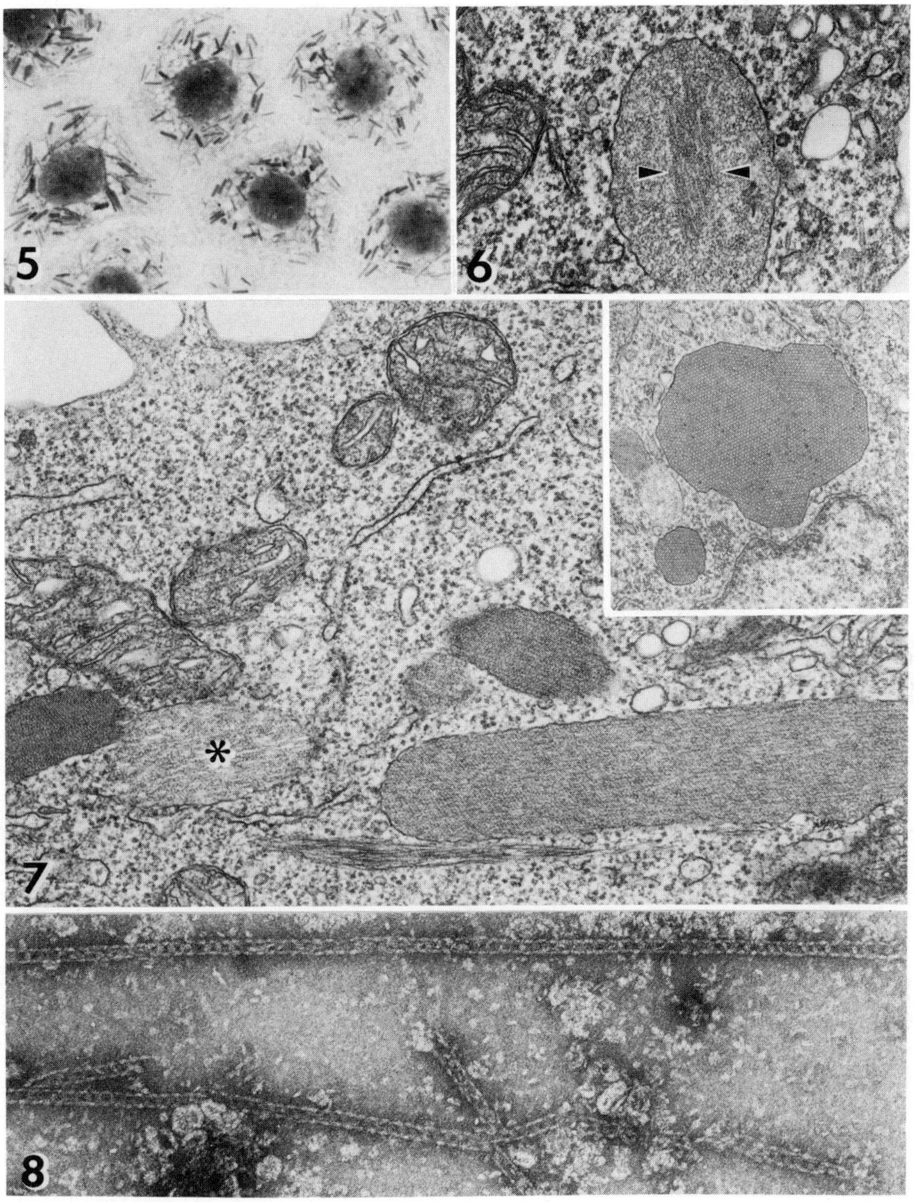

Figs. 5–8. Flat mount of rabbit blastocyst on day 6. Numerous crystalline inclusions pack the perinuclear region of each trophoblast cell while peripheral margins are relatively free of crystals. Fast green FCF (× 440). (6) Rabbit trophoblast. Granular vesicles appear to contain polymerizing, or depolymerizing, elements of crystalline inclusions (between arrowheads) (× 32,000). Hoffman, Olson, Exp Cell Res 127: 1. (7) Profiles of membrane-bounded crystals of rabbit trophoblast. In longitudinal section (lower right) the linear elements with angled cross-striations are shown. Loosely packed inclusion at (*) (× 33,600). Hoffman, Olson, Exp Cell Res 127: 1. Inset: Cross section of inclusions showing hexagonal packing of tubular units (× 28,000). (8) Negatively-stained preparation of individual tubular units of rabbit crystals from lysed trophoblast. The regular spacing along tubules is interpreted as a double helix of coating material; this gives rise to the ladder-like effect (× 92,000). Hoffman, Olson, Exp Cell Res 127: 1.

Furthermore, crystals appear in the uterine glandular epithelium at the same time (day 4 p.c.) as those of the blastocyst and this is near the stage of pregnancy when rabbit endometrial tubulin reaches its peak level (47). There is, however, no reason to believe that the uterine inclusions are actually secreted.

At present there is little experimental evidence which bears on the possible developmental significance of the crystals. There is the suggestion that the inclusions decrease in number during ovo-implantation (41). Furthermore, along with Steer (44), we have witnessed apparent depolymerization of crystalline inclusions in the syncytial trophoblastic knobs during implantation. These knobs fuse with, then

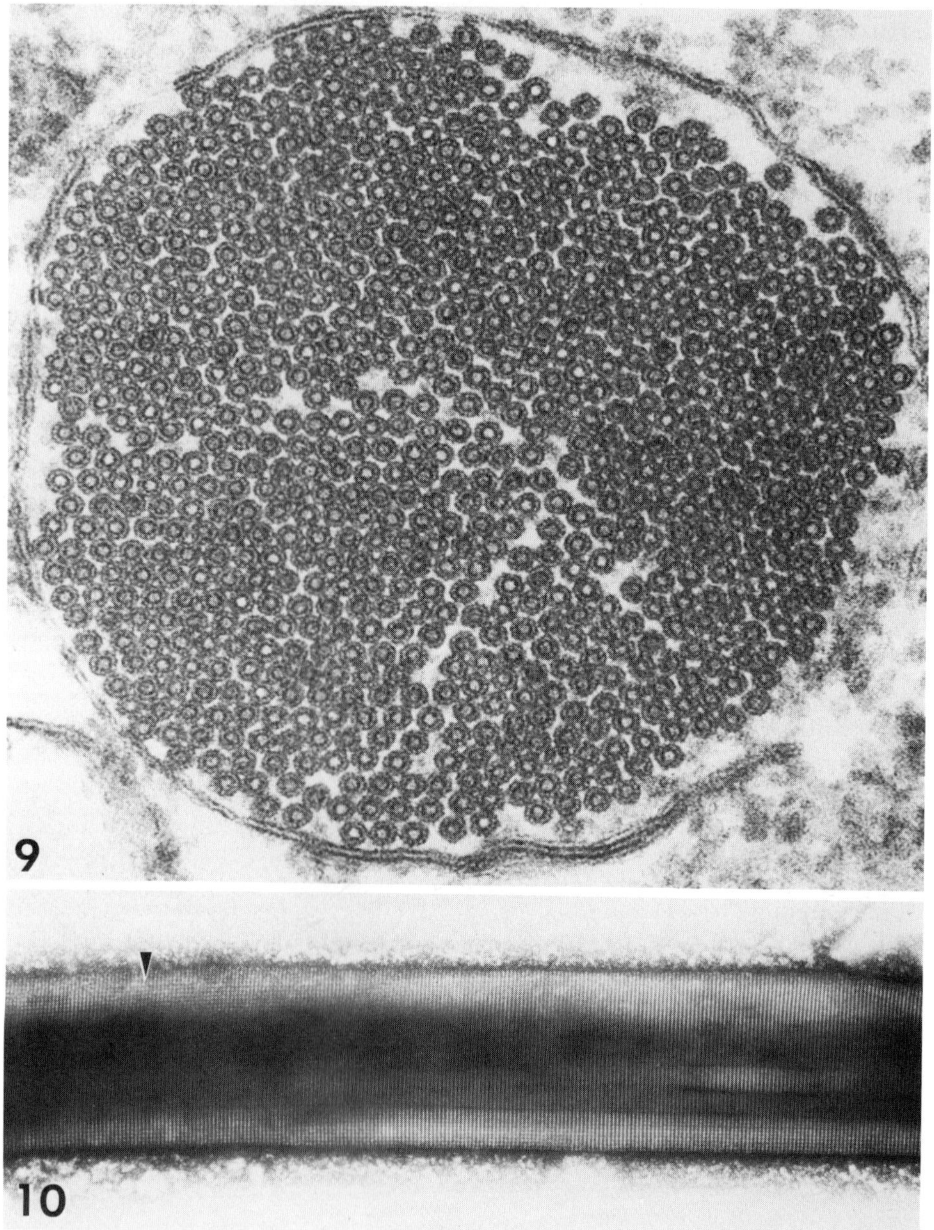

Fig. 9–10. (9) Cross section of rabbit trophoblast crystal from blastocyst preserved in glutaraldehyde/tannic acid. Tubular walls are composed of 11–14 electron-lucent subunits (× 164,000). Hoffman, Olson, Exp Cell Res 127: 1. (10) Negatively-stained crystalline inclusion from lysed trophoblast cell of the sheep. Transverse lines with a period of 5–6 nm are apparent. Longitudinal lines are present (e.g. at arrowhead) but less obvious. This inclusion appears to have a central core of non-crystalline material (× 85,600).

invade the endometrium on the antimesometrial and lateral aspects of the implantation chamber beginning on day 7 p.c. (48). Ultimately, they invade uterine blood vessels beneath the epithelium. We have noted relatively large numbers of microtubules in the cytoplasm of these knobs. Their appearance may coincide temporally with depolymerization of crystalline inclusions, although we hasten to add that this is a preliminary observation.

Large crystalline inclusions are also present in sheep blastocysts. These inclusions, the 'Stabchen' of Bonnet (49) are seen as rhomboidal or needle-shaped acidophilic inclusions of the trophoblast cytoplasm (6). They are also reported to occur in the uterine epithelium during early pregnancy and as a component of the 'uterine milk' (6, 50). The crystals are first seen in chorionic epithelial cells within membranous vesicles ('lysosome-like bodies') on day 10 of

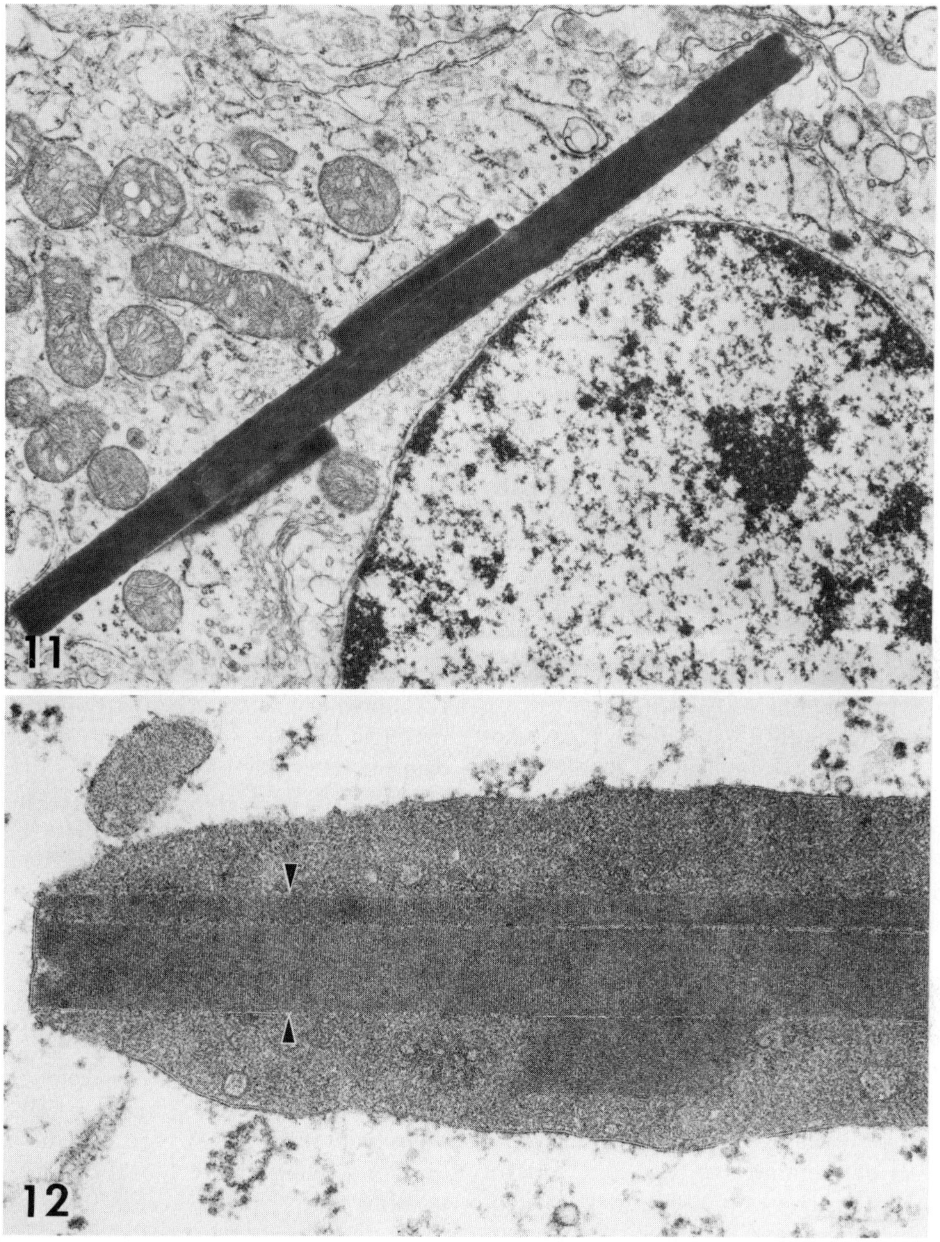

Figs. 11–12. (11) Thin section of needle-shaped crystal from sheep trophoblast. Periodicity is not apparent on this micrograph (× 16,000). (12) Higher magnification of a needle-shaped crystal of sheep trophoblast preserved in tannic acid-containing fixative. The crystal (between arrowheads) in this inclusion is surrounded by a cortex of amorphous material. Transverse lines repeating at 5–6 nm intervals can be seen in the crystal (× 56,000).

gestation according to Wintenberger-Torres and Fléchon (51). These authors noted larger crystals in 14- and 16-day blastocysts and reported a lattice periodicity of ~20 nm. At the time of earliest attachment of trophoblast to the epithelium overlying uterine caruncles, day 16–17, crystal-containing cells were more numerous in abembryonic portions of the chorionic vesicle (51).

We have made ultrastructural observations on the inclusions present in trophoblastic cells of conceptuses from two ewes at relatively early stages of gestation (fetal crown-rump lengths of 6 and 28 mm). A brief description based on these preliminary observations follows.

The sheep inclusions are of two general types, both of which are membrane-limited and are occasionally found in aggregates within a common vesicle which may be comparable to the lysosome-like bodies of

Wintenberger-Torres and Fléchon (51). One type is cubic or rhomboidal while the other is more needle-like in form (Fig. 11). A crystalline lattice is more apparent in the needle-shaped inclusions and our description is based primarily on this type inclusion. Whether or not the larger, rhomboidal, structures represent a somewhat different arrangement of comparable material is at present unexplored.

The crystalline structure within the inclusion membrane frequently shares the enclosed space with amorphous material. Such material may fill a core-like area, separate adjacent crystals or, in some cases, be arranged as a cortex around the crystal (Fig. 12). The lattice shows a transverse series of lines with a repeat of 5–6 nm in thin-sectioned preparations. The lattice, however, is more easily studied in isolated crystals prepared from homogenates of sheep trophoblast. Negatively-stained preparations of this material (Fig. 10) show the transverse dense lines, spaced at 5.5 nm, and a longtitudinally arranged series of lines, also having a period of $\sim$5.5 nm.

The crystals of sheep trophoblast have been referred to as lysosome-like (51), however, we are unaware of evidence indicating acid hydrolase activity. Boshier (52) used histochemical methods for demonstration of acid phosphatase activity in the sheep placentome. No evidence was given in that study to indicate significant quantities of this lysosomal enzyme in the trophoblast. Nevertheless, there would appear to be a structural resemblance between these crystal-containing inclusions in the sheep and the specific granules in eosinophil leukocytes of mammals (see for e.g. Fig. 12). The crystal-containing eosinophil granules are regarded as being lysosomal (53). Perhaps biochemical or histochemical tests for other enzymes present in eosinophil granules (e.g. cathepsin, peroxidase, B-glucuronicase) would be instructive in determining the functional role of the 'Stäbchen'. We have witnessed no consistent association of the inclusions with particular cytoplasmic organelles although many micrographs suggest an association of the needle-shaped inclusions with cytoskeletal filaments.

We are aware of only a single indication for crystalline inclusions in primate blastocysts; this being a brief reference and a single illustration of such a structure in the baboon (54).

5. Placenta and fetal membranes

The crystalline inclusions just described for the sheep blastocyst are present, at least through early gestation, as part of the placentome (55). Heterogeneous inclusions containing cylindrical arrays of myelin-like lamellae are present in trophoblast of interplacentral areas in late pregnancy (56). These appear to be somewhat different structures from the crystalline inclusions described in the previous section. The lattice structures in various rodents and the larger crystalline inclusions of the mouse blastocyst are no longer apparent after implantation and those of the rabbit blastocyst appear not to be present in trophoblastic components of the chorio-allantoic placenta. The latter structures are occasionally encountered in obplacental giant cells, at least through mid-pregnancy.

A crystalline arrangement of cytoplasmic membranes has been described in the yolk sac endoderm of the bat, *Tadarida*, during later gestation (57). There is apparent continuity of the multiple membranes comprising the inclusion with elements of the endoplasmic reticulum. Ollerich (58) reported the occurrence of intramitochondrial crystalline structures in the rat placenta. The crystals had a well ordered, gridlike lattice of symmetrical quadrilateral components which had center-to-center spacing of 37 nm. The crystals were seen only in mitochondria of element III (syncytiotrophoblast) of the chorioallantoic placenta and were encountered more frequently near the end of gestation. Ollerich discussed the possibility that the crystals are associated with an age-related, degenerative process in this region of the placenta.

6. Summary

In this review we have concentrated on certain crystalline structures of mammalian oocytes and preimplantation embryos on which sufficient structural and, occasionally, biochemical data is available to allow tentative conclusions to be made regarding their developmental significance. These have included the lattice structure of rodent oocytes, crystalline inclusions of mouse blastocysts, rod-shaped crystals of rabbit trophoblast and the 'Stäbchen' of sheep trophoblast. All of these have been thought of as storage forms of material to be utilized by the developing embryo. The rodent lattice structures have been designated as yolk material by a variety of authors (11, 12, 15). It seems apparent that the lattices are indeed a depot of material to be employed in early development, although they may not conform in the strictest sense to definitions of 'true yolk' such as that given by Waddington (59). In spite of the number of articles identifying the rodent inclusions as ribosomal in nature (14, 21, 11, 23, 28), we find the evidence available at present insufficiently compelling to consider the matter settled. While there is evidence that the lattices contain protein and may be susceptible to ribonuclease digestion, considering them to be storage forms of ribosomes seems premature. Definitive evidence as to their composition should include isolation of the lattices by cell fractionation and more rigorous biochemical and structural analysis. The larger and more complex crystalline bodies appearing in mouse cleavage stage embryos and blastocysts are even less well characterized. The close association of these crystals with the granular ER and the possibility that their absence is somehow correlated with alterations in protein synthesis in mouse parthenotes and certain lethal mutants imparts to them potential significance for early development.

Of the crystalline inclusions discussed here the only one which has been at least partially isolated for biochemical analysis is the trophoblastic inclusion of rabbit blastocysts. Although questions as to their origin and ultimate fate remain as exciting problems for future inquiry, the available evidence strongly suggests that they are composed primarily of aggregates of microtubules. These crystals occupy a significant portion of the trophoblastic cytoplasm and, considering that cytoplasmic microtubules are likewise abundant, the trophoblast must contain relatively large quantities of microtubular protein. Weisenberg (60) has reported the presence of polymorphic aggregates of tubulin in eggs of the surf clam which break down just before formation of the meiotic spindle. Although the mitotic rate is high in the rabbit blastocyst (61), there is no direct evidence that the crystals are utilized as a source of tubulin for mitotic spindles (41). It is of interest that in many cell types tubulin synthesis decreases rapidly in response to agents which increase the cytoplasmic monomer pool of tubulin (62). This decline in synthesis is associated with a decrease in amounts of both α- and β-tubulin mRNAs; thus, the monomer pool may regulate the rate of tubulin mRNA transcription. We have speculated that large amounts of microtubules may be associated with the process of trophoblastic knob penetration of endometrial tissue during early ovo-implantation in the rabbit. Perhaps in the rabbit blastocyst the isolation of large quantities of microtubular protein as membrane-enclosed aggregates allows the cytoplasmic monomer pool to remain at a lower level, the ultimate effect of which would be a relatively massive accumulation of total microtubule protein for utilization both in continued mitosis and in implantation-related functions of the trophoblast.

The numerous electron-dense crystalline inclusions characterizing sheep trophoblast and, presumably, certain endometrial cells during early gestation likewise remain structures without a known function. Here too, we would expect isolation and detailed structural/biochemical analysis to be necessary in elucidating their composition. We have cited superficial similarities with lysosomal bodies such as found in certain leukocytes and suggest that initial studies into the nature of the crystals be directed toward their possible content of acid hydrolases.

Acknowledgements

We are grateful to Virginia Winfrey for skillful technical assistance, to Dr. Charles Torbit for aid in collecting mouse oocytes and blastocysts, and to Vera Henley for typing the manuscript.

References

1. Nakao K, Meyer CJ, Noda Y: Progesterone-specific protein crystals in the endometrium: An electron microscopic study. Amer J Obstet Gynecol 111: 1034–1038, 1971.
2. O'Malley BW: Hormonal regulation of nucleic acid and protein synthesis. Trans NY Acad Sci 31: 478–503, 1969.
3. Beier HM, Petry G, Kuhnel W: Endometrial secretion and early mammalian development. In: Mammalian Reproduction. Gibian H, Plotz EJ (eds), New York, Springer-Verlag, 1970, pp 264–287.
4. Wynn RM: Histology and ultrastructure of the human endometrium. In: Biology of the Uterus, Wynn RM (ed), New York, Plenum, 1977, pp 341–376.
5. Hoffman LH, Davies J, Long VD: Hormone-induced crystals and intramitochondrial lamellae in uterine epithelium. In: Electron microscopic concepts of secretion. Hess M (ed), New York, John Wiley & Sons, 1975, pp 99–111.
6. Wimsatt WA: Observations on the morphogenesis, cytochemistry and significance of the binucleate giant cells of the placenta of ruminants. J Anat 89: 233–282, 1951.
7. Watzka M: Das Ovarium. In: Handbuch mikr Anat Mensch. 7/3, Möllendorff W, Bargmann W (eds), Berlin, Springer- Verlag, 1957.
8. Merkow LP, Slifkin M, Acevedo HF, Pardo M, Greenberg WV: Ultrastructure of an interstitial (hilar) cell tumor of the ovary. Obstet Gynecol 37: 845–859, 1971.
9. Nagano T, Ohtsuki I: Reinvestigation of the fine structure of Reinke's crystal in the human testicular interstitial cell. J Cell Biol 51: 148–161, 1971.
10. Enders AC, Schlafke S: The fine structure of the blastocyst: Some comparative studies. In: Preimplantation stages of pregnancy. Wolstenholme GEW, O'Connor M (eds), Boston, Little, Brown & Co, 1965, pp 29–59.
11. Szollosi D: Changes of some cell organelles during oogenesis in mammals. In: Oogenesis. Biggers JD, Schuetz AW (eds), Baltimore, University Park Press, 1972, pp 47–64.
12. Nilsson BO: Comparative ultrastructure of the yolk material in preimplantation stages of the hamster, mouse, and rat embryos. Gamete Res 3: 369–377, 1980.
13. Schlafke S, Enders AC: Cytological changes during cleavage and blastocyst formation in the rat. J Anat 102: 13–32, 1967.
14. Mazanek K: Submikroskopische Veränderungen während der Furchung eines Säugetiereies. Arch Biol Liège 76: 49–85, 1965.

15. Dvorák M, Trávnik P, Stanková J: A quantitative analysis of the incidence of certain cytoplasmic structures in the ovum of the rat during cleavage. Cell Tiss Res 179: 429–437, 1977.
16. Weakley BS: Investigation into the structure and fixation properties of cytoplasmic lamellae in the hamster oocyte. Z Zellforsch mikr Anat 81: 91–99, 1967.
17. Weakley BS: Comparison of cytoplasmic lamallae and membranous elements in the oocytes of five mammalian species. Z Zellforsch mikr Anat 85: 109–123, 1968.
18. Weakley BS: Initial stages in the formation of cytoplasmic lamellae in the hamster oocyte and the identification of associated electron-dense particles. Z Zellforsch mikr Anat 97: 438–448, 1969.
19. Hadek R: Cytoplasmic whorls in the golden hamster oocyte. J Cell Sci 1: 281–285, 1966.
20. Calarco P, Brown H: An ultrastructural and cytochemical study of preimplantation development of the mouse. J exp Zool 171: 253–284, 1969.
21. Zamboni L: Ultrastructure of mammalian oocytes and ova. Biol Reprod Suppl 2: 44–63, 1970.
22. Burkholder GD, Comings DE, Okada TA: A storage form of ribosomes in mouse oocytes. Exp Cell Res 69: 361—371, 1971.
23. Garcia RB, Pereyra-Alfonso S, Sotelo JR: Protein-synthesizing machinery in the growing oocyte of the cyclic mouse. Differentiation 14: 101–106, 1979.
24. Bachvarova R, DeLeon V: Stored and polysomal ribosomes of mouse ova. Devel Biol 58: 248–254, 1977.
25. Schultz GA, Clough JR, Braude PR, Pelham HRB, Johnson MH: A reexamination of messenger RNA populations in the preimplantation mouse embryo. In: Cellular and molecular aspects of implantation. Glasser SR, Bullock DW (eds), New York, Plenum, 1981, pp 137–154.
26. Pikó L, Clegg KB: Quantitative changes in total RNA, total poly (A), and ribosomes in early mouse embryos. Devel Biol 89: 362–378, 1982.
27. King BF, Tibbitts FD: Ultrastructural observations on cytoplasmic lamellar inclusions in oocytes of the rodent, Thomomys. Anat Rec 189: 263–272, 1977.
28. Kang Y-H, Anderson WA: Ultrastructure of the oocytes of the Egyptian spiny mouse (Acomys cahirinus). Anat Rec 182: 175–200, 1975.
29. Biagini G, Simoni P, Maraldi NM, Bersani F, Barbieri M: Ribosome crystallization. IV. Biochemical and ultrastructural investigation on the uptake of uridine-^{3}H in ribosomes of hypothermic chick embryos. J Submicr Cytol 4: 95–100, 1973.
30. Bleier WJ: Crystalline structure in the ova and early embryological stages in a leaf-nosed bat, Macrotus californicus. J Mammalogy 56: 235–238, 1975.
31. Hope J: The fine structure of developing follicles of the Rhesus ovary. J Ultrastruct Res 12: 592–610, 1965.
32. Wartenberg H, Stegner HE: Uber die elektronenmikroskopisch Feinstrucktur des menschlichen Ovarialeis. Z Zellforsch mikro Anat 52: 450–474, 1960.
33. Hillman N, Tasca RJ: Ultrastructural and autoradiographic studies of mouse cleavage stages. Amer J Anat 126: 151–174, 1969.
34. Izquierdo L, Lopez T, Marticorena P: Cell membrane regions in preimplantation mouse embryos. J Embryol exp Morph 59: 89–102, 1980.
35. McReynolds HD, Hadek R: A comparison of the fine structure of late mouse blastocysts developed *in vivo* and *in vitro*. J exp Zool 182: 95–117, 1972.
36. Moskalewski S, Sawicki W, Gabara B, Koprowski H: Crystalloid formation in unfertilized mouse ova under influence of cytochalasin B. J exp Zool 180: 1–12, 1972.
37. Van Blerkom J, Runner MN: The fine structural development of preimplantation mouse parthenotes. J exp Zool 196: 113–123, 1976.
38. Calarco PG, Brown EH: Cytological and ultrastructural comparisons of t^{12}/t^{12} and normal mouse morulae. J exp Zool 168: 169–186, 1968.
39. Hadek R, Swift H: A crystalloid inclusion in the rabbit blastocyst. J Biophys Biochem Cytol 8: 836–841, 1960.
40. Daniel JC, Kennedy JR: Crystalline inclusion bodies in rabbit embryos. J Embryol exp Morph 44: 31–43, 1978.
41. Hoffman LH, Olson GE: Crystalline inclusions in the rabbit blastocyst. Evidence for microtubular aggregates. Exp Cell Res 127: 1–14, 1980.
42. Van Blerkom J, Manes C, Daniel JC: Development of preimplantation rabbit embryos *in vivo* and *in vitro*. Devel Biol 35: 262–282, 1973.
43. Hesseldahl H: Ultrastructure of early cleavage stages and preimplantation in the rabbit. Z Anat Entwickl-Gesch 135: 139–155, 1971.
44. Steer HW: The trophoblastic knobs of the preimplanted rabbit blastocyst: a light and electron microscopy study. J Anat 107: 315–325, 1970.
45. Van Blerkom J, Motta P: The Cellular Basis of Mammalian Reproduction. Baltimore, Urban & Schwarzenberg Inc, 1979.
46. Gulyas B: Nuclear extrusion in rabbit embryos. Z Zellforsch mikr Anat 120: 151–159, 1971.
47. Fujimoto GI, Saldana LR, Gaskin F: Uterine tubulin production during early pregnancy in the rabbit. Endocrine Res Commun 3: 219–229, 1976.
48. Enders AC, Schlafke S: Penetration of the uterine epithelium during implantation in the rabbit. Amer J Anat 132: 219–240, 1971.
49. Bonnet R: Präparate und Zeichnungen zur Entwicklungsgeschichte des Schafes. Arch Anat Physiol anat Abt pp 1–106, 1888.
50. Amoroso EC: Placentation. In: Marshall's physiology of reproduction, Vol 2. Parkes AS (ed), Boston, Little, Brown & Co, 1952, pp 127–311.
51. Wintenberger-Torrès S, Fléchon JE: Ultrastructural evolution of the trophoblast cells of the pre-implantation sheep blastocyst from day 8 to day 18. J Anat 118: 143–153, 1974.
52. Boshier DP: A histological and histochemical examination of implantation and early placentome formation in sheep. J Reprod Fert 19: 51–61, 1969.
53. Zucker-Franklin D: Eosinophil function and disorders. Adv Int Med 19: 1–25, 1974.
54. Panigel M, Kraemer DC, Kalter SS, Smith GC, Heberling RL: Ultrastructure of cleavage stages and preimplantation embryos of the baboon. Anat Embryol 147: 45–62, 1975.
55. Davies J, Wimsatt WA: Observations on the fine structure of the sheep placenta. Acta anat 65: 182–223, 1966.
56. Lawn AM, Chiquoine AD, Amoroso EC: The development of the placenta in the sheep and goat: an electron microscope study. J Anat 105: 557–578, 1969.
57. Stephens RJ, Easterbrook N: Ultrastructural differentiation of the endodermal cells of the yolk sac of the bat, Tadarida brasiliensis cynocephala. Anat Rec 169: 207–242, 1971.
58. Ollerich DA: An intramitochondrial crystalloid in element III of rat chorioallantoic placenta. J Cell Biol 37: 188–191, 1968.
59. Waddington CH: Principles of embryology. London, George Allen & Unwin, Ltd, 1962, p 36.
60. Weisenberg RC: Changes in the organization of tubulin during meiosis in the eggs of the surf clam, Spisula solidissima. J Cell Biol 54: 266–278, 1972.
61. Moog F, Lutwak-Mann C: Observations on rabbit blastocysts prepared as flat amounts. J Embryol exp Morph 6: 57–67, 1958.
62. Cleveland DW, Lopata MA, Sherline P, Kirschner MW: Unpolymerized tubulin modulates the level of tubulin mRNAs. Cell 25: 537–546, 1981.

Authors' address:
Department of Anatomy
Vanderbilt University School of Medicine
Nashville, TN 37232, USA

CHAPTER 20

Cellular aspects of implantation

DANIEL J. CHÁVEZ

1. Introduction

Implantation is an essential and integral component of the reproductive process in all mammals except the *Monotremes* (egg layers). Several reviews treating morphological (19, 68, 25), comparative (83), and evolutionary (1) aspects of implantation have appeared in recent years. This chapter focuses on cellular and molecular interactions between the trophoblast and uterus of *Metatherian* (placental) mammals. Most of the present knowledge about implantation has been derived from experimentation with laboratory and domestic species, with occasional observations from wild animals and humans. There are fundamental similarities of implantation that are common to all mammals, and these similarities are discussed herein. A few examples of variation are cited for reasons of contrast, but for the most part, atypical variations are excluded.

2. The pre-implantation period

Embryos of the elephant shrew (*Elephantulus myurus*) initiate the implantation process prior to the blastocyst stage (83) and the chorionic vesicle of the pig reaches one meter in length before beginning implantation. However, although considerable variation exists among species studied, the general pattern of cellular differentiation is fundamentally similar. The majority of mammalian embryos begin implantation at the blastocyst stage. All mammalian blastocysts are organized similarly and consist of two phenotypically divergent cell types. An outer investing layer of trophectoderm cells, destined to become the ectoderm of fetal membranes and the placenta, surrounds an excentrially placed inner cell mass, destined to become the embryo proper, and a fluid-filled blastocyst cavity.

2.1. The oviductal phase of development

There are both wide ranges in the length of time during which embryos remain in the oviduct and wide differences in the developmental stages attained before the embryo arrives in the uterus. In most mammals, the oviductal period is relatively short, and, with the exception of a few species of bats, embryos in the majority of mammals reach the uterus in preblastocyst stages. The greater part of pre-implantation development occurs in the uterus, for reasons probably reflecting the different nutritive capacities of the oviduct and uterus for the developing embryo. In all species in which specific examples have been obtained, except man, retention of embryos within the oviducts, beyond the time when implantation normally would begin, results in embryonic death. This finding indicates that the oviduct is either not capable of providing the embryo with nutritive material sufficient for growth and development beyond the blastocyst stage or that the stroma of the oviduct does not permit implantation to occur. That embryos of certain bats are able to develop to the blastocyst stage within the oviduct is an indication that the oviducts of these animals are capable of providing a supportive milieu for growth and development. In this regard, secretory activity has been reported to occur in the bat during the time the embryos are in the oviduct (14, 61).

2.2. Shedding of the zona pellucida

The first association between blastocyst and uterus involves attachment of the blastocyst to the luminal epithelium of the uterus. This stage is subdivided into an initial stage of apposition and a subsequent stage of adhesion (68). Whether the stage of apposition occurs by expansion of the blastocyst to increase the area of contact with the wall of the uterus or by closure of the uterine lumen around the blastocyst ('clasping'), the blastocyst first must shed its zona

Van Blerkom, J. and Motta, P.M. (eds.), Ultrastructure of reproduction. ISBN 0-89838-572-5.

pellucida. The exact mechanism for this shedding is not clear. Although experimental evidence in mice has implicated the action of an estrogen-dependent uterine factor in zona lysis, other experiments have indicated that embryos from mice and rabbits cultured from the oviductal stage in vitro are capable of escape from the zona pellucida without uterine derived factors.

Shedding of the zona pellucida by mouse blastocysts *in vitro* appears to be facilitated by mechanical stress exerted by regular pulsations of expansion and contraction of the blastocyst (13). However, Cole (13) reported that the initial breakage of the zona appeared to be due to the formation by the trophoblast of specific cellular processes which extended and contracted, and not to a simple increase in blastocoelic pressure. It is possible that the cellular processes observed by Cole served to penetrate localized regions of weakness in the zona and that subsequent escape of the blastocyst from the zona occurred at the point of initial penetration.

Evidence for uterine participation in the shedding of the zona pellucida by mouse blastocysts was presented by McLaren (44, 45, 46) and Mintz (47), who reported that unfertilized oocytes and dead morulae escape from the zona pellucida when they were placed in the uterus of a pseudopregnant mouse. Potts and Psychoyos (59) published a photomicrograph showing a zona-encased unfertilized oocyte in the same uterus which contained blastocysts in delay of implantation that were free of zona. Considered together, these results indicate that uterine-dependent zona lysis is delayed greatly during the stage of delay of implantation. Escape from the zona pellucida has been shown to be delayed by at least a full day during ovariectomy-induced (43) and lactational delay of implantation (42, 64). Escape from the zone pellucida has been reported to occur in the oviducts of mice (53), but is delayed by one day, when compared to blastocysts that reach the uterus on schedule.

Collectively, these data indicate that the mouse blastocyst is capable of shedding the zona pellucida both *in vitro* and in ectopic sites (27, 53), but that the timing is delayed when compared to the shedding of the zona *in utero*. The uterine factor is estrogen-dependent and is therefore absent after ovariectomy; in this situation in embryo sheds the zona pellucida independent of the presence of estrogen.

The uterine involvement in facilitating zona lysis may be either acidification of the uterine secretions or by production of a proteolytic enzyme (47), or both. Proteolytic activity has been reported to increase in the uterus of mice at about the time of zona lysis and the initiation of implantation (54). In this respect, a gelatin-dissolving proteinase is located also in the trophoblast and at the surface of implanting rabbit blastocysts (15). This indicates that escape from the zona pellucida by rabbit embryos may be facilitated by the action of a proteolytic enzyme produced by the blastocyst and not by the uterus.

Whether escape from the zona by primate blastocysts is facilitated by zona lysis or by mechanical stress exerted by contraction and expansion has not been investigated as thoroughly as it has for mice and rabbits. However, Enders and Schlafke (25) observed in the Rhesus monkey several split and torn zonae during their attempts to recover unimplanted blastocysts. This observation indicates that Rhesus blastocysts are capable of escape from undigested zona pellucidae.

Escape from the zona is not necessarily followed immediately by either apposition of the blastocyst or attachment to the uterine epithelium. During delayed implantation, for example, the zona-free blastocyst remains unattached and is recovered easily from the uterus by flushing with medium. Nor does enzymatic removal of the zona result in premature attachment of mouse embryos *in vitro* (73). This finding indicates that the embryo must ecquire adhesiveness before attachment and that the zona does not simply prevent premature attachment.

2.3. Positioning of blastocysts in utero

The position of blastocysts within the uterus has three aspects. First, in species that given birth to more than one offspring at a time (polytocous) the blastocysts must be positioned with respect to each other. Second, the blastocysts of most species become positioned according to the anatomical structure of the uterus, as for example, with respect to the mesometrium, and third, the orientation of the ICM is normally constant within species.

2.3.1. Spacing

In polytocous species, the embryos generally are spaced evenly within the uterus. Although spacing is presumed to be necessary to prevent local overcrowding, the uniformity of spacing varies among species (43, 83). For example, Boving (7) has observed that rabbit conceptuses are generally more uniformly spaced than rat conceptuses although the latter are not spaced at random. McLaren (43) has suggested that the uniformity of spacing varies in relation to the size of the blastocysts. Boving (7) suggested that since rabbit blastocysts expand some 10,000 fold in volume prior to attachment, the stimulus for spacing in the rabbit is expansion of the blastocyst. However, since nonrandom spacing also occurs in species such as the rat, that do not experience such dramatic expansion, different stimuli must be considered in these species. Thus, the expansion hypothesis does not totally account for nonrandomness of spacing. It is generally accepted that spacing and transport of the blastocysts is accomplished by peristaltic waves derived from contraction of circular myometrium.

2.3.2. Position of attachment of the blastocyst to the uterus

Both the site of attachment of the blastocyst to the uterine epithelium with respect to the mesometrium,

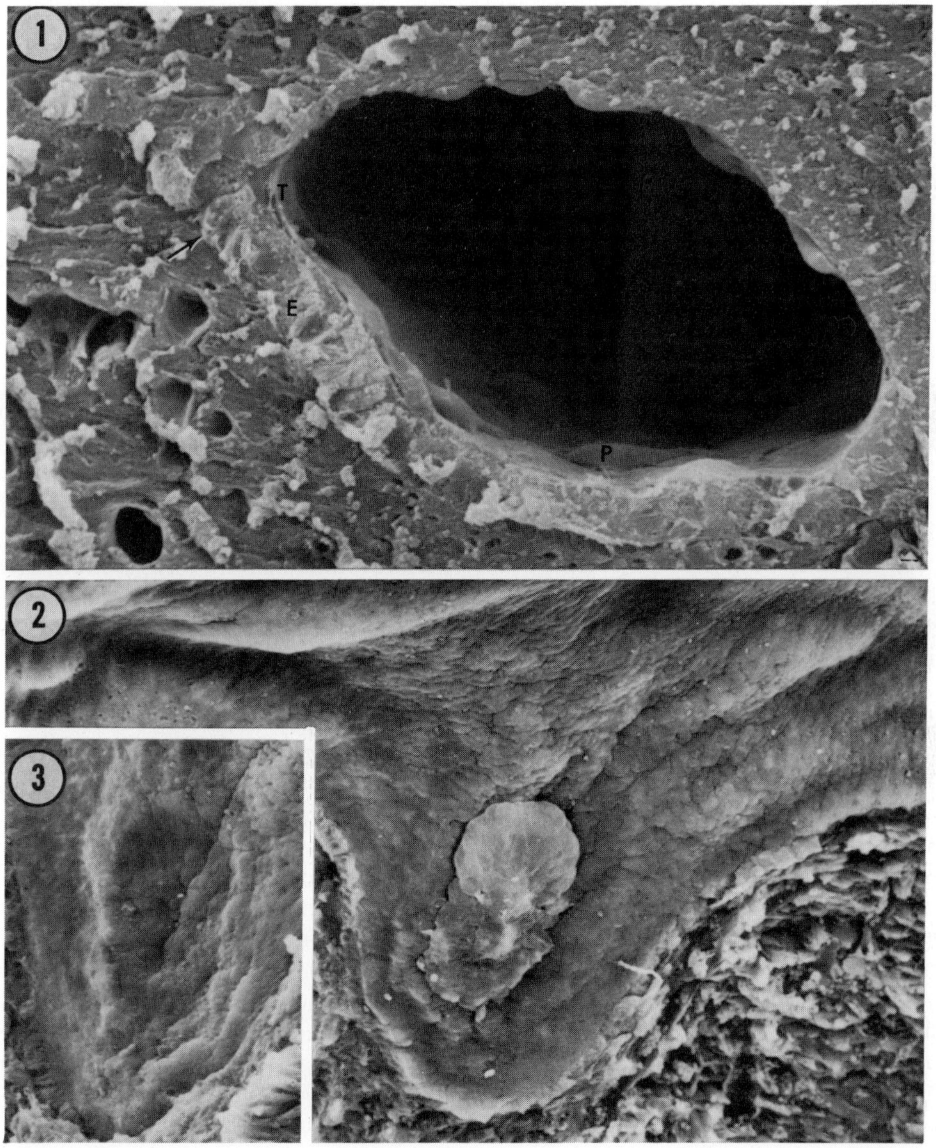

Figs. 1–3 (1) Scanning electron micrograph of a mouse blastocyst implanting *in situ*. In this specimen the uterus was cut longitudinally. The blastocyst is clasped tightly within the implantation chamber and the extremely thin trophoblast cells (T) are closely apposed to the uterine luminal epithelium (E). The basement membrane of the uterine luminal epithelium is indicated by the arrow and parietal endoderm cells (P) may be seen within the cavity of the blastocyst (× 960). (2) Scanning electron micrograph of a split implantation chamber of a mouse. The uterine lumen extends horizontally at the top of the picture and the implantation chamber containing the blastocyst extends antimesometrially below the lumen of the uterus. The long axis of the blastocyst is mesometrially-antimesometrially oriented and the inner cell mass is at the mesometrial pole of the blastocyst (× 400). (3) Scanning electron micrograph of the contralateral aspect of the implantation chamber in Figure 2. The blastocyst was not adherent to the uterine epithelium and was easily separated from it thus leaving an imprint (× 400).

and the manner with which site specific attachment occurs varies greatly among mammals (48). The cross sectional shape of the uterine lumen may by important is providing a limited choice of sites against which the blastocyst may lodge. In cross section, the shape of the uterine lumen may be dorsoventrally compressed, as in the simplex uteri of primates, including man, laterally compressed, as in rodents, or symetrically radiate, as in carnivores, pigs and rabbits. Some uteri, e.g., those of ruminants, possess glandular areas that form specialized zones to which blastocysts attach (implantation pads or caruncles) (48). Other species e.g., many rodents, form special 'implantation chambers' as the result of localized edema, which is presumably stimulated by the presence of the blastocyst. (Figs. 1, 3). Still other species, such as certain bats, form modified vascular patterns of specific implantation sites prior to the arrival of the blastocysts (83) and yet others posess preformed implantation sites, e.g., elephant shrew. Whether the blastocyst is actually attracted to specific implantation sites or arrives at the site of attachment purely by chance is an important aspect that requires more clarification.

2.3.3. Orientation of the blastocyst

The third aspect of positioning the blastocyst within the lumen of the uterus concerns the specific location of the inner cell mass (ICM). In the mouse, the first site of trophoblast attachment is antimesometrial, and the ICM is oriented mesometrially (Fig. 2). It has been suggested by Kirby, Potts, and Wilson (40) that the ICM may rotate to its definitive orientation after the first attachment. The mouse blastocyst possesses a basement membrane between the ICM and trophectoderm (20, 40) and this basement membrane has been suggested to permit migration of the ICM. However, Gardner (28) marked the location of the ICM in preimplantation mouse blastocysts by injecting melanin granules into trophectoderm cells overlying the ICM, (polar trophectoderm) and transferred the blastocysts to the uterus of foster mothers where they subsequently implanted. Examination of sections of the implantation sites indicated no case where the position of the ICM varied with respect to the marked trophoblast cells. Therefore, the blastocyst must somehow become arranged with its long axis mesometrially-antimesometrially oriented and the ICM at the mesometrial pole. There have been indications that under certain conditions, the abembryonic trophoblast binds lectins in different pattern than does the polar trophoblast (8, 52), and these subtle differences in the surface properties of the trophoblast cells may somehow facilitate the orientation of the blastocyst within the uterus.

The importance of position and orientation of the blastocyst within the uterus lies in the fact that the first orientation generally will determine the location where the placenta ultimately will form. One need only consider the frequency and clinical complications of placenta previa in man to recognize the importance of proper position and orientation. However, there have been some indications that placenta previa may result from deficient vascular development at the uterine fundus and not abnormal orientation.

3. The implantation period

3.1. Preparation of the endometrium for implantation

Although some species require a copulatory stimulus to elicit preparation of the endometrium for implantation, in the majority of mammals cyclic changes occur in the endometrium that are anticipatory for implantation with every ovulatation. The changes in the endometrium include increased vascularization of the stroma, and proliferation and differentiation of component cells. These changes are collectively termed decidualization (see section 3.1.3.), and are primarily under the influence of ovarian hormones (55, 69). Additionally, there may be localized changes which are elicited by the presence of the blastocyst.

3.1.1. Uterine luminal epithelium

Functionally, there are three aspects of preparation of the uterine luminal epithelium for implantation. First, there must be modification of the milieu within the uterine lumen to make it compatible with growth and development of the blastocyst. This function is largely dependent on glandular secretions and is dealt with in the next section. The second function is to make the luminal epithelium receptive to attachment of the blastocyst. Third, the epithelium transmits the signal(s) which initiates the decidual reaction of the stroma and elicits vascular changes in the implantation site.

In some species, the functionalis layer of the endometrium is cyclically lost through menstruation. In these species, for implantation to occur, the epithelium must first reestablish itself. The source of tissue for this reconstruction of the functional layer is considered to be the basal layer (5). Both the sloughing that occurs during mestruation and the subsequent proliferation of the functionalis of the endometrium are widely known to be under the influence of ovarian steroid hormones.

The ultrastructure of the epithelium varies among species. The epithelium of all uteri is a simple collumnar type with numerous small microvilli (Figs. 10, 11). The uteri of some species (man and rabbits) posess epithelial cells with numerous cillia (Fig. 10). Under the influence of progesterone, the epithelium undergoes an increased rate of mitosis and the cells may become multinucleate (cow) or form a syncitium (rabbit). Further ultrastructural modifications include the accumulation of lipid droplets (mouse) or rich glycogen deposits (monkeys and man).

3.1.2. Uterine glands
During the peri-implantation period, the composition of the uterine secretions has been reported to change both qualititatively and quantitatively (77). These changes have a direct role in determining whether conditions within the uterine lumen are compatible with implantation and continued growth and development of the embryo (2, 81, 82).

Uterine glands are located in the stroma of the uterus as inpocketing of the luminal epithelium, and as such are morphologically similar to the luminal epithelium. However, the secretory and metabolic activity of glandular epithelium is significantly different from those of the luminal epithelium. The function of uterine glands is secretion into the uterine lumen. Although histochemical methods have shown that the secretion product is rich in mucopolysaccharides, particularly glycogen, biochemical evidence has been obtained from several species that indicates that these glands are also responsible for the synthesis and secretion of uterine specific proteins (4, 29, 77). The presence of uterine specific proteins and their suggested functions have been reported in mice, rabbits, roe deer, cow, pig, sheep, macaque and baboon (4).

Uterine glandular secretions may be important in providing nourishment for the blastocyst (histotrophic). In species where implantation is superficial, histotrophic nourishment might be an important factor in the maintenance of the embryo. However, in species such as the mouse and man, where implantation is interstitial, uterine secretions are probably of less importance. This is especially obvious when considering that embryos grow and develop in ectopic sites. This indicates that although there might be uterine-specific secretions, these secretions are not necessarily essential for embryonic development to occur.

3.1.3 The uterine stroma
Differentiation of the uterine stroma in preparation for implantation has been studied widely at the ultrastructural, biochemical, and physiological levels. Uterine stromal differention is termed decidualization and involves proliferation and differentiation of fibroblasts. Decidualization occurs as a normal part of the proliferative phase in women, is under the influence of the blastocyst in many species, and in some species does not occur at all. In some monkeys, e.g., Rhesus and Macaque, epithelial plaque formation occurs rather than decidualization.

The uterine stromal cell population is not static and shifts in the relative percentage of cell types take place during decidualization (54). During decidualization, fibroblasts enlarge, become polyhedral, and due to endoreplication accumulate as much as 64 times the haploid amount of DNA (17). A dense accumulation of collagen and extracellular matrix material occurs in the spaces between stromal cells. Tight junctions are common and junctional complexes shared between processes of the same cell have been observed. During this time, in addition to stromal cell differentiation, there is also extensive infiltration of mast cells and lymphocytes. This observation indicates a localized inflamatory response and suggests the possibility of an immunological reaction to the implanting blastocyst.

The function of the decidua presumably is to protect the maternal tissue from excessive invasion by the trophoblast. There has also been some discussion that the decidua may serve to protect the embryo from rejection as allogenic tissue. This lack of rejection was specifically demonstrated in the rat where syngenic grafts of epidermal cells grew in a decidualized uterus but were rejected in the absence of decidualization (3).

4. Preparation of the blastocyst for implantation

Prior to attachment to the uterine luminal epithelium, the cell surface of mouse trophoblast has long and rather uniformly spaced microvilli projecting from the apical cell surface (57) (Fig. 11). After the blastocyst has become oriented within the uterine lumen and establishes contact with the luminal epithelium (period of apposition), the surfaces of both the uterine epithelium and trophoblast become smoother and less microvillous (Figs, 4, 6). The remaining microvilli are reduced to small, irregular projections (49). This smoothness is probably the result of the close physical contact between the apposing cell surfaces (Fig. 7).

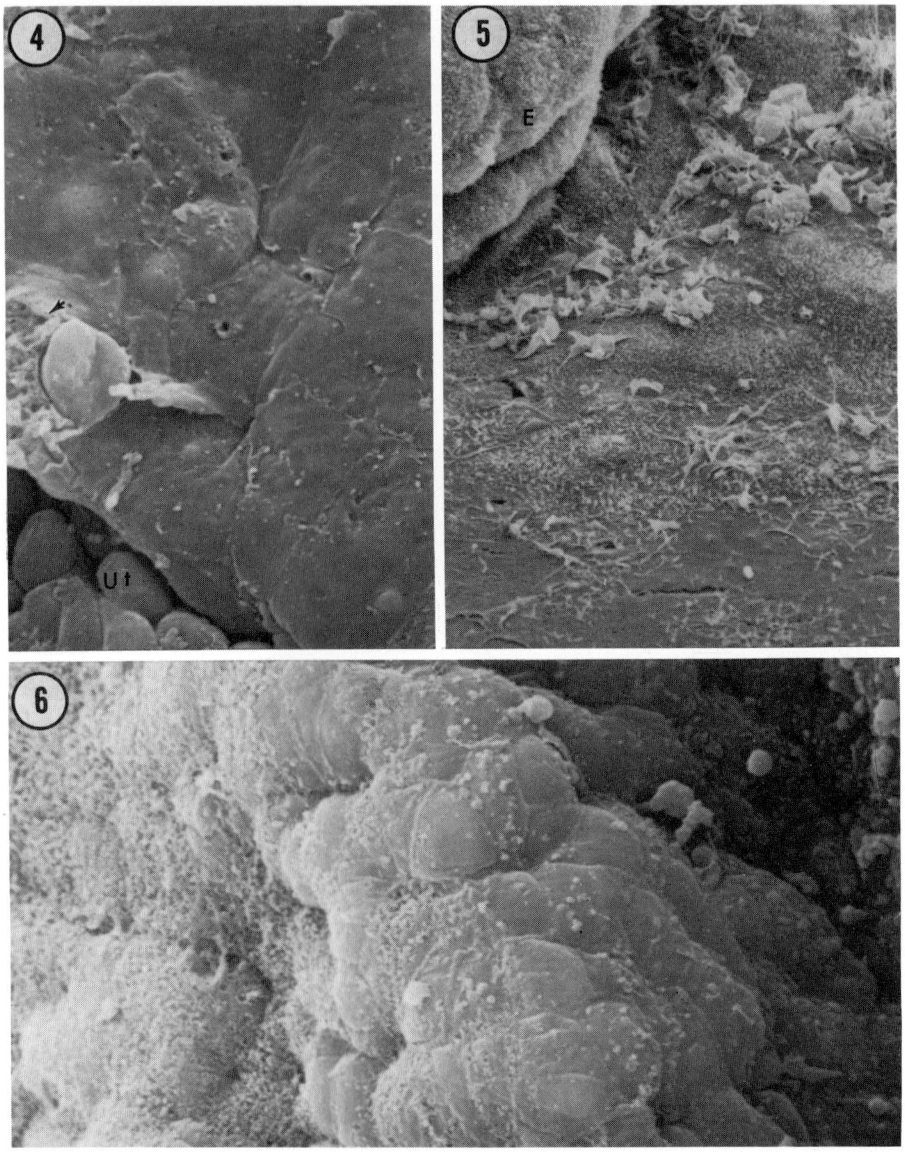

Figs. 4–6. (4) Scanning electron micrograph of an implanting mouse blastocyst. The surface of the trophoblast had begun to adhere to the uterine luminal epithelium and a portion of a trophoblast cell (arrow) was torn as it was pulled away from the uterus in preparation for microscopy. The adhesive trophoblast cells have lost their microvilli and are smooth in appearance. Ut, uterine epithelial cells (× 2,000). (5) Scanning electron micrograph of a portion of a mouse blastocyst during initial attachment to a monolayer derived from a primary explant of mouse uterine epithelial cells. The embroynic surface (E) is extensively microvillous and the surface of the outgrowing trophoblast cells displays extensive ruffles and lamellipodia. The cells derived from the uterine epithelium (foreground) are easily distinguishable by their smooth appearance (× 2,400). (6) Scanning electron micrograph of the margin of an implantation chamber in a mouse uterus. Whereas the cells in the chamber at the right of the picture were formerly in contact with the blastocyst and display a smooth microvilli-free surface, the cells lining the lumen of the uterus retain their microvillous surface (× 2,160).

4.1. Acquisition of adhesiveness

Little is known about the molecular basis of the acquisition of adhesiveness by the trophoblast. The plasmalemma of the trophoblast contains molecules which have been associated with specific fundamental processes. The majority of these molecules are glycosylated proteins or glycosylated lipids; the carbohydrate moiety is termed the glycocalyx. Surface glycoconjugates have been implicated as cell surface markers that enable the cells to signal their own identity and to recognize the identity of cells they come in contact with, thus allowing them to acquire 'positional information'. In addition, surface glycoconjugates may provide receptors for hormones and other molecules. The presence of a glycocalyx on the surfaces of both trophoblast and uterus of several species has been demonstrated (24), and the presence of glycocalyx at the blastocyst-epithelium interface has been noted to persist in the mouse (50). In mouse blastocysts, the acquisition of adhesiveness has been associated with specific changes in the composition of the glycocalyx. For example, the overall negativity of the trophoblast surface decreases, presumably as the result of decreased amount of sialic acid at the trophoblast cell surface (39, 51). Specific alterations in mono- and disaccharides occupying a terminal position in the glycocalyx have been reported (9, 10). Such changes at the trophoblast cell surface could be the result of either incorporation of newly synthesized glycopeptides or modification of pre-existing glycopeptides. Modification of trophoblast surface could be the result of activity of transglycosyltransferases located on the uterine luminal epithelium (63). The altered trophoblast surface coat could interact with the glycocalyx of the uterine luminal epithelium to bring about adhesiveness.

Other hypothesis regarding cell adhesiveness account for the presence in surface membranes of nonprotein components such as glycosaminoglycans and lipids. Basement membrane components such as fibronectin and laminin have been suggested to be important in the attachment and spreading of fibroblasts *in vitro*, however, fibronectin is not present between trophoblast and uterine epithelium at the time of attachment in the mouse (78). The presence of fibronectin at attachment sites in other species has not been investigated.

4.2. Penetration of the uterine luminal epithelium

Perhaps the most perplexing aspect of implantation is the penetration of the uterine epithelium by trophoblast cells. In all species where implantation is interstitial, the trophoblast penetrates the epithelium and underlying basement membrane and becomes inbedded within the underlying stroma. Categorizing penetration in some species occasionally is difficult, as a result of the transiency of this period in the implantation process. In other species, as in rats and mice, it occurs over a period of days and so is much more amenable to investigation. Schlafke and Enders (68) have considered this subject in detail and have classified three methods of accomplishing epithelial penetration: a) by penetration of trophoblast cells between intact and apparently healthy epithelium, termed *intrusive implantation;* b) by sloughing of the epithelium away from the underlying basement membrane and the space being subsequently occupied by the trophoblast, termed *displacement implantation;* and c) by fusion of syncytial trophoblast to epithelial cells, termed *fusion implantation.*

4.2.1. Intrusive implantation

Intrusive implantation occurs both in species with cellular trophoblast and in species with syncytial trophoblast. Enders and Schlafke described penetration of uterine luminal epithelium in the ferret (23) where the first areas of penetration were by processes of syncytial plaque which penetrated between apical ends of epithelial cells lining the lateral antimesometrial wall. In this species, the trophoblast processes displaced the uterine apical junctions and subsequently developed junctions with the uterine cells. The shared complexes between trophoblast and uterine epithelial cells resembled those that were displaced by intrusion. Mechanisms by which the trophoblast subsequently broaches the basement membrane of the epithelium are, as yet, unknown, however, it is possible that initial penetration of the basement membrane is by the decidualizing stromal cells (25) rather than the trophoblast.

4.2.2. Displacement implantation

Displacement implantation occurs when the uterine epithelium becomes dislodged from the basal lamina and is replaced by trophoblast cells. The dislodged cells are frequently phagocytized by trophoblast cells. Epithelial cells are stimulated to detach themselves from the basement membrane, either by the implanting blastocyst or by the underlying decidualizing stroma. Displacement implantation occurs in mice and rats. In these species, during the apposition phase, the anchoring villi of the luminal epithelium become less numerous and penetrate less deeply into the basement membrane (Chávez, unpublished observations). Displacement of epithelial cells can also occur in the absence of an implanting blastocyst, such as in mice, if the decidual cell reaction (deciduoma) is elicited artificially, for example, by an intraluminal injection of oil. In this case, however, it is likely that it is the trauma induced by the oil rather than the underlying decidua which elicits the sloughing response. Sloughing and subsequent death of epithelial cells may be independent of lysosomal secretion by the trophoblast, but may occur by fragmentation (apoptosis) and self-digestion (18).

4.2.3. Fusion implantation

Fusion implantation occurs in rabbits. Prior to fusion, both trophoblast and uterine epithelium become syncytial. The syncytium of the trophoblast is limited to discrete knobs on the abembryonic pole of the blastocyst. These knobs fuse with the syncytium of the uterine expithelium on the anti-mesometrial pole of the uterus. Once fusion is accomplished, the blastocyst and uterus share syncytial areas containing nuclei from both uterine epithelium and blastocyst (22, 75). Maternal nuclei are removed by ingestion by lysosomes and the tissue is then under control of embryonic nuclei. After assuming 'control' of epithelial cytoplasm the syncytiotrophoblast penetrates the basal lamina into the underlying stroma.

5. Maternal recognition of pregnancy

The basic question of how an animal knows that it is pregnant is one of the most intriguing in reproductive biology. From a purely morphophysiological viewpoint, the question addresses the changes that occur in the uterus in preparation for implantation. These changes have been previously described (section 3). In the rat and mouse for example, the timing of these changes is under control of steroid hormones (37).

It has been shown that, in rats and mice, the uterus must receive progesterone 'priming' for at least 48 h before it is 'sensitized' to respond to estradiol. If implantation does not occur in these species, the uterus enters a refractory state during which it is unresponsive to estradiol and the environment it provides may become lethal to the preimplantation blastocyst (60).

In spirte of considerable work concerning the role of steroid hormones during pregnancy, it has not been conclusively demonstrated that estrogen is necessary for implantation to occur in most species. The hypothesis has been advanced that the blastocyst itself secretes minute quantities of estrogen necessary for implantation (16). This hypothesis has not gained universal acceptance.

There can be little doubt that the implanting blastocysts of most species do exert considerable direct influence on the uterus, and may additionally, affect steroid synthesis by the ovary in many species. In the wallaby, the presence of an embryo stimulates an increase in both the weight and protein synthesis of the pregnant horn, as compared to the contralateral non-pregnant horn (62), but does not extend the life of the corpus luteum or interrupt the estrus cycle. In the Rhesus monkey, the presence of an implanting conceptus is required to stimulate increased secretion of progesterone and estrogen by the corpus luteum, and progesterone, in turn, is required for maintenance of the conceptus until the placenta develops the capacity for progesterone synthesis (6).

In the human, the situation is similar to that in the Rhesus. In a recent study by Lenton, Sulaiman, Sobowale, and Cooke (41), it was found that during the cycle in which conception occurred, women had significantly higher plasma progesterone levels that during either their own non-pregnant control cycles, or in nonpregnant women at the same times after ovulation. Therefore, these authors suggested that the higher progesterone concentrations may constitute a preimplantation component of the maternal recognition of pregnancy in women.

6. Experimental approaches to the study of implantation

In recent years, several methods have been developed to study the various aspects of implantation outlined in this chapter. Most of the attempts to elucidate the process of implantation have focused on specific facets of the process. For although the process occurs as a series of inter-related events, many events may be defined as occurring independently. These events collectively compose the overall process of implantation. For example, the morphological aspects of adhesion of the embryo to the uterine epithelium are distinguishable from the physiological response of the epithelium and underlying stroma to the adhering blastocyst. Factors which mediate the specific facets of implantation may be characterized as 1) embryonic factors; 2) maternal factors, and 3) maternal-embryo interactions. To elucidate the various aspects of these factors, various model systems have been devised utilizing both *in vivo* and *in vitro* techniques. For example, trophoblast outgrowth has been used as a model for studying the intrinsic ability of the embryo to grow and differentiate in the absence of maternal factors (11, 73, 79). The artificial induction of decidualization has been used to simulate the uterine response normally observed with an implanting blastocyst (31, 67, 72). This method could also be used to study the effects of the decidualized uterus in the absence of an embryo, on the lifespan and secretions of the corpus lutuem and also the feedback between the pituitary, hypothalamus and ovary. In attempts to study the interaction of the embryo and uterine tissue, in the absence of serological and immunological factors, uterine fragments and blastocysts have been placed in co-culture (32), and blastocysts have been co-cultured with monolayers of tissue culture cells (30), or primary cultures derived from uterine epithelium (11,

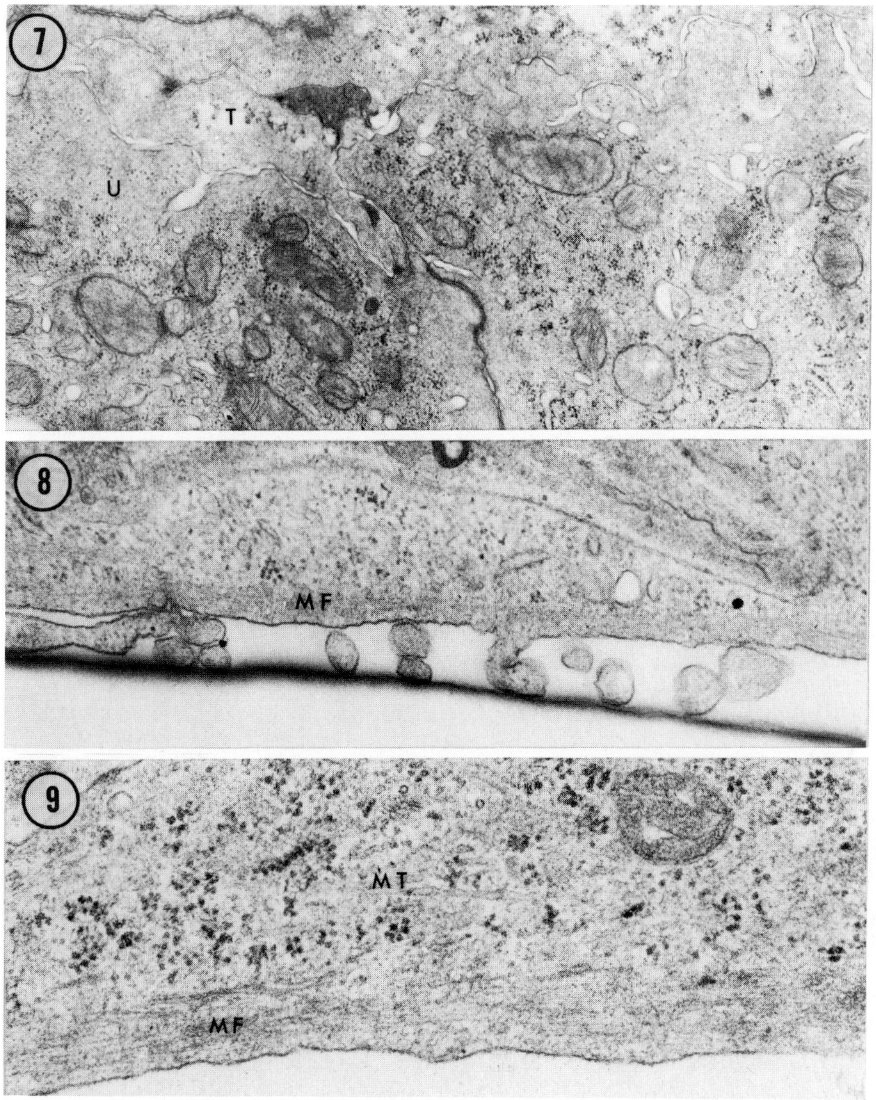

Figs 7–9. (7) Transmission electron micrograph of the early stages of apposition of blastocyst and uterus in the mouse. The trophoblast (T) has formed extensive contact with the uterine epithelial cells (U) and the plasma membranes of the two cell types are closely interdigitated. A tongue of trophoblast cell is shown penetrating slightly between the apical portion of the two epithelial cells (× 3,120). (8) Transmission electron micrograph of a portion of a mouse trophoblast cell during the early stages of outgrowth *in vitro*. The microvilli are shown here as being blunted by physical contact with the surface of the culture dish. Microfilaments (MF) are beginning to form in arrays parallel to the plasma membrane. Compare with the absence of microfilaments in the trophoblast at a similar stage of attachment depicted in A (× 3,760). (9) Transmission electron micrograph of a portion of a mouse trophoblast cell outgrown *in vitro*. In this picture are extensive arrays of microfilaments (MF) arrayed parallel to the plasma membrane and surface of the culture dish. This is a significant deviation from implantation *in vivo* and probably represents an adaptation of the trophoblast cells to two-dimensional outgrowth *in vitro*. (MT, microtubules) (× 4,400).

73) (Fig. 5). These studies have provided some information about cell to cell contacts, but have not yielded much information about invasion. The method seems especially suited for studies dealing with adhesion.

6.1. Trophoblast outgrowth

The ability of trophoblast of many species to proliferate and form monolayer outgrowth *in vitro* has been exploited in numerous studies that have sought to determine 1) the temporal relation of developmental events relative to the developmental potential and biochemical activities of trophoblast and inner

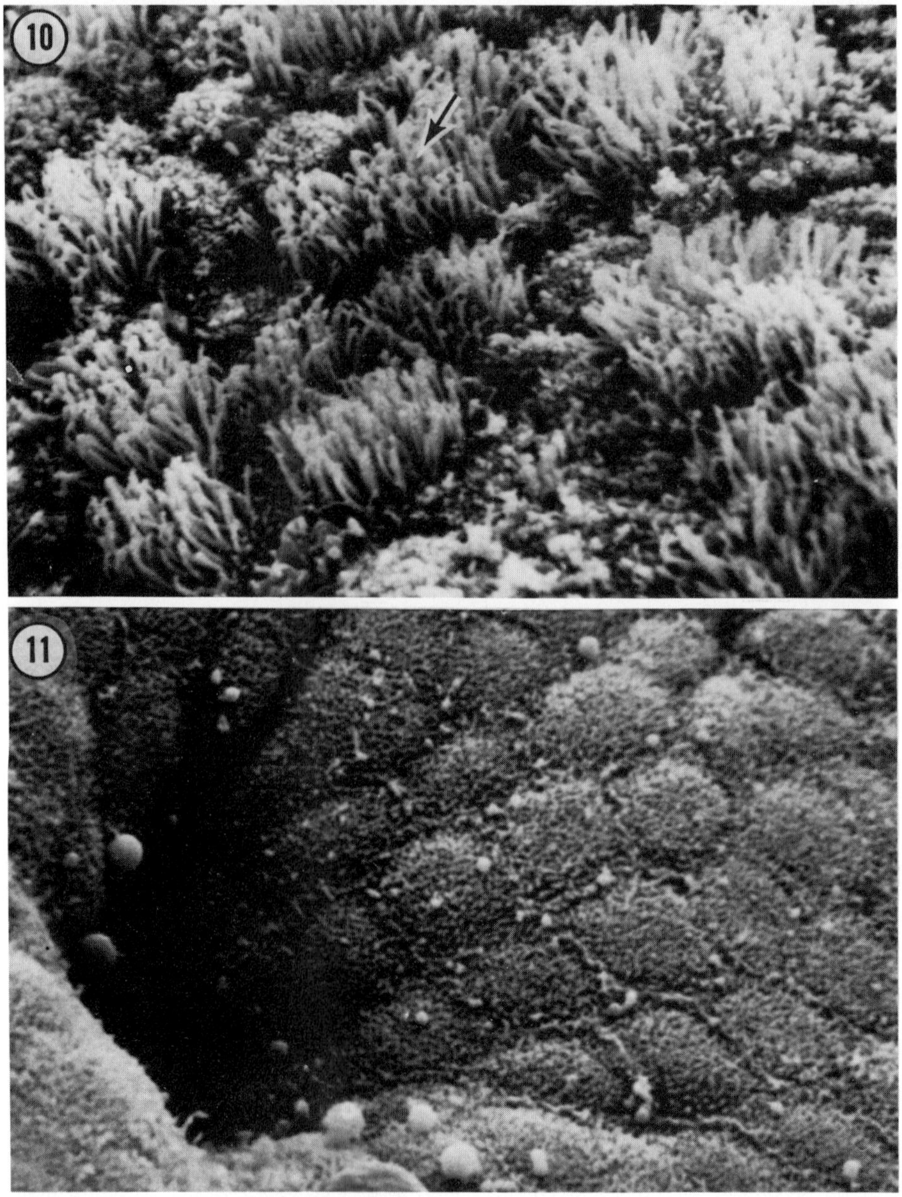

Figs. 10–11. (10) Scanning electron micrograph of an area of the fundus of a human uterus during the mid-secretory phase of the menstrual cycle. Both ciliated (arrow) and now ciliated cells can be seen in this micrograph. Abundant secretory products were present over the surface of the cells. This sample was taken from a hysterectomy specimen of a uterus from a 43-year-old pre-menopausal woman (× 2,500). (11) Scanning electron micrograph of uterine luminal epithelial cells surrounding a glandular pore of a mouse uterus. This uterus contained blastocysts in ovariectomy-induced delayed implantation. The uterus was taken 12 h after estradiol and progresterone injection to reactivate the blastocysts from delay. The uterine cells are extensively microvillous but uterine epithelial cells of the mouse poses no cilia (× 2,400).

cell mass (33, 70) and, 2) the mechanism by which the trophoblast acquires properties that permit it to adhere to and penetrate the uterine luminal epithelium (8, 69, 71). The mouse has been the most widely used model for these studies, and mouse trophoblast outgrowth *in vitro* seems to be a reliable and reproducible approach to the analysis of many molecular and cellular events that occur during the early stages of the peri-implantation period, though not as a model for the implantation process per se. Furthermore, under appropriate culture conditions, normal germ layer differentation and organ formation has been achieved, albeit with a relatively low success rate (34, 35, 36).

These studies have indicated that the timing of trophoblast adhesion and the molecular events associated with adhesion correspond very closely *in vivo* and *in vitro* (30, 33, 74, 79). The morphology of outgrowing trophoblast has been examined and there appear to be numerous differences in ultrastructure at the cell surface and also beneath the cell surface of blastocysts outgrowing *in vitro* and those implanting *in vivo* (11, 21, 79) (Figs. 7–9). It has been suggested that the elaboration of lamellipodia primarily reflects accomodations to *in vitro* culture conditions and not fundamental deficiencies in the culture system (11).

Results from several studies using the *in vitro* model have indicated that the timing of the phases of blastocyst apposition, adhesion, and trophoblast invasion are fundamentally similar *in vivo* and *in vitro* (30, 33, 70, 71, 74, 79). The molecular composition of the trophoblast cell surface that gives the blastocyst adhesive properties appears to be functionally similar in cultured and normally implanting mouse blastocysts (11, 74). Further exploitation of this method may provide insight into the mechanisms of cell-to-cell attachment as well as possible means of interfering with the implantation process.

6.2. *The decidual cell response*

Following sensitization with progesterone, followed by estrogen stimulation, the uterus of many animals will respond to stimuli (e.g., oil trauma) and undergo morphological changes such as fibroblast enlargement, increased vascularization and accumulation of intercellar material that are similar to changes which occur during normal pregnancy. Thus stimulated, a non-pregnant uterus exhibits many characteristics of pregnancy in the absence of stimuli and interaction from an implanting blastocyst. Rats and mice have been the most extensively used animals in studies that have sought to investigate the nature of 1) the chronology and 2) endocrinology of the uterine response to decidualizing stimuli, as well as the nature of various 3) biochemical and 4) morphological aspects of the stimulated uterus (31, 66, 67, 72). Studies using castrated animals have resulted in clarification of the sequence of the action of progesterone and estradiol on the preparation of the uterus for decidualization (31). These studies have also been used to determine the responsiveness of the uterus to various inducing agents (72) in the hopes of clarifying the action of the blastocyst.

7. Summary

Motives for studying the implantation process vary considerably. For example, knowledge gained may enable constructive interference with reproductive processes in order to increase production of economically important domestic animals or to preserve endangered species. Alternatively, destructive interference may be desireable in order to limit the population of our own species, or other species which may be considered a nuisance in some geographical areas. Regardless of the motives, it is apparent that there remains much to be learned about the implantation process. Although a considerable amount of knowledge is derived from studying implantation in rats and mice, very little insight can be gained by extrapolating this knowledge of interstitial implantation to other species such as the cow, where the blastocyst is non-invasive. As there are extremes in the degree of invasion of the stroma of the uterus there are variations in which invasion is achieved. These variations must be borne in mind when selecting a suitable model for the study of implantation and especially when attempting to draw generalized conclusions about the overall process.

The adaption of *in vitro* methods will probably yield useful information about the acquisition of adhesiveness by the blastocyst. Such studies may lead to the discovery of new antifertility agents. For some studies however, the *in vitro* model is unsuited. The use of cell monolayers to represent the uterine luminal epithelium has proved disappointing, for, even cells derived from primary cultures of uterine epithelium do not retain their characteristic (i.e., tall columnar and microvillous) morphology when placed in tissue culture. It is probably correct to assume that the biochemistry of these cells is also altered *in vitro* for we cannot duplicate the physiological hormonal conditions that they encounter *in vivo*.

Techniques are constantly evolving, however, and improved methods in molecular and cell biology, as well as standard methods of careful observation will continue to accumulate more knowledge about the mechanisms involved in the implantation process.

References

1. Amoroso EC: Viviparity. In: Cellular and Molecular Aspects of Implantation. Glasser S, Bullock (eds) Plenum Press, New York and London, 1981, pp 3–25.
2. Aitken RJ: Embryonic diapause. Devl in Mammals 1: 307–359, 1977.
3. Beer AE, Billingham RE: Implantation, transplantation and epithelial-mesenchymal relationships in the rat uterus. J Exp Med 132: 721–736, 1970.
4. Beier HM: The role of uterine proteins in the establishment of the receptivity of the uterus. Prog Reprod Biol 7: 158–172, 1980.
5. Blandau RJ: The female reproductive system. In: Histology. Weiss Greep RO (eds) McGraw-Hill, 4th ed, 1977, pp 881–924.
6. Booher C, Enders AC, Hendrickx AG, Hess DL: Structural characteristics of the corpus lutem during implantation in the Rhesus Monkey (Macaca mulatta) Am J Anat: 160: 17–36, 1981.
7. Boving BG: Biomechanics of implantation. In: Biology of the Blastocyst. Blandau RJ (ed) The University of Chicago Press, Chicago and London, 1971, pp 423–442.
8. Chávez DJ: Polar differences in trophoblast cells of peri-implantation mouse blastocysts. J Cell Biol 95: 143A, 1982.
9. Chávez DJ, Enders AC: Temporal changes in lectin binding of peri-implantation mouse blastocysts. Devel Biol 87: 267–286, 1981.
10. Chávez DJ, Enders AC: Lectin binding of mouse blastocyst: Appearance of Dolichos biflorus binding sites on the trophoblast during delayed implantation and their subsequent disappearance during implantation. Biol Reprod 26: 545–552, 1982.
11. Chávez DJ, Van Blerkom J: In Vitro Attachment and Outgrowth of mouse Trophectoderm. In: Cellular and Molecular Aspects of Implantation, 1981.
12. Chen LT, Hsu YC: Development of mouse embryos *in vitro:* Preimplantation to the limb bud stage. Science 218: 66–68, 1982.
13. Cole RJ: Cinemicrographic observations on the trophoblast and zona pellucida of the mouse blastocyst. J Embryol exp Morph 17: 181–490, 1967.
14. De Bunilla H, Rasweiler JJ: Breeding activity, preimplantation development, and oviduct histology of the short-tailed fruit bat, Carollia in captivity. Anat Rec 179: 385–403, 1974.
15. Denker H-W: Role of proteinases in implantation. Prog Reprod Biol 7: 28–42, 1981.
16. Dickmann Z, Dey SK, SenGupta J: A new concept: Control of early pregnancy by steroid hormones originating in the preimplantation embryo. Vitamins and Hormones 34: 215–242, 1976.
17. Dupont H, Esnault C, Duluc AJ, Mayer G: Evolution de la polyploidie dans le deciduome experimental chez la ratte en grossesse unilaterale. Etude cytophotometrique. Cr hebd Seanc Acad Sci (Paris) 279: 501–504, 1974.
18- El-Sherdhaby AM, Hinchliffe JR: Epithelial autolysis during implantation of the mouse blastocyst an ultrastructural study. J Embryol exp Morph 33: 1067–1080, 1975.
19. Enders AC: Anatomical aspects of implantation. J Reprod Fert, Suppl 25: 1–15, 1977.
20. Enders AC: The fine structure of the blastocyst. In: The Biology of the Blastocyst Blandan RJ (ed) the University of Chicago Press, 1971, pp 71–88.
21. Enders AC, Chávez DJ, Schlafke S: Comparison of Implantation in utero and *in vitro*. Cellular and Molecular Aspects of Implantation. Glasser, Bullock (eds) Plenum, 1981, pp 365–382.
22. Enders AC, Schlafke S: Penetration of the uterine epithelium during implantation in the rabbit. Am J Anat 132: 219–240, 1971.
23. Enders AC, Schlafke S: Implantation in the ferret: epithelial penetration. Am J Anat 133: 291–313, 1972.
24. Enders AC, Schlafke S: Surface coats of the mouse blastocyst and uterus during the preimplantation period. Anat Rec 180: 31–46, 1974.
25. Enders AC, Schlafke S: Comparative aspects of blastocyst-endometrial interactions at implantation. Maternal Recognition of Pregnancy CIBA Foundation Symposium 64 (new series) Heap B (ed) Excerpta Medica, 1979, pp 3–22.
26. Enders AC, Schlafke S: Differentiation of the blastocyst of the Rhesus Monkey. Am J Anat 162: 1–21, 1981.
27. Fawcett DW, Wislocki GB, Waldo CM: The development of mouse ova in the anterior chamber of the eye and in the abdominal cavity. Amer J Anat 81: 413–433.
28. Gardner RL: Analysis of determination and differentiation in the early mammalian embryo using intra and interspecific chimeras. In: The Developmental Biology of Reproduction. -3rd Symposium of the Society for Developmental Biology. Markert EL, Papaconstantinon J (eds) Academic Press, Inc, 1975, pp 207–236.
29. Given RL, Enders AC: Mouse uterine glands during the peri-implantation period. II Autoradiographic studies. Anat Rec 199: 109–127.
30. Glass RH, Spindle AI, Pedersen RA: Mouse embryo attachment to substratum and interaction of trophoblast with cultured cells. J Exp Zool 208: 327–336.
31. Glasser SR Clark JH: A determinant role for progesterone in the development of uterine sensitivity to decidualization and ovo-implantation. The developmental Biology of Reproduction. Markert CL, Papaconstantinou J (eds) Academic Press, 1975, pp 311–345.
32. Glenister TW: Embryo-endometrial relationships during nidation in organ culture. J Obstet Gynaecol Br Commonw 69: 809–814, 1962.
33. Gonda MA, Hsu Y-C: Correlative scanning electron, transmission electron, and light microscopic studies of the *in vitro* development of mouse embryos on a plastic substrate at the implantation stage. J Embryol exp Morph 56: 23–39, 1980.
34. Hsu, Y-C: Post-blastocyst differentiation *in vitro*. Nature, London 231: 100, 1971.
35. Hsu Y-C: Differentiation *in vitro* of mouse embryos beyond the implantation stage. Nature, London 239: 200, 1972.
36. Hsu Y-C, Baskar J, Stevens LC, Rash JE: Development *in vitro* of mouse embryos from the two-cell egg stage to the early somite stage. J Embryol Exp Morph 31: 235–245.
37. Humphrey KW: The induction of implantation in the mouse after ovariectomy. Steroids 19: 591–600, 1967.
38. Jenkinson EJ: The *in vitro* blastocyst outgrowth system as a model for the analysis of peri-implantation development. In: Development in Mammals, Vol. 2 Johnson M (ed) Amsterdam, 1978. pp 151–172.
39. Jenkinson EJ, Searle RF: Cell surface changes on the mouse blastocyst at implantation. Exp Cell Res 106: 386–390.
40. Kirby DRS, Potts DM, Wilson LB: On the orientation of the implanting blastocyst. J Embryol Exp Morph 17: 527–532, 1967.
41. Lenton EA, Sulaiman R, Sobowale O, Cooke ID: The human menstrual cycle: plasma concentrations of prolactin, LH, FSH, oestradiol and progesterone in conceiving and non-conceiving women. J Reprod Fertil 65: 131–139, 1982.
42. McLaren A: Delayed loss of the zona pellucida from blastocysts of suckling mice. J Reprod Fert 14: 159–162, 1967.
43. McLaren A: A study of blastocysts during delay and subsequent implantation in lactating mice. J Endocr 42: 453–463, 1968.
44. McLaren A: The fate of the zona pellucida in mice. J Embryol exp Morph 23: 1–19, 1970a.
45. McLaren A: Early embryo-endometrial relationships. Ovo-Implantation. Human Gonadotropins and Prolactin. Karger, Basel. 1970b. pp 18–73.
46. McLaren A: Blastocyst activation. In: The Regulation of Mammalian Reproduction Segal SJ, Corzier R, Corfman PA, Condliffe PG (eds) Thomas, Springfield, Ill, 1973, pp 321–334.
47. Mintz B: Control of embryo implantation and survival. Adv Biosciences 6: 317–342, 1971.
48. Mossman HW: Orientation and site of attachment of the blastocyst: A comparative study. In: The Biology of the Blastocyst Blandau RJ (ed) The University of Chicago Press, 1971, pp 49–58.
49. Nilsson O: Ultrastructure of the trophoblast-epithelial junction at blastocyst implantation in the mouse. Exp Cell Res 94: 434–436, 1975.
50. Nilsson O: Ultrastructure of trophoblast-epithelium relations during implantation in the mouse. In: Reproductive Endocrinology. Proteins and Steroids in Early Mammalian Development (ICE Satellite Symposium, Aachen, 1976) Beier HM, Karlson P (eds) Springer-Verlag, New York, 1979.
51. Nilsson O, Lindqvist I, Ronquist G: Decreased surface charge of mouse blastocysts at implantation. Exp Cell Res 83: 421–423, 1973.
52. Nilsson O, Naeslund G, Curman B: Polar differences of delayed and implanting mouse blastocysts in binding of alcian blue and Concanavalin A. J Exp Zool 214: 117–180.
53. Orsini MW, McLaren A: Loss of the zona pellucida in mice, and the effect of tubal ligation and ovariectomy. J Reprod Fert 13: 485–499, 1967.
54. Padykula HA: Shifts in Uterine Stromal Cell populations during pregnancy and regression. In: Cellular and Molecular Aspects of

Implantation. Glasser SR, Bullock DW (eds) Plenum Press, 1980, pp 197–216.
55. Perrota CA: Initiation of cell proliferation in the vaginal and uterine epithelia of the mouse. Am J Anat 111: 195–204, 1962.
56. Pinsker MC, Sacco AG, Mintz B: Implantation-associated proteinase in mouse uterine fluid. Devel Biol 38: 285–290, 1974.
57. Potts M: The attachment phase of ovimplantation. Am J Obstet Gynec 96: 1122–1128, 1966.
58. Potts DM: The ultrastructure of implantation in the mouse. J Anat 103: 77–90, 1968.
59. Potts M, Psychoyos A: L'ultrastructure des relations ovoendometriales au cours du retard experimental de nidation chez la Souris. C R Acad Sci Paris 264: 956–958, 1967.
60. Psychoyos A, Casimiri V: Factors involved in uterine receptivity and refractoriness. Prog Reprod Biol 7: 143–157, 1980.
61. Rasweilier JJ: Reproduction in the long-tongued bat Glossophaga soricina. I. Preimplantation development and histology of the oviduct. J Reprod Fertil 31: 249–262, 1972.
62. Renfree MD: Influence of the embryo on the Marsupial uterus. Nature (London) 240: 475–477, 1972.
63. Roth S, White D: Intercellular contact and cell surface galactosyl transferase activity. Proc Nat Acad Sci USA 69: 485–489, 1972.
64. Rumery EE, Blandau RJ: Loss of zona pellucida and prolonged gestation in delayed implantation in mice. Biology of the Blastocyst. Blandau R (ed) p 115. University of Chicago Press, Chicago, 1966.
65. Salomon DS, Sherman MI: Implantation and invasiveness of mouse blastocysts on uterine monolayers. Exp Cell Res 90: 261–268, 1975.
66. Sartor P: Cell proliferation and decidual morphogenesis. Prog reprod Biol 7: 115–124, 1980.
67. Serra MJ, Baggett B, Rankin JC, Ledford BE: The artificially stimulated decidual cell reaction in the mouse uterus. Studies of RNA polymerases and histone modifications. Cellular and Molecular Aspects of Implantation. Glasser SR, Bullock DW (eds) Plenum Press, 1981, pp 291–306.
68. Schlafke S, Enders AC: Cellular basis of interaction between trophoblast and uterus at implantation. Biol Reprod 12: 41–65, 1975.
69. Schmidt IG: Proliferation in the genital tract of the normal mature guinea pig treated with colchicine. Am J Anat 73: 59–80, 1980.
70. Sellens MH, Sherman MI: Effects of culture conditions on the developmental programmes of mouse blastocysts. J Embryol exp Morph 56: 1–22, 1980.
71. Shalgi R, Sherman MI: Scanning electron microscopy of the surface of normal and implantation-delayed mouse blastocysts during development *in vitro*. J Exp Zool 210: 69–80, 1979.
72. Shelesnyak MC: Decidualization: The decidua and the deciduoma. Perspect Biol Med 5: 503–518, 1962.
73. Sherman MI, Salomon DS: The relationships between the early mouse embryo and its environment. In: The Developmental Biology of Reproduction. Markert CL, Papaconstantinou J (eds) 33rd Symposium of the Society for Developmental Biology, Academic Press, NY, 1975 pp 277–310.
74. Sherman MI, Shalgi R, Rizzino A, Sellens MH, Gay S, Gay R: Changes in the surface of the mouse blastocyst at implantation. In: Maternal Recognition of Pregnancy. Ciba Symposium 64 (New Series) Heap B (ed) Elsevier/North-Holland, 1979, p 48–52.
75. Steer, HW: Implantation of the rabbit blastocyst: The adhesive phase of implantation. J Anat 109: 215–228, 1971.
76. Surani MAH: Radiolabeled rat uterine luminal proteins and their regulation by estradiol and progesterone. J Reprod Fertil 50: 289–296, 1977.
77. Surani MAH: Cellular and molecular approaches to blastocyst uterine interactions at implantation. Devel in Mammals 1: 245–305, 1977.
78. Van Blerkom J, Chávez DJ, Bell H: Molecular and cellular aspects of facultative delayed implantation in the mouse. In: Maternal Recognition of Pregnancy. Heap B (ed) Excepta Medica, New York, 1979, p 141–172.
79. Van Blerkom J, Chávez DJ: Morphodynamics of mouse trophoblast outgrowth in the presence and absence of a uterine epithelium monolayer. Am J Anat 162: 143–155, 1981.
80. Wartiovaara J, Leivo I, Vaheri A: Expression of the cell surface-associated glycoptrotein, fibronectin, in the early mouse embryo. Devel Biol 69: 247–257, 1979.
81. Weitlauf HM: Effect of uterine flushings on RNA synthesis by 'implanting' and 'delayed implanting' mouse blastocysts *in vitro*. Biol Reprod 14: 566–571, 1976.
82. Weitlauf HM: Factors in mouse uterine fluid that inhibit the incorporation of ^{3}H-uridine by blastocysts *in vitro*. J Reprod Fertil 52: 321–325, 1978.
83. Wimsatt Wa: Some comparative aspects of implantation. Biol Reprod 12: 1–40, 1975.

Author's address:
Department of Anatomy
School of Medicine
Southern Illinois University
Carbondale, IL 62901, USA

CHAPTER 21

The fine structure and function of mouse and rat trophoblast during delayed implantation

B. OVE NILSSON

1. Introduction

The trophoblast cells of the blastocyst form a layer, the trophectoderm, which encloses the blastocoele cavity. This cavity is filled with fluid and contains the inner cell mass (the embryoblast) or future embryo. The blastocoele fluid constitutes the microenvironment of the embryoblast, delivering its nutrients and draining its waste products. The composition of the fluid is therefore crucial for normal development of the embryo. The trophectoderm controls the qualitative and quantitative nature of the blastocoele fluid.

The preimplantation blastocyst resides freely in the uterine cavity and also has a microenvironment, namely the uterine secretion (1). The biochemical composition of this fluid, which is governed by the secretory activity of the uterine epithelium, is critical for the function of the trophectoderm. But the trophectoderm also influences the uterine epithelium; among other things, the trophoblast cells are assumed to release signal substances to the endometrium. Thus, the trophectoderm is active in both maintaining an optimal composition of the blastocoele fluid and interacting with the endometrium during the various stages of implantation.

The duration of the blastocyst stage is normally about two days. During this period the blastocyst undergoes continuous development (2). The blastocyst loses its zona pellucida and prepares for attachment onto the uterine epithelium; attachment is followed later by invasion into the uterine tissues. This sequence of events an be interrupted, if a phenomenon termed delayed implantation is initiated. The blastocyst then will lie freely in the uterine cavity for days or weeks with subnormal metabolic activity, arrested cell divisions and low DNA synthesis. Under appropriate conditions, however, the blastocyst will be activated to resume its development in the form of blastocyst attachment and implantation.

Implantational delay occurs naturally as an obligate delay in some animals, like the roe dear and the mink. Under some conditions, a facultative delay can occur, for instance, in mice and rats at suckling of a concurrent litter. A facultative delay can also be brought about experimentally in the rat and mouse. This delay is obtained by ovariectomy a few days after fertilization followed by injections of progesterone (3). The treatment will keep the blastocysts in delay of implantation, but the blastocysts will resume their development as soon as an injection of estrogen is given. Then they will be implanting within a day.

Delayed implantation induced experimentally in the mouse and rat by ovariectomy is therefore an appropriate system for studying changes which take place in the blastocyst at implantation. The system offers a homogeneous population of blastocysts and the implantation changes, all of which are initiated by the injection of estrogen, are well synchronized in time. This chapter describes first the general architecture of the trophectoderm and then correlates the ultrastructure of trophoblast cells with their functions during blastocyst delay, activation, and attachment. The results derive from experiments primarily with the mouse but some rat experiments are also included.

2.1. The trophectoderm and the blastocoele cavity

The ovoid blastocyst has a long diameter of about 150 μm, the blastocoele fluid amounts to about 0.2 nl, and the trophectoderm contains about 80 trophoblast cells (4, 5). The cells of the embryonal pole (polar trophoblast), where the inner cell mass is located, have a greater proliferative capacity than those of the abembryonal pole (mural trophoblast) (6). The latter cells, however, are the first ones to attach onto the uterine epithelium at implantation.

The intercellular space between the trophoblast cells has a convoluted course, formed by the numerous extensions of the lateral membrane of the

Van Blerkom, J. and Motta, P.M. (eds.), Ultrastructure of reproduction. ISBN 0-89838-572-5.

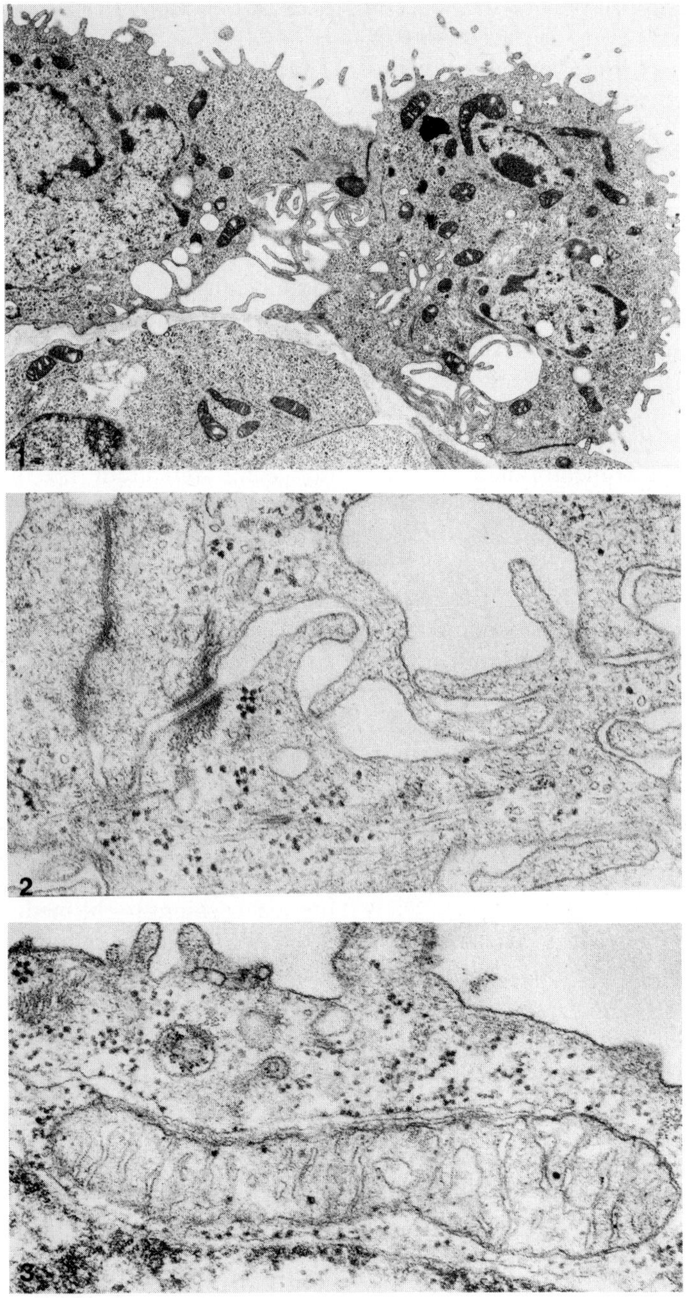

Figs. 1–3. (1) A portion of trophectoderm of a blastocyst, activated for 4 h. The lateral cell membranes of the trophoblast cells are convoluted, which gives the intercellular spaces an irregular shape. The basal cell surfaces are smooth. The tropectoderm rests on a basement membrane, which also limits the blastocyst cavity containing blastocele fluid and embryoblast. Portions of inner cell mass also can be observed (× 6,000). (2) Intercellular space of the trophectoderm of a blastocyst, activated for 4 h. The intercellular space begins at the left of the blastocyst surface with a tight junction. A desmosome follows and then the intercellular space widens. In the cytoplasm, groups of microsomes (polysomes) and microtubules are observed. Also, profiles of an irregular system of channels, a tubulo-cisternal endoplasmic reticulum (TER), can be observed (× 39,000). (3) Peripheral portion of the trophoblast of a blastocyst, activated for 8 h. The cell surface possesses short microvilli. A mitochondrion lies closely to the nucleus and contains dark granules. Profiles of a rough endoplasmic reticulum (RER) can be observed surrounding part of the mitochondrion. Close to the cell membrane, profiles of a tubulocisternal endoplasmic reticulum (TER) are present (× 39,000).

trophoblast (Fig. 1). The upper part of the space is sealed by tight junctions (7) (Fig. 2) and separated from the blastocoele cavity by a basement membrane. Thus, a baso-lateral space intervenes between the blastocoele fluid and the trophectoderm.

Various analyses of the blastocoele fluid have suggested that it is produced by active transport (8, 9). The tight junctions are not leaky, as shown by tests with lanthanum ions (Fig. 6). Because a paracellular route of transport in the trophectoderm thus seems to be excluded, a transcellular transport route should exist. In some other types of cell, this route is evident as a tubulo-cisternal endoplasmic reticulum (TER), which makes up a continuous intercellular system in proximity to the apical surface and forming cisternae along the lateral cell membrane (10). This system can be observed by conventional transmission microscopy, but can be visualized more clearly by a potassium ferricyanide staining technique. The trophoblast cells also seem to exhibit a TER system (Figs. 2, 3), which thus might be the structural basis of the cellular transport. The transport out of the trophoblast occurs at the basolateral cell membrane, and is facilitated by the many extensions of the lateral membrane into the intercellular space, which result in a greatly increased surface area.

A blastocyst left undisturbed in culture is known to undergo pulsating contractions at intervals of several hours (11). Within a few minutes it collapses and then expands slowly over a period of some hours. Such contractions can also be induced by mechanical or environmental disturbances during handling of the blastocyst in experiments *in vitro*. The contractive elements are the various types of fibrils in the cytoplasm (Fig. 4), but the trigger of the contractions in an undisturbed blastocyst is unknown. Probably the normally occurring contractions assist in keeping the composition of the blastocoele fluid optimal for the function of the inner cell mass. It is possible that when the ionic balance of the fluid has become inappropriate as a result of metabolic activity of the embryoblast, most of the fluid drains out of the cavity through the paracellular route by contractions of the trophectoderm. The blastocoele fluid then is replaced by means of activity transport, probably through the transcellular route of the trophoblast. This assumption is strengthened by the finding that contracted blastocysts have an increased capacity for taking up radioactivally labelled α-aminoisobuturic acid (12). Moreover, their baso-lateral cell membranes are very irregular, which promotes transport.

2.2. The trophoblast in delay of implantation

The blastocyst in delay of implantation resides in the uterine cavity in a small amount of secretory material, which in the mouse is so small that the material is hardly noticed in sections of the uterus (13). Functionally, the blastocyst has a low metabolic activity. The mechanisms by which the blastocyst remains inactive – whether there is a growth-impeding substance in the uterine secretion or an absence of optimal conditions for growth-continues to be debated (14, 15).

Structurally, the trophoblast in delay shows signs of being inactive (Fig. 4): the ribosomes are distributed as monosomes, only few profiles of the endoplasmic reticulum are present, and the Golgi apparatus is not well developed (16). Yet, the trophoblast takes up substances from the uterine secretion by endocytosis (Fig. 5). This uptake is receptor-mediated, which implies that the trophoblast in delay of implantation has specific requirements, either those related to the suggested presence of a growth-impeding substance or those related to the nutritional qualities of the secretion.

The nature of the putative growth-impeding substance is as yet unknown. Various experiments have suggested that it is secreted by the uterine epithelium during delay of implantation. It has been possible to impede the growth of blastocysts *in vitro* by adding uterine washings to the medium (17), but the effect of this treatment at the cellular level is not yet known. A blastocyst, which has been in delay of implantation and subsequently activated, reverts to its delayed state if replaced in the uterine cavity of a mouse in delay of implantation (16). Simultaneously, the trophoblast will attain the ultrastructural characteristics of an inactive blastocyst, among other things, its ribosomes will reoccur in the form of monosomes. Perhaps an ultrastructural study of the effect of uterine washings on developing blastocysts *in vitro* can provide information as to the mechanism of action of inhibitory substances in the washings by revealing, for instance, the distribution pattern of the trophoblast ribosomes.

A lack of optimal conditions for blastocyst growth implies any condition which restricts either the blastocyst uptake of substrates or the availability of substrates in the uterine secretion. For instance, a deficiency of sodium ions in the secretion should block sodium-dependent transport in the trophoblast membrane (18–20). Using X-ray microanalysis to evaluate samples of uterine secretion absorbed by single Sephadex beads (21), we have found that Na^+ and K^+ were present in the secretion during delay

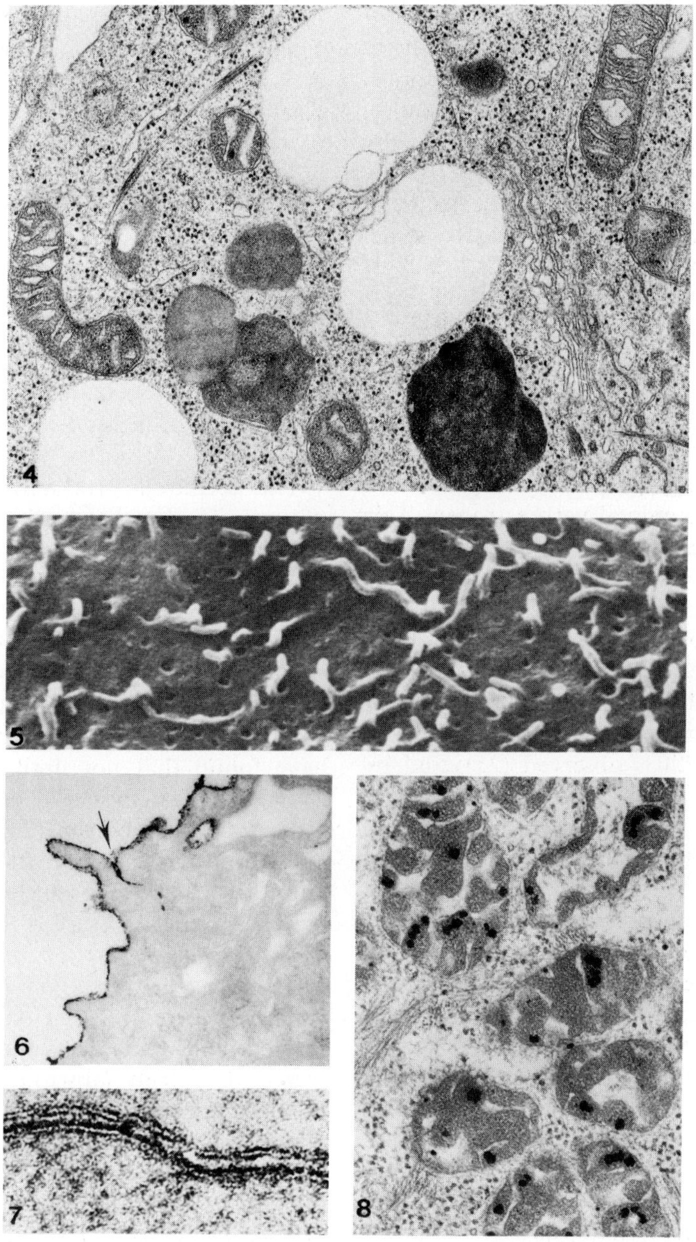

Figs. 4–8. (4) Cytoplasm of a trophoblast cell of a blastocyst in delay. The Golgi apparatus (lower right corner of the micrograph) is small, ribosomes are present as monosomes, and few profiles of a rough endoplasmic reticulum (RER) are visible. The mitochondria contain dilated spaces. Various types of granules are demonstrated, the light areas being dissolved lipid granules and some of the dense areas probably representing lysosomes (× 24,000). (5) Surface of the trophoblast of a delayed blastocyst. Among the microvilli several endocytic pits can be observed (× 9,000). (6) Trophectoderm of an attaching blastocyst, 18 h after induction of implantation. The permeability of the intercellular spaces has been determined by the lanthanum-osmium technique. The stain covers the trophoblast surface but does not penetrate the tight junction of the intercellular space (arrow) which thus is sealed from the outside. The space is located in the upper portion oft the micrograph (× 12,000). (7) Attachment of trophoblast membrane (above) to epithelial cell membrane (below), 24 h after induction of implantation. The two leaflets of each cell membrane can be observed. The outer leaflet of the epithelial cell membrane is denser than the inner one (× 78,000). (8) Mitochondria of a trophoblast cell of an attaching blastocyst, 18 h after induction of implantation. The blastocyst has been prepared according to a potassium pyroantimonate-osmium technique which reveals the presence mainly of calcium and magnesium ions. The many precipitates in the mitochondrial matrix could indicate an increased amount of mitochondrial calcium (× 30,000).

(22). Sodium- and potassium-dependent carrier systems of the trophoblast cells membrane could therefore be able to function also under delay conditions. However, a trophoblast deficiency of, for instance, amino acids can still occur, if they are not secreted by the epithelium and thus are not available in the uterine secretions. Another substrate whose degree of availability could be rate-limiting for blastocyst activity is glucose, because it is known that a lack of glucose in a medium for blastocyst culture will arrest the growth of the blastocysts (23, 24) (see below).

2.3. The trophoblast at activation

The blastocyst in an experimental delay of implantation can be made to begin prepare for implantation by injecting the mother animal with estrogen. This injection initiates a series of structural and functional changes in the blastocyst within a few hours, and the blastocyst is said to be activated for implantation.

The ultrastructure of the activated trophoblast differs in a few respects to that of the inactive trophoblast in delayed implanted blastocysts. By scanning electron microscopy a more irregular surface with more microvilli is observed (25), and by transmission electron microscopy polysomes and glycogen granules are seen (16) (Fig. 8). These changes reflect increased synthesis related to the early stage of activation.

The amount of uterine secretion during delay of implantation is very small, but a few hours after activation it has increased markedly (13, 26). It is not unreasonable to assume that this surge of substances into the microenvironment of the blastocyst could markedly affect its function such that changes of the secretion could work as triggers for blastocyst activation. Among the various mechanisms, which are conceivable, we have tested two of them: a shift in the ionic balance as a cause for changes of the transport properties of the cell membranes, and an influx of glucose into the secretion as a cause for an increased metabolic activity of the blastocyst.

The ionic composition of the uterine secretion in delayed and activated rats have been assessed with the Sephadex technique (Nilsson and Ljung, in preparation). The beads have been left in the uterus to absorb secretion under standardized conditions and were then coated with a thin layer of gold. Using the relations of the elemental peaks of Na^{+}, K^{+} or Ca^{++} to the peak of gold, a ratio has been obtained for each element in each bead. These ratios or normalized element values can be used for comparisons among the beads. We have found that the concentrations of potassium and calcium double while that of sodium remains rather unchanged at activation. This alteration of the ionic balance in the secretion at activation might well frigger changes of the trophoblast function.

Absolute values for the concentrations of sodium, potassium and calcium have been calculated by fitting normalized element values from uterine beads into linear regression plots obtained by taking know concentrations of rat serum and the gold-normalized values of the serum ions as coordinates. Using these reference curves, we have estimated the delay and activation figures for sodium to 117 and 125 mEqv/l, for potassium to 3.0 and 8.3 mEqv/l, and for calcium to 1.4 and 2.3 mEqv/l (Nilsson and Ljung, in preparation).

The secretion produced at activation contains glucose (27). Furthermore, if a conventional *in vitro* medium is depleted of glucose and/or arginine and leucine, the growth of a delayed blastocyst in culture will remain impeded, and not until the substances are added will growth resume (24). It thus seems that the blastocyst is dependent upon glycolysis to become activated. Determination of the oxygen consumption of blastocysts has confirmed this view, because we have found that not until about four hours after activation blastocysts did begin to consume oxygen for aerobic metabolism (28). After determining the mitochondrial capacity for oxidative phosphorylation by a histochemical technique for cytochrome oxidase, we have observed that this enzyme is nearly absent during delay of implantation, while it is clearly present four hours after activation (28). We conclude, therefore, that one contributory factor in the mechanisms of activation is a surge of uterine secretions rich in substrates for glycolysis. Because evidence indicates that the capacity for oxidative phosphorylation is low in the trophoblast in delay of implantation, we believe that glycolysis is utilized by the trophoblast in delay of implantation to obtain sufficient energy to activate its cytochrome oxidase system. At present a morphometric analysis is being undertaken to define what structural changes of the mitochondria are associated with their increased capacity for oxidative phosphorylation (Cieciura, Nilsson, Pietrzkowska and Wandachowicz, in preparation). The activation of the cytochrome oxidase system can be accomplished by activation of enzyme precursors or by translation of preformed RNA templates. Our current experiments with cycloheximide and choramphenicol as metabolic blockers indicate that the cytoplasmic units of the enzyme have to be synthetized while the mitochondrial ones are already present (Nilsson and Magnusson, in preparation).

If the increase in the amount of glucose in the uterine secretion is a trigger of blastocyst activation, it might be expected that experimental conditions that affect the utilization of glucose in the uterine secretion would influence activation. However, neither in animals with experimentally induced diabetes, which have an increased content of glucose in the uterine secretion (28), nor in animals injected with deoxyglucose, which should replace glucose in the secretion, have we noticed any interference either ultrastructurally or functionally with the process of activation. If the lack of effect on the blastocyst is not, for instance, a problem of substrate competition, it can be assumed that in addition to glucose, other growth-controlling substances are present, for example, the already mentioned blastocyst-impeding uterine factor or the amino acids arginine and leucine.

Arginine and leucine are known to be required for obtaining blastocyst activation *in vitro*, but unless the medium also is depleted of glucose only a temporary delay of implantation is obtained (24). If this finding is relevant to *in vivo* conditions, then amino acids must be expected to pass into the uterine secretion very early at activation. This passage seems to be the case, because it is known that radiolabelled α-aminoisobuturic acid (^{14}C-AIB), a synthetic amino acid analogue, is accumulated specifically in uterine tissues at blastocyst activation (29), and that ^{14}C-AIB is taken up rapidly by activated blastocysts lying in the uterine lumen (30). Furthermore blastocysts in delay of implantation, when examined after incubation *in vitro*, take up ^{14}C-AIB to an extent similar to activated blastocysts (12). Thus, it is possible that the availability of amino acids in the uterine secretion and not an ineffeciency in the carrier system in the trophoblast membrane is the factor which limits the uptake of amino acids into uterine blastocysts in delay of implantation.

A signal passes from the blastocyst to the endometrium early during activation. The signal is meant for inducing such endometrial changes which are necessary for a normal course of implantation. The first response observed in the endometrium is a nucleolar change of the stromal cells, occurring about 8 h after the injection of estrogen (31). The nature of the blastocyst signal is still unknown, but among substances discussed are hormonal factors, such as estrogens or prostaglandins, and metabolic products, originating from trophoblast cells.

2.4. The trophoblast at attachment to the uterine epithelium

Attachment to the uterine epithelium, which occurs about 18–20 h after initiation of implantation, is a morphological event characterized by close contact between the cell membrane of the trophoblast and that of the cells of the uterine epithelium. Preliminary results have shown that the distance between the two cell membranes seems to decrease during attachment (32). Occasionally, an asymmetry in thickness of the cell membrane leaflets of the uterine epithelium has been noticed (Fig. 6). However, further high resolution studies of these and other changes are needed in order to characterize the ultrastructural changes of the cell membranes during attachment.

Whether the change in adhesiveness of the blastocyst is due to a change in the uterine secretion surrounding the blastocyst or to a continuation of the activation process of the trophoblast remains to be determined. A change in the blastocyst environment in the form, for instance, of a shift of the calcium ion balance could be implicated as a mechanism for increasing trophoblast adhesiveness. An examination of the ionic content of the uterine secretion by the Sephadex technique have revealed a decrease in the calcium concentration from about 2.3 mEqv/l at activation to about 1.0 mEqv/l at attachment. Within the trophoblast cells, after a potassium pyroantimonate reation, a larger amount of mitochondrial precipitate has been noted in attaching blastocysts than in delayed implanted ones (Fig. 8). Because this reaction demonstrates the presence of some ions, like calcium, it may be assumed that the mitochondrial pool of calcium has increased simultaneously with the decrease of calcium in the secretion. But it has not been established yet how the calcium changes in the uterine secretion, trophoblast cytoplasm, trophoblast mitochondria, and blastocoele fluid are related.

A programmed sequence of development as a cause of the increased adhesiveness of the trophoblast during attachment is also conceivable. Blastocysts transferred to culture chambers or to ectopic sites in the body will attach to the artificial support or to the foreign tissue within a fairly defined period of time (33). This finding also implies that there is little specificity in the trophoblast-epithelial interaction, not even in the uterus because mouse trophoblast has been showed to attach to rat uterine epithelium (34, 35).

Alterations in cellular adhesiveness most probably involve changes of the glycoproteins and glycolipids of the surface coat (the glycocalyx) (36–38). The glycocalyx can be visualized at the electron microscopic level by means of, for example, ruthenium red and Alcian Blue. These markers are nonspecific, cationic stains which stabilize the glycocalyx and facilitate its demonstration with osmium staining. The stains give some morphologic information, like occurrence of glycocalyx among various types of cell and changes in coat thickness at various functional states of cells. However, the surfaces of trophoblast cells have not been found to change at implantation when examined with a ruthenium red procedure (39).

Specific markers of saccharides in the glycocalyx are lectins, which have been used for studies of trophoblast changes (40–43). The lectins have been traced by radioactivity and fluorescence or by coupling to particles such as erythrocytes, hemocyanin, latex beads or ferritin, followed by examination of binding pattern on the blastocyst surface. Since different functional states of blastocysts and different pretreatments of blastocysts have been used, no safe conclusions can be drawn from the results. However, to avoid artifacts in these types of studies is difficult,even if artifactual findings can be lessened by using blastocysts in delay of implantation (since they lack zona pellucida and have defined functional states), by using unfixed material (since fixatives affect glycocalyx in an unknown way), and by being cautious at interpreting the results (since none of the markers is really reliable).

A marker for anionic sites of a cell surface is given by cationic ferritin or a positively charged colloidal iron solution. When applied to blastocysts, a decrease in the number of negative sites of the trophoblast has been observed during attachment (44). Quantification of the negative sites by electrophoresis has shown that the net negative charge is diminished by 30 per cent during implantation, probably due to a partial loss of sialic acid from the blastocysts (45). This change also occurs *in vitro* without any induction by the endometrium, thus indicating the presence of an intrinsic or scheduled program of development. The less negative charge at implantation is regarded to support blastocyst attachment by the mere decrease in negative charge repulsion. However, I believe rather that the prime changes in the glycoproteins and glycolipids, which in turn results in a decrease of the net negative charge, are the important ones.

The trophectoderm contains a genome with both maternal and paternal components, yet it can invade the maternal tissues without evoking any graft rejection. The cause for this is unknown. One line of research attempts to specify the histocompatibility antigens of the developing blastocyst using monoclonal antibodies raised against various embryonal tissues. Monoclonal antibodies specifically raised against the blastocyst surface, however, are cumbersome to obtain due to the rather high number of blastocysts required for a conventional immunization. Further, current methods for detecting an immunological reaction at the blastocyst surface either have a rather low sensitivity (like radioimmunology and immunofluorescence) or require time-consuming processing for sectioning and electron microscopy (like immunocytochemistry). Therefore we have attempted to design techniques for both immunization and antibody detection which require only a restricted number of blastocysts.

Our current procedure (Svalander, Grönvik and Nilsson, in preparation) involves immunization with a few radiation- inactivated blastocysts of a defined functional state by transplanting them under the

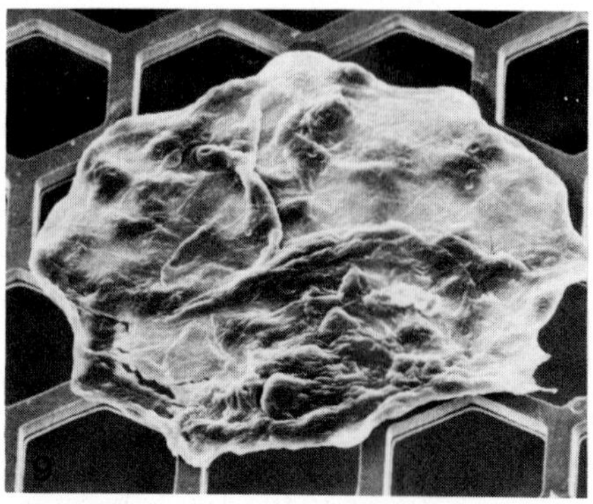

Figs. 9–10. (9) Blastocyst, air-dried onto a hexagonal grid and prepared for scanning electron microscopy. This type of specimen is used in the transmission electron microscope for detecting a presence of protein A-colloidal gold complexes (see Fig. 10) (× 60). (10) Part of an air-dried blastocyst, prepared to have protein A-colloidal gold complexes bound to immunoglobulins on the trophoblast surface. The number of dark particles (gold complexes) is roughly proportional to the number of sites with an antigen against which antibodies have been raised (× 6,000).

spleen capsule. We screen for possible antibodies by a protein A-colloidal gold technique (46), which we have designed for transmission electron microscopy of whole, air-dried blastocysts (Figs. 9, 10). These procedures facilitate the immunization, probably by delivering the antigen directly into the spleen, and they allow for an economic and rapid search for specific antibodies by using single and easily processed blastocysts.

In conclusion, the trophectoderm of the preimplantation blastocyst exhibits a sequence of functional changes which are of fundamental importance both for the development of the inner cell mass and for implantation of the blastocyst. All these changes occur within about 24 h. This requires standardized batches of blastocysts for analyses, and blastocysts in delay of implantation or activated for implantation fullfil these requirements.

Acknowledgement

Technical assistants: Mrs Barbro Einarsson, Mrs Marianne Ljungkvist, Mrs Sibylle Widéhn and Mr Leif Ljung.

Financial supports: Swedish Medical Research Council and the 'Expressen' Prenatal Research Foundation.

References

1. Edwards RG, Surani MAH: The primate blastocysts and its environment. Ups J Med Sci 22: 39–50, 1978.
2. Sherman MI, Wudl LR: The implanting mouse blastocyst. In: The cell surface in animal embryogenesis and development. Poste G, Nicolson GL (eds) Elsevier/North-Holland Biomedical Press, 1976, pp 81–125.
3. Bergström S: Experimentally delayed implantation. In: Methods in mammalian reproduction. Daniel JC Jr (ed) Academic press, 1978, pp 419–435.
4. Biggers JD, Borland RM: Physiological aspects of growth and development of the preimplantation mammalian embryo. Annu Rev Physiol 38: 95–119, 1976.
5. Schiffner J, Spielmann H: Fluoremetric assay of the protein content of mouse and rat embryos during preimplantation development. J Reprod Fertil 47: 145–147, 1976.
6. Copp AJ: Interaction between inner cell mass and trophectoderm of the mouse blastocyst. I. A study of cellular proliferation. J Embryol Exp Morphol 48: 109–125, 1978.
7. Ducibella T, Albertini DF, Anderson E, Biggers JD: The preimplantation mammalian embryo: characterization of intercellular junctions and their appearance during development. Dev Biol 45: 231–250, 1975.
8. Gamow E, Daniel Jr JC: Fluid transport in the rabbit blastocyst. Wilhelm Roux Archiv 164: 261–178, 1970.
9. Benos DJ, Biggers JD: Blastocyst fluid formation. In: Fertilization and embryonic development *in vitro*. Mastroianni L, Biggers JD, Sadler W (eds) New York, Plenum Press.
10. Møllgård K, Rostgaard J: Morphologic aspects of transepithelial transport with special reference to the endoplasmic reticulum. In: Ion transport by epithelia. Schultz SG (ed) New York, Raven Press, 1981, pp 209–231.
11. Cole RJ: Cinematographic observations on the trophoblast and zona pellucida of the mouse blastocyst. J Embryol Exp Morphol 17: 481–490, 1967.
12. Lindqvist I, Nilsson O, Ronquist G: The *in vitro* transport of ^{14}C-α-aminoisobuturic acid into blastocysts from mice in delay and after activation for implantation. Acta Physiol Scand 111: 35–42, 1981.
13. Nilsson O: The morphology of blastocyst implantation. J Reprod Fertil 39: 187–194, 1974.
14. Psychoyos A, Bitton-Casimiri V, Brun JL: Repression and activation of the mammalian blastocyst. In: Regulation of growth and differentiated function in eukaryote cells. Talwar GP (ed) New York, Raven Press, 1975, pp 509–514.
15. Weitlauf HM, Kiessling AA, Activation of 'delayed implanting' mouse embryos *in vitro*. J Reprod Fertil 29: 191–202, 1981.
16. Naeslund G, Lundkvist Ö, Nilsson BO: Transmission electron microscopy of mouse blastocysts activated and growth-arrested *in vivo* and *in vitro*. Anat Embryol 159: 33–48, 1980.
17. Weitlauf HM: Factors in mouse uterine fluid that inhibit the incorporation of ^{3}H-uridine by blastocysts *in vitro*. J Reprod Fertil 52: 321–328, 1978.
18. Borland RM, Tasca RJ, Activation of a Na^{+}-dependent amino acid transport system in preimplantation mouse embryos. Dev Biol 30: 169–182, 1974.
19. Surani MAH, Fishel SB: Embryonic and uterine factors in delayed implantation in rodents. J Reprod Fertil 29: 159–172, 1981.
20. Van Winkle LJ, Activation of amino acid accumulation in delayed implantation mouse blastocysts. J Exp Zool 218: 239–246, 1981.
21. Nilsson BO, Ljung L: Electron probe micro-X-ray analyses of electrolyte composition of fluid microsamples by use of a Sephadex bead. Ups J Med Sci 84: 1–2, 1979.
22. Nilsson BO: Electron microscopic aspects of epithelial changes related to implantation. Prog Reprod Biol 7: 70–80, 1980.
23. Van Blerkom J, Chavez DJ, Bell H: Molecular and cellular aspects of facultative delayed implantation in the mouse. In: Maternal recognition of pregnancy. Ciba Foundation Symposium, 1978, pp 141–163.
24. Naeslund G: The effect of glucose-, arginine- and leucine-deprivation on mouse blastocyst cutgrowth *in vitro*. Ups J Med Sci 84: 9–20, 1979.
25. Bergström S: Scanning electron microscopy of ovo-implantation. Arch Gynecol 212: 258–270, 1972.
26. Nilsson BO, Lundkvist, Ö: Ultrastructural and histochemical changes of the mouse uterine epithelium on blastocyst activation for implantation. Anat Embryol 155: 331–321, 1979.
27. Nilsson BO, Östensson CG, Eide S, Hellerström C, Utilization of glycose by the implanting mouse blastocyst activated by oestrogen. Endocrinologie 76: 82–93, 1980.
28. Nilsson BO, Magnusson C, Widéhn S, Hillensjö T: Correlation between blastocyst oxygen consumption and trophoblast cytochrome oxidase reaction at implantation of delayed mouse blastocysts. J Embryol Exp Morphol 71: 75–82, 1982.
29. Lindqvist I, Nilsson O, Ronquist G: Preferential uptake of ^{14}C-α-aminoisobuturic acid into mouse uterine tissue during early pregnancy. Acta Physiol Scand 99: 37–41, 1977.
30. Lindqvist I, Einarsson B, Nilsson O, Ronquist G: The *in vivo* transport of ^{14}C-α-aminoisobuturic acid into mouse blastocyst during activation for implantation. Acta Physiol Scand 102: 477–483, 1978.
31. Lundkvist Ö, Nilsson BO: Endometrial ultrastructure in the early uterine response to blastocysts and artificial deciduogenic stimuli in rats. Cell Tissue Res 225: 355–364.
32. Nilsson BO: Ultrastructure of trophoblast-epithelium relations during implantation. In: Proteins and steroids in early pregnancy. Beier HM, Karlson P (eds) Berlin, Springer-Verlag, 1982, pp 5–14.
33. Sherman MI: Implantation of mouse blastocysts *in vitro*. In: Methods in mammalian reproduction. New York, Academic Press, 1978, pp 247–257.
34. Håkansson S, Lundkvist Ö, Nilsson O, Alm G: Prolonged survival of implanting rat blastocysts in the uterus of congenitally athymic mice. Scand J Immunol 6: 817–820, 1977.
35. Tachi S, Tachi C: Ultrastructural studies on maternal-embryonic cell interaction during experimentally induced implantation of rat blastocysts to the endometrium of the mouse. Dev Biol 68: 203–223, 1979.
36. Pinsker MC, Mintz B: Change in cell-surface glycoproteins of mouse embryos before implantation. Proc Natl Acad Sci USA 70: 1645–1648, 1973.

37. Johnson LV, Calarco PG: Mammalian preimplantation development: The cell surface. Anat Rec 196: 201–219, 1980.
38. Surani MAH, Kimber SJ, Handyside AH: Synthesis and role of cell surface glycoproteins in preimplantation mouse development. Exp Cel Res 133: 331–339, 1971.
39. Enders AC, Schlafke S: Surface coats of the mouse blastocyst and uterus during the preimplantation period. Anat Rec 180: 31–46.
40. Wu JT, Chang C: Increase in Concanavalin A binding sites in mouse blastocysts during implantation. J Exp Zool 205: 447–453, 1978.
41. Chávez DJ, Enders AC, Temporal changes in lectin binding of peri-implantation mouse blastocysts. Dev Biol 87: 267–276.
42. Nilsson BO, Naeslund G, Curman B, Polar differences of delayed and implanting mouse blastocysts in binding of Alcian Blue and Concanavalin A. J Exp Zool 214: 177–180, 1980.
43. Carollo JR, Weitlauf HM: Regional changes in the binding of (^{3}H) Concanavalin A to mouse blastocysts at implantation: an autoradiographic study. J Exp Zool 218: 247–251, 1981.
44. Nilsson O, Lindqvist I, Ronquist G: Blastocyst surfaçe change and implantation in the mouse. Contraception 11: 441–450.
45. Nilsson BO, Hjertén S: Electrophoretic quantification of the changes in the average net negative surface charge density of mouse blastocysts implanting *in vivo* and *in vitro*. Biol Reprod 27: 485–493.
46. Slot JW, Geuze JH: Sizing of protein A-colloidal gold probes for immunoelectron microscopy. J Cell Biol 90: 533–536.

Author's address:
Department of Human Anatomy Biomedical Centre Box 571
S-751 23 Uppsala, Sweden

Index

Acrosin 170, 181
Acrosome 78, 113, 116
  absence 122
  actin elements 93, 98
  anomalies 122
  band 91
  comparative aspects 86, 95
  formation 86
  membrane changes 80, 84, 113
  reaction 78, 86, 93, 169, 181
  vesicle 89, 94
Actin
  in acrosome 93, 98
Activation
  sperm 169
Adrenal blastema 22
Aminoisobutiric acid 265
Anionic lipids
  spermatozoon 77
Anionic site
  cell surface 266
Annulus
  sperm tail 83
Antrum
  follicle 156
A particles
  virus-like, embryo 224
Area opaca
  chick embryo 13
Area pellucida
  chick embryo 13
Area vasculosa
  chick embryo 15
Arginine 265
Atresia 142, 156
Atretic division 143
Axoneme
  anomalies 123
  proteins 119
  sperm 118

Balbiani body 17
Basal lamina 20
Blastocoele 198, 260
  fluid 198, 260
Blastocyst 199, 247, 260
  early 213
  formation 202
  inclusions 240
  mouse 200, 239
  orientation 250
  position 248
  rat 260
  SEM 200
  signal 265
  spacing 248
  surface features 203
  uterus attachment 249
Blastomeres 186, 193, 206, 209
  polarization 206
  surface features 210, 212
Blood-testis barrier 72
B particles
  virus-like embryo 221
Busulfan 28

Calmodulin 121
cAMP
  oocyte 160
Capacitation 79, 167, 176
  in vitro 167
  membrane changes 80
Carcino-embryonal marker 231
Cationic stains 266
Cauda epididymis 97
Cavitation 194
Cell death 140
Chemotactic substance 15
Chorionic cells 242
Chromatin 147
  fibrils 147
  oocyte 161
Chromosomes
  amphibia 134
  changes in meiosis 161
  lampbrush 134, 152
  oocytes 144
Cleavage 193, 202
Coelomic epithelium 15
Compaction 194, 196, 202

Concanavalin 208
Corpus luteum 27
Cortical
  granules 161, 179, 184
  reaction 184
C particles
  virus-like 230
Crystalline inclusions 235
  embryo 239
  ovary 237
  uterus 235, 236
Cumulus cells 152, 156
  junctions 159
  sperm passage 179
Cytochalasin
  B 238
  D 208

Decidualization 257
Dedifferentiation 20
Dictyosomes 86
Diplotene 135, 143
DNA-bodies
  invertebrates 130
  vertebrates 132
Dynein 120

Embryo
  adhesiveness 252
  carcinoma cells 199, 220
  cell surface 190, 194
  crystals 239
  early development 190, 223
  8-cell 206
  mammals 190
  mosaic organization 205
  mouse 197
  partenogenesis 239
  postimplantation 218
  preimplantation 205, 218, 239
  two-cell 186, 206
  virus particles occurrence 223
Embryoblast 260, 262
Embryogenesis 1, 137
  early 190
Endometrium (see also uterus) 235, 250
Epididymis 97, 176
  proteins 107, 177
  spermatozoa 177
ε particles (virus-like) 219
Equatorial segment (spermatozoon) 78

Fertilization
  human 176
  in vitro 166, 167, 190
  in vivo 172, 173
  mammals 166, 190
  SEM 173
Filipin-sterol complexes 77
Follicle
  antrum 156
  fluid 156
  Graafian 143
  primordial 26, 143, 145
Follicle cells 157
  association with germ cells 157
Folliculogenesis 25, 26

Gap junctions
  metabolic coupling 158
  oocytes 158
Gene products 1, 8
Genetic effect (X rays) 140
Germ cell line 1
Germ cells 4, 12, 25
  ameboid features 15
  chick embryo 12
  degeneration 41
  differentiation 31
  filopodia 17
  genetic regulation 41
  glycogen 13, 31
  intercellular bridges 31, 36, 38, 157
  ionizing radiations 141
  lamellipodia 17
  migration 12
  motility 40
  nucleolus 13
  nutrition 41
  potential 6
  Sertoli relationship 47
  synchronism of maturation 39
  tumors 42
  yolk 13
Germ plasm 1, 17
Germinal crescent 13
Germinal vesicle 156
Glycogen (germ cell) 13, 31
Gonad 12, 19, 26, 27, 29
  anlagen 12
  blastema 22, 29
  indifferentiate stage 26
  morphogenesis 15, 19, 25, 27
  phase (chick) 15
  ridge 19
Gonadotrophins 157
Gonocytes (see also germ cells) 35
Graafian follicle 143
Granulosa cells 156

Histones 162
Hylar cell tumor
  ovary 237

Implantation 247
  activation 264
  delayed 260, 262
  experimental 254

trophoblast 253
Intercellular bridges
elimination 39
function 39
germ cells 31, 36, 38, 39
Ionizing radiations 140
genetic effect 140
in germ cells 141
Irradiation
of oocytes 147, 152

Karyolysis 144
Kinetochore
oocyte 161

Lactice structure 237
embryo 238
oocytes 237
Lectins 104, 266
Leucine 265
Luteinizing hormone 157

Maturation
spermatozoon 97
Meiosis 25, 33, 86, 128, 140
amphibia 128
inducing substance 25
maturation 156
prophase 156
vertebrates 128, 129
Mesenchimal cells 20
Mesonephros 19
blastema 22
cells 20
tubules 22
Mesothelial cells 20
Migration
germ cells 12, 13, 15
Mitosis
oocytes 33, 140
Morphogenesis
gonad 25
Morula 195, 210, 216
cells 210
fate 216
surface 214
Mosaicism
embryo 205
Myoid cells
seminiferous tubule 62

Neoplastic cells
embryo 220
Neutrons 140
Nexin 120
Nuage 4, 17
Nucleolar bodies 134
Nucleolonema 147
Nucleolus
evolution 128
involution 136, 150
multiplication 133
oogenesis 127
ultrastructure 130
vacuolation 150
Nucleosomal beads 162

Oocyte
cAMP 160
chromatin 161
cristalline inclusion 237
DNA 135, 137, 142, 146
gap junction 158
giant 146
human 33, 177
invertebrates 136
irradiation 147, 152
kinetochore 161
lactice structures 238
mammals 156
maturation 156, 178
mice 146
nucleolus 127, 136
nucleus 178
organelles 35, 146, 237
phagocytosis 148
ribosomes 137, 237
RNA 136
Spermatozoon interaction 179, 182
Oogonia
mammals 35, 142
organelles 34
Oolemma 156
sperm fusion with 172
Ovary 31
cords 27
cristalline inclusion 237
differentiation 31
hilar cell tumors 237
Oviduct 247

Pachytene 129
Parergin 120
Partenogenetic embryo 239
Perforatorium 115
PGC (primordial germ cell), see also germ cells 1, 13, 19, 142
Phagocytosis
oocyte 148
Phenotype 210
Placenta 244
Plasminogen 64
Polar cells 213
Polar granule 2
Polarization
blastomeres 207
induction 207
Polymixin-B 77
Polyspermy block 184

Postacrosomal region 79, 98
Post-penetration events 184
Pregnancy 254
Pregranulosa cells 28
Preimplantation 205
  crystals 239
Primitive streak stage 13
Primordial follicles 26, 143, 145
Primordial germ cell (PGC), see also germ cells 1, 13, 19, 142
Proacrosomal granule 89
Progesterone 157
Pronucleus
  female 185
  male 184, 186
Protoms 140
Pseudocleavage 146
Pseudomaturation division 146
Pyknosis 144

Radiation biology 140
Radioresistance 141
Radiosensitivity 140, 141
Reproductive capacity 140
Residual body
  spermatid 51, 60
Rete ovarii 26, 29
Retroviruses 226
  embryogenesis 229
  endogenous 226, 230
  exogenous 227
R particles
  virus like 231

Secretion
  uterus 264
Seminiferous tubule 62, 67
  spermiation 64
Sertoli cells 47, 67
  apical process 52
  culture 64
  cycle 73
  cytoplasm 53, 54, 69
  ectoplasm 53
  germ cell relationship 47, 73
  in vivo 65
  organelles 68, 69
  penetrating process 58
  stalk 52
  steroidogenesis 72
  structure 70, 71
Sex chromosomes 25
Sexual differentiation 25
Sperm (see spermatozoon)
Spermatid 48
  cytoskeleton 62
  disengagement 62
  displacement 54
  Golgi apparatus 86
  movement 48
  residual body 52, 60
  stalk 52
  tubulobulbar complexes 56, 57
Spermatocytes
  intercellular bridges 36, 37
  mammals 35
  organelles 35
Spermatogonia
  mammals 35, 71
  organelles 35
Spermatozoon 75, 97
  activation in vitro 169
  anomalies 121, 122, 124
  axoneme 118
  double 124
  epididymal transit 102
  fertilization capacity 97
  freeze-fracture data 101
  glycoproteins 102, 111
  guinea pig 75, 84
  head 76, 110
  human 110
  in female genital tract 177
  lectins 104
  lipids 77, 102
  macromolecular changes 104
  maturation 97, 100, 102
  membrane organization 75, 84, 97, 98, 100, 110
  movement 121
  neck 116
  nucleus 112
  oocyte interaction 179, 182
  organelles 116
  proteins 119, 120
  radiolabel 105
  surface changes 102, 103
  tail 80, 98, 115, 118
Spermiation 46, 49, 60
  acrosome 50, 51, 54
  myoid cells 62
  seminiferous tubule 64
  Sertoli cell relationship 50
Spermiogenesis 47, 60, 86
Sperm release 46, 49, 60
Sperm-zona association 170, 181
Steroidogenic cells 27
Steroids 26
  metabolism in testis 72

Tail
  fibrous sheath 118
  midpiece 81, 98, 116
  principal piece 82, 98, 117
  proteins 119, 120
  sperm 80, 98
Testis 35
  blood barrier 72
  differentiation 35
Transcription activity

oocyte 164
Triton X100 161
Trophectoderm 199, 215, 260, 261
surface glycoprotein 201
Trophoblast 240, 244
activated 264
adhesiveness 252, 265
delayed 262
fusion 253, 265
implantation 253
inclusion 245
mural 260
outgrowth 255
Tubulin 119, 240
Tumor (germ cell) 42, 43

Uterus 235
aminoacids 265
blastocyst association 249
crystals 235
decidualization 257
endometrium 250
epithelium 236, 250
glands 241, 251
glucose 265
glycocalix 265
implantation 254
secretion 264
stroma 251

Viruses 218
embryo 226
Virus-like particles 218
B-type 218
C-type 218
embryo 220
oocytes 221
Vitellogenesis
amphibia 134

Wolffian duct 21

X-rays 140

Yolk plates 238

Z cell 143
Zona lysin 182
Zona pellucida 156, 247
shedding 247
sperm association 181
Zygote
human 185, 187
in vitro 187